INTERNATIONAL PERSPECTIVES ON GLOBAL ENVIRONMENTAL CHANGE

Edited by **Stephen S. Young**
and **Steven E. Silvern**

INTECHOPEN.COM

International Perspectives on Global Environmental Change
Edited by Stephen S. Young and Steven E. Silvern

Published by InTech

Janeza Trdine 9, 51000 Rijeka, Croatia

Edition 2014

Publishing Process Manager Anja Filipovic

Technical Editor Teodora Smiljanic

Cover Designer InTech Design Team

Additional hard copies can be obtained from orders@intechopen.com

International Perspectives on Global Environmental Change,
Edited by Stephen S. Young and Steven E. Silvern
p. cm.
ISBN 978-953-307-815-1

INTECH

open science | open minds

Contents

Preface IX

Part 1 Climate Change 1

Chapter 1 Two Cultures, Multiple Theoretical Perspectives: The Problem of Integration of Natural and Social Sciences in Earth System Research 3
Diógenes S. Alves

Chapter 2 History and Prediction of the Asian Monsoon and Glacial Terminations, Based on Records from the South China Sea 25
Hong Ao and Guoqiao Xiao

Chapter 3 Climate Change and Health Effects 35
Rajan R. Patil

Chapter 4 Agricultural Technological and Institutional Innovations for Enhanced Adaptation to Environmental Change in North Africa 57
Ali Nefzaoui, Habib Ketata and Mohammed El Mourid

Chapter 5 Possible Evolutionary Response to Global Change – Evolutionary Rescue? 85
Lars A. Bach and Cino Pertoldi

Part 2 Historical Environmental Change 105

Chapter 6 How Did Past Environmental Change Affect Carnivore Diversity and Home-Range-Size in Spain? 107
María T. Alberdi, José L. Prado, Esperanza Cerdeño and Beatriz Azanza

Chapter 7 **Response of Biogenic Silica Production in Lake Baikal and Uranium Weathering Intensity in the Catchment Area to Global Climate Changes 121**
Takuma Murakami, Nagayoshi Katsuta, Takejiro Takamatsu, Masao Takano, Koshi Yamamoto, Toshio Nakamura and Takayoshi Kawai

Chapter 8 **Continental Erosion/Weathering Changes in Central Asia Recorded in the Holocene Sediment from Lake Hovsgol, Northwest Mongolia, by Synchrotron μ-XRF Mapping Analyses 137**
Nagayoshi Katsuta, Takuma Murakami, Yuko Wada, Masao Takano, Masayuki Kunugi and Takayoshi Kawai

Part 3 Biological Responses to Environmental Change 149

Chapter 9 **Primary Succession in Glacier Forelands: How Small Animals Conquer New Land Around Melting Glaciers 151**
Sigmund Hågvar

Chapter 10 **Excess Supply of Nutrients, Fungal Community, and Plant Litter Decomposition: A Case Study of Avian-Derived Excreta Deposition in Conifer Plantations 173**
Takashi Osono

Chapter 11 **Effect of Environmental Change on Secondary Metabolite Production in Lichen-Forming Fungi 197**
Christopher Deduke, Brinda Timsina and Michele D. Piercey-Normore

Part 4 Land Use and Land Cover Change 231

Chapter 12 **Investigating Soils, Vegetation and Land Use in a Lunette Dune-Pan Environment: The Case of Sekoma Lunette Dune-Pan Complex, Botswana 233**
S. Mosweu, J.R. Atlhopheng and M.P. Setshogo

Chapter 13 **Late Quaternary Environmental Changes and Human Interference in Africa 253**
Wolfgang Römer

Chapter 14 **Assessment of the Impact of Land-Use Types on the Change of Water Quality in Wenyu River Watershed (Beijing, China) 275**
Yuanzhi Zhang and Yufei Wang

Part 5 **Indicators of Change 295**

Chapter 15 **Heavy Metals Contamination of a Mediterranean Coastal Ecosystem, Eastern Nile Delta, Egypt 297**
M. F. Kaiser, H.A. Aboulela, H. El-Sereh and H. Ezz El-Din

Chapter 16 **HPLC Fingerprints of Porewater Organic Compounds as Markers for Environmental Conditions 311**
Viia Lepane

Chapter 17 **Management Strategies for Large River Floodplain Lakes Undergoing Rapid Environmental Changes 329**
Giri Kattel and Peter Gell

Part 6 **Management and Policy for Environmental Change 353**

Chapter 18 **Satellite-Based Monitoring of Ecosystem Functioning in Protected Areas: Recent Trends in the Oak Forests (*Quercus pyrenaica* Willd.) of Sierra Nevada (Spain) 355**
M.A. Dionisio, D. Alcaraz-Segura and J. Cabello

Chapter 19 **Linking Sea Level Rise Damage and Vulnerability Assessment: The Case of Greece 375**
Areti Kontogianni, Christos Tourkolias, Michalis Skourtos and Maria Papanikolaou

Chapter 20 **Strengthening Regional Capacities for Providing Remote Sensing Decision Support in Drylands in the Context of Climate Variability and Change 399**
Humberto A. Barbosa and T. V. Lakshmi Kumar

Chapter 21 **Using Fuzzy Cognitive Mapping in Environmental Decision Making and Management: A Methodological Primer and an Application 427**
Elpiniki Papageorgiou and Areti Kontogianni

Chapter 22 **Wind Farming and the Not-in-My-Backyard Syndrome: A Literature Review Regarding Australia's Challenge in Relation to Climate Change and CO2 Emissions 451**
Ingrid Muenstermann

Preface

Almost 150 years ago George Perkins Marsh, in *Man and Nature or the Earth as Modified by Human Action* (1864), took notice on the impact of human activity on the natural environment. Since then, human activities have become a dominant force affecting the functioning of the Earth's biological, hydrological and climatological systems. The use of land, water, air and other natural resources have increased exponentially over the years. With future increases in population, continued technological change and economic development, the demands on the biosphere will continue to grow. With such extensive use, we are now experiencing large scale of transformations that disrupt the functioning of the biosphere and the larger flow of energy and materials on a global scale. We are witnessing significant human-induced impacts on the environment, such as the extensive melting of Arctic sea ice and glaciers around the world, to the depletion of global fish stocks, and the disruption of fresh water ecosystems.

Since Marsh first studied the negative changes associated with agriculture and the development of urban-industrial society, natural and social scientists have continued to explore the local, regional and global dimensions of human-induced environmental change. We now have a much clearer understanding of such adverse human impacts on the environment. Science is increasingly becoming more sophisticated and developing conceptual frameworks and techniques to measure and model environmental changes at all spatial scales. Techniques have emerged such as sediment sampling, ice-core analysis and dendrochronology that help us understand past environmental changes. Geoinformatics with the use of remote sensing, geographic information systems, global positioning systems and information communication technologies enable us to study current and recent changes. Computers and sophisticated modeling techniques are being developed and employed to predict future environmental change.

Our growing scientific knowledge and understanding of the causes and consequences of human activity on the environment is increasingly influential and necessary for humanity's ability to adapt to such changes. Planners, policy-makers and key decision-makers require objective scientific information in order to develop appropriate mitigation plans and policies. For example, computer models of global warming and rising sea levels are being employed to develop plans to protect coastal cities and settlements. Studies of environmental change and transformation are, therefore, critical for risk assessment and reducing uncertainties.

While much of the world has been captivated by global warming and climate change, there are, however, many more dimensions to past and current environmental change that the scientific community is bringing to light. Environmental change is occurring at multiple spatial scales: the local, regional and global scale and across all of the diverse ecosystems and bio-physical environments found on the surface of the planet. Environmental change is thus broad, diverse and multidimensional.

The objective of this book is to advance our scientific knowledge and understanding of some of the many neglected aspects of environmental change. We bring together an international group of experts to fill in the gaps in our knowledge of climate change, historical environmental change, biological adaptation to change, land use changes, indicators of change and management of environmental change. The twenty-two chapters in this book represent a diverse, international set of perspectives on environmental change. The contributors come from different parts of the world and different scientific disciplines. They employ diverse theoretical approaches and scientific methodologies to provide on-the-ground accounts of environmental change around the globe. Taken together as a whole, we hope this text expands the discussion of environmental change beyond Europe and North America to other parts of the world, to include voices of academic researchers whose voices and research is not often heard. The result, we hope, is a text that contributes to building bridges amongst researchers around the world from different fields of study and between researchers and environmental policy makers and decision-makers.

Dr. Stephen S. Young and **Dr. Steven E. Silvern**
Department of Geography,
Salem State University
USA

Part 1

Climate Change

1

Two Cultures, Multiple Theoretical Perspectives: The Problem of Integration of Natural and Social Sciences in Earth System Research

Diógenes S. Alves
National Institute for Space Research (INPE)
Brazil

1. Introduction

The integration of natural and social sciences has been recognized as a key aspect of Earth System (E.S.) research, a cross-disciplinary field involving the study of the geosphere, the biosphere, and society (IGBP, 2006; Leemans et al., 2009; Pfeiffer, 2008; Reid et al., 2010; Young, 2008). Because of societal and political correlates between environmental change and socio-economic development, the study of the Earth System has been increasingly ascribed social and political dimensions emphasizing the need for greater collaboration between the social and natural sciences (Beven, 2011; Kates et al., 2001; Leemans et al., 2009; Reid et al., 2010; Saloranta, 2001; Shackley et al., 1998).

The problem of inter-disciplinary articulation between the social and natural sciences is not specific to E.S. research, and its challenges can be traced back to the very origins of the notions of science and social science (e.g. Comte, 1830-1842; de Alvarenga et al., 2011; Latour, 2000, 2004). To a degree, these challenges could be explained in terms of the increasing gulf between two cultures - those of the sciences and the humanities - as suggested by C.P. Snow (1905-1980) in an instigating essay (Snow, 1990 [1959]), due to the high specialization in science and education, and, not less important, to a "tendency to let our social forms to crystallise" (Snow, 1990: 172). More to the point, the increasing importance attributed to the problem has motivated a growing number of analyses concerning the high level of specialization and fragmentation of science and university education (e.g. de Alvarenga et al., 2011; Moraes, 2005; Snow, 1990), but also the societal and political questions concerning research agendas (e.g. Alves, 2008; Kates et al., 2001; Latour, 2000, 2004; Schor, 2008), the disparities between developed and developing countries not just in affluence level, but also in research capacity (Kates et al, 2001; Pfeiffer, 2008; Schor, 2008), and, finally, from a more methodological point of view, the multiplicity of theoretico-methodological perspectives admitted by the social sciences (e.g. de Alvarenga et al., 2011; Floriani et al, 2011; Giddens, 2001; Leis, 2011; Moraes, 2005; Oliveira Filho, 1976; Raynaut & Zanoni, 2011; Weffort, 2006).

Yet, in the E.S. field the problem of bringing together social and natural sciences has been a permanent and still unresolved challenge (Alves et al., 2007; Alves, 2008; Geoghegan et al.,

1998; Hick et al., 2010; Liverman & Cuesta, 2008), despite its recognized central relevance for E.S. research programs (e.g. Hogan & Tolmasquim, 2001; IGBP , 2006; Leemans et al., 2009; Reid et al., 2010; Young, 2008). In this field, inter-disciplinary articulation is of great interest and importance specially due to the challenges of postulating societal responses to environmental changes attributed to society itself and addressing the considerable level of uncertainty in detecting and predicting E.S. changes as in the case of the Intergovernmental Panel on Climate Change (IPCC) (e.g. Beven, 2011; Bradshaw & Brochers, 2000; Houghton & Morel, 1984; Houghton, 1990; Houghton, 2008; IPCC, 1990, 1996, 2001, 2007; Saloranta, 2001; Shackley et al., 1998; Thatcher, 1990).

The study of the Earth System is the object of a number of research programs that has been generally defined as "the study of the Earth system, with an emphasis on observing, understanding and predicting global environmental changes involving interactions between land, atmosphere, water, ice, biosphere, societies, technologies and economies" (Leemans et al., 2009). It constitutes a cross-disciplinary field of research, including a broad array of disciplines and techniques, for which General Systems Theory (G.S.T.) plays a major role for inter-disciplinary articulation. G.S.T. offers the natural sciences a key, yet conceptually simple method to formulate and solve problems involving a variety of disciplines, and can serve, for the social sciences, as the basis for conceptualizing about social systems by taking into account their functions, reproduction and meaning behind social action (Buckley, 1976; Luhmann, 2010; Rhoads, 1991). At the same time, a number of critical issues concerning environmental change and societal responses to it, including the conditions for the stability of social order, the possibilities for social change, and the role of the knowing human agent (e.g. Giddens, 2001; Habermas, 2000 [1968]; Luhmann, 2010; Rhoads, 1991; Rosenberg, 2010) may need a broader theoretico-conceptual framework extending beyond G.S.T. to be answered.

The main objective of this chapter is to examine inter-disciplinary articulation in E.S. studies, investigating how General Systems Theory and the multiplicity of theoretico-methodological perspectives taken by the social sciences[1] can come together to explore both the "physical" problem of the changing E.S. and the social process of the emergence - for the social world - of the meaning of the changing E.S. problem[2]. The example of the Intergovernmental Panel on Climate Change (IPCC) is taken to illustrate how the problem of climate change may have emerged for the social world. The aim of the chapter is to contribute to broaden the prevailing conceptual model of Earth System studies, in which the technical concepts of observing and modelling are usually better understood and studied, by attempting to complement it with a few reflections about the part played by society.

---

[1]Before addressing the multiplicity of theoretico-methodological perspectives in the social sciences in more detail, it is possible to mention, as examples, the concepts of ideal type (Weber, 2005a [1904]), social fact (Durkheim, 1894), and structure and superstructure (Marx, 1859), which offer different approaches to conceptualize about the social world.

[2]Here it is postulated that in order to recognize and respond to the problem of the changing E.S., the social world needs both to understand the „physical" nature of the environmental changes and to elucidate to itself what such changes might mean. Although natural and social sciences take part in both processes, the emergence for the social world of the meaning of the problem would be seen as the result of social interaction leading to the elucidation of the extent and the consequences of the problem, as well as of possibilities of responding to it. The assumption of the double hermeneutic (Giddens, 2001) described in section 3.1 will help further explore these ideas for the case of the IPCC.

The chapter is organized in three major sections: the first presents an introductory, brief review of the problem of inter-disciplinary articulation and the importance of G.S.T. as a tool for it, the second reassesses the concept of method in the natural and the social sciences, and postulates how the problem of the emergence of meaning of environmental change can be explained within the G.S.T. framework, and the last section examines the workings of the IPCC, postulating the emergence of the ideas of detection and attribution of climate change, and of emission scenarios as shared concepts between the social world and science, that helped the social world to elucidate to itself what climate change might mean.

2. On inter-disciplinary articulation and general systems theory

2.1 A brief account of inter-disciplinary articulation

The question concerning the articulation of scientific knowledge produced by different disciplines has relevance not only for E.S. studies, and includes many different aspects such as the question about the unity of science, the processes leading to disciplinary fragmentation, epistemological differences among sciences, and the varied understandings of the concept of inter-disciplinarity (e.g. Aubin & Dalmedico, 2002; de Alvarenga et al, 2011; Jollivet & Legay, 2005; Jordi, 2010; Leis, 2011; Nowotny et al, 2003; Poincaré, 1968 [1902]; Raynaut & Zanoni, 2011; Schor, 2008; von Bertallanffy, 1950).

The growing importance of this question can be perceived, in particular, following the great achievements of science in the late XVIII and early XIX centuries, and the multiplication of scientific disciplines that started at that time, including the foundation of what would become sociology. In addition to the question of understanding how scientific knowledge could be achieved - which would include enquiries on the nature of scientific knowledge and method - it would be proposed that such knowledge would provide a basis to make society more just and, not less important, to evade social crises such as those of the time of the French Revolution.

One of the key conceptions at that time, one that followed the Galilean tradition, but also reflected new scientific advances in the domains of physics and chemistry, postulated a unifying, analytical view of the world provided by mathematics, as illustrated by the proposition made by the mathematician Marquis de Laplace (1749-1827):

> "We ought [...] to look at the present state of universe as the effect of its previous state, and as the cause of the following one. An intelligence which, for a given moment, would know all the forces animating nature, and the conditions of the beings composing it, if furthermore it would be as immense as to analyze these data, would hold together in the same one formula the movements of the largest bodies in the universe, and those of the lightest atom: nothing would be uncertain for it, and the future as the past, would be before its eyes" [3] (Laplace, 1825: 3-4; my translation)

At about the same time, Auguste Comte (1798-1857) saw the construction of scientific knowledge as needing a more complex logico-theoretical framework. For him, Laplace's

---

[3] *"Nous devons [...] envisager l'état présent de l'univers, comme l'effet de son état antérieur, et comme la cause de celui qui va lui suivre. Une intelligence qui pour un instant donné, connaitrait toutes les forces dont la nature est animée, et la situation respective des êtres qui la composent, si d'ailleurs elle était assez vaste pour soumettre ces données à l'analyse, embrasserait dans la même formule les mouvements des plus grands corps de l'univers et ceux du plus léger atome: rien ne serait incertain pour elle, et l' avenir comme le passé, serait présent à ses yeux."* (Laplace, 1825: 3-4)

ideas would have been presented as a "simple philosophical game" without real consequences not even for the progress of chemistry, and offering no way to achieve a "scientific unity" which might comprehend, for example, "physiological phenomena" (Comte, 1830-1842: 58). Comte envisaged a conceptual interconnection for all scientific knowledge including the new discipline of "social physics" or "sociology", whereas

> "[to determine] the actual dependence of various scientific studies [...] it is possible to organize them among a small number of categories [...] arranged in such fashion that the rational study of each category is founded on the knowledge of the laws [...] of the previous category, and become the foundation for the study of the next one [...] from what follows [the] successive dependency [of observable phenomena]" (Comte, 1830-1842: 77; my translation).[4]

In this conception, the understanding of social phenomena was to contribute to the greatest good of humanity, as a result of "social physics" achieving the same positive stage of the study of astronomical, physical, chemical and physiological phenomena. Such views would not necessarily search for unique, unifying laws encompassing all branches of knowledge, and would leave room for the admission of limits to scientific knowledge at any given moment, but they would nonetheless think of an entire unified scientific building as the result of the juxtaposition of knowledge from the different branches of science.

Throughout the XIX century and early 1900s, a series of developments in physics, mathematics, biology, as well as in the social sciences, motivated lively debates about the nature of science and the construction of scientific knowledge, and, also, about the methods and the role of the social sciences. These debates would have a long list of protagonists, including John Stuart Mill (1806-1873), Charles Darwin (1809-1882), Claude Bernard (1813-1878), Karl Marx (1818-1883), Herbert Spencer (1820-1903), Ludwig Boltzmann (1844-1906), Vilfredo Pareto (1848-1923), Emile Durkheim (1858-1917), Max Planck (1858-1947), Alfred Whitehead (1861-1947), David Hilbert (1862-1943), Max Weber (1864-1920), Bertrand Russell (1872-1970), Albert Einstein (1879-1955), Werner Heisenberg (1901-1976), Kurt Gödel (1906-1978), and many others. These developments would mark the debate on inter-disciplinary articulation, reflecting many different, sometimes opposing, views of the possibilities and methods of science, and producing a long lasting effect on the conception of the inter-relationships among different disciplinary knowledge.

In the field of the physical sciences, in particular, the debate would include a number of issues that have relevance for the field of Earth System research, as can be illustrated by the writings of Henri Poincaré (1854-1912), a prominent French mathematician, physicist and philosopher of science. He was among the several scientists that contemplated the problem of the nature of different sciences and the construction of knowledge in mathematics, mechanics, gas dynamics and other domains. For him, physics would be mainly an experimental science conditioned by the scale of observation; his understanding of an experimental science was based on the idea that every "experimental law is always subjected to revision [and that] we should always expect to see it replaced by another, more precise one". Attuned to the great doubts afflicting his time, Poincaré proposed that neither

---

[4] "*[pour déterminer] la dépendance réelle des diverses études scientifiques [...] il est possible de les classer en un petit nombre de catégories [...] disposées d'une telle manière, que l'étude rationnelle de chaque catégorie soit fondée sur la connaissance des lois [...] de la catégorie précédente, et devienne le fondement de l'étude de la suivante. [...] d'où résulte [la] dépendance successive [des phénomènes observables]*" (Comte, 1830-1842: 77)

space nor time had any absolute sense (1968 [1902]: 116), and that the "science of the numbers" would be "synthetic *a priori*", questioning the validity of the program asserting that mathematics could provide analytical means to "apprehend every truth" of the world (: 32); he also conjectured that Euclidian geometry would be "provisory", while non-Euclidian geometries - like those of Lobatchevsky and Riemann - might prove to be adequate for problems involving "very large triangles or highly precise measurements" (: 74). A very precocious investigator who faced the challenge of "chaotic" behaviour in his studies of the stability of the Solar System, Poincaré would state that:

> "The simplicity of [Johannes] Kepler's [1571-1630] laws [of planetary motion] [...] is nothing but apparent. That should not forbid that they shall be applied [...] to all systems similar to the solar system, yet that should prevent that they'd be rigorously exact" (Poincaré 1968 [1902]: 165; my translation)[5]

These ideas reveal a series of difficulties and impasses verified in the natural sciences at the time, including those that would lead to the relativity and quantum theories, and challenge the efforts of linking atomic theory and the kinetic theory of gases, and the postulates about the foundation of mathematics. This would significantly impact the understanding of what science is, justifying, for example, the proposition of the convention of falsifiability by Karl Popper (1902-1994), the increased perception of incommensurability of scientific knowledge from different disciplines, and the questions concerning the possibilities and the limits of both observation and formal inference.

In this context, the prospect of *a priori* interdependence among scientific disciplines based on shared categories, as conceived by Comte, would fail to provide a consensual, universal framework for scientific articulation, just as Laplace's model would do. At the same time, the natural and social sciences would continue to interact, exploring and borrowing ideas one from another, and investigating problems involving multiple disciplines. This interaction would include, most particularly, the use of analogies, as in the case of V. Pareto, whose concept of social equilibrium was analogue to mechanical equilibrium, and H. Spencer, who extended Darwin's ideas of natural selection to society and thought of society as a social organism formed by different organs, borrowing ideas from mechanics and biology (Buckley, 1971; Rosenberg, 2000). Not less importantly, General Systems Theory ideas of exchange of matter and energy among several elements or systems, as well as the concepts of system reproduction and evolution would provide a valuable investigative framework for a number of problems requiring inter-disciplinary articulation, as examined in the next section.

2.2 General systems theory in the uncertain inter-disciplinary E.S. field

General Systems Theory (G.S.T.) - defined by von Bertallanffy (1950) as a "logico-mathematical discipline [...] applicable to all sciences concerned with systems"- has played a central role in integrating a variety of disciplines in many fields of research (e.g. von Bertallanffy, 1950, 1972; Alves, 2008; Almeida Júnior et al, 2011), and, not less importantly, has been applied to the domain of social systems (Buckley, 1971; Luhmann, 2010; Rhoads, 1991). It has evolved from a series of methods aiming at the representation, simulation

---

[5]"*La simplicité des lois de Képler [...] n'est qu'apparente. Cela n'empêche pas qu'elles s'appliqueront[...] à tous les systèmes analogues au système solaire, mais cela empêche qu'elles soient rigoureusement exactes.*" (Poincaré 1968 [1902] : 165)

and/or control of a broad variety of processes ranging from control theory to biological, ecological and social systems. In Earth System studies, the use of G.S.T. is of key importance as it provides the basic instrumental means to join together the different Earth "sub-systems" for which numerical modelling and simulations are performed.

Here, a system will be understood as an entity formed by interacting elements, whose evolution presupposes exchange of energy and matter with its surrounding environment, at the same time as such entity is capable of maintaining or reproducing itself in this environment. This definition is similar to other system definitions (e.g. Buckey, 1971; Gell-Mann, 1994; Luhmann, 2010), although it could be noticed that it attempts to put as much emphasis on the ideas of system reproduction and evolution as on that of system maintenance. Examples of such entities may be the atmosphere, the oceans and terrestrial ecosystems, that during their entire histories have evolved by continually exchanging energy and matter among themselves.

The atmospheric and the oceanic systems can be considered to be the two central components of Earth System research investigating climate change (e.g. McGuffie & Henderson-Sellers, 2001; Randall et al, 2007), as they are the major entities responsible for heat storage and transport across the globe in climate models. At the same time, the atmospheric-oceanic climate system is connected to other systems, including the terrestrial ecosystems - which can act as a sources or sinks of greenhouse gases, and, not less importantly, transform themselves due to ecological succession in face of climate change. Similarly, social systems – the source of "dangerous anthropogenic interference with the climate system" (United Nations, 1992) – are also expected to evolve, transforming themselves to both mitigate and adapt to climate change.

In the case of the Earth System research, G.S.T. offers a very valuable and unifying framework to join together several different disciplines. Yet, a conceptual understanding of such a system does not imply that the study of the changing Earth System would assure that accurate predictions of climate and environmental change can be achieved, a fact that has had important implications for both seeking legitimacy for E.S. research and for conceiving of how social systems will respond to climate change (e.g. Bevin, 2011; Houghton, 2008; Le Treut et al, 2007; Saloranta, 2001; Verosub, 2010). This state of affairs justifies the need for understanding the different sources of uncertainties[6] in E.S. studies, and here four different uncertainty categories are highlighted:

- uncertainties that are intrinsic to the chaotic nature of some Earth-System processes, significantly affecting the feasibility of long-term prediction of atmospheric and oceanic fluid dynamics (e.g. Lorenz, 1963; Houghton & Morel, 1984);
- uncertainties due to insufficient and incomplete knowledge about key atmospheric, oceanic, and ecosystem processes (e.g. Kesselmeier et al, 2009; Longo et al, 2009; Randall et al, 2007);
- uncertainties resulting from the choices made in implementing numerical models of the Earth System, due to limited computational resources and observational data, and to parameterization in coupling the various E.S. sub-systems (e.g. McGuffie & Henderson-Sellers, 2001; Randall et al, 2007);
- uncertainties arising from the impossibility of actually predicting changes and the evolution in social systems (e.g. Rosenberg, 2000).

[6]For further analises of uncertainty relevant to this context see also Brown (2010), Lahsen (2005) and Shackley et al (1998).

The first three categories can be attributed to the characteristics of the natural sciences objects and methods in Earth System research, which have been the focus of continuous efforts of model improvement and data collection (e.g. Forster et al, 2007; Houghton & Morel, 1984; IPCC, 1990, 1996, 2001; Le Treut et al, 2007; Randall et al, 2007; Solomon et al, 2007). They might be assumed not to be directly relevant to the problem of articulation between natural and social sciences, even though the reader shall keep in mind their potential effects on the reception of E.S. research by the social world (e.g. Beven, 2011; Houghton, 2008; Le Treut et al, 2007; Verosub, 2010).
The assumption of the impossibility of predicting changes in social systems seems to be of greater relevance to analyze the problem of inter-disciplinary articulation involving the social sciences. To address this problem it might be useful to highlight a few perspectives from the social sciences which are relevant to the question concerning environmental change, as attempted below.

2.3 Articulation with the social sciences and environmental change studies

Despite the recognition of existing difficulties in articulation between the natural and the social sciences, environmental change has been the focus of several social science programs and projects with varying degrees of inter-disciplinary articulation with the natural sciences (e.g. Lambin & Geist, 2006; Moran & Ostrom, 2005; Pfeiffer, 2008; Young, 2008). Moreover, the establishment of environmental change as a field of research has contributed to systematizing a number of ideas and perspectives that are helpful to advance the discussions of inter-disciplinary articulation.
First of all, environmental change is frequently assimilated - from a theoretical perspective - to the problem of scarcity or distribution of resources in face of a growing population, usually taking as reference some of the postulates of Thomas Malthus (1766-1834). This theoretical perspective has received attention from several commentators, who discussed the role of technology to answer to population pressure and scarcity of resources (Boserup, 1995 [1965]; Floriani et al, 2011; Hardin, 1968; Mortimore, 1993; VanWey et al, 2005), and its political-economic, ideological and political-philosophic roots (Harvey, 1974; Montibeller, 2008; Walker, 1988). Not less importantly, a number of analyses contributed to refer this debate to questions of inequality among nations and to the development agenda (e.g. Cardoso, 1972; Furtado, 1998 [1974]; Martins, 1976).
More recently, two new fields of study - environmental sociology and political ecology - have offered valuable contributions to the problem of articulation of the natural and social sciences in the context of environmental change, in particular, by systematically reviewing a number of classical issues in the social sciences (e.g. Alimonda, 2002; Alonso & Costa, 2002; Hannigan, 2006).
In these fields, K. Marx (1818-1883), E. Durkheim (1858-1917) and M. Weber (1864-1920) are usually recognized as key references from classical, XIX-century, social theory (e.g. Hannigan, 2006) offering critically relevant, but frequently opposing views to the problem of scarcity and distribution of resources and its relation to social stratification and order. For example, Marx's concepts of structure and superstructure, his attribution of changes in the former to the transformation of the latter, and the assertion that nature is as much a source of value and wealth as labour (Marx, 1859, 1875), assume the pre-eminence of economic relations of production and appropriation of surplus value as sources of both societal contradictions and transformation. Durkheim's definition of social fact, and his distinction

between normal and pathological social phenomena (Durkheim, 1894), his understanding of solidarity (Durkheim, 1893), and anomie (Durkheim, 1897), presuppose the existence of social facts as "things" external to individuals and analyze the role of social norms and practices as something that could help to evade or to understand dysfunctional states and crises in society. Weber's concept of ideal type and his analyses of the nature of the social sciences (Weber 2005a [1904]), the distinction among class, status group and party (Weber 2004 [1922, posthumous]), and his analyses of the German national question (Weber 2005b [1895]) take into account the influence of scientist's values for developing theories and abstractions, and allow to examine social differentiation and stratification beyond the strict limits of economic relations.

This quick, certainly far from comprehensive, recollection of Marx's, Durkheim's, and Weber's ideas is indicative of the different methodological and theoretical perspectives taken by these authors, as well as of their differing logical and philosophical approaches to social phenomena. A number of other classical authors and theories can contribute new perspectives to the context of the problem of the changing Earth System, among which V. Pareto and H. Spencer, for their views of social stratification and competition, and the references to them in the study of social systems (Buckley, 1971); and the XIX-century theories of geographic and biological determinism that have been recognized as being of interest in our context (Bresciani, 2005; Hannigan, 2006).

This multiplicity of methodological and theoretical approaches recognized since the "classical" 1800s has been the cause of continuous and lively debates, in which theories can be tentatively or effectively falsified, and questions concerning the scale and context of their validity can be raised (e.g. Browder et al, 2008; Giddens, 2001; Lambin et al, 2001; VanWey et al, 2005). At the same time, it also represents a critical element of the philosophy, the theory, and the methods of the social sciences, as it is related to the capacity of judgment and intent of the social agent, and to the very question about the possibility of predicting changes in social world (e.g. Arendt, 2010; Giddens, 2001; Rosenberg, 2000).

Here, it is suggested that such multiplicity of approaches is one of the major sources of tension in attempts to articulate the natural and the social sciences in the study of the changing Earth System. Taking into account or ignoring the fact of this multiplicity ends up having important consequences to the very conceptualization of inter-disciplinary articulation, most particularly, in efforts to explore new possibilities of enquiry on how the meaning of environmental change can emerge for the social world, and on the possibilities of articulation with the political field. It is also suggested that exploring the differences in the understanding of the concept of method in the natural and the social sciences can help to better recognize this multiplicity and some of its implications for the study of the changing Earth System.

3. Methodological issues in studies of the Earth System

3.1 Postulating different understandings of the concept of method

By assuming that the study of the changing Earth System requires the articulation between the natural and the social sciences (e.g. IGBP, 2006; Reid et al, 2010), crucial questions about **how** to actually achieve such an articulation will arise, concerning both how to conceive of the investigative process involving multiple disciplines and how to consider the different logical, epistemological, ontological and political perspectives in relation to the problem of changing Earth System (e.g. Alimonda, 2002; Alves, 2008; Geoghegan et al., 1998; Hick et al.,

2010; Liverman & Cuesta, 2008; Moraes, 2005; Oliveira Filho, 1976; Schor, 2008; Shackley et al., 1998). Although these questions have not impeded close collaborative work between natural and social scientists (e.g. Alves, 2008; Lambin & Geist, 2006; Moran & Ostrom, 2005), they can justify a broader examination of the persistent difficulties in approaching the articulation problem (e.g. Alimonda, 2002; Alonso & Costa, 2002; Liverman & Cuesta, 2008; Moraes, 2005; Schor, 2008).

This section's departing point is the different understandings of the concept of method as presented by Moraes (2005), who proposed that methods, in the natural sciences, are understood just as the "technical-instrumental means" of investigation, while, in the social sciences, they in fact represent "logico-theoretical frameworks" for scientific enquiry. This differentiation is summarized in Table 1.

<table>
<tr><th></th><th>Techniques</th><th>Methods</th><th>Theories</th></tr>
<tr><td>Natural sciences</td><td colspan="2">Technical-instrumental means of investigation</td><td rowspan="2">Hypothetico-deductive or inferential systems allowing for interpretation of natural and social phenomena</td></tr>
<tr><td>Social sciences</td><td>Technical-instrumental means of investigation</td><td>Logico-theoretical frameworks for enquiry</td></tr>
</table>

Table 1. Schematization of the different concepts of method in the natural and the social sciences, based on Moraes (2005) and Audi (2005).

In addition to these differences in the concept of method, it is useful to distinguish two different aspects of the methodologico-theoretical problem according to Oliveira Filho (1976): the conception of the process of social investigation and the different logical, epistemological and ontological perspectives that can be found in the field of study. For this author, the process of social investigation can include, for example, functionalism, ethnomethodology, and structuralism, to which it seems appropriate to add possibly different frameworks for data collection, systematization and analysis (e.g Moran & Ostrom, 2005); different logical, epistemological and ontological perspectives can be exemplified by the dialectical, hermeneutical, and pluralistic methods. Here it will be suggested that conceiving of the process of investigation may represent not the largest of the obstacles to collaborative work, provided that the multi-disciplinary team be capable to work towards common investigative problems and questions (see, for example, Alves, 2008; Keller et al., 2009; Moran & Ostrom, 2005; Schor, 2008). On the other hand, different logical, epistemological, and ontological views may be at the origin of a challenge of different nature, in particular, as they can be intertwined with the attribution of different meanings to social phenomena not only by scientists, but also in the social world. Further discussion of the nature of this challenge and its implications can easily expand into the domains of political science, philosophy of science and philosophy of the social sciences (e.g. Arendt, 2010; Latour 2000, 2004; Rosenberg, 2000), possibly creating further barriers for understanding what to expect of inter-disciplinary articulation. Here, this discussion will quickly refer to the concept of double hermeneutics proposed by Anthony Giddens (1938-, e.g. Giddens 2001), which can provide an instrumental reference to conceive of how the meaning of environmental change emerges in the social world, considering, at the same time, the nature of the contribution of science to this process.

The concept of double hermeneutic posits that the social sciences are distinguished from the natural sciences by the fact that the latter "consist of hermeneutic or interpretive efforts [...] [where the interpretation of the natural-science laws] must occur in the domain of theoretical systems" (Giddens, 2001: 101; my translation from the Brazilian edition), while the former are concerned by "[knowing] agents [imbued of intent] that generate and invent concepts, theorize about what they do, as well as about the conditions under which they perform their acts [...] In contrast to the natural science [...] the social sciences entail a double hermeneutic, since the concepts and theories developed in their domain are applied to a world which is constituted of activities performed by individuals who conceptualize and theorize [about their world]" (Giddens, 2001: 111; my translation from the Brazilian edition). Getting back to the distinction between the process of investigation in itself and the different logical, epistemological and ontological ideas permeating different methodological approaches (as in Oliveira Filho, 1976), it can be suggested that the assumption of the double hermeneutics helps further scrutinize the problem of different, frequently opposing logical-philosophical-political views behind the methodological question. In fact, by admitting a "knowing human agent" capable of attributing meaning to the findings of science and to respond to these because he/she is instilled with intent, it also assumes that it is not up to the "social scientist to interpret the meaning of the social world for the social actors therein inserted" (Giddens, 2001: 101; my translation from the Brazilian edition). While stating that, Giddens also proposes that the practical impact of the social science will be found in the social world actually absorbing social sciences concepts, without abdicating from its own capacity of judgment and intent. The concept of the double hermeneutic has been considered in a number of social analyses, ranging from the field of education, to cultural and political-philosophical problems (e.g. Aguiar, 2009; Botelho & Lahuerta (2009); Domingues, 1998, 1999; Magalhães & Stoer, 2002; Rodrigues, ND). As suggested in section 4, it can open new perspectives to assess the role of science by analysing the work of the IPCC. Before concluding this section, it seems appropriate to raise the question about what possible places can be attributed to social systems and to the knowing human agent, in the study of the changing Earth System, where General System Theory plays a central role. This will be explored next.

3.2 The place of social systems in the study of the changing Earth System

The question concerning the effective role of the social sciences in the Earth System field is far from being a consensual one, including different views ranging from the indication that the social sciences "have been reluctant to respond to global-change science" to the proposition that they are "critical in shaping the public discourse on the changing socio-environmental condition". Although this lack of consensus has not prevented collaborative work involving the natural and the social sciences (e.g. Alves, 2008; Lambin & Geist, 2006; Moran & Ostrom, 2005), a fundamental question can be raised about the place of the "social system" in E.S. research, most particularly, if the interest of investigation is how the meaning of environmental change can emerge in the social world, and a "knowing human agent", capable of judgement and intent, is to be recognized.

Here, three aspects of this problem will be referred to, a first one related to the sceptical views about G.S.T. in some domains of the social sciences, a second one pondering the addition of a social component to Earth System models, and a last one discussing how the concept of social systems can be of interest in studies of the changing Earth System.

The sceptical views concerning G.S.T. have their roots in the association of this method with technocratic inclinations, including the postulate that social and economic problems can be resolved based on "objective" knowledge provided by the technocracy, and a tendency to dispense with political constituency and representativeness (see, for example, Habermas, 2000; Leff, 2002; Martins, 1976; Mirowski, 2003; Schwartzman, 1980; Whiteside, 1998). Although this scepticism will not be examined in detail here, it seems pertinent to notice, on one hand, that E.S. researchers should be aware of it, and, on another, that the problem of environmental change is involved in multi-faceted processes that makes the analysis of its political dimensions particularly complex, extending beyond the questions about the technocracy (e.g. Alonso & Costa, 2002; Latour, 2004; Leis, 2011; Raynaut & Zanoni, 2011; Santos & Alves, 2008).

The idea of adding a social sub-system to fully-coupled Earth System models seems to have its roots in the Galilean-Laplacian mathematico-analytical views, in the foundation of cybernetics and modern G.S.T., and, more recently, on agent based models (e.g. Gell-Mann, 1994; Grimm et al, 2005; Holland, 2006; Mirowski, 2003; Parker et al, 2006; von Bertalanffy, 1950, 1972; Whiteside, 1998). Although a complete analysis of this proposal is still to be done, it can be observed that conceiving of a social sub-system as part of a broader system can be instrumental to exercise inter-disciplinary collaboration by taking into consideration social structure and processes. However, reducing the social world to just one element of a larger system may elude key socio-logical and political aspects of the process of emergence of the meaning of environmental change, and contribute to some form of technocratic predisposition concerning the issue of societal responses to the changing E.S. (Mirowski, 2003; Shackley et al, 1998; Whiteside, 1998).

In contrast, the question whether and how the concept of social system could be adopted in Earth System research is suggested to potentially contribute to the approximation of the natural and the social sciences on more conceptual ground. In fact, despite some scepticism concerning the relationship between G.S.T. and the social sciences, a number of authors have examined how different logico-theoretical frameworks can be combined with the concept of social system (e.g. Buckley, 1971; Luhmann, 2010; Rhoads, 1991), pointing to the possibility of taking into account the role of the knowing human agent. Most notably, the contributions by Talcott Parsons (1902-1971), George Homans (1910-1989), and Niklas Luhmann (1927-1998) offer a variety of conceptual frameworks allowing consider the changing, "live" nature of social structure and action, and, in varying degrees, the meanings and intents present in the social world. Although such work does not seem to contribute to the conceptualization and construction of more powerful Earth System models or simulators, it is suggested that they offer important perspectives to bridge the gap between the "two cultures" as they can help to incorporate some key social science issues and categories into the Earth System field debates. Furthermore, it is suggested that by recognizing a "living" social system, it may be possible to re-position some of the questions about the interface between the social world and science.

Such conceptions of the social system presuppose that it is only in the process of social reproduction - including the processes of social interaction and mobilization mediated by social institutions and stratification - that the meaning of environmental change can emerge for the social world. There are two aspects of this proposition that need to be further stressed. First, it does not assume any definite need of incorporating a social sub-model in a single fully-coupled model of the changing Earth System as part of an integrated simulation effort; yet, it still corroborates to the assumption that system theory concepts can serve as a common ground for interaction between natural and social scientists (e.g. Buckley, 1971; Luhmann, 2010). Second, and perhaps most importantly, it re-positions inter-disciplinary research so that it becomes possible to look for questions for scientific investigation outside the strict dynamics of the scientific field, i.e. in the very process of social reproduction, when the meaning of environmental change emerges and evolves continually for the knowing human agent. It is this emergence of questions shared by both the scientific field and the social world that will be postulated and discussed for the case of the IPCC in section 4.

By leaving aside the assumption of the definite need of implementing fully-coupled simulation models including the social world, it is not suggested that there are no relationships between this world and the "physical" Earth System. Rather, it is proposed that the social world may need different concepts and logics than those of the physical system to be elucidated. Thus, it is suggested that social systems can be conceptualized about by recognizing in it a symbolic dimension, in which socially mediated information can be valued by the knowing human agent, and a reproductive-evolutionary dimension in which the meaning of environmental change can emerge and evolve in the process of social reproduction.

Figure 1 illustrates these ideas in the form of a 3-dimensional conceptual space in which physical, ecological, and social systems are represented with the purpose of inter-disciplinary articulation in Earth System studies. The "physico-chemical" axis corresponds to the conceptual dimension of atmospheric-oceanic-biogeochemical "physical-climate" systems, for which the question of exchanging energy and matter, and the idea of positive and negative feedbacks constitute the main elements for conceptualizing and modelling. The figure also suggests that the emergent/reproductive-evolutionary axis together with the physico-chemical one define a 2-dimensional "*conceptual plane*" where the reproduction and evolution of ecological systems can be conceptualized aboutby taking into account both physical processes and ecological succession; in this case, the long term result of ecological succession might be, for example, the emergence of new ecosystems as the product of climate change.

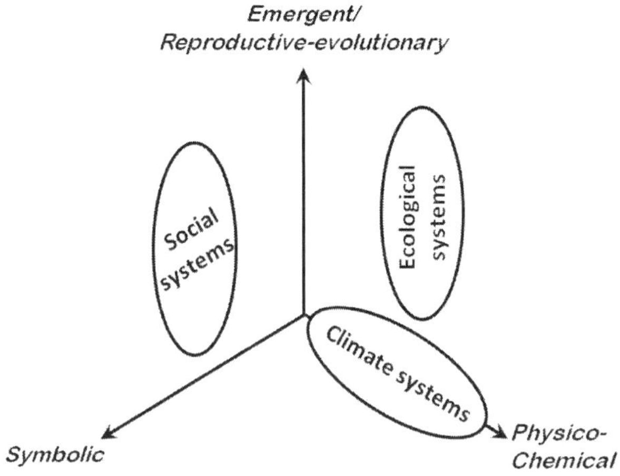

Fig. 1. Schematic representation of three conceptual dimensions for Earth System studies.

Finally, the figure shows a second *"conceptual plane"* defined by the symbolic and the emergent/reproductive-evolutionary axes, which is suggested to be the place for conceptualizing about social systems. It should be noticed that the emergent/reproductive-evolutionary dimension is proposed to highlight the idea of evolution of both the social and the ecological systems, which incorporates both *inheritance* and *random elements*, making the long-term prediction of such systems daring. As noted before, the impossibility of prediction in social systems is also related to the nature of the knowing human agent.

4. An assessment of the Intergovernmental Panel on Climate Change

The Intergovernmental Panel on Climate Change (IPCC) defines itself as an "international body for the assessment of climate change [...] established by the United Nations Environment Programme (UNEP) and the World Meteorological Organization (WMO) to provide the world with a clear scientific view on the current state of knowledge in climate change and its potential environmental and socio-economic impacts". Its constitution assures that it is both "a scientific body [which] reviews and assesses the most recent scientific, technical and socio-economic information produced worldwide relevant to the understanding of climate change" and intergovernmental, in the sense that "governments participate in the review process and the plenary Sessions" (IPCC, ND). The Panel was the recipient of the 2007 Nobel Peace Prize "for [its] efforts to build up and disseminate greater knowledge about man-made climate change, and to lay the foundations for the measures that are needed to counteract such change" (Nobel Foundation, ND).

The Panel's dual constitution as a scientific body in which governments take part has been suggested to offer a new model for the science-policy interface stressing extensive public reviews (e.g. Saloranta, 2001). Yet, the socio-political nature of its procedures seems to be easily eluded when its achievements are seen as the result of the objectivity of its scientific results alone (e.g. Houghton, 2008). The IPCC workings are not immune to the debate involving the problem of making political choices, most notably suggestions of a "technocratic policy orientation to [the] climate change [problem]" (e.g. Shackley et al, 1998). Despite disputes involving the Panel, it is considered a very respectable body since its foundation, as demonstrated by it being awarded the Nobel Peace Prize, the testimonies of several public persons (e.g. Thatcher, 1990), and appraisals of the effectiveness of its contribution to the climate change debate (e.g. Saloranta, 2001).

Notwithstanding the wide public recognition of its "technical" contribution to the climate change debate, it is suggested here that seeing it as a predominantly technical-scientific body can elude the nature of its dual constituency and can be misleading. Indeed, the Panel's mission statement asserting that it "reviews and assesses the most recent scientific, technical and socio-economic information", and the analyses focussing on the nature of its "reviewing" procedure (e.g. Saloranta, 2001) may conceal the fact that it has been built on a privileged relationship with the socio-political world based on sharing the meanings of a number of concepts between the scientific field and the social world. It is proposed that these shared concepts - those of detection and attribution of climate change, and emissions scenarios - played a crucial role in the very institution of IPCC, as well as on the success of public mobilization around the climate change issue. It is further postulated that this can be apprehended based on the double hermeneutic concept of Giddens (2001) summarized in section 3.1.

Indeed, the concepts of detection and attribution[7] - have clear shared meanings for both the political world and the scientific field, following the dialectics assumed by Giddens, in which the impact of the social sciences appears when "socio-scientific concepts [end up being selectively absorbed] by the social world, to which such concepts [become] a constitutive part" (2001: 112; my translation from the Brazilian edition). In addition, it can be observed that attribution of climate change to human action depends on the shared understanding of the uncertainties intrinsic to climate modelling. This dialectics is manifested in the importance attributed to detection and attribution in both the technical and the political discourses, which can be documented by the generalized and continuing use of these concepts, as central elements of both scientific investigation and political deliberation as can be found, for example, in Houghton & Morel (1984), Thatcher (1990), and in the scientific reports of IPCC WG1 (IPCC, 1990, 1996, 2001, 2007).

The role of the concept of emissions scenarios[8] (i.e. scenarios of emissions of greenhouse gases as defined, for example, in IPCC, 2000), can be understood in similar terms. In this case, it also presents particular interest because it offers the prospect of descriptions of the future which would be immediately shared with the social world, but manifestly have not the status of predictions, since the result of the reproduction and evolution of the social world is admittedly uncertain for anyone. Emissions scenarios, more particularly, serve to at least two purposes - as qualitative narratives that can be associated to reference ranges of emissions necessary to parameterize GCM models, and as an idea reflecting the unpredictability of emissions produced by social systems for both the scientific field and the social world itself.

In comparison to other analyses, which put emphasis on the IPCC review procedures, for example, those assuming an "'Extended Peer Community' [where] various stakeholders with various perspectives [...] are brought into the dialogue assessing the input from science to decision-making" (Saloranta, 2001: 492), the double hermeneutic framework may consider that some core issues pertaining to IPCC work are not *a posteriori* deliberated by society. On the contrary, here it is assumed that shared concepts have emerged in a social world in which the scientists are embedded. The success in arriving at some shared concepts does not presupposes consensual, definite and comprehensive responses as a result of the production of "objective" knowledge by science, and the dialectics assumed by the double hermeneutics can potentially recognize situations of conflict, contradictions, and the result of different mobilizations in relation to the environmental change issue (e.g. Alves, 2008; Alves, 2010; Schor, 2008; Shackley et al, 1998).

5. Conclusion

The societal and political aspects of the problem of the changing Earth System have represented a major challenge for both the development of Earth System studies and to consider the question about the societal responses to climate change. In such a context, the articulation between the natural and the social sciences is also seen as a significant challenge

---

[7]Here, detection will be understood as detection of climate change, in particular, temperature, and attribution as attributing climate change to specific causes, in particular, to anthropogenic greenhouse gas emissions (Houghton & Morel 1984; IPCC, 1990).

[8]"Scenario is a plausible description of how the future may develop, based on a coherent and internally consistent set of assumptions about key relationships and driving forces. Note that scenarios are neither predictions nor forecasts." (IPCC, 1995: 33)

involving multiple dimensions, whose solution may be expected to be provided by instrumental means - methods - which would allow to carry Earth Systems studies to a new level, and to formulate new strategies and solutions to face issues like the climate change.
This chapter attempted to examine a few basic differences between that the natural and the social sciences, whose understanding is expected to contribute to the goal of responding to societal and political aspects of the changing E.S. problem by taking into account, in particular, the different understandings of the concept of method. It attempted to show to the reader a variety of perspectives concerning how the social world can be understood, from the point of view of the process of investigation, but, also, admitting that different logical, epistemological, ontological and political perspectives are part of the "logic" of the social world for which the meanings of environmental change have ultimately to emerge.
In conclusion, there are three final points that might be stressed here. First, the proposed approach to analyze the workings of the Intergovernmental Panel on Climate Change attempted to put in evidence that climate change is something relevant for both science and the social world, suggesting that more than just providing assessments of climate change, scientists have been engaged in some kind of dialectical exercise in which the scientists and the social world have ended up sharing a small number of key concepts, and have been similarly conscious of the huge uncertainties facing both science and society in relation to climate change. Second, it is necessary to make it clear that the admission of a variety of perspectives concerning the social world does not attempt to demonstrate that "truth is a relative concept" (e.g. Verosub, 2010), or that political aspects of responding to climate change can be reduced to a matter of supposedly objective cost-benefit analysis (e.g. Beven, 2011). In contrast, it is proposed that the social world is capable of attributing meaning and is imbued of intent, and the more this capacity is recognized and exercised, the greater the likelihood that the social world will respond to the climate change issue, although not without its own contradictions, its own inequities, its own aspirations and intents. Finally, it is suggested that this proposition is part of what the two cultures can attempt to develop as some kind of strategy shared with society towards the problem of the changing Earth System, and understanding the nature of this challenge is one of the key contributions that might be expected from the social sciences and their multiple methods.

6. Acknowledgment

I would like to express my gratitude to the book editors, for the clear indication of the parts of the text that needed revision and clarification to be more intelligible to a larger, multi-disciplinary audience of scientists, and to Josefina Moraes and Roberto Araújo for their comments and suggestions to an earlier version of this chapter. This work is the result of several years of investigating and teaching environmental, climate and land change topics at the National Institute for Space Research (INPE), and I am indebted to my post-graduate students and colleagues that have helped to discuss and develop these ideas.

7. References

Aguiar, J.H.C.M. (2009). Hamlet e Próspero: indivíduo, sociedade e intelectual em Anthony Giddens e Pierre Bourdieu, *Revista Ensaios*, Vol. 2, pp. 1-12. Retrieved from http://www.uff.br/periodicoshumanas/index.php/ensaios/information/readers

Alimonda, H. (2002). Introducción: Política, utopia, naturaleza, In: *Ecologia Política: Naturaleza, sociedad y utopia*, H. Alimonda (Ed.), 7-14, CLACSO, ISBN: 950-9231-74-6, Buenos Aires, Argentina

Alonso, A. & Costa, V. (2002). Por uma sociologia dos conflitos ambientais no Brasil, In: *Ecologia Política: Naturaleza, sociedad y utopia*, H. Alimonda (Ed.), 115–135, CLACSO, ISBN: 950-9231-74-6, Buenos Aires, Argentina

Almeida Jr., A.R.; Molina, S.M.G.; Martirani, L.A; Ballester, M.V.R.; Garavello, M.E.P.E.; Verdade, L.M. & Victoria, R.L. (2011). Interação interdisciplinar : a experiência da Pós-Graduação em Ecologia Aplicada da USP, In : *Interdisciplinaridade em ciência, tecnologia & inovação*, A. Philippi Jr & A.J. Silva Neto (Eds.), 298-324, Manole, ISBN 978-85-204-3046-0, Barueri, Brazil

Alves, D.S.; Becker, B.K. & Costa, W.M. (2007). Apresentação, In : *Dimensões Humanas da Biosfera-Atmosfera na Amazônia*, W.M. Costa, B.K. Becker & D.S. Alves, (Eds.), 7-12, EDUSP, ISBN 978-85-314-0993-6, São Paulo, Brazil.

Alves, D.S. (2008). Taking Things Public : A contribution to address human dimensions of environmental change, *Phil. Trans. R. Soc. B*, Vol. 363, pp. 1903-1909

Alves, D.S. (2010). Nosso progresso é nômade: Ensaio sobre as Dimensões Ambientais da Questão da Terra, In: *AMAZÔNIA: DINÂMICA DO CARBONO E IMPACTOS SÓCIOECONÔMICOS E AMBIENTAIS*, S. M. F. Buenafuente (Ed.), ISBN:69-83, ISBN:9788560125430, EdUFRR, Boa Vista, Brazil.

Arendt, H. (2010). *A Promessa da Política* (3rd edn.) DIFEL, ISBN 9788574320861 , Rio de Janeiro, Brazil

Aubin, D. & Dalmedico, A.D. (2002). Writing the History of Dynamical Systems and Chaos: Longue Durée and Revolution, Disciplines and Cultures, *Historia Mathematica*, Vol. 29, pp. 273-339

Audi, R. (Ed.) (2005). *The Cambridge Dictionary of Philosophy* (2nd edn.) Cambridge University Press, ISBN 978-0-521-63722, New York, USA

Beven, K. (2011). I believe in climate change but how precautionary do we need to be in planning for the future? *Hydrological Processes*, published online 9 feb 2011

Boserup, E. (1995). *The Conditions of Agricultural Growth*, Earthscan, ISBN 85383159X, London.

Botelho, A. & Lahuerta, M. (2005). Interpretações do Brasil, pensamento social e cultura política. *Perspectivas*, Vol. 28, pp. 7-15. Retrieved from http:// http://www.fclar.unesp.br/perspectivas/vol_28_lahuerta_botelho.pdf

Bradshaw, G.A. & Borchers, J.G. (2000). Uncertainty as information: narrowing the science-policy gap, *Conservation Ecology*, Vol. 4, Retrieved from http://www.consecol.org/vol4/iss1/art7/

Bresciani, M.S.M. (2005). *O charme da ciência e a sedução da objetividade*, UNESP, ISBN 978-85-7139-782-8, São Paulo, Brazil

Brown, J.D. (2010). Prospects for the open treatment of uncertainty in environmental research, *Progress in Physical Geography*, Vol. 34, pp. 75-100

Browder, J.O.; Pedlowski, M.A.; Walker, R.; Wynne, R.H.; Summer, P.M.; Abad, A; Becerra-Cordoba, N. & Mil-Homens, J. (2008). Revisiting Theories of Frontier Expansion in the Brazilian Amazon: A Survey of the Colonist Farming Population in Rondônia's Post-Frontier, 1992–2002. *World Development*, Vol. 36, pp. 1469-1492

Buckley, W. (1971). *A Sociologia e a Moderna Teoria dos Sistemas*, Cultrix, São Paulo, Brazil

Cardoso, F.H. (1972). Dependency and Development in Latin America. *New Left Review*, pp. 83-95

Comte, A. (1830-1842). *Cours de Philosophie Positive. 1re et 2e leçons.* Retrieved from. http://www.uqac.uquebec.ca/zone30/Classiques_des_sciences_sociales/index.html

de Alvarenga, A.T.; Philippi Jr., A. ; Sommerman, A.; Alvarez, A.M.S. & Fernandes, W. (2011). Histórico, fundamentos filosóficos e teórico-metodológicos da interdisciplinaridade, In: *Interdisciplinaridade em ciência, tecnologia & inovação*, A. Philippi Jr & A.J. Silva Neto (Eds.), 3-68, Manole, ISBN 978-85-204-3046-0, Barueri, Brazil

Domingues, J.M. (1999). Sociologia da cultura, memória e criatividade social. *Dados* [online], Vol. 42. Retrieved from: http://www.scielo.br/scielo.php?script=sci_arttext&pid=S0011-52581999000200004&lng=en&nrm=iso

Domingues, J.M. (1998). Modernidade, tradição e reflexividade no Brasil contemporâneo. *Tempo Social*, Vol. 10, pp. 209-234. Retrieved from: http://www.fflch.usp.br/sociologia/temposocial/site/images/stories/edicoes/v102/modernidade.pdf

Durkheim, E. (1893). *De la division du travail social.* Retrieved from http://classiques.uqac.ca/classiques/Durkheim_emile/division_du_travail/division_travail.html

Durkheim, E. (1894). *Les règles de la méthode sociologique.* Retrieved from http://classiques.uqac.ca/classiques/Durkheim_emile/regles_methode/regles_methode.html

Durkheim, E. (1897). *Le Suicide*, Retrieved from http://classiques.uqac.ca/classiques/Durkheim_emile/suicide/suicide.html

Floriani, D.; Brandenburg, A.; Ferreira, A.D.D.; Teixeira, C.; Mendonça, F.A.; Lima, J.E.S.; Andriguetto Filho, J.M.; Knechtel, M.R. & Lana, P.C (2011). Construção interdisciplinar do Programa de Pós-Graduação em Meio Ambiente e Desenvolvimento da UFPR, In : *Interdisciplinaridade em ciência, tecnologia & inovação*, A. Philippi Jr & A.J. Silva Neto (Eds.), 342-425, Manole, ISBN 978-85-204-3046-0, Barueri, Brazil

Forster, P.; V. Ramaswamy, V.; Artaxo, P.; Berntsen, T.; Betts, R. ; Fahey, D.W. ; Haywood, J.; Lean, J.; Lowe, D.C.; Myhre, G. ; Nganga, J.; Prinn, R.; Raga, G.; Schulz, M.; & Van Dorland, R. (2007): Changes in Atmospheric Constituents and in Radiative Forcing. In: *Climate Change 2007: The Physical Science Basis. Contribution of Working Group I to the Fourth Assessment Report of the Intergovernmental Panel on Climate Change*, S.D. Solomon, D. Qin, M. Manning, Z. Chen, M. Marquis, K.B. Averyt, M. Tignor & H.L. Miller (Eds.), 129-234, Cambridge University Press, ISBN 0521705967, Cambridge, United Kingdom and New York, NY, USA

Furtado, C. (1998) *O mito do desenvolvimento econômico*, (2nd Edn), Paz e Terra, ISBN 9788521902133, São Paulo

Gell-Mann, M. (1994). Complex Adaptive Systems, In : *Complexity : Metaphors, Models, and Reality*, G. Gowan, D. Pines & D. Meltzer (Eds.), 17-29, Addison-Wesley, Reading, MA, USA

Geoghegan, J; Pritchard Jr., L; Ogneva-Himmelberger, Y.; Chowdhury, R.R.; Sanderson, S. & Turner II, B.L. (1998). «Socializing the Pixel» and « Pixelizing the Social » in Land-Use and Land-Cover Change, In : *People and Pixels, Linking Remote Sensing and Social Science*, D. Liverman, E.F. Moran, R.R. Rindfuss & P.C. Stern, (Eds.), 51-69, National Academy Press, ISBN 0-309-06408-2, Washington, D.C., USA

Giddens, A. (2001). *Em defesa da Sociologia*, UNESP, ISBN: 9788571393639, São Paulo, Brazil

Grimm, V.; Revilla, E.; Berger, U.; Jeltsch, F.; Mooij, W.M.; Railsback, S.F.; Thulke, H-H.; Weiner, J.; Wiegand, T. & DeAngelis, D.L. (2005). Pattern-Oriented Modeling of Agent-Based Complex Systems: Lessons from Ecology, *Science*, Vol. 310, pp. 987-991

Habermas, J. (2000). *La technique et la science comme ideologie*. Paris, France, Gallimard

Hannigan, J. (2006). *Environmental Sociology* (2nd edn.), Routledge, ISBN 9780415355131, New York.

Hardin, G. (1968). The tragedy of the commons, *Science*, Vol. 163, pp. 1243-1248

Harvey, D. (1974). Population, resources, and the ideology of science, *Economic Geography*, Vol. 50, pp. 256-277

Hicks, C.C.; Fitzsimmons, C. & Polunin, N.V.C. (2010). Interdisciplinarity in the environmental sciences: barriers and frontiers, *Environmental Conservation*, Vol. 37, pp. 464-477

Hogan, D. J. & Tolmasquim, M. T. (Eds.) (2001) *Human dimensions of global environmental change – Brazilian perspectives*, Academia Brasileira de Ciências, ISBN: 8585761202, Rio de Janeiro, Brazil

Holland, J. (2006). Studying Complex Adaptive Systems, *Journal System Science & Complexity*, 19, pp. 1–8

Houghton, J. (1990). Preámbulo, In: *Cambio Climatico. Evaluación científica del IPCC*, J.T. Houghton, G.J. Jenkins & J.J. Ephraums (Eds.), v-vi, Instituto Nacional de Meteorología, ISBN 84-7837-068-4, Madrid, Spain

Houghton, J. (2008). Madrid 1995: Diagnosing climate change. *Nature*, Vol. 455, pp. 737-738

Houghton, J. & Morel, P. (1984). The World Climate Research Program (WRCP), In: *The Global Climate*, J. Houghton (Ed.), 1-11, Cambridge University Press, ISBN: 0-521-31256-6, Cambridge, U.K.

IGBP (2006). Science Plan and Implementation Strategy. *IGBP Report No. 55*. IGBP Secretariat, Stockholm, Sweden

Intergovernmental Panel on Climate Change (IPCC) (ND). http://www.ipcc.ch/organization/organization.shtml

IPCC (1990). Resumen dirigido a los responsables de la toma de decisiones, In: *Cambio Climatico. Evaluación científica del IPCC*, J.T. Houghton, G.J. Jenkins, J.J. Ephraums (Eds.), vii-xxxvi, Instituto Nacional de Meteorología, ISBN 84-7837-068-4, Madrid, Spain

IPCC (1995). *Climate Change: A glossary by the Intergovernmental Panel on Climate Change*. Retrieved from http://www.ipcc.ch/pdf/glossary/ipcc-glossary.pdf

IPCC (1996). Summary for Policy Makers, In: *Climate Change 1995. The Science of Climate Change*, J.T. Houghton, L.G. Meira Filho, B.A. Callander, N. Harris, A. Kattenberg & Maskell (Eds.), Cambridge University Press, ISBN 0-521-80493, Cambridge, UK

IPCC (2000). *Special Report on Emissions Scenarios: A Special Report of Working Group III of the Intergovenrmental Panel on Climate Change*, N. Nakicenovic & R. Swart (Eds.), Cambridge University Press, ISBN 0-521-80493, Cambridge, UK

IPCC (2001). *Climate Change 2001: Synthesis Reort. Summary for Policymakers*. Retrieved from http://www.ipcc.ch/pdf/climate-changes-2001/synthesis-spm/synthesis-spm-en.pdf

IPCC (2007). Summary for Policymakers. In: *Climate Change 2007: The Physical Science Basis. Contribution of Working Group I to the Fourth Assessment Report of the Intergovernmental Panel on Climate Change*, S. Solomon, D. Qin, M. Manning, Z. Chen, M. Marquis, K.B. Averyt, M. Tignor & H.L. Miller (Eds.), 1-18, Cambridge University Press, ISBN 978-0521705967, Cambridge, UK

Jollivet, M. & Legay, J.M. (2005). Canevas pour une réflexion sur une interdisciplinarité entre sciences de la nature et sciences socials, *Natures Sciences Sociétés*, Vol. 13, pp. 184-188

Jordi, C. (2010). "The Unity of Science", In: *The Stanford Encyclopedia of Philosophy* (Fall 2010 Edition), E.N. Zalta (Ed.), retrieved from <http://plato.stanford.edu/archives/fall2010/entries/scientific-unity/>

Kates, R.W.; Clark, W.C.; Corell, R.; Hall, J.M.; Jaeger, C.C.; Lowe, I.; McCarthy, J.J.; Schellnhuber, H.J.; Bolin, B.; Dickson, N.M.; Faucheux, S.; Gallopin, G.C.; Grübler, A.; Huntley, B.; Jäger, J.; Jodha, N.S.; Kasperson, R.E.; Mabogunje, A.; Matson, P.; Mooney, H.; Moore, B.; O'Riordan, T. & Svedin, U. (2001). Sustainability Science, *Science*, Vol. 292, pp. 641-642

Keller, M. ; Bustamante, M ; Gash, J. & Dias, P.S. (Eds.) (2009). *Amazonia and Global Change*, American Geophysical Union, ISBN 978-0-87590-476-4, Washington, D.C., USA

Kesselmeier, J; Guenther, A.; Hoffmann, T; Pìedade, M.T. & Warnke, J. (2009). Natural Volatile Organic Compound from Plants and their Roles in Oxidant Balance. In: *Amazonia and Global Change*, M. Keller, M. Bustamante, J. Gash & P.S. Dias (Eds.), 183-206, American Geophysical Union, ISBN 978-0-87590-476-4, Washington, D.C.

Lahsen, M. (2005). Seductive simulations. Uncertainty distribution around climate models. *Social Studies of Science*, Vol. 35, pp. 895-922

Lambin, E.F. & Geist, H.J. (Eds.) (2006). *Land Use Land Cover Change, Local processes and global impacts*, Springer, ISBN 103540322019, Berlin Heidelberg, Germany

Lambin, E.F.; Turner, B.L.; Geist, H.J.; Agbola, S.B.; Angelsen, A.; Bruce, P.S.; Coomes, O.T., Dirzo, R.; Fischer, G.; Folkei, C.; George, P.S.; Homewood, K; Imbernon, J.; Leemans, R.; Li, X.; Moran. E.F., Mortimore, M.; Ramakrishna, P.S.; Richards, J.F.; Skanes, H.; Steffen, W.; Stone, G.D.; Svedin, U.; Veldkamp, T.A., Vogel, C. & Xu, J. (2001). The causes of land-use and land-cover change: moving beyond the myths. *Global Environmental Change*, Vol. 11, pp. 261–269

Laplace, P. (1825). *Essai philosophique sur les Probabilités* (5th edn.), Bachelier, Paris, France. Retrieved from http://www.google.com/books

Latour, B. (2000). When things strike back: a possible contribution of 'science studies' to the social sciences, *British Journal of Sociology*, Vol. 51, pp. 107–123

Latour, B. (2004). *As Políticas da Natureza*, EDUSC, ISBN 9788574601977, Bauru, Brazil

Leemans, R ; Asrar, G. ; Busalacchi, A. ; Canadell, J.; Ingram, J. ; Larigauderie, A. ; Harold Mooney, H. ; Nobre, C. ; Patwardhan, A. ; Rice, M. ; Schmidt, F.; Seitzinger, S.; Virji, H.; Vörösmarty, C. & Young, O. (2009). Developing a common strategy for integrative global environmental change research and outreach: the Earth System Science Partnership (ESSP), Strategy paper, *Current Opinion in Environmental Sustainability*, Vol.1, pp. 4-13

Leff, H. (2002). *Epistemologia Ambiental,* (3rd edn.), Cortez, ISBN 8524907681, São Paulo, Brazil

Leis, H.R. (2011). Especificidades e desafios da interdisciplinaridadenas ciências humanas, In: *Interdisciplinaridade em ciência, tecnologia & inovação,* A. Philippi Jr & A.J. Silva Neto (Eds.), 106-122, Manole, ISBN 978-85-204-3046-0, Barueri, Brazil

Le Treut, H.; Somerville, R.; Cubasch, U.; Ding, Y.; Mauritzen, C.; Mokssit, A.; Peterson, T. & Prather, M. (2007). Historical Overview of Climate Change. In: *Climate Change 2007: The Physical Science Basis. Contribution of Working Group I to the Fourth Assessment Report of the Intergovernmental Panel on Climate Change,* S.D. Solomon, D. Qin, M. Manning, Z. Chen, M. Marquis, K.B. Averyt, M. Tignor & H.L. Miller (Eds.), 94-127, Cambridge University Press, ISBN 0521705967, Cambridge, United Kingdom and New York, NY, USA

Liverman, D.M. & Cuesta, R.M.R. (2008). Human interactions with the Earth system: people and pixels revisited, *Earth Surf. Process. Landforms, Vol.* 33, pp. 1458–1471

Longo, K.M.; Freitas, S.R.; Andreae, M.O.; Yokelson, R. & Artaxo, P. (2009). Biomass Burning in Amazonia: Emissions, Long-Range Transport of Smoke and its Regional and Remote Impacts. In: *Amazonia and Global Change.* M. Keller, M. Bustamante, J. Gash & P.S. Dias (Eds.), 207-232, American Geophysical Union, ISBN 978-0-87590-476-4, Washington, D.C., USA

Lorenz, E.D. (1963). Deterministic Nonperiodic Flow, *Journal of the Atmospheric Sciences*, Vol. 20, pp. 130-141

Luhmann, N. (2010). *Introdução à teoria dos sistemas. Aulas publicadas por Javier Torres Nafarrate* (2nd edn.), Vozes, ISBN 9788532638618, Petrópolis, Brazil

Magalhães, A. & Stoer, S.R. (2002). A nova classe média e a reconfiguração do mandato endereçado ao sistema educativo. *Educação, Sociedade e Culturas.* Retrieved from http://www.fpce.up.pt/ciie/revistaesc/ESC18/magalhaes_stoer.pdf

Martins, L. (1976). *Pouvoir et development economique,* Editions Anthropos, Paris, France

Marx, K. (1859). *A Contribution to the Critique of Political Economy.* Retrieved from http://www.marxists.org/archive/marx/works/download/Marx_Contribution_to_the_Critique_of_Political_Economy.pdf

Marx, K. (1875). *Critique of the Gotha Programme.* Retrieved from http://www.marxists.org/archive/marx/works/download/Marx_Critque_of_the_Gotha_Programme.pdf

Mirowski, P. (2003). What'sKuhn got to do with it? *Soc. Epistemol.*, Vol. 17, pp. 229–239

Montibeller Filho, G. (2008). *O Mito do Desenvolvimento Sustentável* (3th edn.), Editora UFSC, ISBN: 9788532803917, Florianópolis, Brazil

Moraes, A.C.R. (2005). *Meio Ambiente e Ciências Humanas* (5th edn.), AnnaBlumme. ISBN 9788574195483, São Paulo, Brazil

Moran, E. & Ostrom, E. (Eds.) (2005). *Seeing the Forest and the Trees,* MIT Press, ISBN 0-262-63312-4, Cambridge, MA, USA

Mortimore, M. (1993). Population and land degradation, *GeoJournal*, Vol. 31, pp. 15-21

Nobel Foundation (ND). The Nobel Peace Prize 2007. Retrieved from http://nobelprize.org/nobel_prizes/peace/laureates/2007/

Nowotny, H.; Scott, P. & Gibbons, M. (2003). Introduction. 'Mode 2' Revisited: The New Production of Knowledge, *Minerva*, Vol. 41, pp. 179–194

Oliveira Filho, J.J. (1976). Reconstruções metodológicas de processos de investigação social, *Revista de História*, Vol. 54, pp. 263-276

Parker, D.C; Manson,S.M.; Janssen, M.A.; Hoffmann,M.J. & Deadman, P (2004). Multi-Agent Systems for the Simulation of Land-Use and Land-Cover Change: A Review, *Annals of the Association of American Geographers*, Vol. 93, pp. 314-337

Pfeiffer, E. (2008). The Road Ahead: an Introductory Note from the Secretariat, *IHDP Update*, 1.2008, pp. 10-12

Poincaré, H. (1968). *La Science et l'Hypothèse*, Flammarion, Paris.

Randall, D.A.; Wood, R.A.; Bony, S.; Colman, R.; Fichefet, T.; Fyfe, J.; Kattsov, V.; Pitman, A.; Shukla, J.; Srinivasan, J.;. Stouffer, R.J; Sumi, A. & Taylor, K.E. (2007). Climate Models and Their Evaluation. In: *Climate Change 2007: The Physical Science Basis. Contribution of Working Group I to the Fourth Assessment Report of the Intergovernmental Panel on Climate Change*, S.D. Solomon, D. Qin, M. Manning, Z. Chen, M. Marquis, K.B. Averyt, M. Tignor & H.L. Miller (Eds.), 589-662, Cambridge University Press, ISBN 0521705967, Cambridge, United Kingdom and New York, NY, USA

Raynaut, C. & Zanoni, M. (2011). Reflexões sobre princípios de uma prática interdisciplinar na pesquisa e ensino superior, In: *Interdisciplinaridade em ciência, tecnologia & inovação*, A. Philippi Jr & A.J. Silva Neto (Eds.), 143-208, Manole, ISBN 978-85-204-3046-0, Barueri, Brazil

Reid, W.V.; Chen, D.; Goldfarb, L.; Hackmann, H.; Lee, Y.T.; Mokhele, K.; Ostrom, E.; Raivio, K.; Rockcktröm, J.; Schellnhuber, H.J. & Whyte, A. (2010). Earth System Science for Global Sustainability: Grand Challenges, *Science*, Vol. 330, pp. 916-917

Rhods, J.K. (1991). *Critical issues in Social Theory*, Pennsylvania University Press, ISBN: 9780271007533, Philadelphia, USA

Rodrigues, E.V. (ND). Modos e dinâmicas de exclusão social em contexto urbano e periurbano, *Actas dos ateliers do V° Congresso Português de Sociologia.* Retrieved from http://www.aps.pt/cms/docs_prv/docs/DPR4628c4a5f1b41_1.pdf

Rosenberg, A. (2000). Social Science, Philosophy of. In: *A Companion to the Philosophy of Science*, W.H. Newton-Smith (Edt.), 451-460, Blackwell, ISBN 0-631-23020-3, Malden, USA

Saloranta, T.M. (2001) Post-normal science and the global climate change issue, *Climatic Change*, Vol. 50, pp. 395-404

Santos, R.A. & Alves D.S. (2008). Mudanças ambientais na Amazônia e as particularidades da construção institucional, In: *Amazônia: Natureza e sociedade em transformação*, M. Batistella; E.F. Moran; D.S. Alves (orgs.), 221-240, Editora da Universidade de São Paulo, ISBN 9788531411267, São Paulo, Brazil

Schor, T. (2008). Ciência e Tecnologia: O caso do Experimento de Grande Escala da Bioesfera-Atmosfera na Amazônia (LBA), AnnaBlumme, ISBN: 9788574198163, São Paulo, Brazil

Schwartzman, S. (1980). *Ciência, Universidade e Ideologia: a Política do Conhecimento*, Zahar Editores, Rio de Janeiro, Brazil. Retrieved from http://www.schwartzman.org.br/simon/zahar/tecno.htm

Shackley, S.; Young, P.; Parkinson, S. & Wynne, B. (1998) Uncertainty, complexity and concepts of good science in climate change modeling: are GCMs the best tools ? *Climatic Change*, Vol. 38, pp. 159-205

Snow, C.P. (1990). The Two Cultures, *Leonardo*, Vol.23, pp. 169-173

Solomon, S.; Qin, D.; Manning, M.; Alley, R.B.; Berntsen, T.; Bindoff, N.L.; Chen, Z.; Chidthaisong, A.; Gregory, J.M.; Hegerl, G.C.; Heimann, M.; Hewitson, B.; Hoskins, B.J.; Joos, F.; Jouzel, J.; Kattsov, V.; Lohmann, U., Matsuno, T.; Molina, M.; Nicholls, N.; Overpeck, J.; Raga, G.; Ramaswamy, V.; Ren, J.; Rusticucci, M.; Somerville, R.; Stocker, T.F.; Whetton, P.; Wood, R.A. & Wratt, D. (2007). Technical Summary. In: *Climate Change 2007: The Physical Science Basis. Contribution of Working Group I to the Fourth Assessment Report of the Intergovernmental Panel on Climate Change*,S.D. Solomon, D. Qin, M. Manning, Z. Chen, M. Marquis, K.B. Averyt, M. Tignor & H.L. Miller (Eds.), 19-91, Cambridge University Press, ISBN 0521705967, Cambridge, United Kingdom and New York, NY, USA

Thatcher, M. (1990). *Speech at 2nd World Climate Conference.* Geneva (06 November 1990) Retrieved from http://www.margaretthatcher.org/document/108237

United Nations (1992). *United Nations Framework Convention on Climate Change (UNFCCC).* Retrieved from http://unfccc.int/resource/docs/convkp/conveng.pdf

VanWey, L.K.; Ostrom, E. & Merestky, V. (2005). Theories underlying the study of human-environmental interactions. In: *Seeing the Forest and the Trees,* E. Moran & E. Ostrom (Eds.), 23-56, MIT Press, ISBN 0-262-63312-4, Cambridge, MA, USA

Verosub, K.L. (2010). Climate Science in a Postmodern World, *Eos Trans. AGU*, Vol. 91, 291-292

von Bertalanffy, L. (1950). An Outline of General System Theory, *The British Journal for the Philosophy of Science*, Vol. 1, pp. 134-165

von Bertalanffy, L. (1972). The History and Status of General Systems Theory, *The Academy of Management Journal,* Vol. 15, pp. 407-426

Walker, K.J. (1988). The Environmental Crisis: A Critique of Neo-Hobbesian Responses. *Polity*. Vol. 21, pp. 67-81

Weber, M. (2004). *Economia e Sociedade*, Imprensa Oficial do Estado de São Paulo, ISBN 85-7060-253-7, São Paulo, Brazil

Weber, M. (2005a). A objetividade do conhecimento nas ciências sociais. In: *Max Weber*, G. Cohn (Org.), 79-127, Ática, ISBN 8508011458, São Paulo, Brazil

Weber, M. (2005b). O Estado nacional e a política econômica. In: *Max Weber*, G. Cohn (Org.), 58-78, Ática, ISBN 8508011458, São Paulo, Brazil

Weffort, F. (Org.) (2006). *Os Clássicos da Política 1* (14th edn.), Ática, ISBN: 9788508105908, São Paulo, Brazil

Whiteside, K.H. (1998) Systems theory and skeptical humanism in French ecological thought, *Policy Stud. J.*, Vol. 26, 636–656

Young, O. (2008). The IHDP Strategic Plan at Work, *IHDP Update*, 1.2008, pp. 4-8

2

History and Prediction of the Asian Monsoon and Glacial Terminations, Based on Records from the South China Sea

Hong Ao[1] and Guoqiao Xiao[2]
[1]*State Key Laboratory of Loess and Quaternary Geology, Institute of Earth Environment, Chinese Academy of Sciences, Xian,*
[2]*State Key Laboratory of Biogeology and Environmental Geology, China University of Geosciences, Wuhan, China*

1. Introduction

What caused the global ice sheets to come and go? Knowledge of this question is crucial for understanding global climate evolution and predicting future climate changes. Since the 1840s, when geologists firstly noted the expansion and retreat of ice sheets on land, scientists have been trying to solve this question. Although at present it is generally thought that the glacial cycles are driven by changes in solar insolation due to subtle variations in Earth's orbit parameters (Milankovitch, 1941; Hays et al., 1976; Imbrie et al., 1992), the mechanism by which and the degree to which insolation plays a role on the glacial terminations remains unclear. For example, if glacial cycles vary directly in response to insolation, why do glacial terminations not occur at every time of increasing insolation?

The benthic $\delta^{18}O$ in the ocean is known to increase with glaciation and thus can be used to estimate the global ice-volume changes (Hays et al., 1976; Imbrie et al., 1984; Ruddiman, 2003). Therefore, precise timing of the benthic $\delta^{18}O$ records is crucial for testing the exact relationship between glacial terminations and changes in insolation. Generally, a record of benthic $\delta^{18}O$ versus depth was transformed into a record versus time by tuning the benthic $\delta^{18}O$ record to the Earth's orbital parameters (e.g. Imbrie et al., 1984; Ruddiman et al., 1986; Raymo et al., 1989; Shackleton et al., 1990; Lisiecki and Raymo, 2005). However, it is problematic to discuss the linkage between glacial termination and solar insolation based on the astronomical chronology because of the risk of circular reasoning. In the present study, therefore a different procedure independent of orbital tuning was adopted to establish the timescale for the late Quaternary benthic $\delta^{18}O$ record retrieved from Ocean Drilling Program (ODP) Site 1143, southern South China Sea (Fig. 1). On the one hand, Zhang et al. (2007) recently published a high-resolution Asian summer monsoon record over the last 600 kyr using the ratio of hematite to goethite contents (Hm/Gt) from this site. On the other hand, the high-resolution (from orbital down to millennial) variations in Asian summer monsoon in South China over the last 350 kyr are now available from the $\delta^{18}O$ of stalagmites from caves, which were accurately dated by high-resolution U-series analyses (Wang et al., 2001, 2005, 2008; Yuan et al., 2004; Zhang et al., 2008; Cheng et al., 2009). The stalagmite $\delta^{18}O$

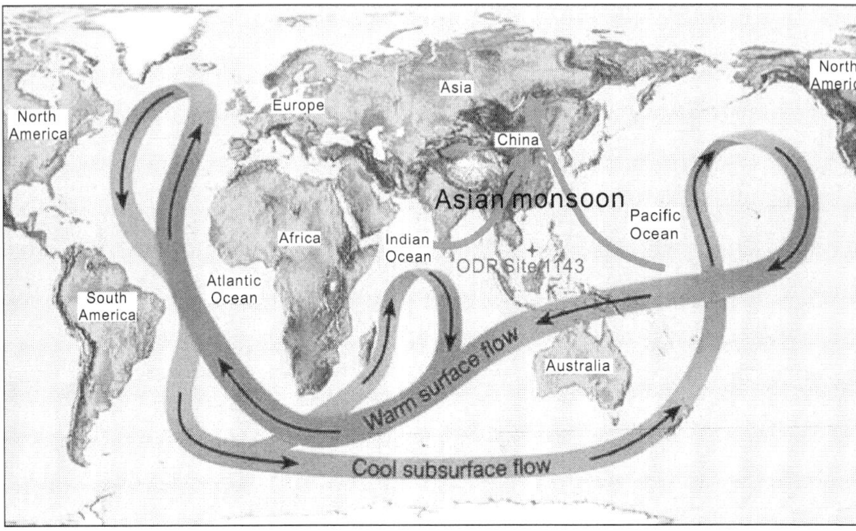

Fig. 1. Map showing the ocean circulation, Asian monsoon and ODP Site 1143 (modified from Friedland (2010)). The orange arrows represent the directions of the Asian summer monsoon.

record is the most accurately dated monsoon record on the relevant 100-kyr time scale, with errors of mere decades. Since both the Hm/Gt record from South China Sea and the stalagmite $\delta^{18}O$ record from South China are good estimates of variations in Asian summer monsoon with similar orbital cycles, we formulate a timescale for ODP Site 1143 over the last 350 kyr by calibrating the Hm/Gt record to the Chinese stalagmite $\delta^{18}O$ record (Wang, et al., 2001, 2008; Cheng, et al., 2009) instead of the orbital parameters as usual. In particular, we test the extent to which the last four terminations as well as the Asian monsoon are linked to solar insolation, based on this orbital-independent timescale without involving in circular reasoning. In addition, the observed late Quaternary relationship between insolation and climate further provide clues for predicting further climate changes.

2. General setting

ODP Site 1143 (9°21.72'N, 113°17.11'E; 2777 m water depth) was drilled in a depression on the carbonate platform that forms the southern continental shelf of the southern South China Sea (Fig. 1). The South China Sea is the largest marginal sea of the western Pacific, covering an area of ~3.5×10^6 km^2. The seasonal reversal of Asian winter and summer monsoon circulations results in cold/dry winters and warm/wet summers over the South China Sea. Due to strong monsoon precipitation and intrusion of low-salinity water from along shore Borneo, the sea surface salinity (SSS) in the southern South China Sea is rather low, ranging from ca. 30‰ to 34‰ (Tian et al., 2004). The SSS in the open western Pacific is as much as 35-35.5‰ throughout the upper 560 m of the water column (Tian et al., 2004). The deposits at ODP Site 1143 mainly consist of terrigenous quartz, feldspar and clay minerals, with only a minor biogenic component (<2%) (Wan et al., 2006).

3. Monsoon proxies and chronology

Hematite (Hm) and goethite (Gt) contents over the last 600 kyr (from 0 to 34 m) were assembled by Zhang et al. (2007) for 315 samples from ODP Site 1143 using a Perkin Elmer Lambda 900 diffuse reflectance spectrophotometer in the Surficial Geochemistry Institute of Nanjing University (China). The average resolution of the iron oxides record is ~2 kyr. For the present study of interest, our age calibration is just based on the interval spanning the last 350 kyr (from 0 to 22.1 m) (Fig. 2). The Hm/Gt ratios of ODP Site 1143 can be used as an indicator of summer monsoon intensity because of the following reasons. (1) During chemical weathering, the relative abundance of goethite to hematite varies with climatic conditions: dry and humid conditions are more favorable for the formation of hematite and goethite, respectively (Curi and Franzmeier, 1984; da Motta and Kampf, 1992; Harris and Mix, 1999; Thiry, 2000; Ji et al., 2004; Zhang et al., 2007). (2) The terrigenous deposits, including hematite and goethite, in ODP Site 1143 are mainly derived from the paleo-Sunda shelf and Mekong Basin through fluvial and marine transportation, with a discharge more than 160×10^6 tons of sediment per year (Milliman and Meade, 1983; Wan et al., 2006). Other rivers such as the Baram River from northwest Borneo and the Chao Phraya River from western Indochina have a combined annual sediment discharge less than 23×10^6 tons to the southwest South China Sea (Wan et al., 2006). (3) Hematite and goethite in ODP Site 1143 are little affected by diagenesis after burial (Zhang et al., 2007, 2009; Ao et al., 2011a). (4) The dry and humid conditions over the South China Sea are mainly modulated by Asian summer monsoon precipitation (Tian et al., 2004, 2005; Wan et al., 2006; Zhang et al., 2007; Clift and Plumb, 2008). Therefore, the strong summer monsoon periods would result in more goethite deposition in the South China Sea, whereas the weak summer monsoon periods would result in more hematite deposition. So, for this region low and high Hm/Gt ratios would imply strong and weak summer monsoons, respectively.

The present timescale for the last 350 kyr as recorded in ODP Site 1143 was established by calibration of the Hm/Gt record to the composited Chinese stalagmite $\delta^{18}O$ record (Wang et al., 2001, 2008; Cheng et al., 2009), because both of them are interpreted as a summer monsoon proxy in South China. This calibration involves downward matches between the Hm/Gt record and the stalagmite $\delta^{18}O$ record (Fig. 2). The strong precession signal in both Hm/Gt and stalagmite $\delta^{18}O$ records guarantees a precise age determination for ODP Site 1143. After our final calibration, the Hm/Gt record was correlated almost cycle-by-cycle with the stalagmite $\delta^{18}O$ record (Fig. 2A–C). Their filtered precession cycles also matched well (Fig. 2D).

4. Discussion

Like the Chinese stalagmite $\delta^{18}O$ record, the Hm/Gt record plotted on our resulted timescale has a good correlation with the solar insolation (Fig. 3 A–C). This is consistent with the response of the Asian summer monsoon in South China to the insolation forcing (Kutzbach, 1981; Wang et al., 2008; Ao et al., 2011b). As indicated by maxima in the benthic $\delta^{18}O$ record from ODP Site 1143, the onsets of the major glacial terminations IV, III, II and I are around 340, 250, 135 and 20 ka, respectively (Fig. 3D). These ages are generally consistent with the recent astronomical estimates for these terminations (Lisiecki and Raymo, 2005). Comparison of the benthic $\delta^{18}O$ record to the summer insolation indicates that all the last four glacial terminations occurred when insolation rose from an outstanding minimum to a prominent maximum (Fig. 3). This is consistent with the primary forcing of glacial

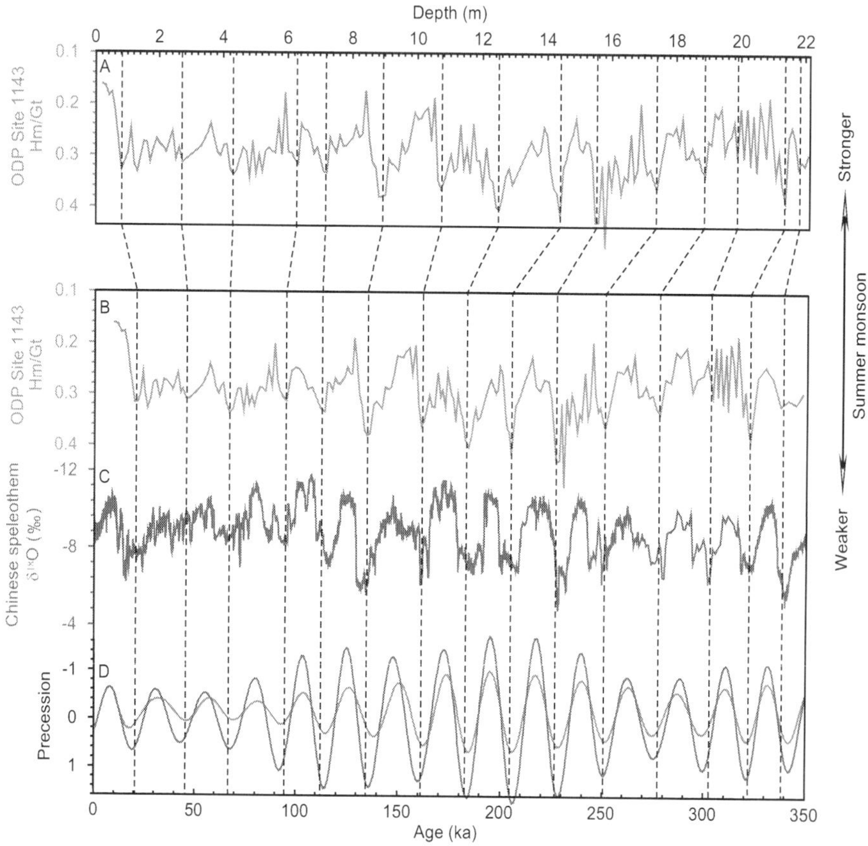

Fig. 2. (A) Hm/Gt record (Zhang et al., 2007) from ODP Site 1143 plotted against depth. (B) Hm/Gt record (Zhang et al., 2007) plotted against our calibrated timescale. (C) Composited Chinese stalagmite $\delta^{18}O$ record (from the Sanbao, Linzhu, Dongge and Hulu caves) (Wang, et al., 2001, 2008; Cheng, et al., 2009). (D) Comparison of filtered precession bands filtered from our calibrated Hm/Gt (orange line) and the composited Chinese stalagmite $\delta^{18}O$ (blue line) records.

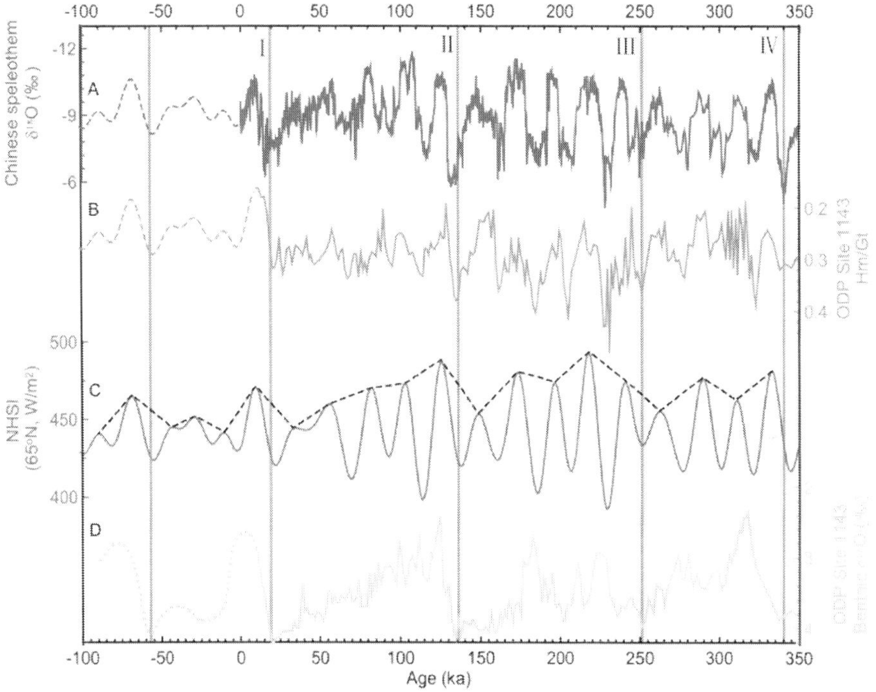

Fig. 3. (A) Composited Chinese stalagmite δ18O record (from the Sanbao, Linzhu, Dongge and Hulu caves) (Wang, et al., 2001, 2008; Cheng, et al., 2009). (B) Hm/Gt record from ODP Site 1143 (Zhang et al., 2007) plotted against the presently calibrated timescale. (C) 65°N summer insolation (Laskar et al., 2004). The black dashed line joins the maximum of insolation. (D) Benthic δ18O record (Tian et al., 2002) from ODP Site 1143 plotted on our calibrated timescale. The vertical shaded lines indicate the onset of terminations revealed by the benthic δ18O record from ODP Site 1143. Terminations are labeled using greek numerals.

terminations by summer insolation, as pointed out by the Milankovitch orbital theory (Milankovitch, 1941). In addition, we noted the following insolation pattern leading up to the glacial terminations: a series of decreased insolation maximum followed by a relatively sharp increase in insolation maximum (Fig. 3). A series of decreased insolation maximum would favor the accumulation of massive ice sheets prior to termination (Broecker, 1984; Peltier, 1994; Raymo, 1997), whose collapse was triggered by a following sharp rise in insolation. This can partly explain why the glacial cycles show a gradual buildup but a rapid collapse. This following insolation maximum, which is generally higher than its nearby insolation maxima, may imply an insolation threshold for triggering a glacial termination (Fig. 3). This is in agreement with the recent view that the amount and rate of insolation rise are important controls on the glacial terminations (Cheng et al., 2009). The insolation maxima may have played a more important role on the ice-age cycles than the insolation minima (Fig. 3C).

In agreement with recent studies (Wu et al., 2005; Cheng et al., 2006, 2009), the Hm/Gt and benthic δ18O records from ODP Site 1143 suggested that each termination occurred when the summer monsoon intensity rose from a minimum to a maximum (Fig. 3). This observation implies that the rising summer monsoon intensity may have played a role in driving the termination to completion to some extent, because the summer monsoon, which transports heat from tropical oceans to the Asian mainland, is expected to lead to a rather warm environment and thus promote the snow-cover meltdowns in Asia (Fig. 1). Furthermore, the summer monsoon would favor the vegetation and wetland covers in Asia, which may in turn produce increased greenhouse gases such as CO_2 and CH_4. The feedback effects of the greenhouse gases are widely regarded as a potentially important player in glacial terminations (Petit et al., 1999; Ruddiman, 2003, 2006; Cheng et al., 2009). Likewise, the ocean circulation may have played an important role on the ice-sheet meltdowns as well, because it is considered as a very important heat transport in Northern Hemisphere (Fig. 1). It transports very warm tropical water to the Northern Hemisphere and warms the air during the transportation. Thus it may have an appreciable impact on the ice-cover meltdowns in Northern Hemisphere. Although the Northern Hemisphere summer insolation intensity is the primary trigger of an initial retreat of northern ice sheets, the modulating impacts from the monsoon system (including not only Asian monsoon, but also African and North American monsoons) and ocean circulation may be much more important than the presently thought (Fig. 1), which should be investigated in detail in future studies of the ice-age terminations.

The observed relationship between insolation and climate during the late Quaternary may provide clues for predicting further climate changes. The insolation from now to the future 100 kyr will be similar to the last 100-kyr insolation behavior (Fig. 3C). Since the Asian summer monsoon is correlated cycle-by-cycle to the insolation during the past 100 kyr, a monsoon behavior similar to the insolation is anticipated to occur during the following 100 kyr. Because outstanding insolation maximum during the Holocene has been over and an insolation minimum is coming, a weak summer monsoon interval may come soon instead of the present Holocene strong summer monsoon period (Fig. 3 A–C). An insolation maximum comparable to the Holocene insolation maximum will appear ca. 70 kyr from now, thus a strong summer monsoon interval comparable to the Holocene strong summer monsoon interval will possibly not occur until then (Fig. 3 A–C). Relatively weakened summer monsoon maxima are likely to occur at ca. 10, 30 and 50 kyr from now, which are correlated to less outstanding insolation maxima of these intervals (Fig. 3 A–C).

As suggested by the benthic $\delta^{18}O$ record from ODP Site 1143, we are presently living within an interglacial period (Fig. 3D). The following insolation maxima at 10 and 30 kyr from now are much lower than these during the last 350 kyr. Thus the present interglacial period will possibly continue shorter than the previous interglacial periods (Fig. 3C, D). This shortened interglacial period will be gradually replaced by a glacial period starting from ca. 10 kyr from now. Subsequently, the next glacial termination will occur at ca. 60 kyr from now when the insolation increases from an outstanding minimum to a prominent maximum, which will be followed by an interglacial period comparable to the present interglacial period. This prediction is consistent with the prediction of Raymo (1997) but in contrast to the predictions of Berger and Loutre (1997) and Ledley (1995). Note that the insolation maximum at ca. 70 kyr from now is much higher than its nearby insolation maxima (Fig. 3C), which should be considered as an insolation threshold for triggering this termination.

5. Conclusions

An orbital-independent timescale for ODP Site 1143 over the last 350 kyr was established by calibration of the Hm/Gt record to the Chinese stalagmite $\delta^{18}O$ record. This resulted timescale enabled a detailed study of the Asian monsoon and glacial terminations and their links with solar insolation during the late Pleistocene. Consistent with the insolation forcing of orbital-scale variability of Asian monsoon in South China suggested by recent studies (Kutzbach, 1981; Wang et al., 2008; Ao et al., 2011b), the Hm/Gt record plotted on our timescale has a good correlation with the solar insolation. The glacial terminations, which are determined by independence of orbital tuned results, generally occurred when insolation rises from an outstanding minimum to a prominent maximum, consistent with a classic summer insolation increase trigger for an initial retreat of northern ice sheets or snow covers. In addition to the primary insolation forcing, the monsoon system and ocean circulation may have played a potentially important modulating role on the glacial terminations as well. Combining the late Quaternary relationship between insolation and climate, we predict that the present warm interglacial periods with strong summer monsoon will gradually develop into a cold glacial period with weakened summer monsoon thousands of years later and the next glacial termination will occur ~60 kyr from now. The next interglacial period with a strong summer monsoon period comparable to that of the Holocene will probably occur ca. 70 kyr from now. If this prediction is true, the coming glacial period with weakened summer monsoon is likely to result in a rather cold period thousands of years later, which then may entirely shut down the present global warming. If so, the presently increasing greenhouse gases are probably helpful for human to adapt the following cold glacial period at a long-term timescale such as millennial and orbital timescales, although the current global warming due to accumulation of greenhouse gases is likely to result in some potentially devastating consequences for humans at a short-term timescale such as the next few hundred years. Therefore, using climate-model simulations to test our climate prediction should be a priority in future investigations on this topic.

6. Acknowledgments

This study was supported by the Key Projects of National Basic Research Program of China (2010CB833400), the International Cooperation program of NSFC (40921120406), National Natural Science Foundation of China (41174057) and the West Light Foundation of Chinese Academy of Sciences.

7. References

Ao, H., Dekkers, M.J., Qin, L., Xiao, G.Q., 2011a. An updated astronomical timescale for the Plio-Pleistocene deposits from ODP Site 1143 and new insights into Asian monsoon evolution. Quaternary Science Reviews, 30, 1560–1575.

Ao, H., Dekkers, M.J., Xiao, G.Q., Yang, X.Q., Qin, L., Liu, X.D., Qiang, X.K., Chang, H., Zhao, H., 2011b. Different orbital rhythms in the Asian summer monsoon records from North and South China during the Pleistocene. Global and Planetary Change, In press.

Berger, A., Loutre, M.F., 1997. Palaeoclimate sensitivity to CO_2 and insolation. Ambio 26, 32–37.

Broecker, W.S., 1984. Terminations. In: Berger, A.L., Imbrie, J., Hays, J.D., Kukla, G., and Saltzman, B. (Eds.), Milankovitch and Climate. Reidel Publishing Company, Dordrecht, pp. 687–698.

Cheng, H., Edwards, R.L., Wan, Y.J., Ko, X.G., Ming, Y.F., Kelly, M.J., Wang, X.F., Gallup, C.D., Liu, W.G., 2006. A penultimate glacial monsoon record from Hulu Cave and two-phase glacial terminations. Geology 34, 217-220.

Cheng, H., Edwards, R.L., Broecker, W.S., Denton, G.H., Kong, X.G., Wang, Y.J., Zhang, R., Wang, X.F., 2009. Ice age terminations. Science 326, 248-252.

Clift, P.D., Plumb, R.A., 2008. The Asian Monsoon: Causes, History and Effects. Cambridge University Press, Cambridge.

Curi, N., Franzmeier, D.P., 1984. Toposequence of oxisols from the Central Plateau of Brazil. Soil Science Society of America Journal 48, 341-346.

da Motta, P.E.F., Kampf, N., 1992. Iron oxide properties as support to soil morphological features for prediction of moisture regimes in oxisols of central Brazil. Pflanzenernahrung und Bodenkunde 155, 385-390.

Friedland, J., 2010. The decade after tomorrow. The Political Climate, http://politicalclimate.wordpress.com/2010/03/15/the-decade-after-tomorrow.

Harris, S.E., Mix, A.C., 1999. Pleistocene precipitation balance in the Amazon Basin recorded in deep sea sediments. Quaternary Research 51, 14-26.

Hays, J.D., Imbrie, J., Shackleton, N.J., 1976. Variations in the Earth's orbit: pacemaker of the ice ages. Science 194, 1121.

Imbrie, J., Hays, J.D., Martinson, D.G., McIntyre, A., Mix, A.C., Morley, J.J., Pisias, N.G., Prell, W.L., Shackleton, N.J., 1984. The orbital theory of Pleistocene climate: support from a revised chronology of the marine $\delta^{18}O$ record. In: Berger, A. L., Hays, J., Imbrie, J. (Eds.), Milankovitch and Climate: Understanding The Response To Astronomical Forcing. Reidel, Dordrecht, pp. 269-305.

Imbrie, J., Berger, A., Clemens, S.C., Duffy, A., Howard, W.R., Kukla, G., Kutzbach, J., Martinson, D.G., Mcintyre, A., Mix, A.C., Molfino, B., Morley, J.J., Peterson, L.C., Pisias, N.G., Prell, W.L., Raymo, M.E., Shackleton, N.J., Toggweiler, J.R., 1992. On the structure and origin of major glaciation cycles 1. Linear response to Milankovitch forcing. Paleoceanography 7, 701-735.

Ji, J.F., Chen, J., Balsam, W., Lu, H.Y., Sun, Y.B., Xu, H.F., 2004. High resolution hematite/goethite records from Chinese loess sequences for the last glacial-interglacial cycle: rapid climatic response of the East Asian Monsoon to the tropical Pacific. Geophysical Research Letters 31, doi:10.1029/2003GL018975.

Kutzbach, J.E., 1981. Monsoon climate of the early Holocene: climate experiment with the earths orbital parameters for 9000 years ago. Science 214, 59-61.

Laskar, J., Robutel, P., Joutel, F., Gastineau, M., Correia, A.C.M., Levrard, B., 2004. A long-term numerical solution for the insolation quantities of the Earth. Astronomy & Astrophysics 428, 261-285.

Ledley, T.S., 1995. Summer solstice solar radiation, the 100 kyr ice age cycle, and the next ice age. Geophysical Research Letters 22, 2745-2748.

Lisiecki, L.E., Raymo, M.E., 2005. A Pliocene-Pleistocene stack of 57 globally distributed benthic $\delta^{18}O$ records. Paleoceanography 20, PA1003, doi:10.1029/2004PA001071.

Milankovitch, M.M., 1941. Kanon der Erdbestrahlung und seine Anwendung auf das Eiszeitenproblem. Royal Serbian Academy, Belgrade.

Milliman, J.D., Meade, R.H., 1983. World wide delivery of river sediment to the oceans. Journal of Geology 91, 1-21.

Peltier, W.R., 1994. Ice age paleotopography. Science 265, 195-201.

Petit, J.R., Jouzel, J., Raynaud, D., Barkov, N.I., Barnola, J.M., Basile, I., Bender, M., Chappellaz, J., Davis, M., Delaygue, G., Delmotte, M., Kotlyakov, V.M., Legrand, M., Lipenkov, V.Y., Lorius, C., Pepin, L., Ritz, C., Saltzman, E., Stievenard, M., 1999. Climate and atmospheric history of the past 420,000 years from the Vostok ice core, Antarctica. Nature 399, 429–436.

Raymo, M.E., Ruddiman, W.F., Backman, J., Clement, B.M., Martinson, D.G., 1989. Late Pliocene variation in Northern Hemisphere ice sheets and North Atlantic deep circulation. Paleoceanography 4, 413–446.

Raymo, M.E., 1997. The timing of major climate terminations. Paleoceanography 12, 577–585.

Ruddiman, W.F., Raymo, M.E., McIntyre, A., 1986. Matuyama 41,000-year cycles: North Atlantic Ocean and northern hemisphere ice sheets. Earth and Planetary Science Letters 80, 117–129.

Ruddiman, W.F., 2003. Orbital insolation, ice volume, and greenhouse gases. Quaternary Science Reviews 22, 1597–1629.

Ruddiman, W.F., 2006. Orbital changes and climate. Quaternary Science Reviews 25, 3092–3112.

Shackleton, N.J., Berger, A., Peltier, W.R., 1990. An alternative astronomical calibration of the lower Pleistocene time-scale based on ODP site 677. Transactions of the Royal Society of Edinburgh 81, 251–261.

Thiry, M., 2000. Palaeoclimatic interpretation of clay minerals in marine deposits: an outlook from the continental origin. Earth-Science Reviews 49, 201–221.

Tian, J., Wang, P.X., Cheng, X.R., Li, Q.Y., 2002. Astronomically tuned Plio-Pleistocene benthic $\delta^{18}O$ record from South China Sea and Atlantic-Pacific comparison. Earth and Planetary Science Letters 203, 1015–1029.

Tian, J., Wang, P.X., Cheng, X.R., 2004. Development of the East Asian monsoon and Northern Hemisphere glaciation: oxygen isotope records from the South China Sea. Quaternary Science Reviews 23, 2007–2016.

Tian, J., Wang, P.X., Chen, R.H., Cheng, X.R., 2005. Quaternary upper ocean thermal gradient variations in the South China Sea: implications for east Asian monsoon climate. Paleoceanography 20, PA4007, doi:10.1029/2004PA001115

Wan, S.M., Li, A.C., Clift, P.D., Jiang, H.Y., 2006. Development of the East Asian summer monsoon: evidence from the sediment record in the South China Sea since 8.5 Ma. Palaeogeography Palaeoclimatology Palaeoecology 241, 139–159.

Wang, Y.J., Cheng, H., Edwards, R.L., An, Z.S., Wu, J.Y., Shen, C.C., Dorale, J.A., 2001. A high-resolution absolute-dated Late Pleistocene monsoon record from Hulu Cave, China. Science 294, 2345–2348.

Wang, Y.J., Cheng, H., Edwards, R.L., He, Y.Q., Kong, X.G., An, Z.S., Wu, J.Y., Kelly, M.J., Dykoski, C.A., Li, X.D., 2005. The Holocene Asian monsoon: links to solar changes and North Atlantic climate. Science 308, 854–857.

Wang, Y.J., Cheng, H., Edwards, R.L., Kong, X.G., Shao, X.H., Chen, S.T., Wu, J.Y., Jiang, X.Y., Wang, X.F., An, Z.S., 2008. Millennial- and orbital-scale changes in the East Asian monsoon over the past 224,000 years. Nature 451, 1090–1093.

Wu, G.J., Pan, B.T., Guan, Q.Y., Xia, D.S., 2005. Terminations and their correlation with solar insolation in the Northern Hemisphere: a record from a loess section in Northwest China. Palaeogeography Palaeoclimatology Palaeoecology 216, 267–277.

Yuan, D.X., Cheng, H., Edwards, R.L., Dykoski, C.A., Kelly, M.J., Zhang, M.L., Qing, J.M., Lin, Y.S., Wang, Y.J., Wu, J.Y., Dorale, J.A., An, Z.S., Cai, Y.J., 2004. Timing, duration, and transitions of the Last Interglacial Asian Monsoon. Science 304, 575-578.

Zhang, P.Z., Cheng, H., Edwards, R.L., Chen, F.H., Wang, Y.J., Yang, X.L., Liu, J., Tan, M., Wang, X.F., Liu, J.H., An, C.L., Dai, Z.B., Zhou, J., Zhang, D.Z., Jia, J.H., Jin, L.Y., Johnson, K.R., 2008. A test of climate, sun, and culture relationships from an 1810-year Chinese cave record. Science 322, 940–942.

Zhang, Y.G., Ji, J., Balsam, W.L., Liu, L., Chen, J., 2007. High resolution hematite and goethite records from ODP 1143, South China Sea: co-evolution of monsoonal precipitation and El Niño over the past 600,000 years. Earth and Planetary Science Letters 264, 136–150.

Zhang, Y.G., Ji, J.F., Balsam, W., Liu, L.W., Chen, J., 2009. Mid-Pliocene Asian monsoon intensification and the onset of Northern Hemisphere glaciation. Geology 37, 599-602.

3

Climate Change and Health Effects

Rajan R. Patil
School of Public Health, SRM University, Chennai
India

1. Introduction

The United Nations Framework Convention (UNFC) on climate change defines climate change as, "a change of climate which is attributed directly or indirectly to the human activity that alters the composition of the global atmosphere and which is in addition to natural climate variability observed over comparable time periods" (UNFCC, 1992). The EU has defined dangerous climate change as an increase in 2 degrees celsius of average global temperatures. Since 1900, global temperatures have risen by 0.7 degrees celsius and are continuing to rise at an estimated rate of 0.2 degrees per decade. If left unchecked, this implies global warming of at least 1.4 degrees celsius (IPCC, 2001).

The United Nations Framework Convention on Climate Change (UNFCCC) was convened in 1992 with an overarching framework to address the challenges of climate change through inter governmental efforts. The objectives of the UNFCCC are: 1. To stabilize greenhouse gas concentrations to levels that prevent dangerous interference with the global climate system; and 2. To achieve these reductions within a time frame that allows ecosystems to adapt naturally to climate change, to ensure that food production is not threatened, and to enable economic development to proceed in a sustainable manner. The Kyoto protocol was developed in 1997 to reinforce the emissions reduction commitments of the UNFCCC. The protocol came into legal force in 2005 when it was ratified by 30 industrialized nations, creating legally binding targets for a 5 percent reduction in emissions below 1990 levels by 2012.

The World Metrological Organization and United Nations Environment Programme (UNEP), in an effort to combat the worsening situation, set up the Intergovernmental Panel on Climate Change (IPCC) in 1988. In recognition of the strong body of evidence that this panel has painstakingly collated, it was honored with the Nobel Peace Prize in 2007. The panel recently released their fourth assessment report which categorically states that the "warming of the climate system is unequivocal, as is now evident from observation of increases in global average air and ocean temperature, widespread melting of snow and ice and rising global average sea level". The fourth assessment report has already identified three areas in which human health has already been affected by climate change. These are: (I) alteration of distribution of some infectious disease vectors, (ii) seasonal distribution of some allergenic pollen species, and (iii) increased heat wave related deaths (Confaloneieri et al 2007).

That climate change impacts health in many ways was highlighted by the World Health Organization (WHO) when it chose to mark World Health Day on April 7 with the theme

"Protecting health from climate change"... The relationship between climate change and human health is multidimensional. The emerging evidence of climate change effects on human health (IPCC 2007) shows that climate change has: altered the distribution of some infectious disease vectors; altered the seasonal distribution of some allergenic pollen species; and increased heat wave-related deaths.

Health effects due to climate change is not a new phenomenon; literate, scholarly systems of medicine dating back more than 3,000 years are available for many parts of the world. Pathological signs in bones, fossil excreta and other items can be studied in archaeological material. Molecular techniques can yield additional information from such remains. In Europe, parish records, the diaries and publications of physicians and other archival material are a rich source of information. Thus, as with climatology, we can turn to a variety of sources for evidence of diseases in past climates (Reiter, 2007). Root cause analysis show that, social and economic developments [driving forces] exert pressure on the environment and, as a consequence, the state of the environment changes. This leads to impacts on e.g. human health, ecosystems and materials that may elicit a societal response that feeds back on the driving forces, on the pressures or on the state or impacts directly, through adaptation or curative action (Griffith, n.d)

The Intergovernmental Panel on Climate Change (IPCC) projected that changes in temperature, precipitation, and other weather variables due to climate change "are likely to affect the health status of millions of people, particularly those with low adaptive capacity" and stated that they had "very high confidence" that climate change is "currently contributing to the global burden of disease and premature deaths" (Paul et al, 2009).

The World Health Organization has concluded that the climatic changes that have occurred since the mid 1970s could already be causing annually over 150,000 deaths and five million disability-adjusted life-years (DALY), mainly in developing countries. The less developed countries are, ironically, those least responsible for causing global warming. Many health outcomes and diseases are sensitive to climate, including: heat-related mortality or morbidity; air pollution-related illnesses; infectious diseases, particularly those transmitted, indirectly, via water or by insect or rodent vectors; and refugee health issues linked to forced population migration. Yet, changing landscapes can significantly affect local weather more acutely than long-term climate change (Partz & Olson, 2006).

2. Health consequences of climate change

Impacts of Climate Change on health are manifested directly due to heat, cold, and injuries or indirectly through changes in environment, agriculture, human behavior and migrations.

2.1 Direct & acute effects

Direct effects on health due to heat, cold, and injuries are some of the acute manifestations resulting due to climate change. These effects can easily be witnessed as a consequence of climate change either in the form of heat and cold waves or direct injuries resulting from heavy rains and wind speeds as witnessed in hurricanes.

2.1.1 Direct effects of extreme events

An increase in the frequency and intensity of extremes of temperature, precipitation and wind speed have clear implications for mortality and morbidity. Flooding and storms increase the risk of deaths and non-fatal injuries. Climate change is expected to increase average temperatures as well as the number and intensity of heat waves. Heat waves are associated with increases in morbidity and mortality in the short term, especially in populations who are not adapted to extremely hot weather. Hot working environments also have non-fatal implications. Heat exposure increases the risk of having accidents. Hot working environments may decrease the ability to carry out physical tasks as well as have implications for mental task ability. Prolonged heat exposure may lead to heat exhaustion or heatstroke. In addition to the implications for health and well-being, climate change may through exposure of workers to heat stress have important direct effects on productivity (Nerlander, 2009).

The Indian metropolitan city of Mumbai was besieged with India's heaviest downpour of the century in July 2005, killing nearly 600 people. According to the Indian Meteorological department, this was the heaviest rainfall ever received in a single day, anywhere in India, recording 94.4 cm in the last 100 years. It broke the record of the previous highest rainfall at one place in India, at Cherrapunjee in Meghalaya (83.82 cm recorded on July 12, 1910). Cherrapunjee in the Northeastern state of Meghalaya is a generally well-known for being the wettest place in the world. Extreme weather changes surpassing their usual statistical ranges and tumbling records in India could be an early warning bell of global warming. Extreme weather events like the recent record setting in the western Indian city of Mumbai, or the all time high fatalities due to the heat wave in southern Indian states, or increasing vulnerability of eastern Indian states to floods could all be a manifestation of climate change in the Asian subcontinent (Patil & Deepa, 2007).

Acute variation in temperature and precipitation, can lead to various Patho-Physiological (Hypo-Hyper thermia, heart stroke, burns, frost bites etc). Extreme weather events such as severe storms, floods and drought can have obvious results such as physical injuries and drowning. Rising sea-levels will also give rise to flooding leading to drowning and population displacement.

2.2 Indirect and chronic effects

There are many indirect effects as: communicable diseases e.g.: vector borne disease, diarrheal diseases; ecological disturbances impacting on agent- host-environment relationships; malnutrition resulting due to agricultural impacts leading to food security issues; environmental health related to air and water quality issues, and human behavior issues such as migrations, and mental health.

2.2.1 Vector borne disease

Climate change is also expected to affect animal, human and plant health via indirect pathways. It is likely that the geography of infectious diseases and pests will be altered, including the distribution of vector-borne diseases, which are highly sensitive to climatic conditions. Extreme weather events might then create the necessary conditions for vector borne disease to expand its geographical range. Strengthening global, regional and national early warning systems is crucial, as are co-ordinated research programs and subsequent prevention and intervention measures (Martin et al,date??). As the ambient temperature of a region rises, the ecology changes and therefore populations of disease carrying animals or insects may increase as well. The rate of replication of the vector itself, or the pathogen (virus, bacteria) within those vectors can be sensitive to temperature. Changes in precipitation patterns can alter the number of breeding sites available leading to explosive epidemics of the following varieties of vector Borne diseases: **Mosquitoes Borne Diseases** e.g., Malaria, Dengue, Chikungunia, Yellow fever, Filaria are some of most climate sensitive diseases in which there is a direct correlation with temperature and rainfall which can be demonstrated. **Rodent-borne diseases** e.g. leptospirosis, are commonly reported in the after-math of flooding. In some areas, drought may reduce the transmission of some mosquito borne diseases, leading to reduction in the proportion of immune persons and therefore a larger amount of susceptible people once the drought breaks. **Pests borne disease**: Pests could become even more important disease vectors as a result of climate change. The spread of Plague, West Nile and Lyme disease are indicative of impact of pests on public health.

2.2.2 Malaria

Climate factors, particularly rainfall, temperature and humidity, interact to greatly affect the development, behavior and survival of mosquitoes transmitting malaria. However, as the Intergovernmental Panel on Climate Change (IPCC) reports, despite known causal links between climate, malaria and transmission dynamics, there is still much uncertainty about the potential impact of climate change on malaria at local and global scales. This is in part due to the complexity and local specificities of malaria transmission. Different mosquito vector species and parasites react differently to various climate conditions. For example, a change in temperature can affect the growth of the parasite within the mosquito and a change in local climate may make it less suitable for one vector. This particularly applies to water habitats for mosquito breeding environmental and institutional factors). However, while there is substantial knowledge on mosquito vectors, there is uncertainty about how climate change may change and influence malaria transmission. Two impacts of climate change at least have to be considered as major factors: temperature and rainfall patterns. The less important, but easiest to model, is the direct effect of temperature. This has effects both on mosquito range and survival, and the period of time it takes for mosquitoes to become infectious following biting an infected individual; the shorter the period, the greater the vectoral capacity. For both reasons, higher temperatures are likely to lead to more malaria, but the effects of this should not be exaggerated, and changes in temperature are unlikely to occur with all other environmental factors remaining constant (DEFID, 2010).
Vector Borne Zoonotic Disease]s [VBZDs: Climate change may affect the incidence of VBZDs through its effect on four principal characteristics of host and vector populations that relate to pathogen transmission to humans: geographic distribution, population density, prevalence of infection by zoonotic pathogens, and the pathogen load in individual hosts and vectors. These mechanisms may interact with each other and with other factors such as anthropogenic disturbance to produce varying effects on pathogen transmission within host and vector populations and to humans. Because climate change effects on most VBZDs act through wildlife hosts and vectors, understanding these effects will require multidisciplinary teams to conduct and interpret ecosystem-based studies of VBZD pathogens in host and vector populations and to identify the hosts, vectors, and pathogens with the greatest potential to affect human populations under climate change scenarios (Mills et al, 2010). Most vector-borne diseases exhibit a distinct seasonal pattern, which clearly suggests that they are weather sensitive. Rainfall, temperature, and other weather variables affect in many ways both the vectors and the pathogens they transmit. For

example, high temperatures can increase or reduce survival rate, depending on the vector, its behavior, ecology, and many other factors. Thus, the probability of transmission may or may not be increased by higher temperatures. The tremendous growth in international travel increases the risk of importation of vector-borne diseases, some of which can be transmitted locally under suitable circumstances at the right time of the year. But demographic and sociologic factors also play a critical role in determining disease incidence, and it is unlikely that these diseases will cause major epidemics in the United States if the public health infrastructure is maintained and improved (Gubler, 2001).

Climate is a major factor in determining: (1) the geographic and temporal distribution of arthropods; (2) characteristics of arthropod life cycles; (3) dispersal patterns of associated arboviruses; (4) the evolution of arboviruses; and (5) the efficiency with which they are transmitted from arthropods to vertebrate hosts. Thus, under the influence of increasing temperatures and rainfall through warming of the oceans, and alteration of the natural cycles that stabilize climate, one is inevitably drawn to the conclusion that arboviruses will continue to emerge in new regions. For example, we cannot ignore the unexpected but successful establishment of chikungunya fever in northern Italy, the sudden appearance of West Nile virus in North America, the increasing frequency of Rift Valley fever epidemics in the Arabian Peninsula, and very recently, the emergence of Bluetongue virus in northern Europe (Gould, 2009)

2.2.3 Chikungunya

Chikungunya is a viral disease that is spread by mosquitoes. It causes fever and severe joint pain. Other symptoms include muscle pain, headache, nausea, fatigue and rash. The disease shares some clinical signs with dengue, and can be misdiagnosed in areas where dengue is common. There is no cure for the disease. Treatment is focused on relieving the symptoms. The proximity of mosquito breeding sites to human habitation is a significant risk factor for Chikungunya.

The Indian capital city of Delhi reported its first ever case of Chikungunya in June 2007.Any new disease in any new region where it was previously not known to occur is certainly a cause of concern, as it is an emergence of a new infectious agent in a hitherto 'virgin' region. It could be a manifestation of disturbed equilibrium in the ecology of a given region. New epidemics in the new regions are a definite signs of an ecological ill health. Hence, if the ongoing climate change can lead to ecological disturbances, it is likely to bring in changes in distribution of vector borne disease like Chikungunya and other vector borne diseases (Patil, 2011)

2.2.4 Lyme disease

Lyme disease, or Lyme borreliosis,is an emerging infectious disease caused by at least three species of bacteria belonging to the genus Borrelia. The disease is named after the town of Lyme, Connecticut, USA, where a number of cases were identified in 1975. Lyme disease is the most common tick-borne disease in the hemisphere. Early symptoms may include fever, headache, fatigue, depression, and a characteristic circular skin rash called erythema migrans. Left untreated, later symptoms may involve the joints, heart, and central nervous system (Ryan, 2004)

Climate change will increase the geographical distribution of Lyme disease. Lyme disease is spread by blacklegged tick bites. A survey conducted from 1992 to 2006 indicates that the incidence of Lyme disease is increasing and rates are highest among children age 5–14 years. The number of reported cases of Lyme disease more than doubled during this time period.19 Children are especially vulnerable to tick bites because they tend to play outside and close to the ground (EPA, u.d) Effect of Climate change on other vector borne diseases **West Nile virus** is spread by infected mosquitoes, and can cause serious, life-altering and even fatal disease. The main route of human infection is through the bite of an infected mosquito. Approximately 90% of West Nile Virus infections in humans are either without any symptoms or very vague symptoms with fever and generalized body pain. The temperature thresholds for WNV survival are not documented, but laboratory studies indicate that the ability of competent vectors to transmit the virus is favored by higher temperatures and the vector's temperature-dependent survival pattern. Climate change may lengthen survival periods of WNV-competent Anopheles) mosquitoes (Table 8) and possibly allow infected hosts (birds) to change their geographic range. These could result in changes in virus prevalence rates and distribution. Therefore, climate change may increase WNV transmission risk. Leishmaniasis. The current environment is conducive to Phlebotomus sandfly survival for several months. Climate change might decrease the number of days suitable for Phlebotomus ariasi. The risk of contracting leishmaniasis may become high. **Mediterranean spotted fever.** The abundant and widespread distribution of the tick as well as the high prevalence of dogs infected with Rickettsia conorii. Because R. sanguineus has a remarkable ability to adapt to its environment, and disease transmission is highest during warmer months, even in harsher arid climatic zones where ambient temperatures exceed 35°C and soil temperatures exceed 45°C In fact, it is possible that climate change may prolong the peak season of MSF cases because of higher temperatures in spring and autumn. **Schistosomiasis:** Environmental conditions can be conducive to Schistosoma transmission, the competent snail population may be infected, and the risk of transmission could be high. Assuming ambient air temperatures as approximations of shallow water temperatures (which affect parasite and vector survival), it is clear that climate change might lengthen parasite Survival periods and vector survival. Focal introduction of the parasite from infected imported human cases to the currently non infected snail population is also possible. If a focal parasite-infected snail population were to occur, if a warmer climate scenario is assumed and that the infected vector population may with time widen its geographic distribution as the favorable temperature period for survival increases significantly, then disease transmission risk may increase toward a medium level (Casimiro, 2006).

3. Food security

Climate change together with other factors can have serious implication on food security consequently resulting in Malnutrition due to following reasons:

Decreased Agricultural Yield: Agricultural production and food security are also linked directly to precipitation patterns – this impacts the nutritional status of the population. Excess or Scarcity of Water resulting from draught, floods, heavy rains can adversely affect agricultural output. Salinization of fertile land : **Rising sea levels** increase the risk of coastal flooding of agricultural land due to sea levels rise leading to decreased yield of crops resulting in malnutrition.**Population Migrations:** Population displacement and also rural to urban migration carries its own health risks e.g., malnutrition and increased risks of communicable diseases. Increased rates of malnutrition as they become more susceptible to other diseases through influx or outpouring of infected population e.g. malaria parasitemia may alter the host and herd immunity leading to increased susceptibility. The vicious cycle between malnutrition and life threatening infectious disease is well demonstrated

3.1 Effect of climate change on malnutrition

About Climate change affects food and nutrition security and further undermines current efforts to reduce hunger and protect and promote nutrition. Additionally, under nutrition in turn undermines the resilience to shocks and the coping mechanisms of vulnerable populations, lessening their capacities to resist and adapt to the consequences of climate change. Climate change further exacerbates the already unacceptably high levels of hunger and under nutrition. Climate change will increase the risk of hunger and under nutrition over the next few decades and challenges the realization of the human rights for health and adequate food. Climate change will affect nutrition through different causal pathways that impact food security, sanitation, water and food safety, health, maternal and child health care practices and many socioeconomic factors. Climate change negatively affects food availability, conservation, access and utilization and exacerbates socioeconomic risks and vulnerabilities. According to the IPCC if current trends continue, it is estimated that 200-600 million more people will suffer from hunger by 2080. Calorie availability in 2050 is likely to decline throughout the developing world resulting in an additional million undernourished children, 21% more relative to a world with no climate change, almost half of which would be living in sub-Saharan Africa. Climate change negatively affects nutrition through its impacts on health and vice versa. Climate change has an impact on water availability and quality, sanitation systems, food safety and on waterborne, food borne, vector-borne and other infectious diseases which eventually both increase nutritional needs and reduce the absorption of nutrients and their utilization by the body. Mitigation is critical to limit impact of climate change on food security and nutrition in low and middle income countries in the future. However, mitigation strategies should not increase food and nutrition insecurity. For example, bio fuel production can have a negative impact on food production and nutrition. Bio fuel production requires large amounts of natural resources (arable land, water, labor, etc.) that might thus be diverted from the cultivation of food crops10 (UNSCN, 2010).

About 46% of the DALYs attributable to climate change were estimated to have occurred in the WHO South-East Asia Region, 23% in countries in the Africa region with high child mortality and very high adult male mortality, and 14% in countries in the Eastern Mediterranean region with high child and adult male mortality. The relative risk estimates for malnutrition, diarrheal diseases, and malaria, respectively, projected for 2030 under the alternative exposure scenarios. The relative risks of malnutrition is directly proportional to underweight; this applies to all diseases affected by underweight (including diarrhea and malaria) (McMichael, 2004).

3.2 Effect of climate change on food security

With "high" or "very high confidence" the IPCC predicts the following, by 2020, in some countries, yields from rain-fed agriculture could be reduced by up to 50%. Agricultural production, including access to food, in many countries is projected to be severely compromised. This would further adversely affect food security and exacerbate malnutrition. According to the IPCC, GCC threatens the health, happiness and even survival of literally hundreds of millions of people, through increased risk of malnutrition and starvation, and increased frequency of deadly weather events (Philos, 2010). In the socio-economics literature on rural livelihoods, it is widely accepted that farming households face three main sources of vulnerability : shocks (unexpected extreme events, for example the sudden death of a family member, or an extreme weather event), seasonal variations (including variations in periodicity and amount of rainfall) and long term trends (such as increases in input prices, or long term changes in mean temperature and rainfall). The problems from all three are likely to increase in intensity, particularly for farmers relying on rain-fed production. Small-scale farming provides most of the food production, as well as employment for 70% of working people. These small-scale producers already face the challenges of climate variability in current climates. For example, intra-seasonal distribution of rainfall affects the timing and duration of the possible cropping season, and periods of drought stress during crop growth. Cropping practices that are often used to mitigate the effects of variable rainfall (Challinor et al, 2007).

Looking at individual sectors, the equity implications of climate change are most pronounced for food security. Low-emission countries are, in general, more adversely impacted (in terms of projected future yield changes of staple crops), more exposed (in terms of the share of agriculture in gross domestic product and labor force), and less able to cope with adverse impacts (in terms of the current level of under nutrition). The analysis for human health also implies that those least responsible for climate change will be most affected by its adverse impacts. Countries with low emissions levels have, on average, a lower current health status (measured by infant mortality and life expectancy), higher socio-economic vulnerability to extreme weather events, and already experience stronger adverse climate impacts on human health (Fussel, 2009).

3.3 Pests

The reproductive success of predators depends, food abundance and population density and their interactions may respond to changes in climatic conditions. Timing of reproduction may increase, during a period of temperature increase. Few studies have investigated how climate change affects predator–prey and parasite–host interactions, although such effects are widely predicted to be key for understanding community level effects of climate change. Theoretical studies suggest that predators and parasites may be particularly susceptible to the effects of climate change due to the direct effects of climate on the distribution and the abundance of prey and host populations, respectively. However, there are only few empirical studies indicating that the ability of hosts to defend themselves against parasites is strongly influenced by environmental conditions. The North Atlantic Oscillation has been shown to affect predator–prey cycles in the Canadian arctic. Studies of the great tit Parus major and its caterpillar prey have shown increasing mal-adaptation of timing of breeding to maximum availability of prey, providing a cause for concern (Nielsen & Moler, 2006).

3.4 Effects of agricultural chemicals and pathogens on human health

Humans may be exposed to agriculturally derived chemicals and pathogens in the environment (i.e., air, soil, water, sediment) by a number of routes, including the consumption of crops that have been treated with pesticides or have taken up contaminants from soils; livestock that have accumulated contaminants through the food chain; fish exposed to contaminants in the aquatic environment; and groundwater and surface waters used for drinking water. Exposure may also occur via the inhalation of particulates or volatiles, or from direct contact with water bodies or agricultural soils (e.g., during recreation). The importance of each exposure pathway will depend on the pathogen or chemical type. The main environmental pathways from the farm to the wider population will be from consumption of contaminated drinking waters and food (Alistair et al, 2009).

3.5 Migration/shift in occupation

At a basic level, for many farmers the challenge will be whether they can continue to farm. Already rural livelihoods at household level are highly diverse, with farming accounting for a lower proportion of disposable income and food security for farming households than 20 years ago. For example, concludes that "diversification out of agriculture has become the norm among African rural populations." There is evidence that households moving out of poverty are those moving either completely or partially out of farming. It is likely that many households will respond to the challenge of climate change by seeking further to diversify into non-farm livelihood activities either in situ or by moving (or sending more family members) to urban centers. For these households, farming may remain as (or revert to) a semi-subsistence activity while cash is generated elsewhere. This would be simply a continuation of a well-established trend towards pluriactive, multi-locational families and the transfer of resources through urban–rural remittances. However, given the acute population and development related challenges faced by most African nations, many households will be forced to remain in the farming sector for livelihood and security for some time to come as the population in Africa undergoes a three-fold increase this century. This will lead to considerable demand for expansion of area under small-farm cultivation for staple crops. Farming for profit, particularly production for international markets, may therefore become more concentrated on fewer farms, as is already happening in the fresh vegetable export market from eastern and southern Africa. Companies with the capital to invest in controlling their production environment through irrigation, netting and crop protection in order to meet stringent quality and bio-safety requirements of European supermarkets are increasing their market share at the expense of smallholders. This should lead to further irrigation development, and contribute to a recommended doubling of irrigated land by 2015.

4. Water borne diseases

4.1 Climate change and water borne disease

High temperatures, water scarcity and water abundance resulting from flooding or heavy precipitation have been shown to be related to diarrheal diseases. Heavy rainfall, even without flooding, may increase rates of diarrheal disease as sewage systems overflow. Increases in soil run-off may contaminate water sources

A lack of availability of water for personal hygiene and washing of food may lead to an increase in diarrheal disease and other diseases associated with poor hygiene. It is important to note that high temperatures in itself an independent risk factor for increased rates of diarrheal diseases, including salmonella and cholera.

Clearly, the health implications of changes to water supply are far-reaching. Currently, more than 3 million people die each year from avoidable water-related disease, most of whom are in developing countries. The effects of climate change on water will exacerbate the existing implications of water shortages on human health (Water Aid u.d), as follows:

Water-borne diseases: result from the contamination of water by human/ animal faeces, or by urine infected with pathogenic viruses/ bacteria, both of which are more likely to occur during periods of flood.

- Water-washed diseases: those resulting from inadequate personal hygiene as a result of scarcity or inaccessibility of water (including many water-borne diseases and typhus).
- Water-based diseases: those caused by parasites that use intermediate hosts living in/near water (e.g. guinea worm).
- Water-related diseases: borne by insect vectors that has habitats in/near water (such as malaria
- Water-dispersed diseases: infections for which the agents proliferate in fresh water and enter the human body through the respiratory tract (e.g. legionella).Climate change and water resources

Climate change may affect the growth and survival of disease-causing organism's related to water- and food-borne illness. The incidence of water- and food-borne illnesses, such as gastroenteritis and infectious diarrhea, is known to increase when outdoor temperature increases, or immediately following storms or floods. Extreme weather can result in the breakdown of sanitation and sewer systems, a loss of power for refrigeration, or inadequate means to thoroughly cook food, increasing the likelihood of water- and food-borne illness. Children are especially susceptible to water- and food-borne illness due to their developing immune systems. In fact, infectious diarrhea is responsible for approximately 1.5 million child deaths per year globally, disproportionately affecting children of developing nations (EPA u.d).

Knowledge about transport processes and the fate of microbial pollutants associated with rainfall and snowmelt is key to predicting risks from a change in weather variability. Although recent studies identified links between climate variability and occurrence of microbial agents in water, the relationships need further quantification in the context of other stresses. In the marine environment as well, there are few studies that adequately address the potential health effects of climate variability in combination with other stresses such as overfishing, introduced species, and rise in sea level. Advances in monitoring are necessary to enhance early-warning and prevention capabilities. Application of existing technologies, such as molecular fingerprinting to track contaminant sources or satellite remote sensing to detect coastal algal blooms, could be expanded. This assessment recommends incorporating a range of future scenarios of improvement plans for current deficiencies in the public health infrastructure to achieve more realistic risk assessments (Bose et al, 2001).

4.2 Harmful algae bloom

A worldwide increase in cyanobacterial (blue-green algae) sources has been observed in both coastal and freshwaters. These harmful algae blooms (HABs), which produce nerve and liver toxins, are longer in duration, of greater intensity, and are suspected of being tied both to increased temperatures due to climate change and nutrient runoff. Exposure to marine toxins has resulted in death and poisonings of California sea lions and Florida alligators. Human exposure is of concern through both drinking water contamination and recreational exposure (English et al, 2009).

4.3 El Niño and severe rainfall/flooding and potential health effects

El Niño is a phenomenon results in the namesake oscillation of wind and ocean currents, usually occurring every three to seven years, there is concern that climate change will increase its frequency or the severity of its consequences. These, in turn, change regional temperatures and precipitation patterns and lead to significantly increased rainfall. Several researchers have established a link between heavy rainfall and flooding—whether resulting from El Niño-associated events or from other meteorological impacts—and subsequent outbreaks of infectious diseases. Extreme meteorological events can easily disrupt water purification and storm water and sewage systems, as well as contaminate uncovered wells and surface water, leading to an increased risk of illness. These risks are even higher when a population lives in a low-lying area, where the land's hydrology causes draining tributaries to meet. Conversely, heavy rains and coastal events can also flush microorganisms into watersheds, affecting those up-coast as well. Nonsustainable development, such as that which contributes to deforestation and soil erosion, influences water contamination by destroying the land's natural ability to absorb runoff, resulting in water-contaminating mudslides(Reiter, 2007).

4.4 Cholera and diarrheal diseases

Climate change can result in increased temperatures in both ocean water and ambient air. Increased sea temperatures have a direct effect on the proliferation of plankton and algae in sea water. Vibrio species organisms, including V. cholera, thrive in particular sea conditions. Among these are warm water, moderate salinity, and number of aquatic invertebrates, all conditions influenced by climate change. In particular, the quantity of vibrio species may increase or the range of the bacteria may extend. Many causative agents of diarrheal disease have a seasonal variability, with peaks in the warmer months. Increased temperatures or higher temperatures for longer times can result in higher than expected diarrhea incidence. Finally, rises in sea level due to increased temperatures can lead to coastal flooding, which can force the use of contaminated water, overwhelm sanitation systems, or prompt migration into areas with insecure water and sanitation availability (Fricas & Tylor, 2007).

Climate change has begun to negatively affect human health, with larger burdens projected in the future as weather patterns continue to change. The climate change-related health consequences like diarrheal diseases (Kristie, 2008). Recent studies examining the potential impacts of climate variability and change on the risks and incidence of water- and food borne illnesses conclude that that the risk of water- and food-borne illness will likely increase with climate change. Studies suggest that extreme precipitation events increase the loading of contaminants to waterways, climate change could increase the risk of illness associated with Cryptosporidium parvum, association between increases in the lagged monthly mean temperature and increases in the number of notifications of salmonellosis infections (Ebi et al, 2006).

5. Air quality and health

5.1 Effect of climate change on air quality

Climate Change also change patterns of air movement and pollution, causing expanded or changed patterns of human exposure and resulting health effects. The formation of many air-pollutants is determined in part by climate factors such as temperature and humidity. In addition the transport and dispersion of air pollutants away from source regions are strongly affected by weather factors. Climate change therefore influence pollutant concentrations, which in turn may affect health as air pollution is related to cardio-respiratory health

Climate and and air quality are closely coupled. Conventional pollutants, such as ozone and particle pollution, not only affect public health but also contribute to climate change. Ozone is a significant greenhouse gas (GHG) and particles can influence the climate by scattering, reflecting, and/or absorbing incoming solar radiation and interacting with various cloud processes. Due to climate change, the IPCC predicted "declining air quality in cities." In summarizing the impact of climate change on ozone and particle pollution, the IPCC concluded that "future climate change may cause significant air quality degradation by changing the dispersion rate of pollutants; the Chemical environment for ozone and particle pollution generation; and the strength of emissions from the biosphere, fires, and dust." Though a great deal of uncertainty remains regarding the expected future impacts of climate change on air quality, recent research suggests that such effects may be very significant, particularly on a local or regional scale (EPA, 2010)

Health effects of air quality the formation of many air-pollutants is determined in part by climate factors such as temperature and humidity. In addition the transport and dispersion of air pollutants away from source regions are strongly affected by weather factors. Climate change may therefore influence pollutant concentrations, which in turn may affect health as air pollution is related to cardio-respiratory health. Exposure to high levels of ground-level ozone, for example, which is formed from the exhaust of transport vehicles, increases the risk of exacerbations of respiratory diseases such as chronic obstructive airways disease and asthma, leading to hospital admissions or increased mortality. The number of forest and bush fires may increase as certain regions face longer periods of extreme dry conditions and such fires can contribute to air-pollution. The direction and magnitude of the effects of climate change on air pollution levels are however highly uncertain and there will be regional variations. National energy policies and transport policies should take into account the health effects of air-pollution40 and early warning systems for levels of air pollution can be implemented. Reducing emission from transport vehicles is a win-win solution contributing both improve health as well as reduce greenhouse gas emissions (Nerlander, 2009).

5.2 Ozone

Because ozone formation increases with greater sunlight and higher temperatures, it reaches unhealthy levels primarily during the warm half of the year. Daily peaks occur near midday in urban areas, and in the afternoon or early evening in downwind areas. It has been firmly established that breathing ozone can cause inflammation in the deep lung as well as short-term, reversible decreases in lung function. In addition, epidemiologic studies of people living in polluted areas have suggested that ozone can increase the risk of asthma-related hospital visits and premature mortality. Vulnerability to ozone effects on the lungs is greater for people who spend time outdoors during ozone periods, especially those who engage in physical exertion, which results in a higher cumulative dose to the lungs. Thus, children, outdoor laborers, and athletes all may be at greater risk than people who spend more time indoors and who are less active. Asthmatics are also a potentially vulnerable subgroup (Ebi, u.d).

6. Climate change and allergies

Pollen allergy currently affects significant proportion of the population. A warmer climate will lead to a longer pollen season and more days with high pollen counts. In addition, a warmer climate increases the risk of proliferation of new plants with well-known allergenic pollens like ragweed, plane tree and wall pellitory. The consequences will be more people with hay fever and pollen asthma, longer allergy seasons and an increase in the severity of symptoms, disease-related costs and demands on health care for diagnosis and treatment of more complex allergies. It is clearly identified that climate change can exert a range of effects on pollen, which might have consequences for pollen-allergic patients. The pollen season might become longer thereby extending the period in which patients suffer from allergy symptoms. This extension of the pollen season could be due to a prolonged flowering period of certain species, e.g. grasses, or the appearance of new species that flower in late summer, e.g. common ragweed. Climate change could cause an increase in heavy thunderstorms on summer days in the grass pollen season, which are known to increase the chance of asthma exacerbations (sommer et al, 2009).
Climate change alters the concentration and distribution of air pollutants and interferes with the seasonal presence of allergenic pollens in the atmosphere by prolonging these periods. The link between climate change and respiratory allergies is most importantly explained by the worsening ambient air pollution and altered local and regional pollen production. Laboratory studies confirm epidemiologic evidence that air pollution adversely affects lung function in asthmatics. Damage to airway mucous membranes and impaired mucociliary clearance caused by air pollution may facilitate access of inhaled allergens to the cells of the immune system, thus promoting sensitization of the airway. Consequently, a more severe immunoglobulin (Ig) E-mediated response to aeroallergens and airway inflammation could account for increasing prevalence of allergic respiratory diseases in polluted urban areas (D'Amato et al, 2010).

6.1 Molds

Aeroallergens that may respond to climate change include outdoor pollens generated by trees, grasses, and weeds, and spores released by outdoor or indoor molds. Because climatologic influences differ for these different classes of aeroallergens, they are discussed separately here. As compared with pollens, molds have been much less studied. This may reflect in part the relative paucity of routine mold monitoring data from which trends might be analyzed, as well as the complex relationships among climate factors, mold growth, spore release, and airborne measurements.63 In addition to potential effects on outdoor mold growth and allergen release related to changing climate variables, there is also concern about indoor mold growth in association with rising air moisture and especially after extreme storms, which can cause widespread indoor moisture problems from flooding and leaks in the building envelope. Molds need high levels of surface moisture to become established and flourish (Kinney, 2008).
The urban heat island effect, a combination of anthropogenic and climatologic heat, can increase urban temperatures as much as 5°C compared with rural locations and further drive the formation of ozone.

Air pollution can interact with pollen grains, leading to an increased release of antigens characterized by modified allergenicity.
Air pollution can interact with allergen-carrying paucimicronic particles derived from plants. The paucimicronic particles, pollen-originated or not, are able to reach peripheral airways with inhaled air, inducing asthma in sensitized subjects.
Air pollution—in particular ozone, PM, and sulfur dioxide—have been shown to have an inflammatory effect on the airways of susceptible subjects, causing increased permeability, easier penetration of pollen allergens in the mucus membranes, and easier interaction with cells of the immune system.
There is also evidence that predisposed subjects have increased airway reactivity induced by air pollution and increased bronchial responsiveness to inhaled pollen allergens.
Some components of air pollution seem to have an adjuvant immunologic effect on IgE synthesis in atopic subjects—in particular, DEPs, which can interact in atmosphere with pollens or paucimicronic particles.

Table 1. The rationale for the interrelationship between agents of air pollution and pollen allergens in inducing respiratory allergy (Shea et al, 2008).

7. Non communicable disease

IPCC expects all parts of the planet to experience more heat exposure in the future (IPCC 2007), while the local extent of heating will vary. Increased heat and climate variability will also influence other exposure routes which are moderated by socio-economic status and other variables. Dehydration increases the concentration of calcium and other compounds in the urine, which facilitates the formation of kidney stones (Cramer and Forrest 2006). In addition to kidney stone disease, there is evidence that during heat waves there is an increase in hospitalizations for acute renal failure and other kidney diseases (Kjellstrom et al, u.d)

7.1 Effect of climate change on non communicable disease

Cardiovascular Disease and Stroke: Association between air quality, especially ozone and particulate burdens, and cardiovascular disease appear to be modified by weather and climate. Ozone is also associated with acute myocardial infarction. Particulate matter is associated with a variety of patho-physiological changes including systemic inflammation, deranged coagulation and thrombosis, blood vessel dysfunction and atherosclerotic disease, compromised heart function, deep venous thromboses,95.Increased burden of PM2.5 is associated with increased hospital admissions and mortality from cardiovascular disease, as well as ischemic heart disease. **Neurological:** climate change on ocean health, resulting in increased risks to neurological health from ingestion of or exposure to neurotoxins in seafood and fresh and marine waters. Neurotoxins produced by harmful algal blooms and other marine microorganisms can cause serious illness and death in humans. The most frequent human exposures are via consumption of seafood containing algal toxins, although some toxins may be present in freshwater sources of drinking water, and others may be aerosolized by surf breaking on beaches and then transported by winds to where they can cause respiratory distress in susceptible individuals who breathe them. Because cooking or other means of food preparation do not kill seafood biotoxins, it is essential to identify contaminated seafood before it reaches consumers. **Human Developmental Effects:** climate change could alter normal human development both in the womb and later in life. Food borne illness and food insecurity, both likely outcomes of climate change, may lead to malnutrition. While adult humans exposed to mild famine usually recover quite well when food again becomes plentiful, nutritional reductions to a fetus in the womb appear to have lasting effects throughout life. Climate change effects on food availability and nutritional content could have a marked, multigenerational effect on human development. Certain commercial chemicals present in storage sites or hazardous waste sites can alter human development. Flooding from extreme weather events and sea-level rise are likely to result in the release of some of these chemicals and heavy metals, most likely affecting drinking and recreational waters. Some of these, including mercury and lead, have known negative developmental effects (IWGCCH, u.d).

8. Cancer

Since last 30 years there has been concern that anthropogenic damage to the earth's stratospheric ozone layer will lead to an increase of solar ultraviolet (UV) radiation reaching the earth's surface, with a consequent adverse impact on human health, especially to the skin. More recently, there has been an increased awareness of the interactions between ozone depletion and climate change (global warming), which could also impact on human exposure to terrestrial UV. The most serious effect of changing UV exposure of human skin is the potential rise in incidence of skin cancers. Climate change, which is predicted to lead to an increased frequency of extreme temperature events and high summer temperatures. This could impact on human UV exposure by encouraging people to spend more time in the sun. While future social trends remain uncertain, it is likely that over this century behavior associated with climate change, rather than ozone depletion, will be the largest determinant of sun exposure, and consequent impact on skin cancer (Diffey, 2004).

9. Mental health

Climate change has potential to influence mental health and behavior. It is observed that those with lower socioeconomic standing are more likely to choose to relocate permanently following a devastating event, often due to limited resources to rebuild property and restore livelihood. In addition, people will continue to experience place-based distress caused by the effects of climate change due to involuntary migration or the loss of connection to one's home environment, a phenomenon called "Solastalgia". (IWGCCH)

Climatic changes may have a significant impact on various dimensions of mental health and well-being. India has been witnessing high incidence of for cotton farmers' deaths/suicides since 1998. The socioeconomic-political factors emerge as very strong determinants of deaths, given the occupational work environment. Also there is decreasing yield of cotton over the years resulting in loss of revenue for the farmers leading them to mental distress. (Patil, 2002)

Violent crime may be exacerbated during heat waves because more stress hormones are released when people are exposed to excessive heat (simister & Cooper, 2004). More alcohol and drugs may be consumed during heat waves, and more people may seek help for their psychiatric problems during these periods (Bulbena et al, 2006). Drought appears to contribute to a variety of mental health effects, including more stress, grief, and hopelessness as well a sense of solastalgia, which describes a palpable sense of dislocation and loss people feel when they perceive changes to their local environment are pervasively harmful (Sartore et al, 2007).Conflict among people may be one of the hallmarks of climate change's severe weather, which can displace thousands or millions and lead to those people competing with others for scarce resources (Abbott, 2008). While many people have short-term reactions to extreme natural disasters—including grief, anger, anxiety, and depression—persistent post-traumatic stress may be the hallmark of climate change, as was demonstrated after Hurricane Katrina (Galea et al, 2007).One study showed that mental illness doubled after Hurricane Katrina (Kessler et al, 2006). One year after Hurricane Katrina, exposed children were four times more likely than before the storm to be depressed or anxious and twice as likely to have behavioral problems (Abramson et al, 2007). Other psychological problems, including family dysfunction, difficulties at work, increased child misbehavior, a sense of lost identity, and more may result from experiences of the extreme disasters that climate change is likely to bring (Bourque et al, 2006). Emotional distress and anxiety will be among the hallmarks of climate change and its effects, and disadvantaged communities are among those to be most harmed (Fritze et al, 2008).

The association between acute psychosis and climatic variation is known, especially in tropical countries. Studies from tropical countries like India suggest an increased prevalence of acute psychosis following viral fever, especially in winter. The hospital admission rates for schizophrenia and "schizoaffective" patients are clearly increased in summer and fall respectively, as reported in an 11-year study from Israel. Schizophrenia patients' mean monthly admission rates correlated with the mean maximal monthly environmental temperature, indicating that a persistently high environmental temperature may be a contributing factor for psychotic exacerbation in schizophrenia patients and their consequent admission to mental hospitals. Around half the children and adolescents exposed to the 'supercyclone' in the state of Orissa in India reported symptoms of the post-traumatic stress disorder (PTSD) syndrome of different severity even after one year. Drought affects family relationships. Stress, worry and the rate of suicide increase. The phenomenon of farmers' suicides in India is a typical example of the consequences of climatic vagaries in poor, predominantly agrarian economies (Chand, 2008)

10. Ethics

Anthropogenic climate change entails important consequences for international equity because both the causes of climate change and its impacts are unequally distributed across (and within) nations. The equity implications of climate change are attracting increasing attention because a comprehensive international agreement on climate change will only be agreed upon if it is considered fair by all parties to the UNFCCC. Therefore, the distribution of mitigation and adaptation costs across countries needs to consider their responsibility for climate change as well as their capacity to act, and the allocation of funds for adaptation need to consider, among others, their vulnerability to climate change. Looking at individual sectors, the equity implications of climate change are most pronounced for food security. Low-emission countries are, in general, more adversely impacted (in terms of projected future yield changes of staple crops), more exposed (in terms of the share of agriculture in gross domestic product and labor force), and less able to cope with adverse impacts (in terms of the current level of under nutrition)... The analysis for human health also implies that those least responsible for climate change will be most affected by its adverse impacts Thus, countries with low (fossil) emissions are not only least responsible for climate change, but they generally have lower socio-economic capacity to cope with adverse impacts of climate change (Fussel, 2009).

Ironically, the most serious victims of climate change are also the ones who do not have a voice in the mitigation of the problem. Therefore, the implementation of policy becomes deeply ethical. Human activity has already resulted in the loss of many thousands of species and the trend will only continue. Going back to the economic arguments, placing an economic value on the existence of a species or an ecosystem is not viable and as such economic arguments fail to be effective. Trying to fix an ethical problem with an economic solution is simply deficient (Helix, 2011). Ethics of global warming emphasizes the need to address concerns about climate change in a responsible way that improves conditions for the poor. The Kyoto climate treaty could cost the world community $1 trillion a year -five times the estimated price of providing sanitation and clean drinking water to poor developing countries, thereby preventing millions of deaths each year (Spencer et al, 2005).

10.1 Mitigation, adaptation, and intergenerational equity

There are three aspects of fairness vis-à-vis climate change: what is a fair cost allocation to prevent further global warming; what is a fair cost allocation to cope with the social consequences of the global warming that will not, in fact, be avoided; and; what is a fair allocation of greenhouse gases emissions over the long term and during the transition to long-term allocation? Helm lists five aspects of equity in climate change ethics: international equity in coping with the impacts of climate change and associated risks; international equity in efforts to limit climate change; equity and social considerations within countries; equity in international processes; and, equity among generations

Bio fuels have been defined as any type of liquid or gaseous fuel that can be produced from biomass and used as a substitute for fossil fuels (Giampietro et al.1997). There have been increasing efforts substitute gasoline and disel by renewable transport bio-fuels that come in the form of ethanol and bio diesel (Davidson, 2003). However in sudden increasing reliance on biofuel in itself can have implication on climate change as follows.

- Emissions may be reduced, but added crop production may affect the ability of the world's poor to feed themselves through increased demand.
- Environmentalists often value low-intensity crop production as it causes less environmental degradation and uses fewer fertilizers and fossil fuels. Higher intensity crop production would allow for greater output and less land transformation.
- Though climate change affects biodiversity, the land use associated with large-scale bio fuel production has the potential to devastate ecosystems, especially in poor countries.
- Finally, a shift to bio fuels will result in rural economic development. This may have implications for the urban economy.
- Should we develop bio fuels if their production could be detrimental to the poor?
- Should we really be developing low intensity energy if it results in the destruction of more natural areas than high intensity energy?
- Should we only be focusing on the ecological after effects of climate change rather than the land impacts created by potential energy systems?
- Should we consider potential effects on rural and urban economies?

10.2 Moral angle to climate change

Philosophers should take the lead in exposing the fallacy that economic growth is any longer the key to human flourishing in wealthy industrial democracies. We should emphasize the need to pursue intellectual/spiritual/personal/relationship growth rather than increased wealth, if we hope to live better lives. Environmental philosophers should also deal honestly with population issues, something we have rarely done in the recent past. At a minimum, we should acknowledge the role population growth plays in environmental destruction, rather than continuing to sweep this unpleasant fact under the rug. We also need to begin to bring "growth is bad" into politics, as well. It is difficult to see how this might be accomplished, however, at least from an American vantage point. For Americans, economic growth is not one goal among many, or a by-product of some more fundamental goal. It is the primary goal of our society, organizing much of our activity, individually and collectively.

Studies have repeatedly shown that while increasing wealth in poor countries does augment happiness, once a society becomes sufficiently prosperous, further increases in wealth no longer boost subjective wellbeing. Throughout the world, the cutoff line seems to be around $10,000, far below the average American income. Meanwhile, psychological studies show that a materialistic outlook is actually an impediment to individuals achieving happiness (Lane 1998, Kasser 2002, Kasser 2006). This is partly because such an outlook interferes with highly valuing people, and good relationships with spouses, friends and co-workers turn out to be very important in securing happiness. All in all, there is little evidence that doubling our wealth will increase Americans' happiness or flourishing. Values and ethics have a strong influence upon the behavioral outcomes that are manifest as the driving forces behind environmental pressures. Although this perspective underplays the structural constraints upon behavior, the influence of beliefs and values can be seen to operate via the configuration of goals, wants, needs, intent and choices. There needs to be consideration of human welfare as the key objective of both the human economy. The misguided nature of existing consumer culture beliefs about what will bring welfare probably represents the core issue in this analysis. Maximum consumption via material good accumulation, and derived services, drives economic and lifestyle choices and is the natural economic (if not the social) outcome of a belief system based on the principle that the external world is the ultimate source of happiness. The accumulation frenzy has required, and resulted in, prodigious natural resource extraction and global labor force exploitation powered largely by the capabilities endowed by fossil fuel energy. The extensive biophysical intervention associated with fossil carbon has led to the looming problems of climate change. (Philos, 2010).

The Middle Way describes the best approach to life as the "golden mean" - a concept shared in various philosophical strands (Marinoff 2007). This is a balanced approach in which basic needs and wants, that genuinely enhance welfare, can and should be satisfied (for all people). This would naturally cover food, clothing, warmth, shelter, and most ecological services as well as psychological security from social and community based needs. Alternately, extremes are avoided and excessive attachment and accumulation is inimical to the three spheres, and individual wellbeing and spiritual progress. The key process is to break and close the endless wants satisfaction circular gap by realization of the heedless nature of clinging to 'tamha' (desire) as a source of wellbeing (Griffith u.d).

11. Conclusion

Climate and weather are two of the most important factors in the emergence of infectious disease in humans. Extreme climate events are expected to become more frequent in the coming years with climate change. The natural history of disease transmission, particularly transmission by arthropods, involves the interplay of a multitude of interacting factors that defy simplistic analysis. The principal determinants are politics, economics, human ecology and human behavior all of which have direct relation to climate change. To detect and respond to the changes in the infectious disease epidemiology caused by the climate change will require strengthening of the public health infrastructure and ensuring increased surveillance for diseases most likely to be influenced by climate with particular attention to those with potentially large public health impacts. Climate change together with other factors can have serious implication on food security consequently resulting in Malnutrition. Agriculture is currently seen by many development experts including economists and policy makers as a sector that can make a significant contribution to the alleviation and mitigation of poverty in the medium term alongside the growth in non-agricultural sectors. The greatest challenges of the climate change in the coming years will be to cater to needs of growing demands to global food in the milieu of climate change.

The risk of non communicable diseases (NCDs) are seen to increase following climate change through number of mechanisms by which increasing population heat exposure and other environmental changes related to global climate change may affect NCDs causing acute or chronic health impacts. Cardiovascular, renal and respiratory diseases may be particularly affected, and people in low and middle income countries are at particular risk due to limited resources for prevention. It follows that in the climate change and health evaluations and action plans a greater focus on NCDs is warranted. The burden of mental health consequences need to be studied from several dimensions: psychological distress per se; consequences of psychological distress including proneness to physical diseases as well as suicide; and psychological resilience and its role in dealing effectively with the aftermath of disasters. When these events happen, people with pre-established mental illnesses often have more extreme difficulty coping than the rest of the population.

Climate change throws larger ethical and moral dilemmas on us as human beings since we have larger responsibility towards our other fellow co-habitants of this lone planet that can support life in the entire universe. While climate changes throws up difficult moral and ethical questions it is important to develop a normative framework of justice for the international-level funding of adaptation to climate change within the United Nations Framework Convention on Climate Change (UNFCCC) architecture. The distribution of power should assure that every party is able to make its interest count in every negotiating stage. According to this principle, the voice of weaker countries in the international regime on adaptation funding must be assured the same weight as that of the developed world. There needs to be guidelines providing for consumption, and hence production, imperatives and choices driving the environmental pressures behind climate change. Climate change may affect our natural resource supplies in terms of quality, quantity and availability. Study after study points to something many people don't want to acknowledge: that we can't continue our present path, and new technologies alone cannot prevent uncontrollable global warming. New thinking and behavior are essential. Without fundamental shifts in our assumptions, beliefs and practices, it is clear we are on a collision course with the planet.

Recognition of the existence of the problem is the first step towards solution, rather than dismissing global climate change as conspiracy theory or hype created by environmentalists. It is important that we have these extreme events on our surveillance radar and verify them for being potential pieces of evidence from India for global climate change. Mitigation measure for reducing health effects due to climate change present phenomenal operational challenges. Unlike in infectious diseases, where there is genuine desire for disease eradication by the affected countries, commitment to efforts to international agreements to reduce green house gas effect give rise to dynamics that are entirely different. There are corporate forces that are working hard to maintain status quo. There are dimensions of economic dependence, politics, fear, suspicion, pressure tactics, intense lobbying, etc, that make commitment to reduction of greenhouse gases very difficult. It's not that the countries that are most likely to be affected due to climate change are not concerned about their health, but their participation in global climate change negotiations is very tentative in nature since their country development and economics is at stake. Therefore it is important that developing countries should strive to strike a balance between economic growth and environmental sustainability.

12. References

Abbott, C. (2008). An uncertain future: Law enforcement, national security, and climate change. London: Oxford Research Group.

Abramson D et al. (2007). The recovery divide: Poverty and the widening gap among Mississippi children and families affected by Hurricane Katrina. Columbia University, New York:

Alistair B.A et al. (2009). Impacts of Climate Change on Indirect Human Exposure to Pathogens and Chemicals from Agriculture. Environmental Health Perspectives, 117, 4

Bourque, L. B et al. (2006). Weathering the storm: The impact of hurricanes on physical and mental health. Annals of the American Academy of Political and Social Science, 604,129-150.

Bulbena, A et al. (2006). Psychiatric effects of heat waves. Psychiatric Services, 57,1519.

Cafaro P. (2010). Economic Growth or the Flourishing of Life:The Ethical Choice Climate Change Puts to Humanity Essays. Philos (2010) 11:44-75

Casimiro E et al. (2006). National Assessment of Human Health Effects of Climate Change in

Challinor A et al.(2007). Assessing the vulnerability of food crop systems in Africa to climate change Climatic Change, 83:381-399

Chand PK,Murthy P. (2008) Climate change and mental health Climate change and mental health. Regional Health Forum, 12, 1.

CLIMATE CHANGE AND AIR QUALITY. Acessed on July 3rd, 2011, Available at : http://www.epa.gov/airtrends/2010/report/climatechange.pdf

Confalonieri U et al. (2007). Human health. Climate Change: Impacts,Adaptation and Vulnerability. In: Parry ML,Canziani OF, Palutikof JP, van der Linden PJ and Hanson CE, editors. Contribution of Working Group II to the Fourth Assessment Report of the Intergovernmental Panel on Climate Change. Cambridge: Cambridge

D'Amato G et al. (2010). Urban Air Pollution and Climate Change as Environmental Risk Factors of Respiratory Allergy: An Update. J Investig Allergol Clin Immunol, 20(2): 95-102

Davidson P. (2003). Deconstructing the Four Pillars of the Climate Change Debate: A Critical Review of the Scientific, Economic,Political, and Ethical Dimensions. Volume 8, Number 8 FES Outstanding Graduate Student Paper Series July 2003 ISSN 1702-3548 (online) ISSN 1702-3521

Diffey B. (2004). Climate change, ozone depletion and the impact on ultraviolet exposure of human skin Phys. Med. Biol. 49 R1 doi: 10.1088/0031-9155/49/1/R01

DEFID. (2010). Malaria: Burden and Interventions. Evidence Overview A Working Paper (Version 1.0) Accessed on July 12, 2011. Available at: http://www.dfid.gov.uk/Documents/prd/malaria-evidence-paper.pdf

Ebi KL et al. (2006).Climate Change and Human Health Impacts in the United States: An Update on the results of the U.S. National Assessment. Environmental Health Perspectives, 114, 9•

Ebi K L et al. Health Effects of Global Change on Human Health SAP 4.6 Chapter 3: Human Health. Accessed on July 3rd, 2011, Available at : http://www.climatescience.gov/Library/sap/sap4-6/public-review-draft/sap4-6prd-ch3-health.pdf

English P B et al. (2009). Environmental Health Indicators of Climate Change for the United States: Findings from the State Environmental Health Indicator Collaborative. Environmental Health Perspectives,117,11, pp

EPA. Learn More About Climate Change U.S. Environmental Protection Agency Climate Change and the Health of Children, Accessed on: July 3rd, 2011,. Available on: http://yosemite.epa.gov/ochp/ochpweb.nsf/content/climate.htm

Fricas J,Martz T.(2007). The Impact of Climate Change on Water, Sanitation, and Diarrheal Diseases in Latin America and the Caribbean water and Sanitation and Their Effects on Diarrheal Illnesses. Accessed on July 3rd, 2011, Available at http://www.prb.org/Articles/2007/ClimateChangeinLatinAmerica.aspx?p=1

Fritze J. G et al. (2008). Hope, despair and transformation: Climate change and the promotion of mental health and wellbeing. International Journal of Mental Health System. Available from http://www.ijmhs.com/content/pdf/1752-4458-2-13.pdf

Füssel HM et al. (2009).The ethical dilemma of climate change: how unequal is the global distribution of responsibility for and vulnerability to climate change?) Climate Change: Global Risks, Challenges and Decisions IOP Publishing IOP Conf. Series: Earth and Environmental Science 6 (2009) 112013 doi:10.1088/1755-1307/6/1/112013.

Galea, S. et al. (2007). Exposure to hurricane-related stressors and mental illness after Hurricane Katrina. Archives of General Psychiatry, 64,1427-1434.

Griffith P D. Climate Change and Buddhist Economic Systems: A Scientific, Ethical Response.

Gould EA et al. (2009). Impact of climate change and other factors on emerging arbovirus diseases. Transactions of the Royal Society of Tropical Medicine and Hygiene Volume, 103, 2, February 2009, 109-121.

Gubler D J et al. (2001). Climate variability and change in the United States: potential impacts on vector- and rodent-borne diseases. Environ Health Perspect. 2001 May; 109(Suppl 2): 223–233.

IWGCCH. ( ) The Interagency Working Group on Climate Change and Health1 A Human Health Perspective On Climate Change. A report outlining the Research needs on the Human Health effects of Climate Change. Environmental Health Perspective of National Institute of Environment Health Science. Accessed on July 3rd, 2011, Available at :http://www.cdc.gov/climatechange/pubs/HHCC_Final_508.pdf

IPCC. (2001). Climate Change: The Scientific Basis

Kessler R. C et al. (2006). Hurricane Katrina Community Advisory Group. Mental illness and suicidality after Hurricane Katrina. Bulletin of the World Health Organization,84, 930-939.

Kjellstrom T et al. ( ) .Public health impact of global heating due to climate change: potential effects on chronic non-communicable diseases. International Journal of Public Health, 55, 2, 97-103

Kinney PL. (2008).The health impacts of climate change Climate Change, Air Quality, and Human Health *American Journal of Preventive Medicine, 35, 5,* 459-467.

Kristie E. (2008). Adaptation costs for climate change-related cases of diarrhoeal disease, malnutrition, and malaria in 2030. Globalization and Health, 4,1.

Martin V et al. ( ). The impact of climate change on the epidemiology and control of Rift Valley fever. vector-borne diseases Rift Valley fever and climate change, 27, 2, 413-426.

McMichael AJ et al. (2004).Global Climate Change.In Comparative Quantification of Health Risks: Global and Regional Burden of Disease due to Selected Major Risk Factors. Edited by Ezzati M, Lopez A, Rodgers A, Murray C. Geneva: World Health Organization; 2004:1543-1649

Mills J N et al. (2010). Potential Influence of Climate Change on Vector-Borne and Zoonotic Diseases: A Review and Proposed Research Plan. Environ Health Perspect, 118(11): 1507–1514.

Nerlander L. (2009). Climate Change and Health. The Commission on Climate Change and Development

Nielsen JT et al. (2006). Effects of food abundance, density and climate change on reproduction in the sparrow hawk. Accipiter nisus Oecologia,149:505–518. DOI 10.1007/s00442-006-0451-y

Patz JA, Olson SH. (2006). Climate change and health: global to local influences on disease risk. Ann Trop Med Parasitol, 100, 535-49.

Patil RR.(2002) Circumstances leading to death of Indian Cotton farmers.International Jounral of Occupational Medicine & Environmental Health, 15 ,4, 405-407.

Patil RR. (2006). Community Based Occupational and Environmental Health Studies: Challenges and Dilemmas. Indian Journal of Occupational and Environmental Medicine , 10, 2, 85-86

Patil RR, , Deepa TM. (2007) Climate change: The challenges for public health preparedness and response- An Indian case study. Journal of Occupational and Environmental Medicine , 11, 3 , 113-115

Reiter P. (2007). Human Ecology and Human Behavior November 2007 International Policy Network, Accessed on 2nd July 2nd,2011, Available from: http://www.csccc.info/reports/report_22.pdf

Rose J B et al.(2009). Climate variability and change in the United States: potential impacts on water- and foodborne diseases caused by microbiologic agents. Environ Health Perspect. 2001 May; 109(Suppl 2): 211–221

Ryan KJ, Ray CG (editors) (2004). *Sherris Medical Microbiology* (4th ed.). McGraw Hill. pp. 434–437. ISBN 0838585299.

Sartore, G et al. (2007). Drought and its effect on mental health—how GPs can help. Australian Family Physician, 36, 990-993\

Shea KM et al. (2008). Climate change and allergic disease. Journal of Allergy and Clinical Immunology, 122, 3,443-453 .

Simister, J., & Cooper, C. (2004). Thermal stress in the U.S.A.: Effects on violence and on employee behavior. Stress and Health, 21, 3-15.

Sommer J, Poulsen LK. (2009). Allergic disease--pollen allergy and climate change. Ugeskr Laeger 6;171(44):3184-7.

Spencer RW. et al. (2005) .An Examination of the Scientific, Ethical and Theological Implications of Climate Change Policy. Interfaith Stewardship Alliance. Accessed on July 3rd, 2011, Available at : www.interfaithstewardship.org

Triple Helix.(2011). Climate Change: An Ethical Perspective On Mitigating Its Impact Source

United Nations Framework Convention on Climate Change. (1992). United Nations, (p. 3)

UNSCN. (2010). CLIMATE CHANGE AND NUTRITION SECURITY.Message to the UNFCCC negotiators.The United Nations System Standing Committee on Nutrition (UNSCN). 16th United Nations Conference of the Parties (COP16) Cancun, November 29th - December 10th, 2010

Venton CC. Climate change and water resources ERM for WaterAid. Environmental Resources Management, London W1G 0ER

4

Agricultural Technological and Institutional Innovations for Enhanced Adaptation to Environmental Change in North Africa

Ali Nefzaoui, Habib Ketata and Mohammed El Mourid
International Center for Agricultural Research in the Dry Areas (ICARDA)
North Africa Program, Tunis,
Tunisia

1. Introduction

North Africa typically is a dry region, comprising the countries of Algeria, Morocco, Tunisia, and Libya, where four subregions may be easily distinguished, namely (i) a northern subhumid coastal subregion, bordering the Mediterranean sea (and the Atlantic Ocean for western Morocco), where average annual rainfall is relatively high, generally above 500 mm and where soils are relatively good for farming; (ii) a semi-arid elevated subregion flanking the first subregion from the southern side, from which it is separated by the Atlas mountains and where rainfall is around 300-500 mm, and soils are light calcareous silt-loam; it is bordered on the southern side by (iii) an arid, lower-altitude subregion, with silt-sandy soils and an average rainfall of 100-300 mm; and (iv) Sahara desert subregion covering the largest part of the countries. Libya is predominantly (90%) desert land, except for a narrow coastal area where some agriculture is practiced. Therefore, reference in this chapter will be mainly made to the 3 countries of Algeria, Morocco and Tunisia.

North Africa is marked by an acute water scarcity, combined with a highly variable Mediterranean climate. While the average world per capita share of fresh water is 7000 cubic meter (m^3), all three North African countries are below the water poverty threshold of 1000 m^3 (Table 1). Agriculture uses the largest share (up to 80%) of available water resources in North Africa where rainfed cropping predominates. The scarcity of natural water resources, combined with the highly variable and generally very low rainfall in most of the region explain in part the low agricultural productivity, especially of key crop commodities, and the reliance of North African countries on food imports to meet their growing national demands; this is especially true for Algeria that has the largest population, and the lowest agricultural contribution to country GDP and to total employment. Water scarcity is further exacerbated by the competition for water from domestic and industrial uses, and the increasing population and urbanization. Cereal crops, mainly wheat and barley, are the major crop commodities grown in North Africa, but their contribution to national food security and household income remains low (Table 1).

Characteristic	Algeria	Morocco	Tunisia
Population (million)	34.4	31.6	10.2
Total area (million ha)	238.1	71.02	16.36
Cultivated area (million ha)	8.4	8.99	5.04
Contribution of agriculture to GDP (%)	8	17	10
Rural population (% total population)	35	44	33
Employment in agriculture (% total employment)	14	45	18
Irrigated area (% cultivated area)	6.9	16.6	8.0
Total annual renewable water resources (km^3)	11.67	29	4.6
Annual per capita renewable water resources (m^3)	339.5	917.5	451.9
Wheat self-sufficiency (%)	29	58	50

Table 1. Selected agricultural characteristics for three North African countries.

To lessen their dependence on highly unpredictable cereal harvests, small-scale farmers may also maintain a small-ruminant (sheep and goats) raising activity that provides them a buffer against poor crop harvest or crop failure in severe-drought years. In fact, the cereal-livestock system forms the backbone of agriculture in the semi-arid zones in contrast to the arid regions where small ruminant raising is the major agricultural activity. Horticultural crops and specific high value fruits (citrus fruits, grapes, etc.) are produced under moisture-favorable conditions in subhumid areas or under irrigation in other areas. Extensive cultivation of olives and other drought tolerant trees are generally produced under rainfed conditions in semi-arid and arid areas. Dates are produced in arid regions or in oases within desert areas.

The future of agriculture in North Africa is further threatened by unfavorable climate change that is expected to drastically affect agriculture productivity and people's livelihoods. The rest of the chapter describes the perceived effects of climate change on natural resources and livelihoods of agropastoral communities in the region. Successful tools and approaches deployed to face climate change are highlighted, including both technological and institutional innovations.

2. Climate change and food security in North Africa

North Africa is widely known for its aridity and dry climate and for rainfall variability. Severe drought indeed has been common in the region, although the causes of such drought were not well understood (El Mourid et al., 2010).

In 2007, The Intergovernmental Panel on Climate Change (IPCC) confirmed (IPCC, 2007) that North Africa is among regions most affected by climate change (CC) with a temperature rise of 1-2°C during the past period 1970-2004, and that it will continue to be affected by global warming at the average rate of 0.2°C per decade for the coming 2 decades. In fact, anthropogenic green house gas (GHG) emissions from within North Africa are very low (Table 2) in comparison to developed countries that have an average emission rate of 14.1 ton CO_2 equivalents (TE-CO_2) and the climate change impacts in North Africa are essentially the result of global GHG emissions. According to the IPCC report, the winter season in North Africa will be shorter, leading to reduced yield and increased diseases and insect outbreaks. Precipitation will undergo a 20% drop by the end of the century, which would reduce crop yield and increase livestock losses. Heat waves also would reduce yield, while expected intense storms will cause soil erosion and damage the crops. High sea level rise will lead to salt water intrusion and salinization of irrigation water (IPCC, 2007). In fact, the frequency of drought in Morocco, for example, has been independently reported (Magnan et al., 2011) to have increased from 1 in 8 years during the period 1940-1979, to 1 in 3 years during 1980-1995, and to 1 in 2 years during 1996-2002. Also, North Africa has been identified as a hot-spot for vulnerability to climate change, based on the analysis of NDVI (Normalized Difference Vegetation Index) data for the period 1982-2000 (De-Pauw, 2008).

GHG emissions	Algeria	Morocco	Tunisia
Total emissions (million TE-CO_2)	103.14	63.34	20.8
Annual per capita emissions (TE-CO_2)	3.0	1.98	2.15
Emission composition (%):			
- carbon dioxide (CO_2)	64.5	67	72
- Methane (CH_4)	29.7	18	14
- Nitrous dioxide (N_2O)	5.9	14	14
Agriculture contribution to total emissions (%)	5.9	25	20

Table 2. Greenhouse gas (GHG) emissions in North African countries (2000).

The livestock sector has been described as a major contributor to global warming, accounting for 18% of the world anthropogenic GHG emissions, namely carbon dioxide (CO_2), methane (CH_4) and nitrous oxide (N_2O) (Koneswaran & Nierenberg, 2008; Steinfeld et al., 2006). Such large contribution of livestock to global warming is primarily the result of the highly intensive livestock system in well endowed, temperate regions of the world. In contrast, the livestock system in North Africa is primarily extensive in nature, where the dominant animals are sheep and goats, essentially raised in open rangeland fields, within the arid and semi-arid areas receiving less than 200 mm of rainfall and no fertilizer, apart from grazing animal manure. Such livestock contributes comparatively little to GHG emissions as compared to intensive livestock systems found in Europe and similar regions. However, rangelands in North Africa are subject to severe degradation, primarily because of cropping encroachment, which is responsible for 50% of rangeland degradation, versus 26% accounted for by overgrazing and 21% by fuel wood utilization. This trend opposes clearly that of the temperate areas, where overgrazing accounts for 70 % of land degradation (Le Houerou, 2000).

The food commodity crisis of 2008 brought-up awareness of the serious threat to food security in many of the world areas, including North Africa, where policy makers realized the importance of food production uncertainty imposed by the vagaries of changing climate and the repercussions it may impose on social and political stability. In all North African countries, swift decisions were taken to encourage farmers and other food producers assure the highest degree possible for self-sufficiency in strategic food commodities. All countries prepared a multi-year plan to boost local agriculture production, taking into consideration climate change and necessary mitigation and adaptation measures. For example, Tunisia developed a national strategy for dealing with climate change (CC) based on the implementation of specific CC studies and a national action plan for adapting to CC. The studies indicated that by year 2020, temperature will have increased by 0.8°C-1.3°C and rainfall dropped by 5-10%, depending on region (Dali, 2008). These effects will impact unfavorably on water resources, ecosystems and agro-systems (including olives, fruit, livestock and rainfed annual crops). Results also indicated a possible 50-cm sea level rise by year 2100, threatening coastal ecosystems and marine biodiversity. Several projects have been developed within the framework of the Clean Development Mechanism (CDM) and are being implemented to reduce GHG emissions and launch actions for sustainable development across the country. In November 2008, Tunisia hosted an International Solidarity Conference on CC strategies for the African and the Mediterranean Regions. In Morocco, the Green Morocco Plan, a 10-year (2010-2020) program, has been started to upgrade the Moroccan agriculture through intensification of key production commodities and combating rural poverty across the country (Badraoui & Dahan, 2011). A new dimension was added to foster the capacity of smallholders cope with the impacts of climate change. Previously, Morocco hosted the Seventh Conference of the Parties to the UN Framework Convention on Climate Change (COP7) at Marrakech, in November 2001. In Algeria, the national 5-year plan (2009-2014) (Cherfaoui, 2009) for the "Renewal of agricultural and rural economy" aims to achieve food security and sustainable development through improved agricultural productivity, enhanced capacity building and employment, and preservation of natural resources, in the context of a changing environment.

3. Technological innovations for enhanced adaptation to climate change in North Africa

The International Center for Agricultural Research in the Dry Areas (ICARDA) has established strong partnership in dryland research jointly with national agricultural research systems (NARS) of North African countries with a focus to reduce food insecurity and enhance sustainable livelihoods of farming communities in the region. During the past 30 years, ICARDA scientists conducted joint dryland research with NARS partners, providing genetic resources and germplasm for selection of cereal and legume crops, and consultancy and training in soil, water and crop management, in integrated pest management, and in rangeland management and small ruminant husbandry. The joint work has been conducted in all North African agro-ecologies under the evolving stresses imposed by the ever-changing climate and the trend of increasing drought and high temperature. The chapter highlights some of the NARS-ICARDA achievements in the area of adaptation to unfavorable environmental changes.

3.1 Soil and water conservation and use

Arable lands cover only a small fraction of total land area in North Africa (Table 1) soils vary widely both in depth and fertility (Matar et al., 1992). Most soils are shallow, with low water-holding capacity, and highly vulnerable to soil erosion. Saline soils are less frequent and apart from those found in desert areas called "sebkha" or "shott", they may be encountered in localized irrigated areas. Over 45% of the agricultural land in North Africa is experiencing some form of degradation. Deep clay soils (Vertisols) are found in certain fertile plains of the North Africa region. However, arid and semi-arid soils are more common, and often are nutrient deficient, generally with low organic matter content (1% or less). These soils are deficient in nitrogen (N) to such an extent that N fertilization has become the norm in cereal cropping systems in North Africa. As most soils also are high in calcareous, and have high pH, they all present P deficiency, which led to deliberate and continuous application of phosphorous by cereal growers, resulting in wasteful application of this mineral. ICARDA researchers have promoted awareness of the necessity for soil analysis as a guide to sound fertilizer application (Ryan & Matar, 1992). In contrast, North African soils are not deficient in potassium for most crops and this mineral is therefore not applied to cereal crops.

In the past, cereal-based cropping systems in North Africa were dominated by the cereal-fallow rotation and the continuous mono-cropping. While the cereal-fallow rotation in semi-arid areas has the advantage of storing some moisture in the fallow season for use by the cereal crop in the following season, the system is inefficient, especially in favorable or moderately favorable environments. ICARDA researchers have advocated and shown the benefits of replacing the fallow with a legume crop, such as vetch, lentil, or faba bean (Ryan et al., 2008). Research results indeed show a favorable effect of legume-based rotations on crop yield and water use efficiency. The introduced legume crop also leads to a beneficial build up of soil N, thus improving soil quality and contributing to sustainability of land use in the semi-arid regions. Cereal-legume crop rotation is now widely adopted by North Africa farmers, especially where annual rainfall is about or above 350 mm.

Because of the dominant aridity and fragile nature of land resources in North Africa, NARS and ICARDA researchers developed efficient technologies for soil and water conservation and management to minimize runoff and soil erosion and improve water retention and infiltration. In arid areas, rainfall is rare, unpredictable, and sometimes comes in unexpected violent bursts causing erosion and floods, and quickly evaporating under the dry and hot conditions of the arid environment. ICARDA has revived, enhanced and promoted an old indigenous practice of collecting (harvesting) the runoff water for subsequent use (Oweis et al., 2001). To retain water, farmers generally use small circular or semi-circular basins or bunds around the trees or the plants. Soil is assembled and raised in such a way as to make a barrier to hold the water, which is therefore collected and made available for agricultural or domestic uses. Water harvesting (WH) proved effective for replenishing the soil water reserve and for the establishment and maintenance of vegetation cover, trees, shrubs or other crops for various uses. Larger catchments are similarly arranged to harvest water and exploited in arid areas by sheep herders to sustain rangeland species. Water harvesting not only provides a much needed additional source of water for drinking or growing plants for feed and food, but it also raises soil moisture, reduces soil erosion and contributes to C sequestration and improved soil quality. In more favorable, semi-arid or wetter regions, and where topography allows, large sloping areas of a few hundred hectares may be targeted for catchments to collect large amounts of water into large ponds or hill reservoirs (or lakes), with a capacity of up to hundreds of thousands cubic meters, requiring more solid, locally-made structures to retain the water (Ben Mechlia et al., 2008). In Tunisia alone, there are about 1,000 hill lakes across the country, contributing to the shrinking water resources. Such large hill lakes are managed with the participation of local communities or organizations for an equitable water distribution among farmers. Although priority is given by governments to strategic crops like wheat, farmers still prefer irrigating summer vegetables instead, for their higher return, although they consume more water. In the semi-arid areas of North Africa, field crops such as wheat and barley are traditionally grown under rainfed conditions, where average yields vary between 1 and 2 t ha^{-1} (Table 3), although in good years, yields can reach up to 3-4 t ha^{-1} and more. The highly variable rainfall and the increased frequency of very dry years in semi-arid regions (Bahri, 2006) induced farmers into irrigating once or twice their winter-grown crops to reduce risk and secure a harvest. Such a practice, referred to as supplemental irrigation (SI) augments the rainwater with some additional water applied to the crop at a critical time during the growing season, when rain fails to come and plants are most vulnerable to water stress (Oweis, 1997).

Crop	North African country		
	Algeria	Morocco	Tunisia
Wheat	1.22	1.24	1.59
Barley	1.23	0.85	0.91

Table 3. Average grain yield (t ha^{-1}) for wheat and barley under rainfed conditions of North Africa (1998-2000).

Our experience shows that candidate critical periods of rain deficiency occur at planting, at stem elongation, and at anthesis, generally not always occurring together in the same season. Supplemental irrigation of wheat in semiarid and subhumid environments of Tunisia resulted in increases of yield and WUE reaching up to 100% and 73%, respectively (Rezgui, et al., 2005). Similarly, a single 60 mm irrigation of winter sown chickpea in Morocco, resulted in lengthened green area duration and associated yield gain (Boutfirass, 1997). Supplemental irrigation of barley during 3 years in coastal areas of Libya with an annual rainfall of 200-300 mm resulted in 50-400% yield increase, depending on rainfall, with larger increases occurring in drier seasons. As compared to rainfed cropping, supplemental irrigation in semiarid Tunisia yielded a net return of 60-170% for cereal crops and up to over 400% for vegetables (Ben Mechlia et al., 2008).

Research conducted in North Africa and similar semi-arid regions shows that SI not only improves and stabilizes grain yield, but it also gives "more crop per drop", i.e. it has a good water return or high water-use efficiency (WUE) or, equivalently, high water productivity (WP), both terms referring to crop return, such as grain yield or value, per unit of consumed water (Oweis, 2010). Unlike land productivity that refers to crop return per unit of land (e.g. "x" Kg ha^{-1}), WP relates the crop return to consumed water, as water is the most important limiting factor, especially in dry regions. In cereal-based cropping systems of semi-arid areas, WP of wheat is generally around 0.35-1 kg grain/m^3 water, but may reach up to over 2.5 kg grain/m^3 water under SI (Oweis, 2010). Therefore SI is a water saving procedure that effectively reduces the impact of drought on farmer's livelihood. However, certain farmers tend to over-irrigate and waste valuable water resources, thinking "the more water, the better". In fact, results of wheat research show that WP is maximum for an optimum level of SI, beyond which it starts decreasing; the optimum SI level is about 1/3-2/3 the level of full irrigation (FI), the latter being equal to the full crop water requirement. Full irrigation is not as efficient as supplemental irrigation in using the water resources (Oweis & Hachum, 2006; Shideed et al., 2005). In fact, in wheat WP for FI is 1 kg/m^3 but it is 2.5 kg/m^3 for SI. In scarce-water conditions, it is therefore more rewarding for the farmer to use SI to optimize WP rather than maximize yield. This approach saves water to grow the crop on a larger area and the farmer ends up with a larger total output, while using water sustainably (Oweis & Hachum, 2006). Also, water productivity can be further improved through proper crop management, including early planting, weed control, fertilizer application, and irrigation at critical times to avoid or minimize detrimental water stress, e.g. at flowering time and fruit or grain formation. For example, supplemental irrigation of wheat, combined with early planting in the Tadla region of Morocco hastened maturity, enabled the crop to escape terminal drought and heat stress, and doubled grain yield and WP (Karrou & Oweis, 2008) The beneficial effect of SI is further enhanced when SI is combined with the use of adapted varieties (Karrou & Boutfirass, 2007).

While water harvesting and supplemental irrigation are effective technologies for augmenting and enhancing the value of fresh water resources, these resources are still too limited to cope with the increasing rural and urban user demands that are further exacerbated by unabating climate change. However, there is a potential for other avenues that could be explored for additional water sources, including brackish water, saline water and treated wastewater. Brackish water and saline water have been used in irrigation with disappointing results in all three countries (ICID, 2003) primarily because of very high evaporative demand in desert or arid regions, and the lack of fresh water and adequate drainage for leaching the salts away. The dry environments in such areas preclude the normal growing of regular crops, but special-purpose, halophytic crop species may be grown successfully, to provide essential oils, folk medicine, biofuel, fodder, shade for animals, or to retain soil and arrest desertification (Neffati et al., 2007; Qadir, 2008). In more favorable semiarid or subhumid areas, brackish water may be successfully used to grow tolerant plants (such as barley) where both fresh water and drainage facilities are more readily available. The use of treated wastewater, although feasible, has been limited so far to less than 20,000 hectares in North Africa, due primarily to unreliable delivery, regulatory exclusion of vegetables growing, and social unacceptance. However, the increasing water scarcity will ultimately make it a *de facto* alternative for fodder, grains and tree cultivation, especially as suitable regulatory frameworks are established and promoted (Qadir, 2008).

Despite all the newly adopted and proposed technologies that make the best use of available water resources, including water harvesting, supplemental irrigation, and utilization of non-conventional water resources, the fact remains that North Africa is a water-deficient region, and will be more so in the future. A more sustainable long-term solution that will not only provide enough fresh water for generations to come, but will also enable reclaiming salinity-degraded land resources, lies in seawater desalination using solar power from the Sahara desert. The Sahara is 9 million km^2 large and North Africa sea coast extends over 4000 km. These two virtually limitless resources for clean energy from the Sahara desert and bountiful fresh water from the sea will make of North Africa a sustainable water-rich region.

3.2 Conservation agriculture

Conservation agriculture (CA), referred to under different labeling (direct seeding, NT or no-till, zero tillage) in different countries, is an agricultural technology that combines minimum or no soil disturbance, direct seed-drilling into the soil, cover crop or residue retention and crop diversification through rotation (Kassam et al., 2010). Now practiced on 117 million hectares worldwide, it covers diverse agroecologies and cropping conditions. In North Africa, CA was introduced about 30 years ago in both Morocco and Tunisia where it now covers 6,000 ha and 12,000 ha, respectively. Algeria's work in CA started only 7 years ago and is gaining momentum (Zaghouane et al., 2006). In addition to the obvious benefits of reduced labor and energy cost, and some yield advantage (generally realized a few years from the start), the most striking effects in semi-arid regions of North Africa is the reduced erosion, especially in sloping areas. CA also presents the advantage of flexibility for the implementation of field crop management that allows timely planting and input application, despite unfavorable field conditions that do prevent such operations in conventional agriculture (e.g. wet soil at planting time). CA prevents soil plowing which has been identified as a major cause for CO_2 emission. Cover crops, residues and crop roots contribute to better soil structure and composition with enhanced build up of organic matter, while crop residues protect the soil and minimize soil evaporation (Angar et al., 2010; Ben Moussa-Machraoui et al., 2010; Gallali et al., 2010; Mrabet, 2006, 2008). CA therefore contributes both to CC mitigation through reduced GHG emissions and enhanced C sequestration, and to adaptation through soil water retention and infiltration, and increased water use efficiency. Therefore, CA based on the NT system is an effective technology to conserve natural soil and water resources while minimizing the drought effect on crop production and contributing to better food security in North Africa.

A difficulty faced in CA is a compaction in the upper soil caused by excessive animal grazing during the wet season (Angar et al., 2010). Major challenges to adoption of CA technology in North Africa are posed by severe drought of rainfed arid regions and the consequent need for fodder resources during the dry season, both of which threaten the maintenance of crop mulch, a key component of CA. In such a situation, partial stubble grazing could offer a compromise. Results in Tunisia indeed show beneficial effects of CA (improved soil organic matter, better soil infiltration, higher wheat yield) despite the low amount of crop residues (1-2 t residue ha^{-1}). Another solution will be some sort of compensation to farmers for environmental services (Lal, 2010) and sustainability of natural resources that will help farmers secure alternative feed resources for the dry season. Other challenges to CA adoption in North Africa are (a) high weed infestation at the initial stage of CA adoption (Dridi et al., 2010), and (b) the unavailability of suitable CA-ready seed-drills (El Gharras & Idrissi, 2006). In fact, the adoption of NT technology in Tunisia is limited to farms of size ≥100 ha, where farmers could afford a high investment for the purchase of NT equipment. ICARDA and collaborating partners are pursuing efforts in North Africa to promote local manufacturing of low-cost NT drills, which will expand CA adoption to small-scale farmers who represent the majority of North African farmers (Requier-Desjardins, 2010). Here is another opportunity for policy makers to encourage farmers reduce the impact of CC, by promoting CA through reduced cost of NT drills.

3.3 Biodiversity and crop variety development

Protracted drought in semiarid regions inevitably leads to disappearance and loss of plant species and varieties in extremely dry or hot years. The likelihood of this happening has increased with climate change. Realizing this risk, researchers around the world make efforts to conserve genetic diversity of plant species in their own environments (*in situ*) where the plants can preserve their specific characteristics while they are living and evolving naturally. Researchers also conserve these indigenous species or varieties under controlled conditions, both in research farms and in genebanks (*ex-situ*). For example, ICARDA maintains over 130,000 accessions of cereals (essentially wheat and barley), legumes (lentil, chickpea, and faba bean) and other species at its Gene bank under cold conditions for medium (up to 30 years) and long-term (100 year) storage, and distributes annually over 30,000 samples to requesting researchers around the world. Seed of requested materials is sent along with associated information on genealogy, special characteristics, and area of origin and adaptation. Such information will assist the user to target the requested genetic resources to specific environments. In return, the user's feedback enriches the information database that gets more valuable as it is accessed by more users.

In addition to the wealth of genetic resources that are characterized and maintained in gene banks, several hundreds of newly-bred crop entries are annually shared with breeders and other researchers around the world through the ICARDA International Nursery Network.

ICARDA and partner breeders use hybridization and selection to develop new germplasm, possessing desirable traits for different purposes and uses in various environments, including tolerance to biotic and abiotic stresses, and good agronomic, nutritional and industrial properties. Breeders work combines the use of both conventional breeding methods based on field and lab manipulations, observations and measurements, and biotechnological tools to speed up germplasm development and identification and transfer of useful traits from varied sources, including alien species. The outcome is a pool of germplasm with a broad genetic base. Of particular interest and relevance to climate change are varieties possessing tolerance to drought and heat, and to major diseases and insect pests prevailing in North Africa. For example, the Moroccan durum wheat varieties INRA-1804 and INRA-1805 are resistant to Hessian fly, a major wheat insect pest, in contrast to the older variety Karim, susceptible to such a level that it yields no grain under heavy infestation, where the resistant varieties yield 1.5 t ha^{-1} under severe drought conditions. In Algeria, the new durum wheat variety Boussalem yields 3.5 t ha^{-1} compared to 2 t ha^{-1} for the widely grown variety Waha, both grown under the same semiarid conditions (El Mourid et al., 2008). In Tunisia, the newly released durum variety Maali is drought tolerant and has an average grain yield of 4 t ha^{-1} compared to 3 t ha^{-1} for the common variety Karim. New barley and wheat varieties are developed with participation of farmers in selection and evaluation on their own farms. In fact, participatory plant breeding where farmers and breeders make independent selections, contributes to maintaining a good level of genetic variability in breeders and farmers selections, which is purposely maintained through generations of continuous selection within heterogeneous populations (ICARDA, 2008; Ceccarelli et al., 2010). Such heterogeneity assures a certain degree of resilience to climate change and ensuing environment variation. Early maturing cultivars are particularly adapted to semi-arid areas of North Africa, where late-season drought is very common. Such cultivars suffer least in dry environments and contribute to lessen the effect on farmer of drought risk and harvest uncertainty. Early maturity, controlled by photo-thermal response genes, is therefore a prime objective in crop breeding of major field crops in North Africa. However, breeders are also investigating other genetic sources of drought tolerance in land races and wheat synthetics and work to incorporate such genes in useful genetic background (ICARDA, 2007; ICARDA, 2010). Wheat synthetic types are derived from crossing durum wheat (*Triticum turgidum var. durum*) with goat grass (*Aegilops tauschii*). Interspecific and intergeneric hybridization generates new genetic variability and contributes to enhancing biodiversity, a valuable asset or 'vaccine' for adaptation and survival and development in erratic environments.

Breeding has been also an effective tool in combating diseases and insect pests and reducing their negative impact on crop productivity and resilience. Genes for resistance to pathogens and pests of wheat, barley, chickpea, faba bean and lentil crops were identified in various crops and wild species and successfully incorporated into commercial cultivars (ICARDA, 2006, 2007, 2008, 2009, 2010). However, there is indication that climate change, with trends of increasing temperature and decreasing rainfall, is favoring the appearance of new pathogen and pest types. Examples include the recent appearance of yellow rust (also called stripe rust) on wheat in relatively warm areas where it was not a problem in the past, as the causal pathogen, the fungus *Puccinia striiformis*, was known to be favored by moderately low temperature; the appearance of the disease in warm areas is likely the result of the appearance of a new race of the pathogen. Similarly, chickpea varieties that were tolerant to the fungal disease *Ascochyta* blight during the period 1979-2000, started showing signs of susceptibility during the period 2001-2007, the two periods differing quite well in rainfall and temperature pattern (Abang & Malhotra, 2008). While researchers pursue exploiting the genetic sources of resistance to diseases and pests, they are also investigating other avenues for an integrated pest management approach that include also crop management techniques and biological control to minimize the recourse to the use of agrichemicals (ICARDA, 2009). The successful development of new improved varieties does not bear fruit unless the new varieties are effectively adopted and grown by farmers. Experience through a durum wheat project in Algeria, Morocco and Tunisia (El Mourid et al., 2008), shows that small-scale farmers in semi-arid areas do not have easy access to new varieties. This is attributed to high prices of certified seed, and inefficient seed multiplication that inhibits its wide distribution across the country's regions. Informal seed production of those varieties through trained community farmers led to rapid seed multiplication and dissemination within the communities and in nearby areas. Although the seed was not certified, it was of very good quality and free of diseases or pests, and could be bought at an affordable price. Such an informal seed production system was especially successful in remote semiarid areas of all three countries, where the new varieties were available to growers within 3 years only. Similar successful examples of village-based seed enterprises in other regions (ICARDA, 2009) confirm the importance of farmer participation in solving local issues and its relevance to food security and community welfare in remote rural areas. The availability of a number of different varieties of various species gives farmers the opportunity to choose. In fact, most farmers choose more than one variety, to increase their odds against poor or no harvest. By so doing, they also contribute to enhancing biodiversity, a powerful tool to adapt to changing climate and associated changes in agro-ecosystems.

3.4 Integrated crop-livestock-rangeland production systems

Although the dominant production systems in North Africa are based on livestock and crops, livestock is still the main source of income of rural populations in the North African countries. Sheep and goat make up the major portion of livestock in North Africa with 30 million and 10 million heads, respectively (Table 4). Several factors including climate change threaten the sustainability of the production systems. There are considerable gaps in our knowledge of how climate change will affect livestock systems and the livelihoods of these populations. Management of the production risk caused by the fluctuation of feed availability is the main problem hampering the development of livestock production in North Africa. Under the framework of Research-for-development project, the Mashreq/Maghreb project, NARS and ICARDA developed over a decade sound technical, institutional and policy options targeting better crop/livestock integration, community development and improvement of the livelihoods of agropastoral communities in 8 countries (Algeria, Iraq, Jordan, Lebanon, Libya, Morocco, Syria, and Tunisia). These options include (i) organization of local institutions to facilitate both collective and individual adaptation and response to climate change, (ii) an innovative approach to their sustainable improvement and management including institutional solutions for access to communal/collective rangelands, (iii) better use of local natural resources with an emphasis on water harvesting and appropriate use of adapted indigenous plant species, such as cactus and fodder shrubs, and (iv) efficient animal feeding involving cost-effective alternative feeds Including feed blocks, and (v) nutrition and health monitoring.

Country	Sheep	Goat
Algeria	17.3	3.2
Morocco	16.7	5.2
Tunisia	6.9	1.4
Total	30.9	9.8

Table 4. Number (million heads) of sheep and goat in three African countries (adapted from Iniguez, L., 2005a)

Two critical trends prevail in the current production context. The first trend involves a crisis in the feed supply reflecting water scarcity, exacerbated by the progressive decline of rangelands' productivity due to overgrazing, cultivation encroachment, or the disruption of institutional arrangements for resource utilization. Moreover, very low ratio of cultivated forages prevails in the cropping systems.
The second trend involves the expansion of market demand for livestock products leading to opportunities for productivity and income improvement.

3.4.1 Participatory collective rangeland management

The pastoral and agropastoral societies in North Africa went through deep mutation during the past few decades. In the mid-20th century, the mobility pattern of the pastoralists was dictated by accessibility and availability of forage and water. With the mechanization of water transportation and the reliance on supplemental feed, animals can be kept continuously on the range, which disturbs the natural balance and intensifies range degradation (Nefzaoui, 2002, 2004). Mechanization profoundly modified rangelands' management in the steppes of North Africa. Water, supplements and other services are brought by trucks to flocks. As a result, families settle close to cities for easier access to education, health, and other services, with only sheepherders moving flocks to target grazing areas (transhumance).
Production systems are intensifying and it is nowadays possible to find in the steppe a continuum between intensive fattening units that are developing in peri-urban areas and along the main transportation routes, mixed grazing-fattening systems, and purely intensive systems based on hand feeding only to provide feed supplements to animals.
Agropastoral societies have developed their own strategies for coping with drought and climate fluctuation. These strategies include (Hazell, 2007; Alary et al. 2007):

- mobile or transhumant grazing practices that reduce the risk of having insufficient forage in any location;
- feed storage during favorable years or seasons;
- reciprocal grazing arrangements with more distant communities for access to their resources in drought years;
- adjustment of flock sizes and stocking rates as the rainy season unfolds, to best match available grazing resources;
- keeping extra animals that can be easily sacrificed in drought conditions, either for food or cash;
- investment in water availability (wells, cisterns, and water harvesting);
- diversification of crops and livestock (agropastoralism), especially in proximity to settlements, and storage of surplus grain, straw and forage as a reserve in good rainfall years;

- diversification among animal species (sheep, goats, cattle, camels, donkeys) and different breeds within species;
- income diversification into non-agricultural occupations, particularly seasonal migration for off-farm employment in urban areas.

However, recent infrastructural and demographic changes as a result of urbanization have made such strategies less effective.

In a recent study conducted within the Mashreq/Maghreb project in Chenini agropastoral community, in Southern Tunisia, perception of drought and livelihood strategies to mitigate drought has been investigated using a "sustainable livelihood approach". Figure 1 translates the perception of agropastoralists of drought and climate change during the past decades, as well as the tools used to adapt to or mitigate climate fluctuation. Indeed, while in the thirties, there was a self reliance on drought coping mainly through transhumance, food and feed storage and goat husbandry, these options shifted gradually towards a significant reliance on government intervention mainly through subsidizing feeds and facilitating feed transport from the North to Southern arid areas.

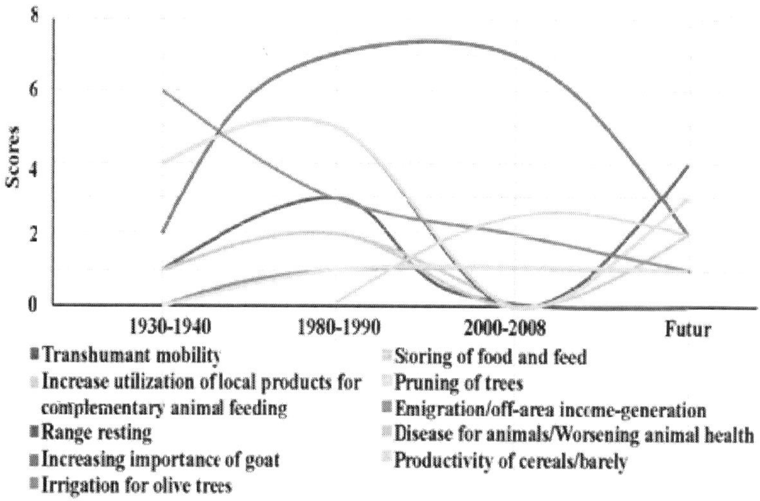

Fig. 1. Tendencies of major drought strategies in Chenini agropastoral community, Southern Tunisia (Nori et al., 2009)

However, science and technology, including climatic adaptation and dissemination of new knowledge in rangeland ecology and a holistic understanding of pastoral resource management are still lacking. Successful adaptation depends on the quality of both scientific and local knowledge, local social capital and willingness to act. Communities should have key roles in determining what adaptation strategies they support if these have to succeed. The integration of new technologies into the research and technology transfer systems potentially offers many opportunities to further contribute to the development of climate change adaptation strategies. Geospatial information, spatial analysis tools, and other decision support tools will continuously play a crucial role in improving our understanding of how climate change will affect livelihoods of pastoral communities. Climate change also offers the opportunity to promote payment to pastoralists for environmental services, as in the case of some livestock keepers in Europe. These services could include watershed management, safeguarding biodiversity, landscape management and carbon sequestration (MacOpiyo et al., 2008).

3.4.2 Matching small ruminant breeds to environments

It is widely recognized that pastoralists and their communities play an important role in conserving domestic animals diversity. In North Africa, seven of the sixteen sheep breeds of the arid regions are at high risk of disappearance (Table 5), either because animals are totally replaced by exotic species or because they are crossbred with more productive breeds. Most of these local breeds (Table 5) are well adapted to harsh environments and their genetic makeup is attracting many western countries that are preparing for similar climate change in Europe. ICARDA has been documenting the status of the diversity and phenotypic characteristics of sheep and goat breeds in the West Asia and North Africa (WANA) region jointly with NARS partners. Many breeds are shared across the region and have important adaptive traits to dryland conditions (Iñiguez, 2005).

Breeds	Average rainfall	Country*	Risk to genetic erosion	Primary purpose
Atlas Mountain breed	500 (mountain)	M	High	Meat + wool + skin
Barbarine	75-500	MATL	High/low	Meat, milk
Barki	150-300	EL	None	Meat, wool
Beni Guil	100-250	M	High	Meat, wool
Berber	450-500 (mountain)	A	High	Meat, milk
Boujaad	300	M	None	Meat, wool
D'man	100 (oasis)	MAT	High	Meat, manure
Farafra	100 (oasis)	E	None	Meat, wool
Hamra	200-250	A	High	Meat, fleece, milk
Ouled Djellal	200-500	AT	none	Meat, fleece, milk
Queue Fine de l'Ouest	200-400	T	None	Meat
Rembi	300	A	Moderate	Meat, fleece, milk
Sardi	300	M	None	Meat
Taadmit		A	Extreme	
Tergui-Sidaou	50	A	Low	Meat
Timahdite	500 (mountain)	M	None	Meat, wool

Table 5. Sheep breeds of non-sedentary (pastoral and semi-pastoral) production systems in North Africa (Dutilly-Diane, 2007). (*) A: Algeria, E: Egypt, L: Libya, M: Morocco, T: Tunisia.

3.4.3 Efficient animal feeding using cost-effective alternative feeds

Managing the production risk caused by the variability of feed availability is the central issue in the small ruminant (SR) production system of the WANA region. Desertification, increased drought frequency and duration, greenhouse emissions, and decreased livestock performance, justify the need for a serious understanding on the readjustment and or the establishment of new feeding strategies targeting the improvement of animal production without detrimental effects on the environment. Moreover, the development of simple and cost-effective techniques such as feed blocks, pellets, and silage (Ben Salem and Nefzaoui, 2003) to valorize local feed resources (e.g. agroindustrial byproducts) help smallholders to better manage livestock feeding throughout the year. Main benefits from these options for the animal, the environment and their impact on farmers' livelihoods are reported in Table 6. Overall, the interesting results on the positive effect on animals of tanniniferous (e.g. in situ protection of dietary proteins, defaunation, reduced emission of methane, anthelmintic activity) and/or saponin (e.g. increased absorption rate of nutrients, defaunation, decreased production of methane) containing forages to improve feed efficiency and to control gastrointestinal parasites, and thus improve the productive and reproductive performance of ruminants should promote plants rich in secondary compounds in grazing systems. These options offer promising solutions to reduce the use of chemicals in livestock production systems to enhance livestock productivity and to decrease emission of methane (Nefzaoui et al., 2011).

Options	Impact on the animal	Impact on the environment	Impact on farmers livelihoods
Feed blocks	- Improved digestion of low quality diets and increased growth and milk production - Improved health conditions due to decreased parasitic load (use of medicated FBs)	- Decreased pollution with perishable AGIBs (olive cake, tomato pulp, etc.) - Decreased pressure on rangelands - Better quality manure	- Decreased feeding cost, increased animal performance and hence higher income - Diversification of farmers' income (sale of FBs) - Employment generation through mechanized unit for FBs making
Cactus (*Opuntia* spp.)	- Improved digestion of low quality forages - Improved animal performance	- Improved soil condition - Decreased pressure on primary resources (water and rangelands)	Added value cash crop (fruit and cladodes sale), and increased animal performance result in increased income
Shrub mixing	- Complementarities between shrub species (nutrients and secondary compounds) increased animal performances	- Combat desertification - Soil protection	Reduced budget allocated for feedstuffs purchasing
Rangelands resting	- Increased feed intake and digestion - Increased productive and reproductive performances	- reduces degradation risk - Protection of plant and animal biodiversity (domestic and wildlife animals)	- reduced feeding cost and increased performances resulting in increased income

Table 6. Productive, environmental and social benefits of some alternative feeding options (Nefzaoui et al., 2011).

Feed blocks (FBs) technology

Cold-processed feed blocks are made of a mixture of one or more agro-industrial by-products (e.g. olive cake, tomato pulp, etc.), a binder (e.g. quicklime, cement and clay), water and common salt, as well as urea with or without molasses. The technique of FB making is well described in the literature (e.g. Ben Salem and Nefzaoui, 2003; Ben Salem *et al.*, 2005a). Some variations in the blocks include the incorporation of polyethylene glycol as a tannin-inactivating agent, which has increased the utilization of tanniniferous browse foliage in ruminant feeding (Ben Salem *et al.*, 2007). Mineral enriched FBs (e.g. with phosphorus, copper, etc.) are fed to animals to mitigate deficiency and improve reproduction in ruminants. Benefits from the integration of FBs in the diet of sheep and goats are reflected by data compiled in Table 7. It is clear that depending on the formula, FBs can partially or totally replace concentrate feeds, thus reducing feeding costs without detrimental effects on livestock performances.

Basal diet	Supplement*	Animals	Growth rate (g/day)	Feeding cost variation	Country
Stubble grazing	Concentrate (250 g/d)	Lambs	95		Algeria
Stubble grazing	Conc. (150 g/d) + FB1	Lambs	136	-81%	Algeria
Wheat straw *ad lib*	Conc. (500 g/d)	Lambs	63		Tunisia
Wheat straw *ad lib*	Conc. (125 g/d) + FB2	Lambs	66	-11%	Tunisia
Acacia leaves	FB4	Lambs	14		Tunisia
Acacia leaves	FB5 enriched with PEG	Lambs	61		Tunisia
Rangeland grazing	Conc. (300 g/d)	Kids	25		Tunisia
Rangeland grazing	FB4	Kids	40		Tunisia

Table 7. Compiled data on the potential use of feed blocks as alternative feed supplements for sheep and goats in the Mediterranean area (Ben Salem *et al.*, 2005a). (*) FB1: wheat bran (10%), olive cake (40%), poultry litter (25%), bentonite (20%), salt (5%); FB2: wheat bran (25%), wheat flour (15%), olive cake (30%), rapeseed meal (10%), urea (4%), quicklime (8%), salt (5%), minerals (1%); FB4: wheat bran (28%), olive cake (38%), wheat flour (11%), quicklime (12%), salt (5%), minerals (1%), urea (5%); FB5: wheat bran (23%), olive cake (31.2%), wheat flour (9%), quicklime (9.9%), salt (4.1%), minerals (0.8%), urea (4.1%), PEG (18%).

Fodder shrubs and trees (FST) in the smallholders farming systems

Trees and shrubs are part of the Mediterranean ecosystem. They are present in most natural grazing lands of the North Africa region. Some species are high in essential nutrients and low in anti-nutritional factors (e.g. *Morus alba*), some others are low in nutrients but high in secondary compounds (e.g. *Pistacia lentiscus*) while some shrubs are high in both nutrients and secondary compounds (e.g. *Acacia cyanophylla, Atriplex* spp.). Such characteristics enable the plants to withstand grazing and to provide ground for selective grazing. In arid and semi-arid North Africa regions where available forage species cannot grow without irrigation, FST could be used as feed supplements. Saltbushes (*Atriplex nummularia, Atriplex halimus* and *Salsola vermiculata*) are planted in dry zones in North Africa and have many advantages because of their wide adaptability to harsh agro-climatic conditions and ability to grow over a longer period. As trees require little care after establishment, the production cost is low (Nefzaoui et al., 2011).

Alley-cropping

This technique consists of cultivating herbaceous crops of both graminae and legume species between rows of trees or shrub species. Among the reasons for the low adoption of pure shrubs planting are the technical design of plantation, mismanagement, and competition for land often dedicated to cereal crops. Alley cropping overcomes some of these disadvantages because it (1) improves soil; (2) increases crop yield; (3) reduces weeds and (4) improves animal performance. Properly managed alley-cropping allows diversification to benefit from several markets. It also promotes sustainability in both crop and livestock production. Benefits from cactus-barley alley cropping system were evaluated in Tunisia (Alary *et al.,* 2007; Shideed *et al.,* 2007). Compared to barley alone, the total biomass (straw plus grain) of barley cultivated between the rows of spineless cactus increased from 4.24 to 6.65 tones/ha and the grain from 0.82 to 2.32 tons ha^{-1}. These results are due to the change of the micro-environment created by alley-cropping with cactus, which creates a beneficial 'wind breaking' role that reduces water loss and increases soil moisture. The barley crop stimulated an increase in the number of cactus cladodes and fruits, while the cactus increased the amount of root material contributing to the soil organic matter. The alley-cropping system with *Atriplex nummularia* proved efficient in the semi arid regions of Morocco (annual rainfall 200-350 mm). Barley was cropped (seeding rate 160 Kg ha^{-1}) between Atriplex (333 plants ha^{-1}) rows. Compared to farmers' mono-cropping system, dry matter consumable biomass yield of Atriplex was significantly higher in the alley-cropping system. The latter system was more profitable than mono-cropping. Indeed, Laamari *et al.* (2005) determined the net benefit from Atriplex monocropping and barley-atriplex alley cropping over 15 years. The cumulative net benefit was 732.18 \$ ha^{-1} and 3,342.53 \$ ha^{-1}, respectively. The economic and agronomic assessment of alley cropping shows that this technology is economically profitable. Therefore, it should be extended on a large scale in the agro-pastoral areas of the North Africa region.

Shrub mixing technique

Most Mediterranean fodder shrubs and trees are either low in essential nutrients (energy and/or digestible nitrogen) or high in some secondary compounds (e.g. saponins, tannins, oxalates). These characteristics explain the low nutritive value of these fodder resources and the low performance of animals. For example, *Acacia cyanophylla* foliage is high in condensed tannins but low in digestible nitrogen. *Atriplex* spp. are low in energy and true protein although they contain high levels of crude protein, fibre and oxalates. Cactus cladodes are considered an energy source and are high in water but they are low in nitrogen and fibre. Moreover, they are remarkably high in oxalates. A wealth of information on the complementary nutritional role of these shrub species and the benefit of shrub mixing diets for ruminants, mainly sheep and goats are reported in the literature (Ben Salem *et al.,* 2002, 2004, 2005b). This technique permits to balance the diet for nutrients and to reduce the adverse effects of secondary compounds and excess of minerals including salt. The association cactus-atriplex is a typical example of shrub mixing benefits. The high salinity and the low energy content of atriplex foliage are overcome by cactus. Some examples of the effects of shrub mixed diets on sheep and goats performance are reported in Table 8. In summary, diversification of shrub plantations should be encouraged to improve livestock production in the dry areas of North Africa.

Basal diet[1]	Supplement[2]	Animal	Daily gain (g)
Acacia (417 g/d)	Atriplex (345 g/d); Barley (280 g/d)	Lambs	54
Cactus (437 g/d)	Atriplex (310 g/d); Acacia (265 g/d)	Lambs	28
Cactus (499 g/d)	Straw (207 g/d); Atriplex (356 g/d)	Lambs	81
Atriplex grazing	Cactus (290 g/d)	Lambs	20
Native shrubland grazing	Cactus (100 g/d); Atriplex (100 g/d)	Kids	60

Table 8. Effect of shrub mixed diets on sheep and goat growth (adapted from Nefzaoui et al., 2011). ([1]) Acacia: *Acacia cyanophylla*; Cactus: *Opuntia ficus indica f. inermis* (cladodes); Atriplex: *Atriplex nummularia*. ([2]) Values between parentheses are daily dry matter intake

3.5 Cactus

The Cactaceae family includes about 1600 species, native to America, but worldwide disseminated. *Opuntia* is the most widely known genus of this family. The species *Opuntia ficus indica* is cultivated in more than 20 countries. Around 900,000 ha of cactus have been planted in North Africa including 600,000 ha in Tunisia. The total area of cactus is estimated at 5 million ha of which 3 million are wild and located in Mexico. Cacti have been consumed by humans for over 9000 years. From underused crop, cacti received an increasing attention during the last few years. Thus, from 1998 to 2000 more than 600 researchers published over 1100 articles on Cacti.

Specific *Opuntia species* have developed phenological, physiological and structural adaptations for growth and survival in arid environments in which severe water stress hinders the survival of other plant species. Among these adaptations stand out the asynchronous reproduction and CAM metabolism of cacti, which combined with structural adaptations such as succulence allow them to continue the assimilation of carbon dioxide during long periods of drought, reaching acceptable productivity levels even in years of severe drought.

3.5.1 Cacti: The perfect candidate to mitigate climate changes in arid zones

CAM plants (Agaves and Cacti) can use water much more efficiently with regard to CO2 uptake and productivity than do C3 and C4 plants (Nobel, 2009). Biomass generation per unit of water is on an average 5 to 10 times greater than C4 and C3 plants (Table 9). In contrast to C3 and C4 plants, CAM plants net CO2 uptake occurs predominantly at night (Nobel, 2009). As stated by Nobel (2009), the key for the consequences between nocturnal gas exchange by CAM plants and C3 and C4 plants is temperature. Temperatures are lower at night, which reduces the internal water vapor concentrations in CAM plants, and results in better water use efficiency. This is the key reason that makes CAM species the most suited plants for arid and semi-arid habitats.

	C3	C4	CAM
WUE*	0.0013-0.005	0.0025-0.010	0.013-0.040
TR**	200-800	100-400	25-80

Table 9. Comparative water use efficiency (WUE) and transpiration rate (TR) for C3, C4, and CAM plants (adapted from P.S. Nobel, 2009). (*) Water-Use Effi ciency (WUE): ratio of the CO2 fixed in photosynthesis to water lost via transpiration. (**) Transpiration rate (TR): amount of water lost through transpiration over the CO2 fixed in photosynthesis.

C3 and C4 plants suffer irreparable damage once they lose 30 % of their water content. On the other hand, many cacti can survive an 80 to 90 % loss of their water content and still survive. This is due to the ability of CAM plants to store a lot of water, to shift water around among cells to keep crucial metabolism active, and to tolerate extreme cellular dehydration (Nobel, 2009).

These three abilities stem from the cacti characteristics including the extra thickness of the cuticles providing efficient barrier to water loss, the presence of mucilage and the daytime stomatal closing. In addition, cacti have an asynchronous development of various plant organs, so that even under the worst conditions some part of the plant is not affected. It is well known that cacti grow in desert where temperatures are extremely high. It has been reported by many authors (i.e. Nobel, 2009) that many agaves and cacti can tolerate high temperatures of up to 60 and 70 °C.

A full chapter has been devoted to this aspect by Park Nobel (2009) in his recent book "Desert Wisdom Agaves and Cacti CO2, Water, Climate Change". In view of the specific phenological, physiological and structural adaptations of cacti described above, it can be assessed that they are well positioned to cope with future global climate change. *Opuntia ficus indica*, for example, can generate a carbon sequestration of 20 tons of dry matter (equivalent to 30 tons of CO2) per ha and per year under sub-optimal growing conditions similar to those in North Africa arid regions. In this regard and as stated by Drennan and Nobel (2000) and Nobel (2009), agaves and cacti with their substantial biomass productivitiy and their high WUE should be considered for the terrestrial sequestration of atmospheric CO2 in underexploited arid and semi-arid regions. Such regions, which occupy 30 % of the Earth's land area, are poorly suited to C3 and C4 crops without irrigation.

3.5.2 How cacti can help adapting to climate change in dry areas?

Soil and water conservation

Several methods like water harvesting strips, contour ridges, gully check structures, biological control of rills and small gullies by planting cactus have been tested and have given good results. The contour ridges consisting of parallel stone ridges are built 5 to 10 m apart to stop runoff water (and the soil it carries) from damaging downstream areas. Each ridge collects runoff water from the area immediately upstream, and the water is channeled to a small plantation of fodder shrubs or cactus. Indeed with a suitable combination of well designed ridges and cactus, farmers are able to meet a large proportion of their fodder requirements.

In the countries of North Africa, particularly Tunisia, cactus is successfully associated with water harvesting structures. Planted according to contour lines, cactus hedges play a major

role in erosion control. Soil physical properties are considerably improved under these hedges and in immediate adjacent areas, with an improvement in organic matter and nitrogen as compared to non-treated fields. About 40 to 200% increase in organic matter and nitrogen have been reported. Top-soil structural stability is enhanced, susceptibility to surface crusting, runoff and erosion are reduced, while permeability and water storage capability are increased (Nefzaoui and El Mourid, 2009).
Comparing different cultivation systems, such as downhill planting, contour planting, reduced weeding, and intercropping with contour hedges, it was found that soil losses (0.13 to 0.26 t ha^{-1} y^{-1}) are the lowest with the last technique. Cactus planting in contour hedges may help retaining up to 100 t ha^{-1} soil annually (Margolis et al., 1985). Experiments conducted in Brazil and Tunisia show clearly that planting cactus in agroforestry system is more efficient for soil and water conservation than conventional land use (Table 10).

Crop type	Soil preparation phase	Cultivation phase	Harvest until next growing season	Total Soil losses	C factor
Bare soil	7.19	8.20	13.71	29.10	1.000
Cotton	2.42	1.77	6.72	10.91	0.392
Maize	1.51	0.68	3.75	5.94	0.199
Maize + beans	1.36	0.55	2.02	3.93	0.119
Opuntia ficus-indica	0.48	0.02	1.48	1.98	0.072
Perennial grass	0.00	0.02	0.01	0.03	0.001

Table 10. Comparison of soil losses (tons per ha per year) under different crops in semi-arid Northeastern Brazil (Margolis et al., 1985)

Cactus to rehabilitate degraded rangelands
Cacti and Opuntia in particular are some of the best plants for the reforestation of arid and semi-arid lands because they can survive under scarce and erratic rainfall and high temperature. Impressive results are obtained with fast growing shrubs (*Acacia cyanophylla, Atriplex nummularia*) or cactus (*Opuntia ficus indica*) planting in Central Tunisia where average annual rainfall is 200-300 mm (Table 11).

Rangeland type	Productivity (forage unit per hectare)*
Natural rangeland in Dhahar Tataouine, Tunisia (100 mm rainfall)	35 -100
Private rangeland improved by cactus crop in Ouled Farhane, Tunisia (250 mm rainfall)	800-1000
Cooperative rangeland improved through *Acacia cyanophylla*, Guettis, Tunisia (200 mm rainfall)	400-500

Table 11. Productivity (forage units per hectare) of natural and improved rangelands in Tunisia (Nefzaoui and El Mourid, 2009) (*): One forage unit is equivalent to 1 kg barley grain metabolizable energy

3.5.3 Cacti: A multipurpose crop and a source of income for the rural poor

Cactus crop is easy to establish and to maintain and has various utilizations. It produces good quality fruits for local or international markets; it is an excellent fodder; cactus young cladodes (nopalitos) are used as vegetable; it produces the "perfect red dye" from a cochineal that lives only on a specific type of cactus; and recent research revealed the vast interesting areas of its medicinal and cosmetic uses.

Fruit production: Cactus pear is cultivated for fruit production since the Aztec time. Main producers today include Mexico, Italy, South Africa, Tunisia, Morocco, Peru, and others. Yields are extremely high and reach 25 tons fruits per ha. Harvest and post-harvest techniques are well developed in producing countries (Inglese et al., 2002). The fruit quality is quite similar to orange or papaya. Recent findings show that cactus fruit has a high content of anti-oxidants and other neutraceuticals.

Use of cactus as forage: Cacti present high palatability, digestibility, and reduce the water needs to animals; however, they must be combined with other feedstuff to complete the daily diet, as they are poor in proteins, although rich in carbohydrates and calcium (Nefzaoui and Ben Salem, 2002). Animals can consume large amounts of cladodes. For instance, cattle may consume 50 to 70 kg fresh cladodes per day, and sheep 6 to 8 kg per day. The energy content of cladodes is 3,500 to 4,000 kcal kg^{-1} dry matter, just over half of which is digestible, coming mainly from carbohydrates. In arid and semi-arid regions of North Africa, cereal crop residues and natural pastures generally do not meet the nutrient requirements of small ruminants for meat production. Cladodes can provide a cost-effective supplementation, for raising sheep and goats on rangelands. When cladodes are supplied to grazing goats that have access to alfalfa hay, the milk yield is increased by 45% (to 436 g day^{-1}). When cladodes are associated with a protein-rich feedstuff, they may replace barley grains or maize silage without affecting body weight gains of sheep and cattle. For instance, milk yield for lactating goats supplied with 2.2 kg alfalfa hay day^{-1} is actually slightly higher (1.080 g day^{-1}) when 0.7 kg cladodes replaces an equal mass of alfalfa. Water scarcity can depress feed intake, digestion, and therefore weight gains of sheep and goats. Thus, supplying livestock with water during the summer and during drought periods is crucial in hot arid regions. Animals consume considerable energy to reach water points. Therefore, the high water content of cladodes is a solution to animal raising in dry areas. In fact, animals given abundant supplies of cladodes require little or no additional water (Nefzaoui and Ben Salem, 2002).

Use of cactus as vegetable and other valuable products: It is feasible to industrialize cladodes, fruit, and nopalitos. This potential market deals mainly with concentrated foods, juices, liquors, semi-processed and processed vegetables, food supplements and the cosmetic industry; it is feasible, but it requires sustained effort and investment to develop the market (Saenz, 2002 - 2006). Many brands of jellies, marmalades and dried sweets are prepared and sold in Latin America, South and North Africa. Juice obtained from the strained pulp is considered a good source of natural sweetener and colorants. Pads are widely used as a dietary supplement to increase fiber content in the human diet and for other beneficial purposes such as weight reduction, decrease in blood sugar and the prevention of colon cancer. The world market for pills made from powdered cactus is growing at a fast pace and small-scale producers could well benefit from this trend (Saenz, 2006).

Medicinal uses: There is some experimental research with promising results on the use of "nopalitos" for gastritis; for diabetes due to the reduction of glucose in blood and insulin;

for hypercholesterolemia due to reduction of total cholesterol, LDL cholesterol and triglycerides serum levels; and for obesity (Nefzaoui et al., 2007).
Economic profiles have been developed for the main exploitation alternatives and reveal that these are indeed viable and with adequate investment returns. These projects bring additional benefits, such as the generation of employment, environmental improvement, etc., which do not represent income for investors, but do contribute to humanity.

4. Institutional innovations: Empowering local communities

Development projects in the past failed to adequately address real community issues and concerns in agropastoral dry areas in North Africa. Decision-makers and all research and development partners are increasingly aware that "the heart of the rangeland sustainable management" is linked to institutional issues. Indeed, in the past the situation of rangelands was relatively better not only because population pressure and demand for meat were lower, but also because the management of rangelands was more strictly controlled by traditional institutions (*jmaas in Morocco, Myaad in Tunisia*) that enjoyed effective power. Numerous policy and institutional reforms have been carried out in several countries of North Africa. In most cases, policy and institutional reforms weakened pastoral institutions. These institutional reforms can be classified into three main approaches: state appropriation of rangeland resources, strengthening customary tribal claims, and privatization with titling (Ngaido and McCarthy, 2004).
ICARDA and IFAD (International Fund for Agricultural Development) worked together within the framework of the Mashreq/Maghreb Project to develop and implement a new participatory approach aiming at sustainable development of agropastoral communities in dry areas of the North Africa region. More specifically, it aims to develop participatory methodologies and tools that empower local communities and promote sustainable livelihood and conservation of agropastoral resources in those areas.
A methodology is developed through the joint inputs of all stakeholders including community members, agricultural specialists, extension services, researchers, local institutions, and decision makers. The methodology consists of the following steps: characterization of the community, diagnosis, planning and programming, institutional set-up, implementation, and monitoring and evaluation. The pillar of the methodology is an effective communication where all stakeholders negotiate community development plan (CDP) on an equal basis and where all sources of knowledge are explored, encompassing both indigenous and research-based knowledge. So far there is little integration of indigenous knowledge into development planning, thus concerned communities are becoming more powerless. It is suggested that development agencies should use indicators extracted from local know-how of agropastors to prepare relief instead of just relying on satellite imagery.
This participatory approach has been accepted and embraced by communities and development agencies in Tunisia, Algeria, and Morocco. It has been documented and disseminated through different channels including: a field manual in English and Arabic, linkage with Karianet network, and specific websites (www.icarda.org; www.mashreq-maghreb.org).
Key lessons from this experience include: (i) participatory characterisation of communities is essential for cooperation and trust among stakeholders; (ii) recognition of local know-how is an important step for successful diagnosis; (iii) the preparation of annual and long-term

development plan approved by communities is an efficient tool to mobilize resources and ease project implementation; (iv) not to underestimate the ability of communities to identify appropriate technical solutions, to solve internal conflicts particularly relating to property rights and land use, and the importance of additional-income generating activities; (v) the success and the sustainability of the process depends on the promotion of elected community-based organizations that play a key interface role between communities and other actors (government agencies and decision makers, non government agencies, donors, and other communities).

Promoting community-based organizations and empowerment will support adaptation to climate change (Garforth, 2008) through:

- help building strong institutions that can facilitate both collective and individual adaptation and response to climate change and other external pressures, both in the short and long term;
- platforms for managing conflict over natural resources
- creating and intensifying learning opportunities, to broaden the set of information and knowledge available to farmers and support local innovation: Livestock Field Schools are an example of how this can be done;
- supporting local innovation processes;
- helping livestock keepers identify opportunities to enrich the set of options they have when making livelihood choices: re-thinking how advisory services are provided, particularly to small-scale, relatively poor livestock keepers.

Recent experience of communal rangeland management in Southern Tunisia (IFAD PRODESUD Project) is quite successful. The community-based organizations (GDAs) are built up on socio-territorial units that correspond to the traditional tribe boundaries. They are fully participating in the design and implementation of their integrated local development. The approach involves the real participation of agropastoral communities, in a new bottom-up mode, for the establishment of community development plan (CDP) that reflects the real issues and priority needs of the community. This is developed through the joint inputs of all stakeholders including community members, agricultural specialists, extension services, local administration and state representatives. Best-bet options for technical, institutional and policy issues are jointly identified for implementation, monitoring and evaluation. The community is represented by a formal community-based organization (CBO), directly elected by community members and fully recognized by government authorities as their equal partner for implementation of all actions set out in the jointly developed CDP. This includes such crucial issues as management of communal pasture and rangelands (for example more than 50,000 ha of collective rangelands are put under rest and fully controlled by the communities), as well as the procurement of funds and necessary inputs and facilities, and the independent and transparent contact with all stakeholders and similar CBOs in the region for exchange of relevant information and experiences (Nefzaoui et al., 2007).

5. Conclusions

In this chapter we reviewed some of the agricultural achievements realized in three North African countries where agriculture depends primarily on rainfed production systems dominated by cereal crops and small ruminant livestock. Successful adopted technologies under unfavorable climate conditions include drought tolerant and disease resistant crop cultivars, water harvesting and supplementary irrigation, no-tillage practices, efficient feeding of small ruminants including communal rangeland management, utilization of balanced feed blocks, fodder shrubs and cactus, a marvelous multi-purpose crop, and above all the participation and empowerment of rural communities facing the adverse effects of climate change. Most of these innovations have been outscaled to other countries and regions with similar agro-ecologies. Thus, within the framework of FAO-ICARDA International Technical Cooperation Network on Cactus (FAO-ICARDA Cactusnet), cactus technologies have been disseminated to countries in the Near East (Jordan, Iran, Arabian Peninsula), South East Asia (India, Pakistan), Subsaharan Africa (Mauritania, Mozambique), and East Africa (Ethiopia, Eritrea). Within ICARDA Mashreq-Maghreb project (www.mashreq-maghreb.org), technological options such as feed blocks and fodder shrubs and trees have been adopted in countries of the Near East (Lebanon, Jordan, Syria, Iraq) and South East Asia (India, Pakistan). Institutional innovations including the community approach and community development plan have been applied and largely adopted in Jordan, Syria, Lebanon, Pakistan, and Mauritania.

The adoption of the improved technologies and institutional innovations and of further upgrade thereof will enable the North African rural communities cope with climate change. However, the recent IPCC projections point to an alarming increase in temperature, an important sea level rise, a general trend of rainfall reduction, and frequent occurrence of extreme events including severe drought and floods. These projections call for a firmer commitment and investment of North African countries and concerned international institutions in scientific innovations to address the worsening of natural resources status and the ensuing threat to food security. Issues to be addressed include (i) modeling climate change effect on a local level to develop more accurate warning systems for better coping with and adaptation to climate change; (ii) use of new biotechnological tools to increase the level and stability of animal and plant productivity encompassing tolerance to drought and heat stress as well as diseases and pests; (iii) developing new fresh water resources through desalination of bountiful sea water using limitless solar energy (iv) developing and promoting socially-acceptable policies on insurance against climatic risks and on water pricing and property rights; (v) North African countries should strive for a fair share of the post-Kyoto and other arrangements for access to the world carbon market and payment for environmental services.

6. References

Abang, M. & Malhotra, R. (2008). Chickpea and Climate Change. *ICARDA Caravan,* No.25. (December 2008), pp. 48-50

Alary, V., Nefzaoui A. & El Mourid, M. (2007). How risk influences the adoption of new technologies by farmers in low rainfall areas of North Africa?. In: El-Beltagy, A. M.C. Saxena and Tao Wang (eds). *Human and Nature – Working together for sustainable Development of drylands. Proceedings of the 8th International Conference on Development of Drylands,* 25-28 February 2006, Beijing, China. ICARDA, Aleppo, Syria. pp. 802-810.

Angar, H.; Ben Haj Salah, H. & Ben-Hammouda, M. (2010). Semis Direct et Semis Conventionnel en Tunisie: Les Résultats Agronomiques de 10 Ans de Comparaison. *Actes des 4e Rencontres Méditerranéennes du Semis Direct*, pp. 9-13, ISSN. 1111-1992, Sétif, Algérie, 3-5 Mai 2010

Badraoui, M. & Dahan, R. (2011). The Green Morocco Plan in Relation to Food Security and Climate Change, In: *Food Security and Climate Change in Dry Areas: Proceedings of an International Conference*, Solh, M & Saxena, M. C. (Eds.), pp. 61-70, ISBN 92-9127-248-5, Amman, Jordan, February 1-4, 2010, International Center for Agricultural Research in the Dry Areas (ICARDA)

Bahri, A. (2006). Water Use Efficiency in Morocco, In: *AARINENA Water Use Efficiency Network: Proceedings of the Expert Consultation Meeting,* I. Hamdan; T. Oweis & G. Hamdallah (Eds.), pp. 127-138, Aleppo, Syria, November 26-27, 2006, ICARDA, Aleppo, Syria

Ben Mechlia, N.; Oweis, T.; Masmoudi, M.; Mekki, I.; Ouessar, M.; Zante, P. & Żakri, S. (2008). Conjunctive Use of Rain and Irrigation Water from Hill Reservoirs for Agriculture in Tunisia. On-Farm Water Husbandry Research Report No. 6. ICARDA, Aleppo, Syria

Ben Moussa-Machraoui, S.; Errouissi, F.; Ben-Hammouda, M. & Nouira, S. (2010). Comparative effects of conventional and no-tillage management on some soil properties under Mediterranean semi-arid conditions in northwestern Tunisia. *Soil & Tillage Research,* 106, (2010), pp. 247-253

Ben Salem, H. & Nefzaoui, A. (2003). Feed blocks as alternative supplements for sheep and goats. *Small Ruminant Research*, 49: 275-288.

Ben Salem, H. & Smith, T. (2008). Feeding strategies to increase small ruminant production in dry environments. *Small Ruminant Research* 77, 174-194.

Ben Salem, H., Nefzaoui, A. & Makkar, H.P.S. (2007). Feed supplementation blocks for increased utilization of tanniniferous foliages by ruminants. In: H.P.S. Makkar, M. Sanchez, A.W. Speedy (Eds.): *Feed supplementation blocks. FAO Animal Production and Health paper 164*, FAO Rome. 185-205.

Ben Salem, H., Nefzaoui, A. & Ben Salem, L. (2004). Spineless cactus (*Opuntia ficus indica* f. inermis) and oldman saltbush (*Atriplex nummularia* L.) as alternative supplements for growing Barbarine lambs given straw-based diets. *Small Ruminant Research*; 51: 65-73.

Ben Salem, H., Abdouli, H., Nefzaoui, A., El-Mastouri, A. and Ben Salem, L. (2005b). Nutritive value, behaviour and growth of Barbarine lambs fed on oldan saltbush (*Atriplex nummularia* L.) and supplemented or not with barley grains or spineless cactus (*Opuntia ficus indica f. inermis*) pads. *Small Ruminant Research*; 59, 229-238.

Ben Salem, H., Norman, H.C., Nefzaoui, A., Mayberry, D.E., Pearce, K.L. & Revell, D.K. (2010). Potential use of oldman saltbush (*Atriplex nummularia* Lindl.) in sheep and goat feeding. *Small Ruminant Research*, 91:13-28.

Ben Salem, H., Al-Jawhari, N., Daba, M.A., Chriyaa, A., Hajj Hassan, S., Dehimi, D.L., & Masri, M.S. (2005a). *Feed block technology in West Asia and North Africa.* ICARDA, 111p.

Ceccarelli, S.; Grando, S.; Maatougui, M.; Michael, M.; Slash, M.; Haghparast, R.; Rahmanian, M.; Taheri, A.; Al-Yassin, A.; Benbelkacem, A.; Labdi, M.; Mimoun, H. & Nachit, M. (2010). Plant breeding and climate changes. *Journal of Agricultural Science*, (2010), 148, pp. 627-637

Cherfaoui, M. L. (2009). Le Plan National de Développement Agricole et Rural : Vision, Objectifs Stratégiques, Démarche, Dispositif de Mise en Œuvre- Eléments de Doctrine. Institut National de la Recherche Agronomique d'Algérie (INRAA), Algiers, Algeria

Dali, N. (2008). Principal guidelines for a National Climate Change Strategy: Adaptation, Mitigation and International Solidarity, In: *Proceedings International Conference-Livestock and Global Climate Change*, Rowlinson, P, Steele, M. & Nefzaoui, A. (Eds.), pp. 1-5, ISBN 978-0-9065626-62-8, Hammamet, Tunisia, May 17-20, 2008

De-Pauw, E. (2008). Hot spots to vulnerability to climate change. *ICARDA Caravan* No. 25, (December 2008), pp. 43-44

Dridi, N; Mekki, M; Cheikh M'hamed, H. & Ben-Hammouda, M. (2010). Impact de Deux Modes de Semis (Conventionnel vs. Direct) sur la Flore Adventice des Cultures. *Actes des 4e Rencontres Méditerranéennes du Semis Direct*, pp. 39-47, ISSN. 1111-1992, Sétif, Algérie, 3-5 Mai 2010

Dutilly-Diane, C. (2007). Pastoral economics and marketing in North Africa: A literature review. *Nomadic Peoples*, volume 11, Issue 1, 69-90

El Gharras, O. & Idrissi, M. (2006). Contraintes technologiques au développement du semis direct au Maroc, In: *Options Méditerranéennes, Série A, Numéro 69 :Troisièmes Rencontres Méditerranéennes du Semis Direct/Third Mediterranean Meeting on no-Tillage*, J.L. Arrue & C. Cantero-Martinez, (Eds.), 121-124, Zaragoza, March 2006

El Mourid, M.; De-Pauw, E. & Tahar, F. (2010). Agriculture in the North Africa Region and Constraints to Sustainable Productivity, In: *Explore On-Farm: The Case of North Africa,* El Mourid, M. Gomez-Macpherson, H., Rawson, H. M. (Eds), pp. 7-15, ISBN 978-9973-9992-5-2, Tunis, Tunisia

El Mourid, M.; Ketata, H. & Nefzaoui, A. (2008). Fostering Adoption of Durum Wheat Low-Cost Technologies for Increased Income and Improved Food Security in Less-Favored Areas of West Asia and North Africa. (IRDEN Project Final Report). ICARDA, Aleppo, Syria. 96 pp, ISBN 978-9973-9992-2-1

Gallali, T.; Brahim, N.; Bernoux, M.; Ben Aissa, N. & Ben Hamouda, A. (2010). Carbon sequestration under conventional and conservation agriculture in Mediterranean area: Tunisia case study, In: *Proceedings of the European Congress on Conservation Agriculture-Towards Agro-Environmental Climate and Energetic Sustainability*, pp. 371-376, ISBN 978-84-491-1038-2, Madrid, Spain, October 4-7, 2010

Garforth, C.J. (2008). Impacts on livelihoods. Pages 25-26. In: *Proceedings Livestock and Global Climate Change International Conference* (P. Rowlinson, M. Steele and A. Nefzaoui, eds.), 17-20 May, 2008, Hammamet, Tunisia. pp. 25-26..

Hazell, P. (2007). Managing drought risks in the low-rainfall areas of the Middle East and North Africa. In: "Perinstrup-Anderson and Cheng, eds. *Food policy for developing countries: the role of Government in the global food system*". Cornell University, Ithaca, New York.

ICARDA. (2006). ICARDA Annual Report 2005. *International Center for Agricultural Research in the Dry Areas*, ISSN 0254-8313, Aleppo, Syria

ICARDA. (2007). ICARDA Annual Report 2006. *International Center for Agricultural Research in the Dry Areas*, ISSN 0254-8313, Aleppo, Syria

ICARDA. (2008). ICARDA Annual Report 2007. *International Center for Agricultural Research in the Dry Areas*, ISSN 0254-8316, Aleppo, Syria

ICARDA. (2009). ICARDA Annual Report 2008. *International Center for Agricultural Research in the Dry Areas*, ISSN 0254-8313, Aleppo, Syria

ICARDA. (2010). ICARDA Annual Report 2009. *International Center for Agricultural Research in the Dry Areas*, ISSN 0254-8313, Aleppo, Syria

ICID. (2003). *Saline Water Management for Irrigation*. International Commission on Irrigation and Drainage (ICID), New Delhi, India

Inglese, P., F. Basile, and M. Schirra. (2002). Cactus pear fruit production. In " P.S. Nobel (eds) *Cacti*. California Unversity Press, 163-183.

Iniguez, L. (2005a). *Characterization of small ruminant breeds in West Asia and North Africa. Vol.1.* North Africa. International Center for Agricultural Research in the Dry Areas (ICARDA) , Aleppo, Syria, 462 pp. ISBN:92-9127-164-9

Iniguez, L. (2005b). *Characterization of small ruminant breeds in West Asia and North Africa. Vol.2. North Africa.* International Center for Agricultural Research in the Dry Areas (ICARDA) , Aleppo, Syria, 196 pp. ISBN:92-9127-177-3

IPCC. (2007). Climate Change 2007: Synthesis Report. Contribution of Working Groups I, II, and III to the Fourth Assessment Report of the Intergovernmental Panel on Climate Change [Core Writing Team, Pachauri, R.K and Reisinger, A. (eds)]. IPCC, Geneva, Switzerland

Karrou, M. & Boutfirass, M. (2007). *Gestion intégrée de l'eau en agriculture pluviale.* INRA, Rabat, Morocco

Karrou, M. & Oweis, T. (2008). Climate Change and Water: the Challenges for Dry Areas. *ICARDA Caravan* No. 25. (December 2008), pp. 17-20

Kassam, A.; Friedrich, T. & Derpsch, R. (2010). Conservation agriculture in the 21st century: a paradigm of sustainable agriculture, In: *Proceedings of the European Congress on Conservation Agriculture-Towards Agro-Environmental Climate and Energetic Sustainability,* pp. 19-68, ISBN 978-84-491-1038-2, Madrid, Spain, October 4-7, 2010

Koneswaran, G. & Nierenberg, D. (2008). Global farm animal production and global warming: impacting and mitigating climate change, *Proceedings International Conference-Livestock and Global Climate Change,* Rowlinson, P; Steele, M. & Nefzaoui, A. (Eds.), pp. 164-169, ISBN 978-0-906562-62-8, Hammamet, Tunisia . May 17-20, 2008

Laamari A., Boughlala, M. & Chriyaa, A. (2005). Adoption and impact studies in Morocco. In: H. Shideed Kamil and M. El Mourid (eds). Adoption and impact assessment of improved technologies in crop and livestock production systems in the WANA region. *The development of integrated Crop/Livestock Production in Low Rainfall Areas of Mashreq and Maghreb Regions* (Mashreq/Maghreb Project).ICARDA, Aleppo, Syria, viii + 160 pp. En. pp. 107-118.

Lal, R. (2010). A dual response of conservation agriculture to climate change: reducing CO_2 emissions and improving the soil carbon sink, In: *Proceedings of the European Congress on Conservation Agriculture-Towards Agro-Environmental Climate and Energetic Sustainability,* pp. 3-18, ISBN 978-84-491-1038-2, Madrid, Spain, October 4-7, 2010

LeHouerou, H. N. (2000). Restoration and Rehabilitation of Arid and Semi-arid Mediterranean ecosystems in North Africa and West Asia: A review. *Arid Soil Research and Rehabilitation,* 14: 3-14

MacOpiyo, L., Angerer, J., Dyke, P. & Kaitho, R. (2008). Experiences on mitigation or adaptation needs in Ethiopia and East African rangelands. Pages 64-67. In *Proceedings Livestock and Global Climate Change International Conference* (P. Rowlinson, M. Steele and A. Nefzaoui, eds.), 17-20 May, 2008, Hammamet, Tunisia.

Magnan, N.; Lybbert, T.J.; Mrabet, R. & Fadlaoui, A. (2011). The Quasi-Option Value of Delayed Input Use under Catastrophic Drought Risk: the Case of No-till in Morocco. *Amer. J. Econ.* Vol.93, No.2, pp. 498-504

Margolis, E., Silva, A.B. & Jacques, F.O. (1985). Determinacao dos fatores da equacao universal de perda de solo para as condicoes de Caruaru (PE). *Rev. Bras. Cienc. Solo* 9:165–169.

Matar, A.; Torrent, J. & Ryan, J. (1992). Soil and fertilizer phosphorous and crop responses in the dryland Mediterranean zone. *Adv. Soil Sci.,* Vol.18, pp. 82-146

Mrabet, R. (2006). Soil quality and carbon sequestration : Impacts of no-tillage systems, In: Options Méditerranéennes, Série A, Numéro 69 :Troisièmes Rencontres Méditerranéennes du Semis Direct /Third Mediterranean Meeting on no-Tillage, J.L. Arrue & C. Cantero-Martinez, (Eds.), 43-55, Zaragoza, March 2006

Mrabet, R. (2008). N-Tillage Systems for Sustainable Dryland Agriculture in Morocco. INRA, Fanigraph Edition, Rabat, Morocco

Neffati, M.; Ouled Belgacem, A. & El Mourid, M. (2007). Putting MAPS on the Map. *ICARDA Caravan* No. 24, (December 2007), pp. 35-38

Nefzaoui A. and Ben Salem, H. (2002). Forage, fodder, and animal nutrition. Chapter 12. In: P.S. Nobel (ed.), *Cacti, biology and uses*. University of California Press, 280 pp.

Nefzaoui, A and El Mourid, M. (2010). Cactus pear for soil and water conservation in arid and semi-arid lands. In: Nefzaoui, A. Inglese, P. and T. Belay (Eds.). 2010. Improved utilization of cactus pear for food, feed, soil and water conservation and other products in Africa. *Proceedings of International Workshop, Mekelle* (Ethiopia), 19-21 October, 2009. 265 pp.

Nefzaoui, A. & El Mourid, M. (2010). Coping with climate change and risk management strategies for a sustainable livestock and rangeland systems in the WANA region. In: "Eds. A. El-Beltagy and M.S. Saxena- *Proceedings of the Ninth International Conference on Development of Drylands*"; pp. 630-640.

Nefzaoui, A. & El Mourid, M. (2009). Cacti: A key-stone crop for the development of marginal lands and to combat desertification. *Acta Horticulturae*, 811:365-372.

Nefzaoui, A. El Mourid, M., Alary,V., Ngaido, T., & El Harizi, K. (2007). Empowering rural communities for better management of desert collective rangelands- from concept to implementation. In: El-Beltagy, A. M.C. Saxena and Tao Wang (Eds.). *Human and Nature - Working together for sustainable Development of drylands. Proceedings of the Eighth International Conference on Development of Drylands*, 25-28 February 2006, Beijing, China. ICARDA, Aleppo, Syria. pp : 620-632.

Nefzaoui, A. (2002). Rangeland management options and individual and community strategies of agropastoralists in Central and Southern Tunisia. *International conference on policy and institutional options for the management of rangelands in dry areas*. CAPRi Working paper N° 23; pp.14-16

Nefzaoui, A., Ben Salem, H. & El Mourid, M. (2011). Innovations is small ruminants feeding systems in arid Mediterranean areas. In: "R. bouche, A. Derkimba, F. Casabianca (eds) *New trends for innovation in the Mediterranean animal production*". EAAP publication No 129. Wageningen Academic Publishers, ISBN 978-90-8686-170-5, ISSN 0071-2477. pp. 99-116.

Nefzaoui, A., Nazareno, M. & El Mourid, M. (2008). Review of Medicinal Uses of cactus. *FAO-ICARDA Cactusnet Newsletter*, issue 11: 3-18.

Nefzaoui, A. (2010). Use of cactus as feed: review of the international experience. In: Nefzaoui, A. Inglese, P. and T. Belay (Eds.). 2010. Improved utilization of cactus pear for food, feed, soil and water conservation and other products in Africa. *Proceedings of International Workshop, Mekelle* (Ethiopia), 19-21 October, 2009. 265 pp.

Ngaido, T. & McCarthy, N. (2004). Collective action and property rights for sustainable development. Institutional Options for Managing Rangelands. Vision 2020 for Food, Agriculture, and Environment. *Focus 11*, February, 2004.

Nobel, P.S. (2009). *Desert Wisdom, Agaves and Cacti, CO2, Water, Climate Change*. Universe, New York, Bloomington. ISBN: 978-1-4401-9151-0. 198 pp.

Nori, M., El Mourid, M. & Nefzaoui, A. (2009). Herding in a shifting Mediterranean changing agro-pastoral livelihoods in the Mashreq and Maghreb region. *EUI Working papers, RSCAS 2009/52*. 22 pp. www.eui.eu/RSCAS/Publications/

Oweis, T. & Hachum, A. (2006). From water efficiency to water productivity: Issues of research and development, In: *AARINENA Water Use Efficiency Network, Proceedings of the Expert Consultation Meeting*, I. Hamdan, T. Oweeis & G. Hamdallah, (Eds.), 13-26, ISBN 92-9127-210-4, Aleppo, Syria, November 26-27, 2006, ICARDA, Aleppo, Syria

Oweis, T. (2010). Improving agricultural water productivity: a necessary response to water scarcity and climate change in dry areas. *Proceedings of the Ninth International Conference on Development of Drylands: Sustainable Development in Drylands- Meeting the Challenge of Global Climate Change*, El-Beltagy, A. & Saxena, M. C. (Eds.), 184-196, International Dryland Development Commission., December 2010

Oweis, T. (1997). Supplemental Irrigation: A Highly Efficient Water-Use Practice. ICARDA, Aleppo, Syria

Oweis, T.; Prinz, D. & Hachum, A. (2001). *Water Harvesting: Indigenous Knowledge for the Future and the Drier Environments*, ISBN 92-9127-116-0, ICARDA, Aleppo, Syria.

Qadir, M. (2008). Putting bad water to good use. *ICARDA Caravan* No. 25, (December 2008), pp. 45-47

Requier-Desjardins, M. (2010). Impacts des changements climatiques sur l'agriculture au Maroc et en Tunisie et priorités d'adaptation. *Les Notes d'analyse du CIHEAM*, No.56, (Mars 2010), CIHEAM

Rezgui, M.; Zairi, A; Bizid, E. & Ben Mechlia, N. (2005). Consommation et Efficacité d'Utilisation de l'Eau chez le Blé Dur (Triticum durum Desf.) Cultivé en Conditions Pluviales et Irriguées en Tunisie. *Cahiers Agricultures*, Vol.14, No.4, (Juillet-Août 2005), pp. 391-397

Ryan, J. & Matar, A. (1992). Fertilizer Use Efficiency under Rainfed Agriculture. *Proceedings, Fourth Regional Soil Test Calibration Workshop, Agadir, Morocco. May 5-11, 1991.* ICARDA, Aleppo, Syria.

Ryan, J.; Masri, S.; Singh, M.; Pala, M.; Ibrikci, H. & Rashid, A. (2008). Total and Mineral Nitrogen in a Wheat-Based Rotation Trial under Dryland Mediterranean Conditions. *Basic and Applied Dryland Research* Vol.2, (2008), pp. 34-36

Sáenz, C. (2002). Cactus pear fruit and cladodes: A source of functional components for foods. *Acta Horticulture* 581: 253-263.

Sáenz, C. 2006. Utilizacion agroindustrial del nopal. *Boletin de servicios agricolas de la FAO*, 162: 164pp.

Shideed, K., Alary, V., Laamari, A., Nefzaoui, A. & El Mourid, M. (2007). Ex post impact assessment of natural resource management technologies in crop-livestock systems in dry areas of Morocco and Tunisia. In: *International Research on Natural Resource Management* (Waibel, H. and D. Zilberman, eds). CAB International.

Shideed, K.; Oweis, T.; Gabr, M. & Osman, M. (2005). *Assessing On-Farm Water-Use Efficiency: A New Approach*, ISBN 92-9127-163-X, ICARDA, Aleppo, Syria

Steinfeld, H.; Gerber, P.; Wassenaar, T.; Castel, V.; Rsales, M. & de Haan, C. (2006). *Livestock's Long Shadow: Environmental Issues and Options*. Food and Agriculture Organization of the United Nations. Rome, Italy

Zaghouane, O.; Abdellaoui, Z. & Houassine, D. (2006). Quelles perspectives pour l'agriculture de conservation dans les zones céréalières en conditions algériennes?, In: *Options Méditerranéennes, Série A, Numéro 69 : Troisièmes Rencontres Méditerranéennes du Semis Direct/Third Mediterranean Meeting on no-Tillage*, J.L. Arrue & C. Cantero-Martinez, (Eds.), 183-187, Zaragoza, March 2006

5

Possible Evolutionary Response to Global Change – Evolutionary Rescue?

Lars A. Bach and Cino Pertoldi
University of Aarhus
Denmark

1. Introduction

1.1 Climate-induced environmental changes

With a pace that is higher than observed in the past 10,000 years global warming is currently changing the global and local environments. On average, the global temperature has increased by 0.7 degree over the past century and future projections show an acceleration of global temperature rise (Walther et al., 2002) which produces climate-induced environmental changes (CIEC). Increasing the mean temperature furthermore corresponds to an increasing range between the minimum and the maximum temperatures due to a pure scaling effect of the variance with the mean (Pertoldi et al., 2007a). Additional factors may then add even more to the increased range of temperatures combined with increased variability in precipitation patterns. An increased temperature range is translated into a fluctuating selective regime for natural populations and amplified environmental variability (σ^2e) which have several consequences at different levels of organization.

In order to understand what limits the ability of species to adapt to CIEC, we need to integrate (local) short-term and (local) long-term changes and to increase our knowledge on the importance of genetic and environmental components on phenotypic variability (σ^2p) (Pertoldi et al., 2005). A notorious debate between ecologists and geneticists concerns the relative importance of genetic and ecological factors for the persistence of populations. There is a need for a deeper understanding of how genetic measures can be used to indicate causal processes, including the genetic signature of population declines or expansions due to CIEC. Evolutionary biologists and ecologists have increasingly turned to molecular genetics to study the demographic and genetic consequences of CIEC on populations. However, this approach has some serious limitations: 1) many different population processes lead to similar patterns of genetic structure and 2) population genetic models most commonly applied to these systems are based on the assumption of equilibrium conditions typically not found in nature and surely not in disturbed ecosystems.

1.2 A natural experiment from the past and experimental investigations on the consequences of climate-induced changes

Detailed knowledge on how CIEC have shaped the genetic composition and the present geographic distribution of species can help us to better comprehend the possible future consequences of climatic changes. The biotic effects of Pleistocene glaciations exemplify how climatic changes influence species distributions by alternately inducing southward range contractions with northward expansions. The geographic patterns resulting from these processes differ with the varying dispersal abilities and ecological requirements of species. The geographical distribution of genetic diversity in species may be used to reconstruct historical biogeographies (Avise, 1998). CIEC do not only affect the distribution of organisms, through changing the abiotic environment, they also change the patterns of biotic interactions between species, and their morphology. More emphasis should therefore be given to morphometrical investigation, which can unravel ecological patterns that are undetectable using neutral molecular markers (Pertoldi et al., 2003). The joint application of different molecular genetic and morphometric methods may prove useful in the description of population structure and help in identifying factors that shape the observed demographical, morphometrical, geographical and genetic structure (Røgilds et al., 2005; Plejdrup et al., 2006).

In particular, studies have been conducted in the attempt to obtain more detailed knowledge on the potential of $\sigma^2 p$ in an evolutionary context. A number of investigations have shown that $\sigma^2 p$ is positively associated with the level of genetic and environmental stresses that individuals experience (Kristensen et al., 2004; Røgilds et al., 2005). Several studies have also tried to elucidate the effect of genetic variability ($\sigma^2 g$) on $\sigma^2 p$ (e.g. Pertoldi et al., 2003). These studies include analyses of differences in $\sigma^2 p$ between males and females of haplo-diploid taxa, or parthenogenetic and sexually reproducing individuals (Andersen et al., 2002). Pertoldi et al., (2006b) have suggested several methods to split-up the different components of $\sigma^2 p$ (canalization, plasticity and developmental homeostasis), developing algoritms and suggesting the use of clonal organisms to remove the effect of $\sigma^2 g$ and its interaction (GXE) with $\sigma^2 e$ by means of admixture analysis (Pertoldi et al., 2006b).

Several investigations have also been conducted in order to resolve the controversies existing about the causal relationships between molecular genetic variation and phenotype-based measures of success. Pertoldi et al. (2006a) recently suggested that greater clarity would be achieved by partitioning genetic diversity into two components: that arising from adaptive evolution and that resulting from long-term historical isolation. The former can be estimated through analysis of phenotypic variation, while the latter is readily assayed through molecular phylogeography. Both approaches have their place, but measure different components of intraspecific diversity. Pertoldi et al., (2007a) suggested that a proper comparison between genetic variability using neutral molecular markers and genetic variability detected in quantitative and fitness related traits could significantly add to the open debate among evolutionary biologists on the correlation between these two measures. Recent genetics studies are beginning to broaden in scope and impact by attempting to correlate genetic, demographic and phenotypic properties of the same populations (Plejdrup et al., 2006). Furthermore, recent progress in biostatistics and mathematics (e.g. theory of coalescence, Bayesian statistics, individual-based population dynamics, algorithms for efficient simulation and sampling of complex processes), have strengthened our potential to infer population genetic processes of neutral and non-neutral genes via the development of theoretical models (Randi et al., 2003; Pertoldi et al., 2007a).

Modelling techniques having the capacity to incorporate explicit genetic variables linked to important life history traits can also be constructive for the identification of the factors (and their interactions) which are affected by CIEC and can be used as complementary tools (Strand, 2002; Bach et al., 2007). Simulation models can also easily accommodate different global change scenarios, which may not be readily accomplished by mathematical analysis.

Stochastic genetic models may mimic events at individual loci, so-called finite loci or allelic models, or may be parameter based, unfolding the average genetic effects according to quantitative genetics theory (Verrier et al., 1990; Wang, 1996).

1.3 Consequences of CIEC on biodiversity

Determining the biodiversity impacts of climate change is a great challenge (Schwenk et al., 2009). The major consequences of CIEC for biodiversity at various scales include: distributional range of species, phenology, community structure and species interactions (Walther et al. 2002). The demographic context of σ^2p has considerable significance to the process of adaptation. Not only does dispersal among patches influence the evolution of traits and their plasticity, but the changing meta-community also plays a role in determining how populations respond to change (Angilletta, 2009; Mitchell & Angilletta, 2009). Given this situation, predictions at the community level seem either pointless at present or unworthy of pursuit (Ricklefs, 2008), especially since initial conditions, instabilities, and model errors should greatly affect the impact of climate change on ecological communities. Substantial shifts in the ranges and phenologies of species from an array of groups have occurred in response to climate change (Steltzer & Post, 2009). This emphasises the importance of mitigating such shifts through e.g. corridors or by securing large coherent areas with suitable habitats for wildlife. Without such initiatives many populations may become extinct due to combined effects of environmental stress, lack of evolutionary potential and inbreeding depression.

1.4 Shifts in the ranges and phenologies of the species as a consequence of climate-induced changes

Biologists no longer doubt that biological systems have already responded to the current global anthropogenic changes in climate. Many studies have demonstrated substantial shifts in the ranges and phenologies of species from a broad array of taxa, indicating a coherent fingerprint of climate change (e.g. Chen et al., 2009; Steltzer & Post, 2009; Knudsen et al., 2011). Given the substantial evidence of shifting ranges and phenologies, and of substantial range shifts in the past (Davis & Shaw, 2001), much attention has also been given to forecasting the likely effects of ongoing climate change on species distributions and ecosystems from these perspectives (e.g. Kearney et al., 2008). A large, and often contentious, literature has developed about how changes in species' ranges should be modelled and how biotic interactions mechanisms might be incorporated to generate novel insights (e.g. Jeschke & Strayer 2008; Keith et al., 2008).

2. Exploiting population variation and molecular techniques

Although environmental variation is not necessarily reflected in transformed vital rates, such as growth rate, interplay between environmental variation and population dynamics has been shown in a variety of species Stenseth et al., 2002). Understanding the consequences of demographic stochasticity in populations requires information of local fluctuations in population size, extinction probability and colonisation potential as well as reproductive success, which can be gained from population dynamics analyses. DNA analyses are progressively used to estimate the extent and organization of genetic diversity in populations in order to infer the causes of spatio-temporal dynamics (Schwartz et al.,

2007). Such assessment is performed by investigating the degree of neutral genetic variation, which is informative in inferring ancient or recent historical dynamics of populations. Information on the genetic composition of a populations prior to environmental perturbation is now accessible thanks to the recent progress in biostatistics and mathematics (e.g. theory of coalescence, Bayesian statistics, individual-based population dynamics, algorithms for efficient simulation and sampling of complex processes), which have greatly improved the possibility to infer population genetic processes through the development of theoretical models (Stephens & Balding, 2009). Going beyond plain parameter estimation is possible in applying a Bayesian approach, which can integrate both genetic and non-genetic data and hence test hypotheses about the factors that control demographic and genetic changes. In particular, the development of Bayesian models aimed to infer historical population dynamics and population parameters are particularly promising (Riebler et al., 2008).

The causal relationship between molecular genetic variation and phenotype-based measures of success are associated with some debate. Part of this incongruity stems from confusing the levels of organization at which genetic variation and phenotypic accomplishment have been conceptualized (Coulson et al., 2006). Further, molecular markers cannot identify the likelihood of loss of genetic variance in traits of ecological significance, as the correlation between molecular diversity (which is per definition neutral) and ecologically relevant traits (which are per definition non-neutral) is weak and becomes even weaker in expanding or declining populations. However, the attempt to correlate neutral and non-neutral variability can be made by using a promising new tool in conservation genetics consisting of the single nucleotide polymorphisms (SNPs). It is at present viewed as the richest polymorphic genetic marker in many genomes and may get round some of the problems related to microsatellites because of the enhanced resolution of genetic variation. In natural populations SNPs hold the potential to expand our ability to survey both neutral (non-coding region) variation as well as genes under selection (coding region), while also providing wider genome coverage compared to microsatellites (Morin et al., 2004). Further, moving the genomic methodology from lab-model organisms to non-model organisms is now becoming achievable, allowing genomic analysis in a population- and species wide fashion (Mitchell-Olds et al., 2008). Until recently, the genomic tools and resources have unfortunately been limited when it came to key ecological species as opposed to models species with plenty of genomic approaches readily available.

Recent identification of functional genes and genes linked to quantitative traits are opening the way to the analysis of functional genes and components of genetic control of physiological processes and are therefore expected to contribute to the understanding of local adaptation (Marsano et al., 2010). Population genomics will very soon add important contributions to these issues, delivering substantial amounts of data on regulatory polymorphisms on a genomic scale. Moreover, we may address the question of whether the regulatory variation per se cause adaptation to local conditions and whether it is able to significantly alter life-time reproductive success.

Quantitative genetic analyses are important in the assessment of the extinction risk since this approach can give information on the amount of non-neutral genetic variability present for a given trait. This information enables us to scrutinize fitness components on various genetic and environmental backgrounds, producing information on the fate of genetic diversity and the force of selection acting on the populations. Note however, that in practice we are thus limited to manageable organisms with short generation times. Nevertheless our ultimate aim is to determine how much a response of a given trait to environmental change is due to plastic and/or evolutionary response. Such information is becoming very relevant for evolutionary biology as there is a need for detailed studies on how variation at the level of genes translates, through developmental and physiological processes, into phenotypic variation for ecologically significant traits (Coulson et al., 2006).

Quantitative genetic investigations have thus far often been limited to laboratory conditions and the neutral molecular markers in natural populations are not necessarily relevant to understand the evolution of functional genes subject to selection, which point to the potential adaptability of a population to environmental changes. In natural populations it is difficult to show selection (let alone to quantify). However, genome scans and association studies are increasingly promising due to new statistical methods with improved power (Stephens et al., 2009). Although identifying selected and functionally important genes is no easy task, genome scans offer the possibility of finding genomic domains with selective value, which in turn is a first step in separating selection from the background of random genetic drift. This would make way for describing how changing environments (and fragmentation) can affect different domains of the genome. Hence, finding genomic domains under selection may be at least as useful as gene finding per se. A combination of ecological genomics and quantitative genetics will therefore lead to a greatly increased understanding of ecological responses, starting from genetic variation in natural populations to the description of shifts in phenotypes as a result of evolutionary responses to environmental changes (Luikart et al., 2003).

3. Theoretical approaches

The development of theoretical models and the use of computer simulations have also contributed significantly to the understanding of the consequences of CIEC. These models include stochastic environmental effects, allowing us to make probabilistic predictions that can be reasonably precise when we consider averages over large scales. Considerable progress has been achieved in incorporating age- or stage-structure into population genetic models, mostly in the context of life history evolution and estimation of the effective population size (N_E) of large and stable populations (Engen et al., 2010). However, knowledge on the interaction between age- or stage-structure and other factors, such as variance in reproductive success, temporal fluctuations in population size, is still fairly limited. Although attempts have been made to combine ecological and genetics theory, there is still insight to be gained from integrating the disciplines further.

Deterministic simulations are based on algebraic equations that predict the likely outcome of sampling, while stochastic (Monte Carlo) simulation models mimic random processes. Although being transparent and analytically tractable, deterministic predictions cannot deal with the same level of complexity over many generations as stochastic simulations. The benefit of combining these approaches is evident from simulations used to verify the accuracy when prediction equations are developed. Stochastic simulations are relevant for the design of risk estimates and there are no inherent limitations excluding representation of the genetic level.

The study objects, such as populations or individuals, do not necessarily comply with the mean field assumptions that all units are organised as uniform masses and interactions are unconditioned and can be averaged. In such cases the individual-based models (IBM) or agent-based approaches can be appropriate ways to allow variation in many aspects of the individual's characteristics as well as variable and conditional interactions (Travis et al., 2009). Likewise the geospatial implementations of IBM can account for specific spatial effects. This approach can be especially relevant for heterogeneous populations of higher animals in spatiotemporally heterogeneous environments with behaviour depending on its own state, the state of conspecifics, or the specific states of the environment (Bach et al., 2006; Bach et al. 2007). In other words the individual in an IBM does not perceive and interact with 'the average individual' of an abstract averaged population according to an average encounter rate and it does not experience the average environment. However, as entities, interactions and environment can be freely defined it follows that the extreme flexibility can become a challenge when designing simulations to address simple questions. In terms of genetics, another advantage of IBM is the straightforward implementation of genotypes, representing either neutral or selected genes where the latter permit the agents to adapt to changing environments. Such models are often referred to as complex adaptive systems (CAS) (DeAngelis & Mooij, 2005). Also the fact that events in IBM simulations are inherently stochastic may prove an advantage when the goal is to obtain probabilities. Much depends on the specific question and available data.

4. Developments in geographical ecology for understanding the consequences of climate-induced environmental changes and its interactions with other biotic and abiotic factors

Given that human impacts in terms of both anthropogenic climate warming, habitat loss and fragmentation, are likely to increase over the 21st century (Smith et al., 2009), the consideration of geographical ecology research is an important new avenue of research.
Therefore, the inclusion of new developments in geographical ecology towards much improved quantification of the determinants of species distributions and diversity patterns will be interesting (Guisan & Zimmermann, 2000). Notably the role of geographic variation in environmental factors such as climate creates an important basis for predicting responses to future climate change (e.g. Thomas et al., 2004). Furthermore, climatically-driven global geographical variation in metabolic rates may both be of fundamental importance to biodiversity and ecosystems and a determining factor in organism sensitivity to stressors (Dillon et al., 2010). Another motivation to look towards geographical ecology is the question of ascertaining effects of habitat destruction and fragmentation on species distribution changes from the separate effects of stressors, as well as their interactions (as fragmentation may affect exposure and susceptibility to environmental stressors (Gandhi et al., 2011).

5. Demographic and genetic consequences of CIEC

Environmental factors and their changes are to a large extent mirrored in the genetic composition of affected populations, which in turn impact the potential for adaptation to future selective forces such as CIEC. Even small alterations of environmental conditions can affect the genetic composition of populations, both via demographic and selective responses (Lande & Shannon, 1996; Björklund et al., 2009). Adaptation is one of the core principles in evolutionary biology and natural selection is universally regarded as the primary cause of evolutionary changes (Vermeij, 1996). The effects of rapid environmental changes, such as global warming, can cause problems particularly for small, isolated populations. Small populations may lack the genetic diversity that would allow adaptation to a new environment, and thus might risk extinction (Spielman et al., 2004). Further, genetic drift in small populations (Gilpin and Soulè, 1986) leads to loss of genetic diversity, further depressing the evolutionary potential and thereby the ability to respond to changing environments (see Lynch 1996). Additionally, in small populations the chance of mating among relatives is increased due to the limited number of individuals, which causes inbreeding and further decreases mean fitness (Spielman et al., 2004). The increased probability of mating among relatives and the accelerated rate of loss of genetic variability in populations are strongly associated with a reduction of NE which is the size of an "ideal" (stable, random mating) population that results in the same degree of genetic drift as observed in the actual population (Wright, 1931). Due to the numerous ways in which natural populations can deviate from the "ideal" population, N_E may be only a fraction of the population census size (N) size (Lande and Barrowclough, 1987). The N_E of a population can predict its capacity to survive in a changing environment more reliably than the census size and/or the amount of genetic variability (Nunney, 2000).

Global scale environmental change may affect the local N_E in several ways that may not be entirely independent. Firstly, as environmental changes accelerate, the demand for rapid adaptation becomes more pronounced, as in the simplest case where an optimum mean trait value shifts as a result of e.g. a rise in mean temperature. This requires a certain 'standing crop' of genetic variation in order for the population to track the moving optimum. Failing to do so, the populations may suffer demographically from the load of being maladapted. Secondly, the variance of environmental conditions may increase putting its toll on genetic variance by lowering the harmonic mean (HM) through the population dynamic response to environmental fluctuations. Theoretical models predict that fluctuation in population size is one of the dominant causes of reduction of N_E and the low N_E/N ratios (Kalinowski and Waples, 2002). If generations are non-overlapping, N_E can be approximated as the HM of the population census size N (Caballero, 1994).

The expected heterozygosity (H_e), a measure of genetic variability, can provide an indication of the immediate evolutionary potential of a population, but it has no necessary relationship to longer term potential (Nunney, 2000). This is particularly true when the environment of the population is changing. The notion of N_E can therefore be viewed as a bridging point between ecology and genetics, with the ecological characteristics including life history traits, social structure and population dynamics determining N_E and hence the rate of loss of genetic variation (Caballero, 1994). Likewise, environmental factors and changes thereof are mirrored in the genetic composition of affected populations. Moreover, recent work points to the impact of altered environmental variability on the variation of vital rates, which in turn obviously affects the demography and therefore N_E (see Boyce et al., 2006 and references therein). The effective population size is therefore related to the temporal variability of the population, which is a fundamental property of the ecological system. Theoretical studies have established that both statistical and biological mechanisms have the potential to influence the temporal variability of populations (Tilman, 1999). Statistical averaging and mean variance rescaling are predominantly statistical mechanisms, while species interactions and contrasting responses of different species to environmental fluctuations are primarily biological mechanisms. These mechanisms may very well be interdependent, and some have both statistical and biological elements (Tilman, 1999).

Most ecologists are familiar with the general propensity of the variance (δ^2) to increase with the mean (μ) which is why ecological data are often log-transformed prior to statistical analysis. For populations experiencing constant per capita environmental variability, the regression of log δ^2 versus log μ gives a line with a slope of 2 and this positive relationship between δ^2 and μ can be described in terms of Taylor's power relation (Taylor, 1961),

$$\delta^2 = K\bar{\mu}^{\beta} \quad (1)$$

where, K is a constant, and β is the scaling coefficient, which here is equal to 2. Larger values of β indicate that the variance increases more rapidly with μ than expected. Values of $\beta > 2$ are not uncommon, and several authors have suggested that β may lie anywhere in the range of 0.6 to 2.8 (Taylor and Woiwod, 1982). Taylor and Woiwod (1980) estimated β for 97 aphid species and for 31 of these species β was found to be above 2.

Several authors showed that environmental stochasticity (σ^2_e) can lead to a substantial extinction risk also for large populations, not merely small ones, and especially so if the population growth rate is low (Lande, 1993; Foley, 1994).

Mean time to extinction is a function of the carrying capacity (K_c) raised to the power of $(K_c)^{\omega}$ where

$$\omega = \frac{2r}{\sigma^2 e} - 1,$$

and where σ^2_e is the environmental variance due to environmental stochasticity, which is the most instantaneous effect on the risk of extinction; and r is the mean growth rate of the population, which is affecting the long-term persistence of populations, (Saltz et al., 2005). There is at the present a general accord that in a stable environment the mean time to extinction of a local population grows with the carrying capacity K (Lande, 1993) whereas under adequately strong, uncorrelated environmental stochasticity, the dependence is characterised by a power law (Foley 1994). Hence, large populations should practically never go extinct for the duration of ecological timescales. The main reason for the discrepancy between this prediction and reality is that real populations are also exposed to deleterious processes other than demographic stochasticity (σ^2_d) which with small population size, is playing a large role on the probability of extinction. Although the time to extinction is expected to increase with population size, other factors influence the dynamics of populations as e.g. mechanisms of density dependence and population growth rates (Sæther & Engen, 2003). Fluctuations in population size is a factor that strongly affects the extinction risk of a population, because larger fluctuations increase the probability that one of these excursions in population size reaches zero with extinction as a result (Boyce et al., 2006; Pertoldi et al., 2008).

Intuitively, for a given average abundance, one expects the risk of extinction to increase with temporal variability; however, many studies conducted on long-term data from natural populations have found a contradictory result (Pimm, 1993). These studies use temporal variability as a direct proxy for population vulnerability, where population variability measures are calculated from time series data as standard deviations (sd), log$N$ or coefficient of variation (CV). The reasons for the discordant results obtained in these correlational studies have been the subject of a debate and the relative importance of density dependence process on population dynamics has been compared to the relative importance of environmental variability (Turchin, 1998), which is probably, the most important of stochasticity affecting population viability (Drake & Lodge, 2003). An additional complication consists of considering the effect of the colour of the environmental noises (i.e. temporal environmental autocorrelation which can be negative, positive or uncorrelated) (Ranta et al., 2008; Björklund, 2010). Theoretical studies, however, have produced conflicting results even when predicting the sign of the effect of the different kind of noises (Halley & Inchausti, 2002), depending on interactions between the environmental noise and demographic processes (Ruokolainen et al., 2009) and on the time scale at which the amplitude of environmental noise is measured (Heino et al., 2000).

Extinction risk can also be deeply influenced by the community context (Guichard, 2005) and/or spatial structure (Engen et al., 2002). Currently, it is possible to simulate the realistic and complex population dynamics and hence quantify extinction risks (Schodelbauerova et al., 2010), and the predicted extinction risk can be a more objective measure rather than many other metrics (Fujiwara, 2007). Modelling approaches for quantifying extinction, such as population viability analyses, are however often faced with so many levels of uncertainty that their utility has been questioned by some researchers (Fieberg & Ellner, 2000). Additionally, before the analyses of the more complex models, it would be natural to understand the fate of a single local population in absence of the various possible biotic interactions (Hakoyama & Iwasa, 2005).

There is a general consensus among ecologists that assuming an initial population size which is large enough for the population to avoid a rapid initial extinction, the distribution of extinction times is exponential in almost any kind of population model, including very complex individual-based models (Grimm & Wissel, 2004). Thus, in this case, the mean time to extinction is a sufficient proxy for predicting the full distribution of extinction times. As previously mentioned, population fluctuations also act to reduce HM of the population census size estimated over time (Pertoldi et al., 2007b). Pertoldi et al., (2007b, 2008) proposed a simple model to estimate the risk of extinction and population persistence based on a description of the HM, defined by the two parameters of the scaling equation (Pertoldi et al., 2007b). The risk of intercepting zero is highly dependent on the way the variance of the population size relates to its mean and Pertoldi et al., (2007b) demonstrated that the minimum population size required for a population not to go extinct can be determined by a scaling equation relating the variance to the arithmetic mean. Pertoldi et al., (2007b) showed that for values of β >2 the relation between μ and HM remains non-linear and non-monotonic as with increasing HM first increase, followed by a domain of decreasing HM with increasing. Therefore, it can be deduced from the model that for certain values of K and β a population will become extinct even if its population size is sufficiently large to restrict the impact of σ^2_d. This description allows a separation of the domains of population persistence versus those of extinction and hence allows the identification of populations on the verge of extinction. The method also presents the estimated minimum population size required for population persistence in the presence of different levels of σ^2_e. To sum up, the model shows that maximizing the population size may not always reduce the extinction risk. Additionally, increasing population size is not always equivalent to an increasing N_E, but may decrease and hence lower the adaptive potential critical to the evolutionary response to changing environments.

At the same time some factors can increase β above 2 and therefore it would be interesting in relation to the application of the following model described below: Pertoldi et al., (2007b; 2008) has shown that environmental stochasticity either increase or reduce the amplitude of the population fluctuations depending on the sign of the correlation between population size and environmental fluctuations as:

$$\sigma^2_{tot} = \sigma^2 + \sigma^2_e + 2rp(\sigma\sigma_e)\ , \tag{2}$$

where σ^2_{tot} is the variance of the population size in the presence of environmental noise, σ^2 is the variance of the population size in absence of environmental noise, σ^2_e is the environmental noise and $r(\sigma\sigma_e)$ is the covariance between the environmental noise and the population fluctuation. The covariance is given by two times the product of rp and the sd of the population size and the environmental fluctuations (σ and σ_e respectively). Hence, a negative correlation ($rp < 0$) between environmental stochasticity and population fluctuations will decrease the fluctuations of the population size, with $\sigma^2 tot < \sigma^2$, whereas in case of a positive correlation ($rp > 0$) we will observe an increase in population fluctuations in the presence of environmental stochasticity ($\sigma^2_{tot} > \sigma^2$). Clearly if σ^2_e overwhelm σ^2 of the population dynamics, it will be the main determinant of the amplitude of oscillation of σ^2_{tot} whatever the correlation between σ^2_e and σ^2 is.
Hence, a population near the carrying capacity with β near the value of 2 should be more prone to extinction, as when an environmental stochastic event is added β will become larger than 2, which means an increased risk of extinction. The fact that β depends on the density of the population, makes it quite evident that β and K should be considered when interpreting the fluctuations of a population. As demonstrated by Pertoldi et al., (2007b), extinction risk of populations can only increase with increasing population size only if the β values can reach values above 2 ($\beta > 2$). Pertoldi et al., (2007b) showed that for certain combinations of β and K ($\beta > 2$ and $K < \mu^{(2-\beta)}$, the following equation:

$$\bar{\mu} - \sqrt{K\bar{\mu}^{\beta}} = 0 \tag{3}$$

predicts the largest mean population size allowed before extinction is expected. Furthermore, Pertoldi et al., (2008) obtained after several rearrangements, the following inequality:

$$K^{\frac{1}{(2-\beta)}} < \bar{\mu} \tag{4}$$

Where μ represents the minimum viable population size necessary for the population to persist. Values of $\beta > 2$ are not uncommon, and several authors have suggested that β may lie anywhere in the range of 0.6 to 2.8 (Taylor and Woiwod, 1982). Factors increasing β above 2 could therefore be interesting in relation to the model. Some possible scenarios where it can be speculated that β values could reach values above 2 could be for example when two species interact in a predator-prey interaction, or there is a primary consumer of a resource which fluctuates with time.
Another factor potentially affecting the σ^2 is temperature fluctuations and there is currently general concurrence that global warming is affecting animal and plant populations in multiple ways (Parmesan, 2006). Different degrees of σ^2_e and their correlation with the dynamic of the fluctuations of the population can allow the population to reach values of β above 2 and change the K values. Note also that the risk of β values above 2 is increasing when the population is approaching its carrying capacity.

Given that HM is mainly dominated by the minimum value reached in a fluctuating population, it must be kept in mind that even if the environmental stochasticity does not have a constant period of fluctuation (and is not synchronised with the population fluctuations), what will be important for the determination of HM is the maximum positive value of rp and/or the minimum peak of the environmental stochasticity fluctuation reached in a given time interval. More precisely it will be the maximum rp observed when the population size values are below and when the first derivative is negative ($f^i < 0$). In this interval the population will attain its smallest value, which in turn strongly influences the HM. Without evoking the correlations between environmental noise and population fluctuations it seems rather intuitive that if the noise is positively autocorrelated (reddened), especially in the time interval where the population size is below the average of the fluctuations and f^i is negative ($f^i < 0$), the probability to attain the minimum value of the population size and consequently the minimum HM is increased and this phenomenon should be taken into account due to evidence that long-term ecological data sets demonstrate reddened spectra (Halley & Inchausti, 2002).

Temporal variability estimated using sd, logN or CV as a direct proxy for population vulnerability, could be misleading as such measures of variability only should be used if the variance scales proportionally to the square of the mean ($\beta = 2$), and we have illustrated how β often differs from 2. Consequentially, there is a call for for detecting regime shifts in the dynamic behaviour of populations as changes in the global environment begin to accelerate and it would be interesting to establish a method which allows an estimation of the importance of σ^2_e on the two parameters β and K and on how much alteration of the parameters will push the population towards the extinction threshold. There are, however several other complications associated with the preservation of biodiversity and/or genetic variability: An enduring debate in ecology has also been how the diversity affects the temporal stability of biological systems. The ecological consequences of biodiversity loss have gained growing attention over the past decade (Bangert et al., 2005; Reusch et al., 2005). Current theory suggests that diversity has divergent effects on the temporal stability of populations and communities (Tilman, 1996). Theoretical work suggests a paradoxical effect of diversity on the temporal stability of ecological systems: increasing diversity should result in decreased stability of populations, while the community stability enhances (Tilman, 1996). While empirical work corroborates that community stability tends to increase with diversity, investigations of the effect of diversity on populations have not exposed any clear patterns. This consideration, together with the observation that changes in vital rate may have opposing effects on growth rate and N_E, is of key importance, as it can produce disagreement about the optimal management strategies.

It is well known that demographic instability in a population is translated into fluctuations of N and a reduced N_E which is close to HM of the varying N values (Vucetich et al., 1997). Therefore, a management strategy with the goal of preserving biodiversity on the community level could theoretically lead to a reduction of N_E in single populations. In the same way, the attempt of increasing growth rate in a population by modifying some of the vital rates can also produce a reduction of N_E. An increase in growth rate will increase N and therefore reduce the demographic stochasticity which is related to the population's risk of extinction, but may simultaneously lead to a reduction of the genetic diversity.

Another question emerges from considering both short- and long-term adaptability in a changing environment and whether genetic variability is always beneficial. This is not always the case as for example in constant environments genetic variability in a quantitative character creates a segregational load each generation due to stabilizing selection against individuals that deviate from the optimum phenotype (Lande & Shannon, 1996). Consider a presumably ordinary situation where natural selection acting on quantitative characters favours intermediate phenotypes. In an intermediate-optimum model, the genetic variability may be either beneficial or detrimental, depending on the pattern of environmental change (the frequency, the amplitude and the degree of autocorrelation of the environmental oscillations) (Lande & Shannon, 1996, Björklund et al., 2011).

The genetic consequences of CIEC can be subdivided in two main categories, namely consequences in small populations and consequences in large populations:

In small populations, random genetic processes (genetic drift) lead to loss of genetic variability, which may depress the evolutionary potential and thus the ability to respond to changing environments (Pertoldi et al., 2006a). It is also anticipated that populations only persist if the rate of adaptive evolution at least matches the rate of environmental change since the evolutionary response of quantitative traits to selection necessitates the presence of genetic variability (Burger and Lynch, 1995). In fact, this is the case even in the presence significant capacity to respond plastically, including adaptations in behaviour, physiology, morphology, growth, life history and demography. The rate of loss of genetic variability in populations is associated to a reduction of N_E. Reduction of N_E due to amplified population fluctuations, reduce the evolutionary potential, by reducing the additive genetic variance (σ^2_a) and the heritability (h^2) of the traits, which in turn is inversely related to σ^2_e.

In large populations, the regime of alternating selective pressures has the potential to increase the average population fitness, selecting for genes implicated in the expression of plasticity. Various modelling approaches have shown that to optimize fitness, phenotypic plasticity evolves by trading the adaptation to acquire resources against the costs of maintaining the potential for plasticity (Ernande & Dieckman, 2004). Plastic responses include changes in behavior, physiology, morphology, growth, life history and demography, and can be expressed either within the lifespan of an individual or across generations (Pertoldi et al., 2005; Røgilds et al., 2005). Two ways of adapting to environmental changes are therefore possible, by evolutionary or by plastic responses, including maternal transmission (trans-generational plasticity). Hence, the survival of populations relies on genetic variation and/or phenotypic plasticity. Populations with small N_E and/or little genetic variability have mainly the option of adapting in a plastic way, therefore the importance of plasticity is quite evident (Pertoldi et al., 2007b).

6. Conclusions

6.1 Future directions

In conclusion, to test the robustness of the theoretical foundations of evolutionary and ecological genetics, three main categories of questions should be answered, using a multidisciplinary approach consisting of: (A) experimental population genetics, (B) collection and analysis of empirical data, and (C) computational population genetics combined with ecological information as for example life history characteristics of the study organism.

6.1.1 Experimental population genetics

Numerous experiments should be conducted including different model organisms. Many of these experiments should be based on innovative methods accounting for the experimental errors due to unpredictable environmental components (for applications see Kristensen et al., 2004). Clonally reproducing strains should be used to study the extent of adaptive phenotypic plasticity, and maternal effects, including the effect of parental ageing. The use of clonal strains will allow us to exclude the genetic components and their interactions with the environment. Therefore, unbiased estimates of genetic and environmental canalization, plasticity, developmental homeostasis and σ^2_e, will be obtained. A more correct interpretation of the interplay between these parameters will provide important contributions to: 1) The evolutionary importance of phenotypic plasticity, maternal effects, environmental and genetic stressors. 2) The consequences of outbreeding on population fitness and phenotypic plasticity, and 3) The selective effects of fluctuating selective regimes on plasticity genes.

In order to investigate in which way environments fluctuating with different intervals can affect the mean population average fitness, and to quantify the costs and benefits of genetic variability in fluctuating environments, sexually reproducing strains ought to be utilised, creating fluctuating temperature environments and making truncated selection experiments in which the extreme phenotypes at the two tails of the phenotypic distribution are selected away. Important information could in this way be obtained about the extent of the environmental information that will be transmitted to the offspring, and to what extent it can enlarge the plastic response of a trait when selecting for plasticity genes.

6.1.2 Collection and analysis of empirical data

Molecular and quantitative genetics studies should be conducted on several species with different ecological characteristics and with different demographic history, such as recent and ancient population decline or expansion. Changes in population size and range are frequent consequences of CIEC, and examples include habitat fragmentation and rapid colonization or recolonization processes. Extensive collections of several species provide the opportunity to analyse large numbers of samples on a temporal scale and directly document changes in genetic diversity. The results of these analyses will improve our understanding of the historical dimension of population change, and provide important data for the interpretation of genetic diversity studies in an ecological and evolutionary context. The possibility of amplifying ancient DNA from old museum specimen (Pertoldi et al., 2005b), should also be used. Furthermore, a phylogeographic approach should be carried out. The innovative aspect of this approach consists of the fact that different molecular and quantitative genetics techniques should be employed simultaneously.

To document the range of genetically based morphological variation within and among populations, a comparison of the degree of quantitative genetics distance (QST) with neutral genetic distance (FST) should be made (Mckay & Latta, 2002). Comparisons of morphometrical (for example, size and shape) and life-history variability (for example, longevity and fecundity) of populations and their crosses with molecular variability (using microsatellites) could present important information about the influences of environmental and genetic components in a non-genetic-equilibrium situation. Furthermore, it will provide important information about the extent to which crosses between different strains affect the various components of σ^2_P (plasticity, developmental homeostasis, canalization and σ^2_e).

The combination of ecological models of the distribution of the species investigated with both mitochondrial DNA (mtDNA) data and synthetic genetic maps constructed from multivariate analysis of microsatellites and morphometric data will allow us to discuss hypothesized historical biogeographic scenarios. By directly dating and quantifying changes in genetic diversity, these investigations will allow examination of postulated causes of population decline, including habitat loss and temperature increase.

The genetic data obtained from the investigations mentioned above provides the information on postglacial history as well as on current demographic threats (fragmentation, relict populations, marginal populations; levels of inbreeding). The results of this approach will provide important information about: 1) The genetic consequences of population fragmentation and rapid recolonization caused by climate change. 2) The extent of the genetic diversity of modern populations compared to that in the late Pleistocenic environment, and 3) Pattern of species recolonization in Europe and their response to environmental change after the last glaciation.

6.1.3 Computational population genetics

Stochastic simulation tools, based on a quantitative infinitesimal model, where the size of N_E can be varied, should be developed. In the simulation models, each phenotype should be considered to be the sum of independent genetic components (σ^2_a, dominance and epistasis) and σ^2_e, and the σ^2_p in a population should be described as the sum of independent variances for each of these effects. Several questions could be answered:

1. Understanding how different environmental scenarios can affect both genetic and demographic parameters.
2. Understanding how much difference in life history between ecologically similar species can cause substantial differences in N_E and σ^2_a, and to what degree fluctuations in vital rate parameters induced by environmental change can alter N_E,
3. Quantifying the interactions of each particular life history parameter with other factors (sensitivity analysis), and
4. Quantifying the effects and the interactions that factors such as N_E, inbreeding, gametic phase disequilibrium, plasticity, and developmental homeostasis have on the speed at which a population can react to a selective pressure.

Given that the information obtained from the computational approach can be combined with empirical data, obtained from approaches 6.1.1 and 6.1.2 the model will be a powerful tool for understanding complex dynamics and to make predictions concerning the possible effects of CIEC and their interactions with other factors.

6.2 Expected yield of the multidisciplinary approach

The establishment of such an approach which integrate experimental, theoretical and applied ecological and evolutionary genetics, will create synergistic effects and contribute to the understanding of the consequences of CIEC and the questions addressed will provide important contributions to general ecology and conservation genetics as there is a requirement for detailed studies on how variation at the level of genes translates through developmental and physiological processes, into phenotypic variation for ecologically important traits. Further scientific progress will be achieved by merging and complementing recent efforts in evolutionary and ecological genetics by: 1) Collecting informative genetic and environmental data sets in natural populations and from preserved specimens, 2) Merging taxonomic, ecological and genetic databases, 3) Using molecular data in combination with quantitative traits and environmental data, and **4)** Unravelling the distribution of variation at functional vs. non-coding sequences in natural populations.

7. References

Andersen H. D.; Pertoldi C.; Scali V. & Loeschcke V. (2002) Intraspecific Hybridisation, Developmental Stability and Fitness in *Drosophila mercatorum*. Evolutionary Ecology Research 4: 603-621.

Angilletta, M. J. (2009). Thermal adaptation: a theoretical and empirical synthesis. Oxford University Press, Oxford.

Avise, J. C. (1998) The history and purview of phylogeography: a personal reflection. Molecular Ecology 7: 371-379.

Bach, L. A.; Thomsen, R.; Pertoldi C. & Loeschcke V. (2006). Evolution of density-dependent dispersal in an Individual-based adaptative metapopulation model including explicit kin competition and demographic stochasticity. Ecological modelling 192: 658-666.

Bach, L. A.; Ripa J. & Lundberg P. (2007). On the evolution of conditional dispersal under environmental and demographic stochasticity. *Evolutionary Ecology Research* 9(4): 663-673.

Bangert, R.; Turek R. J.; Martinsen G. D.; Wimp G.M.: Bailey J. K. & Whitham T. G. (2005). Benefits of conservation of plant genetic diversity to arthropod diversity. *Cons. Biol*, 19: 379-390.

Boyce, M. S.; Haridas, C. V.; Lee C. T. & NCEAS Stochastic Demography Working Group. (2006). Demography in an increasing variable world. - *Trends Ecol. Evol*. 21: 141-148.

Björklund, M.; Ranta E.; Kaitala, V.; Bach, L. A.; Lundberg, P. & Stenseth, N. C. (2009). Quantitative trait evolution and environmental change. *PLos One* 4(2), (e4521).

Björklund, M.; Ranta E.; Kaitala V.; Bach L. A. & Lundberg P. (2011). Environmental Fluctuations and Level of Density-Compensation Strongly Affects the Probability of Fixation and Fixation Times. *Bulletin of Mathematical Biology*. 73(7): 1666-1681.

Burger, R. & Lynch, M. (1995). Evolution and extinction in a changing environment: a quantitative-genetic analysis. Evolution 49:151-163

Caballero, A. (1994). Developments in the prediction of effective population size. - *Heredity* 73: 657-679.

Chen, I. C.; Shiu H.; Benedick S.; Holloway J. D.; Chey V. K.; Barlow H. S.; Hill J. K. & Thomas C. D. (2009). Elevation increases in moth assemblages over 42 years on a tropical mountain. *Proc Natnl Acad Sci USA* 106:1479-1483

Coulson, T. et al. (2006). Estimating individual contributions to population growth: evolutionary fitness in ecological time. *Proceedings of the Royal Society B: Biological Sciences* 273(1586): 547-555.

Dillon, M. E.; Wang. G. & Huey R. B. (2010). "Global metabolic impacts of recent climate warming." *Nature* 467(7316): 704-706.

DeAngelis, D. L. & W. M. Mooij. (2005). Individual-based modeling of ecological and evolutionary processes. Annual *Reviews of Ecology and Evolutionary Systematics* 36:147-168.

Drake, J. M. & Lodge D. M. (2003). Effects of environmental variability on extinction and establishment. *Ecology Letters* , 7: 26–30.

Engen, S.; Lande R.; Saether B.-E. & Gienapp P. (2010). Estimating the ratio of effective to actual size of an age-structured population from individual demographic data. *Journal of Evolutionary Biology* 23(6): 1148-1158.

Ernande, B. & Dieckmann, U. (2004). The evolution of phenotypic plasticity in spatially structured environments: Implications of intraspecific competition, plasticity costs, and environmental characteristics. *Journal of Evolutionary Biology* 17: 613–628

Fieberg, J. & Ellner, S. P. (2000). When is it meaningful to estimate an extinction probability?, *Ecology* 81: 2040–2047.

Foley, P. (1994). Predicting extinction times from environmental stochasticity and carrying-capacity, *Conserv. Biol.* 8: 124–137.

Fujiwara M. (2007). Extinction-effective population index: Incorporating life-history variations in population viability analysis, *Ecology* 88: 2345–2353.

Gandhi Nilima et al. (2011). Implications of geographic variability on Comparative Toxicity Potentials of Cu, Ni and Zn in freshwaters of Canadian ecoregions. *Chemosphere* 82(2): 268-277.

Gilpin M. E. & Soulé M. E. (1986). Minimum viable populations: processes of species extinction. - In: Soulé, M. E. (ed). *Conservation biology: the science of scarcity and diversiy*. Sinauer Associates, Inc., Sunderland, Massachusetts, pp. 19-34.

Grimm, V. & Wissel C. (2004). The intrinsic mean time to extinction: a unifying approach to analysing persistence and viability of populations, *Oikos* 105: 501–511.

Guichard, F. (2005). Interaction strength and extinction risk in a metacommunity, *Proc. R. Soc. Lond., Ser. B: Biol. Sci.* 272: 1571–1576.

Hakoyama, H. & Iwasa, Y (2005). Extinction risk of a meta-population: aggregation approach, *J. Theor. Biol.* 232: 203–216.

Halley, J. M. & Inchausti P. (2004). The increasing importance of 1/f-noises as models of ecological variability, *Fluct. Noise Lett.* 4 (2004), pp. R1–R26.

Heino, M. *et al.*, (2000). Extinction risk under coloured environmental noise, *Ecography* 23: 177–184.

Jeschkem J. M. & Strayerm D. L. (2008). Usefulness of bioclimatic models for studying climate change and invasive species. *Ann NY Acad Sci* 1134:1-24

Kalinowski, S. T. & Waples, R. S. (2002). Relation of effective to census size in fluctuating populations. - *Conserv. Biol.* 16: 129-136.

Kearney M, Phillips B. L. Tracy C. R. Christian K. A. Betts G. Porter W. P. (2008). Modelling species distributions without species distributions: the cane toad in Australia under current and future climates. *Ecography* 31:423-434

Keith, D. A.; Akçakaya H. R.; Thuiller, W.; Midgley G. F.; Pearson R. G.; Phillips S. J.; Regan H. M.; Araújo M. B. & Rebelo T. G. (2008). Predicting extinction risks under climate change: coupling stochastic population models with dynamic bioclimatic habitat models. *Biol Lett* 4:560-563

Knudsen, E.; Linden, A.; Both, C.; Jonzen, N.; Pulido, F.; Saino, N.; Sutherland, W. J.; Bach, L. A.; Coppack, T.; Eron, T.; Gienapp, P.; Gill, J. A.; Gordo, O.; Hendenström, A.; Lehikoinen, E.; Marra, P. P.; Møller, A. P.; Nilsson, A. L. K.; Peron, G.; Ranta, E.; Rubolini, D.; Spartks T. H. Spina F. Studds C. E. Sæther S. A. Tryjanowski P. & Stenseth N. C. (2011). Challenging claims in the study of migratory birds and climate change. *Biological Reviews* (86(4): 928-946).

Kristensen T. N. Pertoldi C. Pedersen L. D. Andersen D. H. Bach L. A. & Loeschcke V. (2004) The increase of fluctuating asymmetry in a monoclonal strain of collembolans after chemical exposure--discussing a new method for estimating the environmental variance. Ecological Indicators 4: 73-81.

Lande, R. & Barrowclough G. F. (1987). Viable populations for conservation. - In: Soulé, M. E. (ed). *Effective population size, genetic variation and their use in population management*. Cambridge University Press, Cambridge, pp. 87-123.

Lande, R. & Barrowclough G. F. (1987). Viable populations for conservation. - In: Soulé, M. E. (ed). *Effective population size, genetic variation and their use in population management*. Cambridge University Press, Cambridge, pp. 87-123.

Lande, R, & S Shannon. (1996). "The Role of Genetic Variation in Adaptation and Population Persistence in a Changing Environment." *Evolution* 50(1): 434-437.

Luikart, Gordon, Phillip R England, David Tallmon, Steve Jordan, & Pierre Taberlet. (2003). "The power and promise of population genomics: from genotyping to genome typing." *Nature Reviews. Genetics* 4(12): 981-994.

Lynch, M. (1996). A quantitative genetic perspective on conservation issues. - In: Avise, J. C. & Hamrick, J. C. (eds). *Conservation Genetics: Case Histories from Nature* Chapman & Hall, New York, pp. 457-501.

Marsano, Francesco et al. (2010). "Effects of Mercury on Dictyostelium discoideum: Proteomics Reveals the Molecular Mechanisms of Physiological Adaptation and Toxicity." *Journal of Proteome Research* 9(6): 2839-2854.

Mckay, J. K. & Latta, R. G. (2002). Adaptive population divergence: markers, QTL and traits. *Trends in Ecology and Evolution*. 17: 285-291.

Mitchell WA, Angilletta MJ Jr (2009) Thermal games: frequency-dependent models of thermal adaptation. *Funct Ecol* 23:510-520

Mitchell-Olds, T, M Feder, & G Wray. (2008). "Evolutionary and ecological functional genomics." *Heredity* 100(2): 101-102.

Morin, Phillip, Gordon Luikart, Robert Wayne, and The. (2004). "SNPs in ecology, evolution and conservation." *Trends in Ecology & Evolution* 19(4): 208-216.

Nunney, L. (2000). The limits to knowledge in conservation genetics. The value of effective population size. - In: Clegg et al. (eds.). *Evolutionary Biology 32, Kluver Academic/Plenum Publishers*, pp. 424-450.

Parmesan, C. (2006) Ecological and evolutionary responses to recent climate change. *Annual Review of Ecology, Evolution and Systematics*, 37, 637-669.

Pertoldi, C. Bach, L. A. Madsen, A.B. Randi E & Loeschcke, V. (2003). Developmental instability of the Eurasian badger Meles meles in Danish populations. *Journal of Biogeography* 30(6): 949-958.

Pertoldi, C. Andersen D. H. Røgilds A. & Loeschcke V. (2005) Heat induced maternal effects in *Drosophila mercatorum* and its evolutionary consequences. *Evolutionary Ecology Research* 7: 203-217.

Pertoldi, C. Givskov J. S. David J. R. & Loeschcke V. (2006a). Lerner's theory on the genetic relationship between heterozygosity, genomic coadaptation and developmental instability, *Evolutionary Ecology Research*, 8: 1487-1498.

Pertoldi, C. Kristensen, T. N. Andersen, D. H. & Loeschcke, V. (2006b). Review: Developmental Instability as an estimator of genetic stress. *Heredity* 96: 122-127.

Pertoldi C. García-Perea R. Godoy A. Delibes M. & Loeschcke V. (2006c).. Morphological consequences of range fragmentation and population decline on the endangered Iberian lynx (*Lynx pardinus*). *Journal of Zoology* 268: 73-86.

Pertoldi, C. & Bach, L. (2007a). Evolutionary aspects of climate induced changes and the need for multidisciplinarity *Journal of Thermal Biology*, 32: 118-124.

Pertoldi, C.; Bach, L. A.; Barker, J. S. F.; Lundberg, P. & Loeschcke, V. (2007b). The consequences of the variance-mean rescaling effect on effective population size. *OIKOS*, 5: 769-774.

Pertoldi C. Bach L. A. & Loeschcke V. (2008). On the brink between extinction and persistence. *Biology Direct*, 3: 47

Plejdrup J. K. Simonsen V. Pertoldi C. Schøyen M. & Bayley M. (2006). Genetic and morphological diversity in populations of Nucella lapillus (L., Neogastropoda) in response to tribultyl. *Ecotoxicology and Environmental Safety* 64: 146-154.

Randi E. Davoli F. Pierpaoli M. Pertoldi C. Madsen A. B. & Loeschcke V. (2003). Genetic structure in European otter (*Lutra lutra*) populations: Implications for conservation. *Animal Conservation* 6: 1–10.

Ranta, E., Kaitala, V., Björklund, M., Lundberg, P., Bach, L.A. & Stenseth N. Chr. (2008). Environmental Forcing and Genetic Differentiation in Subdivided Populations. *Evolutionary Ecology Research* 10(1): 1-9.

Reusch, T. B. H., Ehlers, A., Hämmerli, A. & Worm, B. (2005): Ecosystem recovery after climatic extremes enhanced by genotypic diversity. *Proc. Natl. Acad. Sci. USA*. 102: 2826-2831.

Ricklefs, R.E. (2008). Disintegration of the ecological community. *Am Nat* 172:741-750

Riebler, A., Held, L & Stephan, W. (2008). "Bayesian Variable Selection for Detecting Adaptive Genomic Differences Among Populations." *Genetics* 178(3): 1817-1829.

Ruokolainen L. *et al.*, (2009). Ecological and evolutionary dynamics under coloured environmental variation, *Trends Ecol. Evol.* 24: 555–563.

Røgilds, A., Andersen, D. H., Pertoldi, C., Dimitrov, K. & Loeschcke, V. (2005) Maternal and grandmaternal age effects on developmental instability and wing size in parthenogenetic *Drosophila mercatorum*. Biogerontology 6: 1-9.

Sæther BE, Engen S (2003) Routes to extinction. In *T. Blackburn & K. Gaston, editors. Macroecology (Blackwell Publishing, Oxford, UK)*. 218–236.

Saltz, D., Rubenstein, D. I. & White G. C. (2005). The impact of increased environmental stochasticity due to climate change on the dynamics of Asiatic wild ass. - *Cons. Biol.* 20: 1402-1409.

Schwartz, M. Luikart, G. & Waples, R. (2007). "Genetic monitoring as a promising tool for conservation and management." *Trends in Ecology & Evolution* 22(1): 25-33.

Schwenk K, Padilla, D.K, Bakken, G.S, Full, R.J. (2009). Grand challenges in organismal biology. *Integr. Comp. Biol.* 49:7-14

Schodelbauerova I *et al.*, (2010) Prediction vs. reality: Can a PVA model predict population persistence 13 years later?, *Biodivers. Conserv.* 19: 637–650.

Smith, J.B., Schneider, S.M., Oppenheimer, M., Yohe, G.W., Hare, W., Mastrandrea, M.D., Patwardhan, A, Burton I, Corfee-Morlot J, Magadza, CHD, Füssel, H-M, Pittock, A.B., Rahman, A, Suarez, A, van Ypersele, J-P (2009). Assessing dangerous climate change through an update of the Intergovernmental Panel on Climate Change (IPCC) "reasons" for concern. *Proc. Natnl Acad Sci USA* 106:4133-4137

Spielman, D., Brook, B. W. & Frankham, R. (2004). Most species are not driven to extinction before genetic factors impact them. - *Proc. Natl. Acad. Sci. USA* 101: 15261-15264.

Steltzer, H, Pos,t E (2009). Seasons and life cycles. *Science* 324:886-887

Stenseth, N.C., Mysterud, A., Ottersen, G., Hurrell, J.W., Chan, K-S. & Lima M. (2002). Ecological effects of climate fluctuations. *Science* 297:1292-1296

Stephens, M., & Balding, D. J. (2009). Bayesian statistical methods for genetic association studies. *Nature Reviews Genetics* 10(10): 681-690.

Strand, A. E. (2002). metasim 1.0: an individual-based environment for simulating population genetics of complex population dynamics. *Molecular Ecology Notes* 2: 373–376.

Taylor, L. R. (1961). Aggregation, variance and the mean. *Nature* 189: 732 735.

Taylor, L. R. & Woiwod I. P. (1980). Temporal stability as density-dependent species characteristic. - *J. Anim.Ecol.* 49: 209-924.

Taylor, L. R. & Woiwod, I. P. (1982). Comparative synoptic dynamics. 1. Relationships between interspecific and interspecific spatial and temporal variance mean population parameters. - *J. Anim.Ecol.* 51: 879-906.

Thomas C. D. Cameron A. Green RE, Bakkenes M, Beaumont LJ, Collingham YC, Erasmus BFN, de Siqueira MF, Grainger A, Hannah L, Hughes L, Huntley B, van Jaarsveld AS, Midgley GF, Miles L, Ortega-Huerta MA, Peterson AT, Phillips OL, Williams SE (2004). Extinction risk from climate change. *Nature* 427:145-148

Tilman, D. (1999). The ecological consequences of changes in biodiversity: a search for general principles. *Ecology* 80: 1455-1474.

Travis, J. M. J., Mustin, K., Benton, T. G. & Dytham, C. (2009). Accelerating invasion rates result from the evolution of density-dependent dispersal *Journal of Theoretical Biology* 259(1): 151-158.

Turchi,n P. (1999). Population regulation: a synthetic view. *Oikos.* 84: 153–159.

Vermeij, G. J. (1996). Adaptation of clades: resistance and response. - In: (Rose, M. R. & Lauder, G. V. (eds.). *Adaptation.* Academic Press Inc, San Diego, California, USA. pp. 363-380.

Verrier E. Colleau, J. J. & Folley J. L. (1990). Predicting cumulative response to directional selection in finite panmictic populations. *Theoretical Applied Genetics* 79: 833–840.

Vucetich, J. A. Waite T. A. & Nunney L. (1997). Fluctuating population size and the ratio of effective to census population size (N_e/N). *Evolution* 51: 2017-2021.

Walther G-R. Post E. Convey P. Menzel A. Parmesan C. Trevor J. Beebee C. Fromentin, J. M. Hoegh-Guldberg O. & Bairlein F. (2002). Ecological responses to recent climate change. *Nature* 416: 389-395.

Wang. J. (1996). Deviation from Hardy-Weinberg proportions in finite populations. Genetical Research Cambridge 68: 249–257.

Wright. S. (1931). Evolution in Mendelian populations. - *Genetics* 16: 97-159.

Part 2

Historical Environmental Change

6

How Did Past Environmental Change Affect Carnivore Diversity and Home-Range-Size in Spain?

María T. Alberdi[1], José L. Prado[2],
Esperanza Cerdeño[3] and Beatriz Azanza[4]
[1]*Departamento de Paleobiología, Museo Nacional de Ciencias Naturales, CSIC, Madrid,*
[2]*INCUAPA, Departamento de Arqueología, F.C.S., Universidad Nacional del Centro, Olavarría,*
[3]*Departamento de Paleontología, IANIGLA, Centro Científico Tecnológico CONICET, Mendoza,*
[4]*Departamento de Ciencias de la Tierra, Universidad de Zaragoza, Zaragoza,*
[1,3]*Spain*
[2,4]*Argentina*

1. Introduction

An important aspect of evolutionary biology is to know the faunal diversity changes through time. There is a general agreement that the knowledge of evolutionary patterns through time from an ecological perspective provides information that is not available from ecological studies on extant faunas (Vrba, 1985, 1995a; Behrensmeyer et al., 1992; Jablonski, 2005). Iberian Cenozoic basins provide an exceptional record of fossil mammals and continental environments, giving a great opportunity to evaluate the ecological and evolutionary responses of mammalian communities to climatic changes through the last millions years (Azanza et al., 1999, 2000; Hernández Fernández et al., 2007; Van Dam et al., 2006). This knowledge is essential for linking the dynamics of biotic change from ecological to evolutionary time scales and for understanding the processes that transform ecosystems over geologic times (Badgley et al., 2008).
Previous works have proposed different explanations for biodiversity changes through time, emphasizing the influence of both physic and biotic factors (e.g., Van Valen, 1973; Janis, 1989; Stucky, 1990; Vrba, 1995a, 1995b, 2000; Prothero, 1999, 2004; Alroy, 2000; Alroy et al., 2000; Barnosky, 2001, 2005; Vrba & DeGusta, 2004). The response of mammals throughout the late Cenozoic has been often reflected by migrations or variation of their area of distribution, related to the vegetation cover and latitudinal displacement of biomes. Patterns of change in the home range size (HR, the size of the minimum area that can sustain the individual's energetic requirements) through time can provide important insights into the ecological and evolutionary responses of mammalian communities to new environmental conditions. In a lesser degree, mammals can also respond evolving into new species. These events could modify the structure of mammalian communities, triggering new internal

dynamics. In this sense, different authors such as Kolfschoten (1995) and Koenigswald (2002, 2003) have postulated an alternation of "temperate faunas" and "cold faunas" in interglacial and glacial periods, respectively, in Central Europe; Spassov (2003) for Eastern Europe; and Vrba (1995a), Turner (1990, 1999), Behrensmeyer et al. (1997), McKee (2001) and Werdelin & Lewis (2005) have discussed the African faunal turnovers.

Among biotic factors that may alter the biodiversity, intra- and interspecific competition and niche occupancy seem to be related to the pre-existing community structure, while prey/predator interactions may differ from area to area (Abrams, 2000; Chesson, 2000). It seems reasonable to assume that several factors contributed in different ways to modify mammalian guilds. In the case of carnivores, the dynamics might depend more on prey-predator relationships and to the availability of prey than on climatic factors (Carbone & Gittleman, 2002). Consequently, one could assume that the diversity of carnivores would be less related to climatic-environmental conditions than that of herbivores.

A previous paper (Palombo et al., 2009) has tried to evaluate the progressive changes in the composition of carnivore guilds in north-western Mediterranean area as a direct response to climatic changes. As in other Mediterranean areas, Iberian Peninsula was influenced by the Plio-Pleistocene climatic oscillations that provoked migrations of taxa from East and Central Europe, as well as from Africa, across the Levantine corridor (Alberdi et al., 1997; Azanza et al., 1997, 2000, 2004; Palombo & Valli, 2005; Palombo, 2004, 2007a, 2007b). The present chapter summarizes an updated study on the relationships of biodiversity and HR size variation with climatic changes for the Spanish Plio-Pleistocene Carnivora.

2. Material and methods

Most data for the present study are based on the selected and taxonomically updated faunal lists of large predatory mammals that have been compiled from previous analyses on the Plio-Pleistocene mammals of the North-Western Mediterranean region (Alberdi et al., 1997; Azanza et al., 1999, 2000; Palombo et al., 2009), herein restricted to the Spanish area (Figure 1). The considered biochronological scale has followed discrete time intervals following the mammal ages broadly accepted for western Europe, the European Mammal Neogene System, or MN "zones" (Mein, 1975, 1990; Agustí et al., 2001; van Dam, 2001), which mainly concern Miocene-Pliocene times. In turn, four MmQ "zones" have been differentiated for the Spanish Early-Middle Pleistocene mammalian faunas (i.e., Agustí et al., 1987). From this general scheme, we have adapted thirteen biochronological units (BU; Figure 2), established by multivariate procedures (Azanza et al., 1999, 2004) and defined as "lapses of time during which faunas have certain taxonomic homogeneity, the discontinuity between them corresponding to faunal reconfigurations associated with major changes in environmental conditions". Some localities count with absolute dating that allow estimating the duration of each unit (Palombo, 2007a; Palombo et al., 2009).

The comparison of carnivoran guilds implies some difficulties related to the usual lesser representation of this group with respect to other mammals, for instance herbivore ungulates, with the subsequent underestimation of their taxonomic diversity. In any case, the whole sample of a site is always a partial reflect of the actual diversity of the community that once lived there. Trying to avoid this bias, we consider faunal complexes gathering several local faunal assemblages, and assume that these complexes allow checking the main faunal turnovers through time (Palombo et al., 2009).

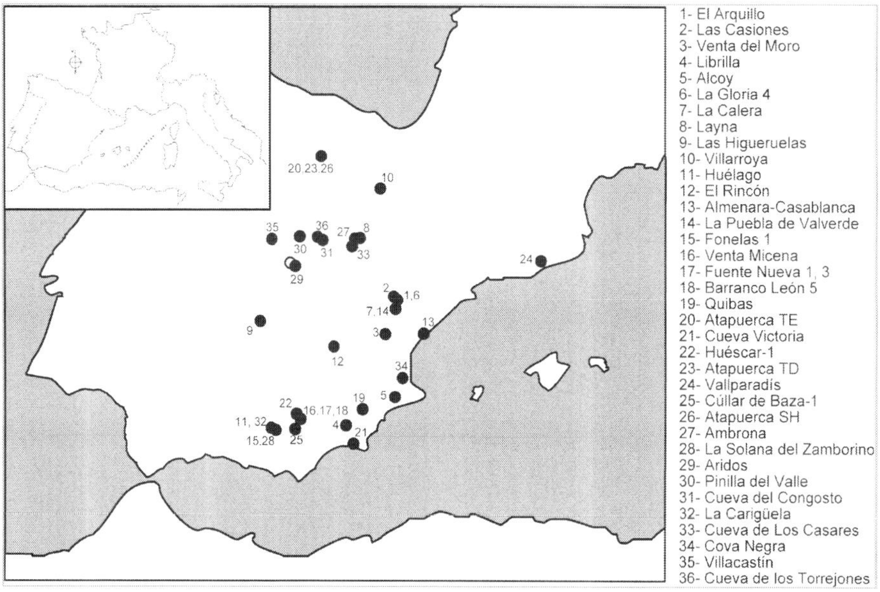

Fig. 1. Geographical distribution of the selected localities from Spain.

Van Valkenburgh (1985, 1988, 1989) confined large terrestrial carnivore guilds to predators weighting 7 kg or more (jackal size and larger), proposing this size threshold because of the evidence in extant species for strong competitive interactions among carnivores larger than 7 kg, being weaker among smaller carnivores; in addition, the representation of large carnivores in the fossil record is better. On the other hand, because larger predators achieve a higher net gain rate by concentrating on large prey, Carbone et al. (2007) predicted the threshold of 14.5 kg, where predators switch from small to large prey. We follow this criterion, and restrict our study to species of the three families that include top terrestrial predators: Canidae, Felidae and Hyaenidae. Ursidae are excluded due to their omnivorous feeding behaviour that rarely includes meat.

For each species, we compiled data of its presence in fossil sites, body-size, diet and preferred habitat. We used a taxon-free characterization by means of two ecological criteria of classification. According to the feeding behavior, species were classified in two trophic categories: (1) Carnivore (C): hypercarnivores with a diet that consists of 70% or more flesh meat, bone-eaters, bone-crushers and scavenging bone-crackers; and (2) Carnivore-Omnivore (OM): including flesh-eaters (with less than 10% flesh in their diet), taxa feeding on invertebrates, and occasionally on fruit. Concerning the preferred habitat, three major ecological categories were considered: (1) Forest dwellers (FH): taxa inhabiting forest, closed woodland, bushland, Mediterranean "macchia", open woodland, and miscellaneous woodland; (2) Ubiquitous (MXH): including more flexible taxa, inhabitants of shrubland or woodland, as well as open landscape, or at the edge of both; and (3) Open landscape dwellers (OH): including taxa inhabiting grassland, steppe or savanna (Figure 2).

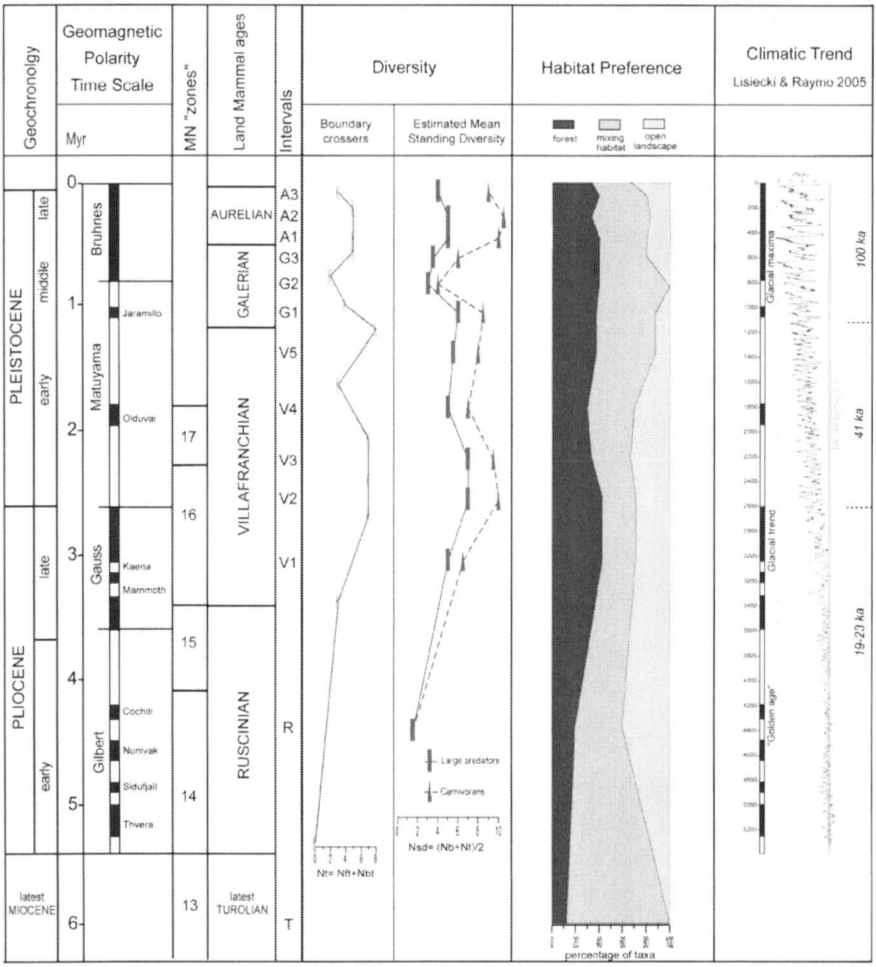

Fig. 2. Biochronological framework. Carnivoran standing diversity, habitat preferences over the Plio-Pleistocene and climatic trend.

The HR-size of fossil species was estimated using the regression for each family, habitat preference and trophic categories. The relationship between HR-size and body size (or body weight, BW) has been stated by McNab (1963) using the allometric equation HR = a BW^b. This author considered two categories: hunters and croppers, the former having a greater allometric coefficient (a), a fact he attributed to the relatively low density of their preferred food items. Posterior studies, such us that by Gittleman & Harvey (1982), have established that carnivorous species had larger HR-size than herbivorous species of similar body mass, and that the distribution of food resources is more homogeneous for herbivores than for carnivores. However, other studies (Kelt & Van Vuren, 1999) defend that scaling relations of

HR-size are not statistically different between carnivorous and herbivorous or omnivorous species. Among terrestrial mammals, carnivore guilds have the broadest range of HR-sizes. Some of the variation in the HR-size of Carnivora can be explained by diet.

Paleobiodiversity refers to the richness of species, measured as the total number of taxa in a time interval. To avoid over estimation of taxa, we use the "Estimated mean standing Diversity" (N_{sd}), following the methodology developed by Foote (2000). The N_{sd} has been calculated for each time interval using both the initial data base and the data base resulting from the taxon-free characterization (following Palombo et al., 2009).

The quality of the fossil data is crucial for validating results and conclusions, because the fluctuating frequencies of coexisting species may reflect actual changes in diversity or may depend simply on sampling bias. Incomplete preservation is usually estimates by the proportion of range-through taxa (or Lazarus taxa) supposing that their existence is mainly a consequence of deficient sampling. Sampling adequacy was explored using the completeness indices ($CI = [(N_{tot} - N_{rt})/N_{tot}] \times 100$ and $CI_{bda} = (N_{bda}/N_{bt}) \times 100$) proposed by Maas et al. (1995).

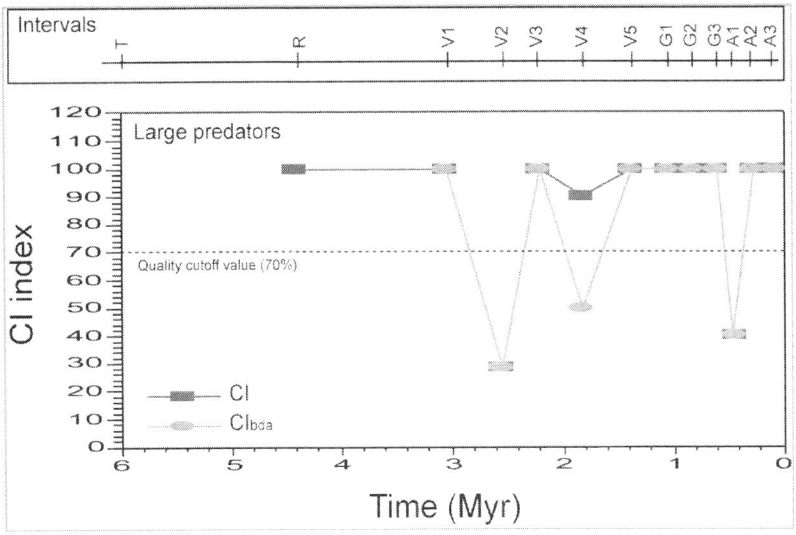

Fig. 3. Quality assessment of the carnivoran data from Spain.

The values of the completeness indices are affected by BU duration, because increasing the length of a time span means that it can potentially include a more extensive fossil record and also reduces the number of the Lazarus taxa. We use as cut-off threshold the value of 0.14 that results of dividing the minimum acceptable index value of 70% by the canonical span time (500 ka) suggested by Maas et al. (1995).

3. Results

Figure 3 shows the values of completeness indices for the data of large predators. The more conservative CI_{bda} index has a value below 70% in three BUs: V2, V4 and A1, whereas if the more generous CI index is used, the quality of the record is only deficient in two (V2 and A1 BUs), indicating that data from these BUs should be treated circumspectly (Maas et al., 1995).

The diversity trends (N_t and N_{sd}) show three "increase plus decrease" pulses, with peaks at the Plio-Pleistocene boundary (V2 to V3 BUs), at the beginning of the Middle Pleistocene (G1 BU) and at the Middle-Late Pleistocene boundary (A1 to A2 BUs). This general pattern is also observed in Italy and France, but in non-synchronous pulses (Palombo et al., 2009). Despite these fluctuations, diversity increases during the Plio-Pleistocene, but large predators show a more moderate increase than other carnivores. The lowest diversity outstandingly corresponds to the Early Pliocene (R BU) that is, by far, the longest BU. This can be related to the most relevant extinction event in Iberian Peninsula, occurred at the Mio-Pliocene transition, when all the Turolian large predators became extinct. These trends roughly correlate with the general global climate signal (Lisiecki & Raymo, 2005, see Figure 2). The first peak is coincident with the onset of Northern Hemisphere Glaciations and the following decrease in diversity began when glacial-interglacial cycles of 41 Ka were clearly installed. The second peak coincides with the intensification of glacial cycles that change from 41 to 100 ka periodicity. The third maximum, during the Middle Pleistocene, correlates with a hot period (Holstein stage) between Mindel and Riss glaciations. The first and third peaks are herein considered to be more reliable, because the diversity of the second one (V2 and A1 Bus) could be under estimated (Palombo et al., 2009).

The HR size quantifies the animal's inherent ability to move. Thus, dispersal distance and HR-size co-vary across mammal species when considered independently of body size (Bowman et al., 2002). Figure 4 documents the patterns of HR-size change through time. Reference lines represent the predicted threshold of 14.5 kg where predators switch from small to large prey (Carbone et al., 2007). In general, the HR-size for Canidae decreases during the Pliocene, independently of which estimation of HR-size is considered (family, habitat preference and trophic categories). The Canidae species are clearly separated by the predicted threshold of 14.5 kg. Those with carnivorous diet have a larger HR-size than the omnivorous species. Around 8 Ma ago, Beringia allowed the dispersal of Canidae throughout Eurasia. The first Eurasian record is *Canis cipio*, only present in the Iberian Peninsula during the Latest Miocene. From 7.2 to 5.3 Ma the genus *Eucyon* was represented by *E. monticinensis*. The genus probably dispersed again in Spain at the end of the Early Pliocene (Sotnikova & Rook, 2010).

Figure 4 shows that HR-size decreases in *Eucyon* during the Pliocene. The small HR-size diversified with the occurrence of the genus *Nyctereutes* and later with *Vulpes*. All these taxa are small predators and remain in the omnivore niche. The main change in HR-size structure among canids occurred at the Early Pleistocene when the glacial-interglacial cycles were well established. A new large HR-size appears for Canidae related to the switch on carnivore niche. These canids are coyote-like and wolf-like forms with hypercarnivore diets. The dispersal of hypercarnivore canids in Iberian Peninsula during the Early Pleistocene is evidenced by an increasing number of fossils not only of large wild dogs (*Lycaon falconeri* as cited in Martínez-Navarro & Rook, 2003) but also Villafranchian wolves (*Canis* ex gr. *C. etruscus*). This is the traditionally called "Wolf Event" due to the massive appearance of these canids in the fossil record (Azzaroli, 1983; Sardella & Palombo, 2007; Rook & Torre, 1996). Even the *C. accitanus* coyote-like dogs increased the HR-size of omnivore canids at this time. This event integrates a faunal renewal that occurred at approximately 2 Ma, strongly related to climatic and environmental changes. Wolves and large hypercarnivorous canids progressively dispersed through the Iberian Peninsula and occupied new niches, may be related to the widespread of open habitats (Figure 2). By increasing the HR-size, they could have expanded their feeding area and eaten the same diet for longer periods. However, it should be taken into account that the evolution of predator size is likely to be influenced by changes in prey size, and a significant trend toward larger size has been documented for large northwestern Mediterranean herbivores (Rodríguez et al., 2004; Prado et al., 2004). The diversity of canids decreases at the end of the Middle Pleistocene when the extant lineages (*Vulpes vulpes* and *Canis lupus*) appear. The diversity increases again with the sporadic presence of *Cuon alpinus* during the Late Plistocene.

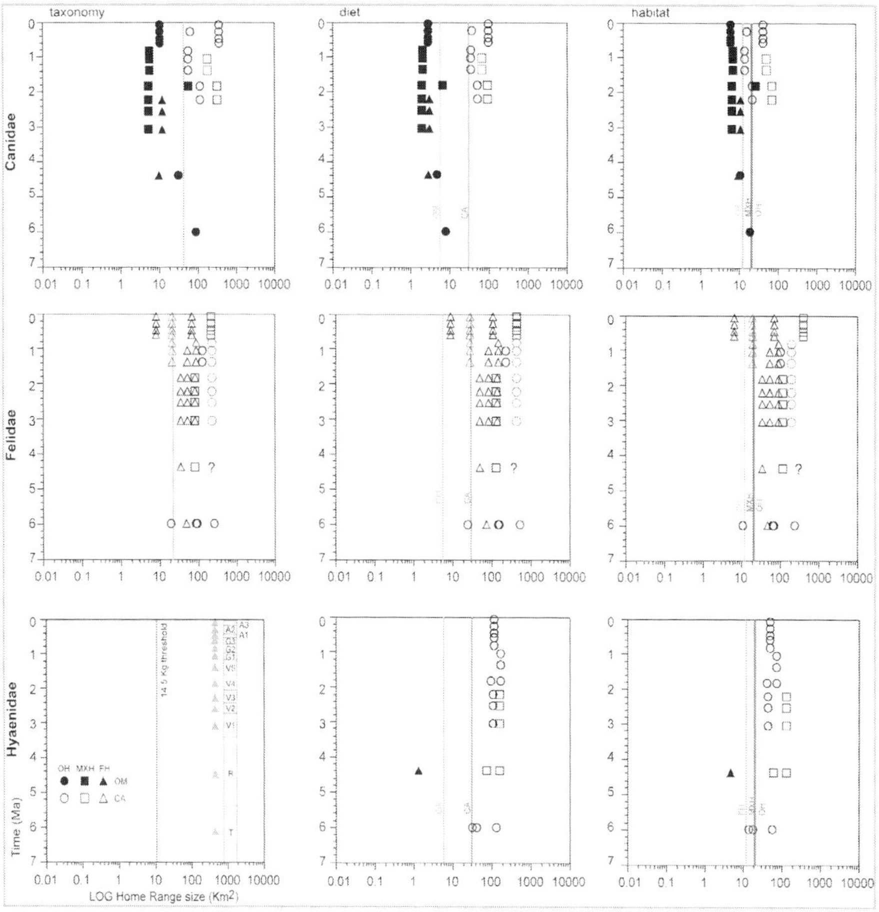

Fig. 4. Home range size of selected species of Canidae, Felidae and Hyaenidae from Spain during the last 7 Ma. The symbols are plotted at the mid-point of the time interval. Vertical axis represents time from past to present.

The Felidae HR-size structure is rather different from that of Canidae. There are four groups of HR-size clearly distinguished from Middle to Late Pleistocene (Figure 4). The smallest HR-sizes correspond to the latest Miocene and the latest Middle to Late Pleistocene. During the Late Pliocene, the intermediate HR-sizes overlap. This moment was characterized by the predominance of open and mixed habitats. In the present African landscape, most of the large predators are abundant in open savannas and savannas-woodlands, coincident with the highest ungulate densities. Rodriguez et al. (2004) showed an increase of size for the ungulates of that time, and without small-sized herbivores. This pattern remains a more close fit between prey and predator. New species appear during Lower Pleistocene and seem to begin a competitive displacement of the four groups of HR-size. After the "Wolf Event" (latest Early Pleistocene), the intermediate niche is segregated into two. This pattern suggests competitive size displacements as the main influence that might have caused HR-size changes.

The Hyaenidae pattern is similar to that of Canidae (Figure 4). The distribution of HR-size is conditioned by the diet, but this pattern was acquired previously and the presence of omnivorous species is occasional. The Hyaenidae were more diverse during the Late Miocene. Together with a precocious bone-eater hyaenidae, *Adcrocuta eximia,* two thalassictine forms were recorded. The "thalassictine group" is basically canid-like and establishes the general trend toward hypercarnivory (Werdelin & Solounias, 1991). All these taxa disappeared at the end of the Miocene and were replaced by a new hyaenid fauna. The canid-likehyaenids show a decrease in diversity that is strongly correlated with the arrival of Canidae into Eurasia during the Late Miocene-Early Pliocene. However, the only hunting hyaenid, *Chasmaportetes lunensis,* continues successfully during the Pliocene. Its disappearance in the Early Pleistocene seems to be related to the "Wolf Event". *C. lunensis* was the only cursorial, meat and bone eater hyaenid (Turner et al., 2008), adapted to open habitats. The Pliocene shows the full emergence of an organised guild of large carnivores in Europe. The appearance of *Pliocrocuta* at the R BU, followed by the gigantic *Pachycrocuta* at Late Pliocene-Early Pleistocene, shows that hyenas were able to obtain a consistent living from scavenging when necessary. In these taxa, the bone cracking component of the dentition is developed at the expense of the shearing component. Scavenging and hunting form part of a spectrum of behaviours. Living hyenas show a range of food-obtaining strategies that go through various aspects of their skeletal morphology (Wederlin & Solounias, 1991; Turner et al., 2008). The "Wolf Event" did not represent any important change despite the immigration of *Pachycrocuta brevirostris,* the largest true hyena ever recorded. Only one species of hyaenid was recorded from this event to latest Pleistocene, *Crocuta crocuta.*

4. Remarks

The analysis of Spanish Carnivora diversity trends shows that the hypothesis of a direct influence by climate change is partially supported by the data. There are two main moments of maximum diversity. The first one occurred during the late Pliocene, followed by a detached decrease during the Early Pleistocene. This lapse of time coincides with the progressive development of the Mediterranean double seasonality (related to the emergence of cold winters) that culminated with the start of the Pliocene glacial trend at 2.7-2.6 Ma (Suc et al., 1995; Suc & Popescu, 2005). The establishment of glacial cycles of 41 Ka is consistent with the diversity decrease during the Late Pliocene–Early Pleistocene. The second increase happened in the middle Pleistocene and was followed by a minimum during the last Glacial. These diversity changes can be correlated with the general global climate signal (Shackleton, 1995); the beginning of this phase is roughly correlated with the shift of the 100 Ka glacial cycles at 1.0 Ma.

Among Plio-Pleistocene Carnivora, only a few are specialized species; many others were ubiquitous or generalist, occupying broad niches. This fact would favor that a high number of carnivoran taxa survives major environmental changes. Comparing diversity trends and shifts in the percentage of carnivorans in each of the habitat categories, it seems that trends in diversity are roughly correlated with the variations of the relative abundance of forest habitat through time. This rough correspondence should be explained as an ecological response, given that the forest-dwelling species were generally more dependent on environmental conditions and shifts of the vegetation cover. On the other hand, the relative increase of ubiquitous taxa during the Pleistocene could be related to the environmental fluctuations associated with the glacial cycles, because these taxa are expected to be more able to adapt to environmental changes. Indeed, forest carnivores seem to be much more sensitive to climate and humidity changes.

In sum, climatic changes altered the ecological equilibrium of palaeocommunities. The effect this had in shaping local patterns of carnivoran diversity was possibly significant. Even during the recent past, environmental conditions have differentially affected migration and dispersal of large mammals towards and across the Iberian Peninsula. Carnivora that progressively dispersed into this region did not occupy preexisting free niches, but new ones that became available as a consequence of the environmental change (Palombo et al., 2007a, 2009). Consequently, there organization of guilds seems to be more closely related to the time and mode of dispersal events of the Carnivora than to the major climatic changes.

5. Acknowledgement

Work supported by a Projects CGL2007-60790/BTE and CGL2010-19116/BOS from the Dirección General de Investigación Científica y Técnica of Spain and Projects AECID A/023681/09 and A/030111/10; grants from the Universidad Nacional del Centro de la provincial de Buenos Aires (http://www.unicen.edu.ar) and the Project ANPCYT (http://www.agencia.gov.ar) PICT 07-01563, Argentina. The funders had no role in study design, decision to publish, or preparation of the manuscript.

6. References

Abrams, P.A. (2000). The Evolution of Predator-Prey interaction: Theory and Evidence. *Annual Review of Ecology and Systematics*, Vol.31, pp. 79–105, ISSN 0066-4162

Agustí, J.; Cabrera, L.; Garcés, M.; Krijgsman, W.; Oms, O. & Parés, J.M. (2001). A calibrated mammal scale for the Neogene of Western Europe. State of the art. *Earth-Science Reviews*, Vol. 52, pp. 247-260, ISSN 0012-8252

Agustí, J.; Moyà-Solà, S. & Pons, J. (1987). La sucesión de mamíferos en el Pleistoceno inferior de Europa: proposición de una nueva escala bioestratigráfica. *Paleontologia i Evolutió*, Vol. 1, pp. 287-295, ISSN 0211-609X

Alberdi, M.T.; Azanza, B.; Cerdeño, E. & Prado, J.L. (1997). Similarity relationship between Mammal faunas and biochronology from latest Miocene to Pleistocene in the Western Mediterranean area. *Eclogae Geologicae Helvetiae*, Vol. 90, pp. 115-132, ISSN 0012-9402

Alroy, J. (2000). Understanding the dynamics of evolutionary trends within evolving lineages. *Paleobiology*, Vol. 26, pp. 319–329, ISSN 0094-8373

Alroy, J.; Koch, P.L. & Zachos, J.C. (2000). Global climate change and North American mammalian evolution, In: *Deep Time: Paleobiology's Perspective*, D.H. Erwin & S.L. Wing (Eds.), 259–288, Allen Press, ISBN 0-9677554-2-5, Lawrence, Kansas, USA

Azanza, B.; Alberdi, M.T.; Cerdeño, E. & Prado, J.L. (1997). Biochronology from Latest Miocene to Middle Pleistocene in the Western Mediterranean Area. A Multivariate Approach, In: *Actes du Congrès BiochroM'97*, J.L. Aguilar, S. Legendre & J. Michaux (Eds.), 567-574, Mémoires des Travaux de l'Institut de Montpellier, ISBN 0-335-8178, France.

Azanza, B.; Alberdi, M.T. & Prado, J.L. (1999). Large mammal diversity and turnover patterns during the Plio-Pleistocene in Western Mediterranean Area. *Revista Sociedad Geológica de España*, Vol. 12, No. 1, pp. 113-122, ISSN 0214-2708

Azanza, B.; Alberdi, M.T. & Prado, J.L. (2000). Large mammal turnover pulse correlated with late Neogene glacial trends in the Northwestern Mediterranean region, In: *Climates: Past and Present*, M.B. Hart (Ed.), Geological Society, Special Publications, Vol. 181, pp. 161-170, ISBN 1-86239-075-4, Oxford UK

Azanza, B.; Palombo, M.R. & Alberdi, M.T. (2004). Large mammal turnover pulses and palaeoclimate changes from the Miocene to the Late Pleistocene in Italy. *Rivista Italiana di Paleontologia e Stratigrafia*, Vol. 110, No. 2, pp. 531-545, ISSN 0035-6883

Azzaroli, A. (1983). Quaternary mammals and the "End-Villafranchian" dispersal event - A turning point in the history of Eurasia. *Palaeogeography, Palaeoclimatology, Palaeoecology*, Vol. 44, pp. 117-139, ISSN 0031-0182

Badgley, C.; Barry, J.C.; Morgan, M.E.; Nelson, S.V.; Behrensmeyer, A.K.; Cerling, T.E. & Pilbeam, D. (2008). Ecological changes in Miocene mammalian record show impact of prolonged climatic forcing. *Proceedings of the National Academic of Sciences, USA*, Vol. 105, pp. 12145-12149, ISSN 0027-8424

Barnosky, A.D. (2001). Distinguishing the effects of the Red Queen and Court Jester on Miocene mammal evolution in the Northern Rocky Mountains. *Journal of Vertebrate Paleontology*, Vol. 21, pp. 172–185, ISSN 0272-4634

Barnosky, A.D. (2005). Effects of Quaternary Climatic Change on Speciation in Mammals. *Journal of Mammalin Evolution*, Vol. 12, No. 1-2, pp. 247-264, ISSN 1064-7554

Behrensmeyer, A.K.; Damuth, J.D.; DiMichele, W.A.; Potts, R.; Sues, H.-D. & Wing, S.L. (1992). *Terrestrial Ecosystems Through Time. Evolutionary Paleoecology of terrestrial Plants and Animals ETE. The Evolution of terrestrial Ecosystems Consortium*, The University Press of Chicago Press, ISBN 0-226-04155-7, Chicago USA

Behrensmeyer, A.K.; Todd, N.E.; Potts, R. & McBrinn, G.E. (1997). Late Pliocene faunal turnover in the Turkana Basin, Kenya and Ethiopia. *Science*, Vol. 278, pp. 1589–1594, ISSN 0036-8075

Bowman, J.; Jaeger, J.A.G. & Fahrig, L. (2002) Dispersal Distance of Mammals Is Proportional to Home Range Size. *Ecology*, Vol. 83, No. 7, pp. 2049-2055, ISSN 0012-9658

Carbone, C.; Teacher, A. & Rowcliffe, T.M. (2007). The costs of carnivory. *PlosBiology*, Vol. 5, No. 2, e22, pp. 0363-0368, eISSN 1545-7885

Carbone, C. & Gittleman, J.L. (2002). A Common Rule for the Scaling of Carnivore Density. *Science*, Vol. 295, pp. 2273-2276, ISSN 0036-8075

Chesson, P. (2000). Mechanisms of maintenance of species diversity. *Annual Review of Ecology and Systematics*, Vol. 31, pp. 343-366, ISSN 0066-4162

Foote, M. (2000). Origination and extinction components of taxonomic diversity: general problems. *Paleobiology*, Vol. 26, No. 4, pp. 74–102, ISBN 0-9677554-2-5

Gittleman, J.L. & Harvey, P.H. (1982). Carnivore Home-Range Size, Metabolic Needs and Ecology. *Behavior Ecology and Sociobiology*, Vol. 10, pp. 57-63, ISSN 0340-5443

Hernández Fernández, M.; Álvarez Sierra, M.A. & Peláez-Campomanes, P. (2007). Bioclimatic analysis of rodent palaeofaunas reveals severe climatic changes in Southwestern Europe during the Plio-Pleistocene. *Palaeogeography, Palaeoclimatology, Palaeoecology*, Vol. 251, No. 3-4, pp. 500-526, ISSN 0031-0182

Jablonski, D. (2005). Mass extinctions and macroevolution. *Paleobiology*, Vol. 31, No. 2, pp. 192–210, ISSN 0094-8373

Janis, C.M. (1989). A climatic explanation for patterns of evolutionary diversity in ungulate mammals. *Palaeontology*, Vol. 32, pp. 463–481, ISSN 0031-0239

Kelt, D.A. & Van Vuren, D. (1999). Energetic constratints and the relationship between body size and Home range area in Mammals. *Ecology*, Vol. 80, pp. 337-340, ISSN 0012-9658

Koenigswald, W. von (2002). Migrations and extinctions in the Quaternary faunas of Central and Western Europe. *Annales Géologiques Pays Helléniques*, Vol. 39, pp 327-335, ISSN 1105-0004

Koenigswald, W. von (2003). Mode and causes for the Pleistocene turnovers in the mammalian fauna of Central Europe, In: *Distribution and migration of Tertiary mammals in Eurasia. A volume in honour of Hans de Bruijn*, J.W.F. Reumer & W. Wessels (Eds.), *Deinsea*, Vol. 10, pp. 305-312, ISSN 0923-9308, Rotterdam, Netherland

Kolfschoten, T. (1995). On the application of fossil mammals to the reconstruction of the palaeoenvironment of Northwestern Europe. *Acta Zoologica Cracoviensia*, Vol. 38, pp. 73-84, ISSN 0065-1710

Lisiecki, L.E. & Raymo, M.E. (2005) A Pliocene-Pleistocene stack of 57 globally distributed benthic δ18O records. *Paleoceanography*, Vol. 20, No. PA1003, doi: 10.1029/2004PA001071, ISSN 0883-8305

Maas, M.C.; Anthony, M.R.L.; Gingerich, P.D.; Gunnell, G.F. & Krause, D.W. (1995) Mammalian generic diversity and turnover in the Late Paleocene and Early Eocene of the Bighorn and Crazy Mountains Basins, Wyoming and Montana (USA). *Palaeogeography, Palaeoclimatology, Palaeoecology*, Vol. 115, pp. 181-207, ISSN 0031-0182

Martínez-Navarro, B. & Rook, L. (2003). Gradual evolution in the African hunting dog lineage. Systematic implications. *Comptes Rendus Palevolution*, Vol. 2, pp. 695-702, ISSN 1631-0683

McKee, J.K. (2001). Faunal turnover rates and mammalian biodiversity of the late Pliocene and Pleistocene of eastern Africa. *Paleobiology*, Vol. 27, pp. 500–511, ISSN 0094-8373

McNab, B.K. (1963). Bioenergetics and the determination of home range size. *American Naturalist*, Vol. 47, pp. 133-140, ISSN 1537-5323

Mein, P. (1975). Résultats du groupe de travail des vertébrés: bizonation du Néogène méditerranéen à partir des Mammifères. *Report Activity Regional Committee on Mediterranea Neogene Stratigraphy working groups*, pp. 78-81

Mein, P. (1990). Updating of MN zones, In: *European Neogene Mammal Chronology*, E.H. Lindsay, V. Fahlbusch & P. Mein (Eds.), 73-90, Plenum Press, ISSN 0-306-43391-5, New York, USA

Palombo, M.R. (2004). Guilds of large mammals from the Pliocene to the Late Pleistocene in Italian peninsula, En: *Homenaje a Emiliano Aguirre*, E. Baquedano & S. Rubio (Eds.), 372-391, Zona Archeologica, Vol. 4, No. 2 Paleontologia, ISBN 84-451-2654-7, Madrid

Palombo, M.R. (2007a). Which boundary for the Quaternary period and Pleistocene epoch? The contribution to the debate given by turnover patterns in large mammalian complexes from North-Western Mediterranean region. *Quaternaire*, Vol. 18, No. 1, pp. 35-53, ISSN 1142-2904

Palombo, M.R. (2007b). Climate changes versus biotic interaction: a case study of large mammal faunal complexes on the Italian Peninsula from the Pliocene to the Late Pleistocene: new methodological approaches. *Courier Forschungsinstut Senckenberg*, Vol. 259, pp. 13-46, ISSN 0341-4116

Palombo, M.R.; Alberdi, M.T.; Azanza, B.; Giovinazzo, C.; Prado, J.L. & Sardella, R. (2009). How did environmental disturbances affect Carnivora diversity? A case study of the Plio-Pleistocene Carnivora in North Western Mediterranean. *Evolutionary Ecology*, Vol. 23, pp. 569-589, ISSN 1573-8477

Palombo, M.R. & A.F.M. Valli. (2005). Large mammal faunas turnover at Early-Middle Pleistocene transition in France, In: *Early–Middle Pleistocene transitions: the land–ocean evidence*, M.J. Head & P.L. Gibbard (Eds.), 277-286, Geological Society London, Special Publication, No. 247, ISBN 1-86239-181-5

Prothero, D.R. (1999). Does climatic change drive mammalian evolution?. *Geological Society of American, Today*, pp. 1-7, ISSN 1052-5173

Prothero, D.R. (2004). Did impacts, volcanic eruptions, or climate change affect mammalian evolution? *Palaeogeography, Palaeoclimatology, Palaeoecology*, Vol. 214, pp. 283- 294, ISSN 0031-0182

Prado, J.L.; Alberdi, M.T.; Azanza, B. & Rodríguez, J. (2004). Patterns of body-size change in mammals during the late Cenozoic in the Northwesterns Mediterranean. En: *Miscelánea en homenaje a Emilano Aguirre*, E. Baquedano & S. Rubio (Ed.), 464-479, Zona Arqueológica, Vol. 4, No. 2 - Paleontología, Museo Arqueológico Regional, ISBN 84-451-2654-7, Madrid

Rodríguez, J.; Alberdi, M.T.; Azanza, B. & Prado, J.L. (2004). Body size structure in north-western Mediterranean Plio-Pleistocene mammalian faunas. *Global Ecology and Biogeography*, Vol. 13, pp. 163-176, ISSN 1466-822X

Rook, L. & Torre, D. (1996). The wolf-event in Western Europe and the beginning of the Late Villafranchian. *Neues Jahrbuch für Geologie und Paläontologie Monatshefte*, Vol. 8, pp. 495–501, ISSN 0028-3630

Sardella, R. & Palombo, M.R. (2007). The Pliocene-Pleistocene Boundary: Which Significance for the so called "Wolf Event"? Evidences from Western Europe. *Quaternaire*, Vol. 18, No. 1, pp. 65-71, ISSN 1142-2904

Shackleton, N.J. (1995). New data in the Evolution of Pliocene Climatic Variability, In: *Paleoclimate and Evolution with Emphasis on Human Origins*, E.S. Vrba, G.H. Denton, T.C. Partridge & L.H. Burckle (Eds.), 242-248, Yale University Press, ISBN 0-300-06348-2, New Haven, USA

Sotnikova, M. & Rook, L. (2010). Dispersal of the Canini (Mammalia, Canidae: Caninae) across Eurasia during the Late Miocene to Early Pleistocene. *Quaternary International*, Vol. 212, pp. 86–97, ISSN 1040-6182

Spassov, N. (2003). The Plio-Pleistocene vertebrate fauna in South-Eastern Europe and megafaunal migratory waves from the east of Europe. *Revue de Paléobiologie*, Vol. 22, No. 1, pp. 197-229, ISSN 0253-6730

Stucky, R.K. (1990). Evolution of land mammal diversity in North America during the Cenozoic. *Current Mammalogy*, Vol. 2, pp. 375–432, ISBN 0-306-42430-4

Suc, J.P.; Bertini, A.; Combourieu-Nebout, N.; Diniz, F.; Leroy, S.; Russo-Ermolli, E.; Zheng, Z.; Bessais, E. & Ferrier, J. (1995). Structure of West Mediterranean vegetation and climate since 5.3 Ma. *Acta Zoologica Cracoviensia*, Vol. 38, No. 1, pp. 3-16, ISSN 0065-1710

Suc, J.P. & Popescu, S.M. (2005). Pollen records and climatic cycles in the North Mediterranean region since 2.7 Ma, In: *Early-Middle Pleistocene transitions: the land-ocean evidence*, M.J. Head & P.L. Gibbard (Eds.), 147-158, Geological Society, Special Publication, ISBN 1-86239-181-5, London, UK

Turner, A. (1990). The evolution of the guild of larger terrestrial carnivores during the Plio-Pleistocene in Africa. *Geobios*, Vol. 23, pp. 349–368, ISSN 1777-5728

Turner, A. (1999). Evolution of the African Plio-Pleistocene mammalian fauna: correlation and causation, In: *African biogeography, climate change and human evolution*, T.G. Bromagem & F. Schrenk (eds), 76– 87, Oxford University Press, ISBN 0-195-11437-X, Oxford, UK

Turner, A.; Antón, M. & Werdelin, L. (2008). Taxonomy and evolutionary patterns in the fossil Hyaenidae of Europe. *Geobios*, Vol. 41, No. 5, pp. 677-687, ISSN 0016-6995

Van Dam, J.A. (2001). The Upper Miocene mammal record from the Teruel-Alfambra region (Spain). The MN system and continental Stage/Age concepts discussed. *Journal of Vertebrate Paleontology*, Vol. 21, No. 2, pp. 367-385, ISSN 0272-4634

Van Dam, J.A.; Abdul Aziz, H.; Álvarez Sierra, M.A.; Hilgen, F.J.; van den Hoek Ostende, L.W.; Lourens, L.; Mein, P.; van der Meulen, A.J. & Peláez-Campomanes, P. (2006). Long-period astronomical forcing of mammal turnover. *Nature*, Vol. 443, No. , pp. 687-691, ISSN 1476-4687

Van Valen, L. (1973). A new evolutionary law. *Evolutionary Theory*, Vol. 1, pp. 1-30, ISSN 0093-4755

Van Valkenburgh, B. (1985). Locomotor diversity between past and present guilds of large predatory mammals. *Paleobiology*, Vol. 11, No. 4, pp. 406-428, ISSN 0094-8373

Van Valkenburgh, B. (1988). Trophic diversity in past and present guilds of large predatory mammals. *Paleobiology*, Vol. 14, No. 2, pp. 155-173, ISSN 0094-8373

Van Valkenburgh, B. (1989). Carnivore dental adaptations and diet: a study of trophic diversity withinguilds, In: *Carnivore behavior, ecology, and evolution*, J.L. Gittleman (Ed.), Vol. 1, pp. 410–436, Cornell University Press, ISBN 0-412-34350-9, Ithaca, NY, USA

Vrba, E.S. (1985). Environment and evolution: alternative causes of the temporal distribution of evolutionary events. *South African Journal of Science*, Vol. 81, pp. 229-236, ISSN 0038-2353

Vrba, E.S. (1995a). On the Connections between Paleoclimate and Evolution, In: *Paleoclimate and Evolution with Emphasis on Human Origins*, E.S. Vrba, G.H. Denton, T.C. Partridge & L.H. Burckle (Eds.), 24-45, Yale University Press, ISBN 0-300-06348-2, New Haven, USA

Vrba, E.S. (1995b). The fossil record of African Antelopes (Mammalia, Bovidae) in relation to human evolution and paleoclimate, In: *Paleoclimate and Evolution with Emphasis on Human Origins*, E.S. Vrba, G.H. Denton, T.C. Partridge & L.H. Burckle (Eds.), 383-424, Yale University Press, ISBN 0-300-06348-2, New Haven, USA

Vrba, E.S. (2000). Major features of Neogene mammalian evolution in Africa, In: *The Cenozoic of southern Africa*, T.C. Partridge & R.R. Maud (Eds.), 277– 304, Oxford University Press, ISBN 0-7923-0266-4, New York, USA

Vrba, E.S. & DeGusta, D. (2004). Do species populations really start small? New perspectives from the Late Neogene fossil record of African mammals. *Philosophical Transactions of the Royal Society, London*, Vol. B359, pp. 285–293, ISSN 0080-4622

Werdelin, L. & Lewis, M. (2005). Plio-Pleistocene Carnivora of eastern Africa: species richness and turnover patterns. *Zoological Journal of the Linnean Society*, Vol. 144, pp. 121–144, ISSN 0024-4082

Werdelin, L. & Solounias, N. (1991). The Hyaenidae: taxonomy, systematics and evolution. *Fossils and strata*, Vol. 30, pp. 1-104, ISSN 0300-9491

7

Response of Biogenic Silica Production in Lake Baikal and Uranium Weathering Intensity in the Catchment Area to Global Climate Changes

Takuma Murakami[1,6], Nagayoshi Katsuta[2], Takejiro Takamatsu[3], Masao Takano[1], Koshi Yamamoto[1], Toshio Nakamura[4] and Takayoshi Kawai[5]

[1]*Graduate School of Environmental Studies, Nagoya University,*
[2]*Faculty of Education, Gifu University,*
[3]*National Institute for Environmental Studies,*
[4]*Center for Chronological Research, Nagoya University,*
[5]*Association of International Research Initiative for Environmental Studies,*
[6]*Low Level Radioactivity Laboratory, Institute of Nature, Environmental Technology, Kanazawa University,*
Japan

1. Introduction

Lake Baikal, in southeast Siberia, is a structural basin in the Baikal rift valley (Fig. 1a) and is the largest lake on earth in terms of fresh water volume (23,000 km^3). With a surface 454 m above sea level (asl), it covers an area of 31,500 km^2 (length, 636 km; maximum width, 80 km) and has a maximum depth of 1,620 m. The vegetation around the lake is characteristic of the steppe and taiga. And the annual mean temperature and rainfall around the lake are -2.2 °C and 400-500 mm per year, respectively. The catchment area of the lake is 540,000 km^2, extending from northwest Mongolia to southeast Siberia (Fig. 1b), of which 83% constitutes the drainage basin of the Selenga River (Fig. 1b and c). This is the largest river flowing into the lake and its water inflow makes up 50% of the total riverine input. The climate in the Lake Baikal region is influenced by westerly wind weather systems (Mackay, 2007). Therefore, most of the atmospheric moisture in southeast Siberia is from the North Atlantic Ocean and the Arctic Ocean.

The bottom sediment of Lake Baikal documents the long-term history of environmental changes in the Asian continental interior (southeast Siberia), showing the shift in climate on various time-scales. The main proxy records obtained from the sediment are based on the concentration of diatom frustules and biogenic silica (bioSi) (Colman et al., 1995; Kashiwaya et al., 2001; Mackay, 2007; Prokopenko et al., 1999, 2001, 2002; Williams et al., 1997) and the amount of pollen fossils (Shich et al., 2007; Tarasov et al., 2005) in the sediment indicating the bioproductivity in the lake and its surrounding watersheds. The main source of the bioSi in the sediment is diatom frustules (Karabanov et al., 1998). Studies to date using the biological records revealed that the diatom and vegetation changes were correlated with the Milankovitich periods (Kashiwaya et al., 2001; Williams et al., 1997) and were in phase with the glacial-interglacial cycles (Colman et al., 1995; Prokopenko et al., 2001a, 2002; Shichi et al., 2007). Moreover, the variations are found to follow the centennial-to-millennial-scale climate changes that occurred in the North Atlantic region: Bond cooling cycles during the last glacial periods (Prokopenko et al., 2001b); a Younger Dryas cold period for the last glacial/Holocene transition (Prokopenko et al., 1999); and IRD (ice-rafted debris) cooling (Bond) events in the Holocene (Mackay, 2007; Tarasov et al., 2005).

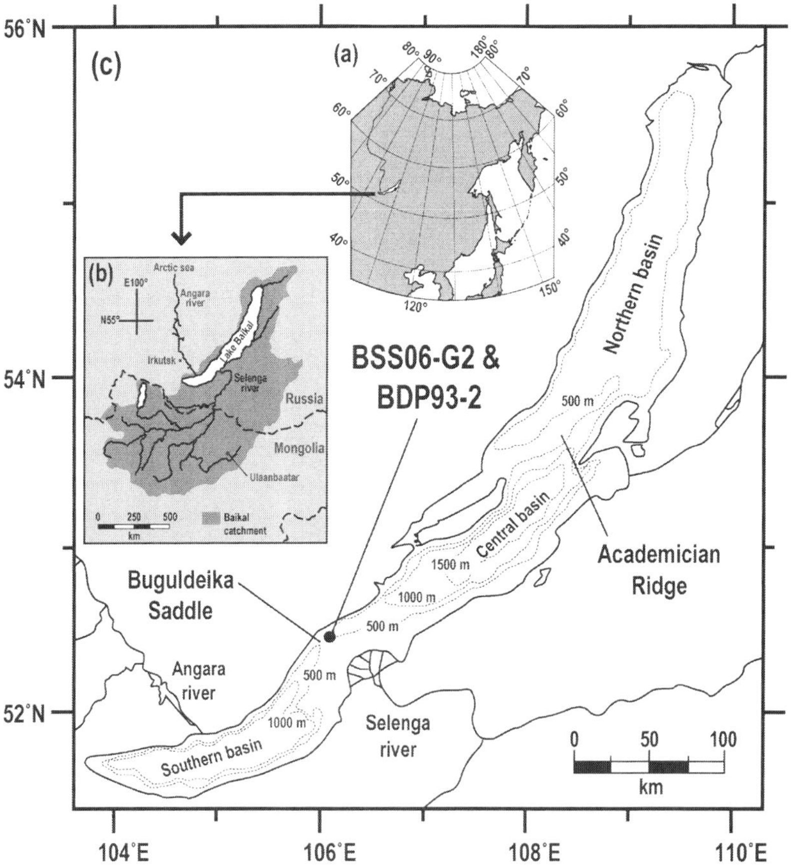

Fig. 1. Maps of (a) northeastern part of continental interior Asia and (b) catchment basin of (c) Lake Baikal. Bathymetric map of the lake showing the collection sites of cores BSS06-G2 and BDP93-2 at Buguldeika saddle.

On the other hand, the uranium record in Lake Baikal sediments as well as the biological records provides information on the environmental changes in the region. The variation in the uranium concentration is thought to be due to the weathering intensity in the Selenga drainage basin associated with changes in rainfall/moistures levels (Goldberg et al., 2010; Murakami et al., under review). The U-Th isotope study of Edgington et al. (1996, 1997) revealed that the uranium in Lake Baikal sediment is composed mainly of authigenic components, which originated in uranium-bearing rocks distributed in Mongolia and southeast Siberia, and that the uranium is transported from the source rock into the lake via the Selenga River. Based on geochemical evidence, Edgington et al. also concluded that the variation of uranium concentration in the sediment resulted from changes in the input from the Selenga River and its tributary. The uranium variations are reported to have corresponded to the Pleistocene glacial-interglacial cycles (Chebykin et al., 2007; Edgington et al., 1996; Goldberg et al., 2010) and the Holocene Bond events (Goldberg et al., 2005).

The purpose of the present study is to investigate the degree of similarity in the variations in the Lake Baikal records of bioSi and uranium and the paleoproxy records of global climate changes on a centennial-to-millennial-scale as well as a glacial-interglacial time scale. An examination of the correlation among these paleoclimate proxy datasets has been conducted for lake sediment from the underwater Academician ridge (Fig. 1) in Lake Baikal (Edgington et al., 1996), focusing on the variations on a glacial-interglacial scale. However, we analyzed the geochemical data of sediment from the Buguldeika saddle in Lake Baikal (Fig. 1). This site is located at the opposite side of the Selenga Delta, where uranium is directly transported to from the watershed of the Selenga River.

In the present study, we used the SPECMAP $\delta^{18}O$ record (Imbrie et al., 1984) and the North Atlantic IRD index (Bond et al., 1997, 2001) as a proxy of global climate change. The SPECMAP was acquired by stacking $\delta^{18}O$ data of planktonic foraminifera collected from five deep-sea cores in low- and mid-latitudes, generally reflecting the continental ice sheet volume. The IRD index represents cooling events in the North Atlantic, which occurred nine times during the Holocene period. The IRD cooling events are referred to as Bond events labeled 8-0 from the past to the present day.

2. Materials and methods

2.1 Sediment cores

In the present study, we used two cores, BDP93-2 and BSS06-G2 taken at Buguldeika saddle in Lake Baikal, southeast Siberia (Fig. 1). Buguldeika saddle is a local elevation that developed at the opposite side of the Selenga Delta, separated from the mouth of the river by a deep trough. Because of these topographic features, the sedimentation of the saddle is controlled by the supply of a fine suspended load from the Selenga River. Seismic surveys and the lithologic features of the drilled cores indicate that the upper 50 m of sediment at the drill sites consists of continuous and sub-parallel layers (BDP-93 Baikal Drilling Project Members, 1997).

Core BDP93-2 was collected in March 1993 at a water depth of 354 m (52° 31′ 3.0″N and 106° 09′ 6.01″E) using a piston corer (BDP-93 Baikal Drilling Project Members, 1997). This core was 102 m long. We measured the chemical components of 228 sediment samples from the core. The samples were collected at intervals of 30 to 80 cm corresponding to the temporal resolution of 2–7 kyr. In the present study, we used the upper 37.2 m of the core, equivalent to the last 250 kyr for data analyses because the concentrations of the chemical components in this section show remarkable patterns on the glacial-interglacial cycles. On the other hand, core BSS06-G2 as well as -G1 was collected in August 2006 at a water depth of 360 m (52°27'27.1"N, 106°07'46.1"E) using a gravity corer. This core was 39 cm long, sliced into 39 sediment samples every 1.0 cm. Each sample from cores BDP93-2 and BSS06-G2 was freeze-dried, then powdered and homogenized with an agate mortar for chemical analyses.

2.2 Analyses

The concentrations of bioSi in the sediment of core BSS06-G2 were determined following the protocol of DeMaster (1981). Each 50-mg dry sediment sample was mixed in a 50-ml polypropylene centrifuge tube with 50-ml of 5% Na_2CO_3 solution. The mixture was then heated at 85°C in a thermostatic bath and the 2-ml samples were collected after 3, 4, 5, and 6 hours. Each digested sample was diluted to 20-ml with distilled water for the analyses. Concentrations of the dissolved Si were analyzed using an ICP-AES (Ultima 2, HORIBA Jobin Yvon) at Gifu Univ. and a UV-Vis spectrometer (Metertik SP-830, Metertech Inc.) at Nagoya Univ. Their analytical precisions were <2% and <3%, respectively. Finally, the concentrations of the bioSi were calculated by determining the intercept of the line through the timed aliquots. With respect to core BDP93-2, we used Figure 3 of Colman et al. (1999), and read the values of concentrations of the bioSi corresponding to the core depth of our samples used in the uranium analysis.

The concentrations of uranium in the bulk-sediment of Lake Baikal were analyzed by two methods. In core BDP93-2, the uranium concentrations were determined neutron-activation instrument analysis (INAA; Koyama and Matsushita, 1980). This experiment was performed at the Nuclear Reactor, Kyoto Univ., Kyoto, Japan. Each 50-mg dry sample was packed in a double polyethylene film bag. It was then placed together with 30-μg of Co as a standard in a capsule. Neutron irradiation of the sample was performed in a pneumatic transfer tube for 50 min. After 1-week of cooling, the gamma-ray spectra were measured using a diode detector system of Ge (Li) coupled with a 4096-channel pulse-height analyzer. In core BSS06-G2, the bulk-U was quantified using a microwave-digestion-ICP-MS (HP4500, Hewlett Packard) at Nagoya Univ. based on the method of Murakami et al. (2010). The analytical precisions of the INAA and ICP-MS were <17% and <3%, respectively.

The radiocarbon (^{14}C) analyses of core BSS06-G2 were conducted using a Tandetron accelerator mass spectrometer (AMS, Model-4130, HVEE) at the Center for Chronological Research, Nagoya Univ, according to the procedures of Watanabe et al. (2009).

2.3 Sediment chronologies

For core BDP93-2, we used the chronology established in Colman et al. (1999). This is based on a combination of the ^{14}C ages and the paleomagnetic relative intensity correlations. The chronology of the depth interval 0–3.68 m in the core was determined by using the ^{14}C dates on the total organic carbon (TOC). Generally, in Lake Baikal, the topmost layer of the bottom sediment is known to give a "non-zero" TOC date. This is mainly caused by contamination by terrigenous organic matter. The apparent surface ages near the Buguldeika saddle range from 1000 to 1600 ka (Colman et al., 1996; Karabanov et al., 2004; Watanabe et al., 2009) and are evaluated to average 1.16 ka for two BDP93 cores (Colman et al., 1996). Colman et al. (1999) applied a 1.16-kyr correction to each date in core BDP93-2. Consequently, the age of the core at the depth interval of 0–3.68 m ranged from 0 to 21.0 ka. The date for the depth interval 3.68–50 m was determined from the correlation of the paleomagnetic relative intensities of the cores between BDP93 and dated marine sediment. Consequently, it ranged from 21.0 to 338 ka.

The chronology of core BSS06-G2 was determined in the present study using five AMS ^{14}C dates on the TOC (Fig. 2). As in core BDP93-2, the topmost layer has a ^{14}C age of 1.418 ka. We therefore applied a 1,418-year correction to the five ^{14}C dates over the entire sequence prior to calibration. The five dates were calibrated using an IntCal04 (Reimer et al. 2004). The age model for core BSS06-G2 was established assuming that the sedimentation rate was constant between the dated levels.

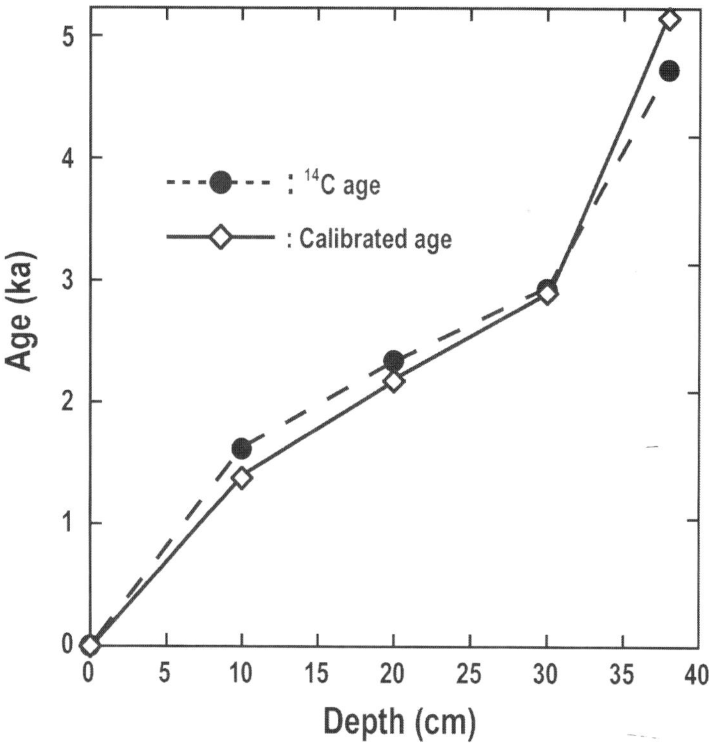

Fig. 2. Plot of age against core depth, based on 5 AMS ^{14}C dates for TOC in core BSS06-G2. The dates were calibrated with CALIB 5.0.1 and IntCal04 (Stuiver et al., 1998; Reimer et al., 2004).

2.4 Cross correlation analysis

Cross correlation analysis was used to examine the time lags between Lake Baikal core BDP93-2 and the SPECMAP $\delta^{18}O$ records. A cross-correlation analysis of such two time-series identifies the time delay that makes the two time-series most similar. The correlation coefficient quantifies the similarity of the two time-series at that time offset. The cross-correlation function for two time-series *f(t)* and *g(t)* is defined as follows:

$$\varnothing fg(\tau) = \frac{1}{T-\tau}\int_0^{T-\tau} f(t)g(t+\tau)dt$$

where τ and T are the time lag and length or period of the time-series, respectively. The SPECMAP $\delta^{18}O$ data is given as a function of age in steps of equally spaced 2 kyr. However, the BDP93-2 datasets are unevenly spaced time-series with an interval of 1.9–6.8 kyr. We therefore used the average interval 2.76 kyr (standard deviation, 1.30 kyr) as a time lag for the BDP93-2 core. The correlation coefficients of cross-correlation between these two time-series were computed after linearly interpolating the neighboring two values in the SPECMAP $\delta^{18}O$, in order to adjust the SPECMAP data points to the BDP93-2 ones.

3. Results and discussion

3.1 Long- and short-term variability in the Lake Baikal records of bioSi and uranium

Figure 3 shows successive 250 kyr-profiles of bioSi and uranium concentration in core BDP93-2, SPECMAP $\delta^{18}O$, and the 65°N June insolation. The concentration of bioSi and uranium in core BDP93-2 increased during the interglacial periods (bioSi, 5.0–45.0%; uranium, 2.5–20.0 ppm), and declined during the glacial cold periods (bioSi, 3.0–10.0%; uranium, 2.5–7.0 ppm). With respect to the variations during MIS 5 and 7, the uranium record has peaks corresponding to the all the substages (Fig. 3b), whereas the bioSi record does not show peaks corresponding to MIS 7a and b (Fig. 3a). A comparison of the BDP93-2 (Fig. 3a and b) and SPECMAP (Fig. 3c) records indicates that the uranium record shows a pattern similar to the SPECMAP $\delta^{18}O$ curve. Especially, the uranium variation during MIS 5 displays a gradual decrease (that is, sawtooth pattern) representative of the late Pleistocene glacial/interglacial cycle observed in the δD record from the Antarctic ice core (EPICA, 2004; Petit et al., 1999) and the SPECMAP $\delta^{18}O$ record.

The relationships between the SPECMAP $\delta^{18}O$ and BDP93-2 records (Fig. 4) show a lower correlation coefficient for the bioSi (R = 0.37) than for the uranium (R = 0.46). These results are consistent with the visual inspections of the variations in the records (Fig. 3). On the other hand, the cross-correlation coefficients of the bioSi are 0.37~0.50 and those of the uranium are 0.46~0.52 at a time-lag range of 0 to 5.52 kyr (Fig. 5). In addition, the bioSi and uranium have a time-lag of 2.76 kyr and 5.52 kyr, respectively, but this difference is not significant because the standard deviation of the time-lag is 1.30 kyr. The results of these cross-correlation analyses indicate that there are no statistically significant differences between the Baikal bioSi and uranium records (Fig. 5); therefore we may regard the bioSi and uranium variations as paleoproxy records having the same degree of response to global climate changes.

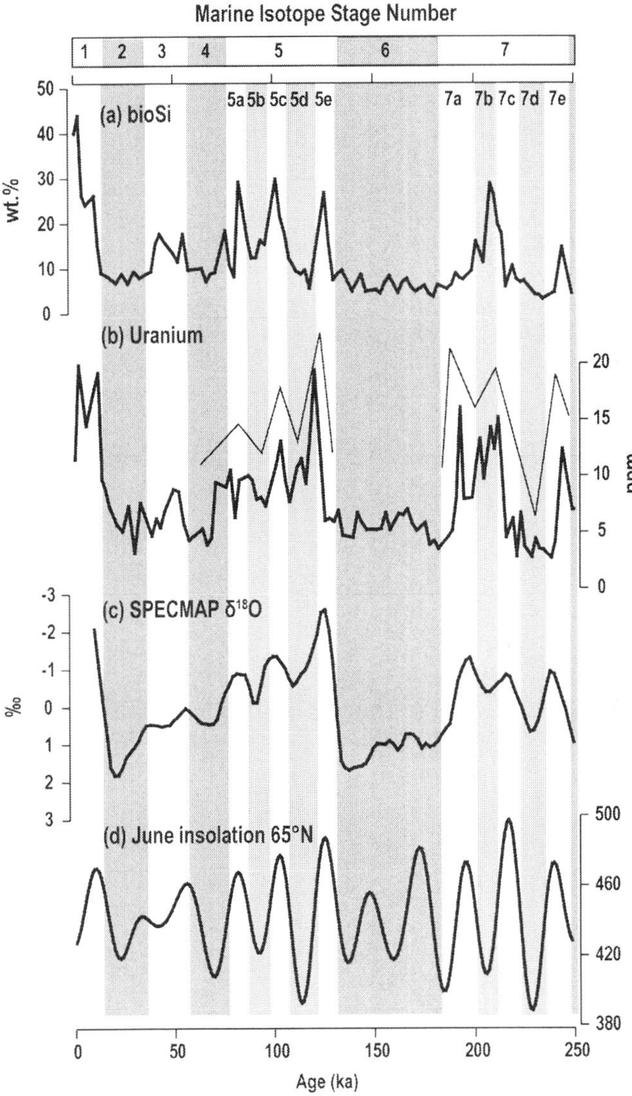

Fig. 3. Comparison of (a) bioSi (Colman et al. 1999) and (b) U in core BDP93-2 with (c) SPECMAP $\delta^{18}O$ data (Imbrie et al. 1984) and (d) 65° N June insolation (Laskar et al., 1993). Shaded and blue bands represent glacial periods and cooling substages in the interglacial periods, respectively. Boundaries between the glacial and interglacial periods were identified by Colman et al. (1999) based on correlations between the Lake Baikal bioSi data and SPECMAP curves.

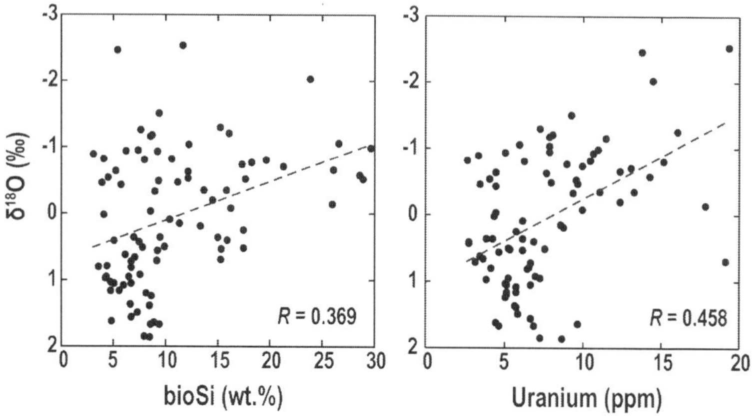

Fig. 4. Diagrams showing (a) relationship between Lake Baikal bioSi and SPECMAP $\delta^{18}O$ and (b) relationship between Lake Baikal uranium and SPECMAP $\delta^{18}O$.

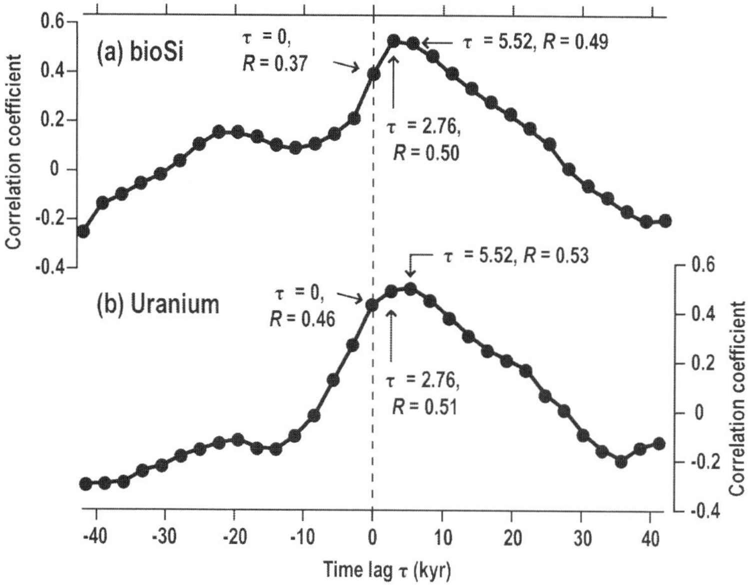

Fig. 5. Correlation coefficients versus time lag in thousand years (kyr) for individual cross correlations between Baikal paleoenvironmental proxies ((a) bioSi and (b) uranium) and SPECMAP $\delta^{18}O$ data. The highest peak represents the time lag at which the two time series were most similar. P-values of correlation coefficient in (a) and (b) are <0.01. This P-value represents >99% confidence level of the obtained correlation coefficient.

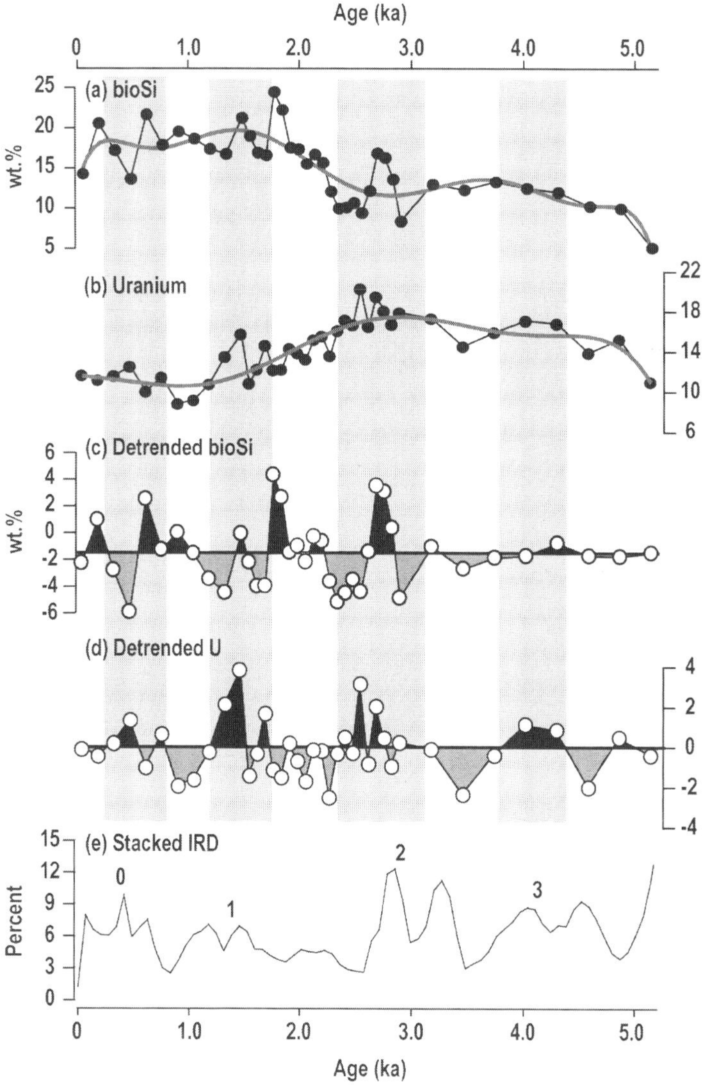

Fig. 6. Comparison of (a-d) Lake Baikal sediment records with (e) stacked IRD index in the North Atlantic (Bond et al., 1997, 2001): concentrations of (a) bioSi, (b) U, (c) the detrended bioSi, and (d) the detrended U in the sediment of core BSS06-G2. Variability in (c) and (d) were extracted by removal of the long-term trend in (a) and (b) (an 8th-order polynomial, red line) from the original curves, respectively. Shaded bands represent intervals of relative high U concentration in (d). Numbers in (e) indicate the Holocene IRD events labeled by Bond et al. (2001).

These tendencies observed in the long-term Lake Baikal records are also identified in the short-term variations in core BSS06-G2. Figure 6 shows the successive profiles of the Baikal bioSi and uranium and the North Atlantic stacked IRD records. A slight decline in the bioSi record, indicating the need to detrend (Fig. 6c), are identified for IRD events 2–0 (Fig. 6e). On the other hand, the uranium records show small peaks corresponding to events 3–0 during the last 5.2 kyr (Fig. 6b and d).

3.2 Relationship between the Lake Baikal records and climates in the Asian continental interior

In core BDP93-2 (Fig. 3), the bioSi and uranium increased in the interglacial period and decreased in the cold glacial periods. These fluctuation patterns can be associated with the bioSi production in the lake and the uranium weathering intensity in the watershed, reflecting the climate changes in the region. Specifically, the bioSi and uranium variations are dependent on changes in temperature and moisture in the region, respectively. This tendency is consistent with the previously reported climate in the Asian continental interior: dry conditions developed during the glacial periods and wet conditions during the interglacial periods.

Vegetation changes reflecting such climatic conditions were observed in the pollen records of sediment from Lake Baikal. Shichi et al. (2007) conducted pollen analyses on two cores from the northern and southern basins in Lake Baikal, and then reconstructed the geographical distribution of vegetation-types in the watershed between the glacial and interglacial periods. According to the analytical results, during the interglacial periods (MIS 1, 5, 7, and 9), the forest, mainly composed of *Pinus* and *Picea*, spread widely in the northern and southern watershed, whereas during glacial periods (MIS 2, 4, 6, and 8), open forest of *Artemisia*, *Betula*, etc., was distributed mainly around the southern region, while open forest and desert greatly expanded in the northern region. Rainfall reconstructed from pollen analyses of Lake Baikal sediment represents about 250 mm at the early stages of the last glaciation (117.5–114.8 ka) and 350–500 mm during the Holocene (MIS 1) and the last interglacial periods (MIS 5) (Tarasov et al., 2007).

The glacial aridity in the Asian interior is considered to have resulted from the expansion of the ice sheet along the north Eurasian continental margin, thus reducing the transport of water vapor from the North Atlantic via the westerly winds. The influence of the continental ice sheet on the glacial climates in the Asian interior has been previously investigated using GCM (general circulation model) simulations (Manabe & Broccoli, 1985; Bush, 2004). According to the GCM study, it is estimated that the annual mean soil moisture at 18 ka decreased by 20~60% compared to present day values (15 cm in soil depth) and that the difference between the annual mean precipitation during the LGM (Last Glacial Maximum) and the present day was 0.05–0.1 cm/day.

In the BDP93-2 records during the isotopic substage 5e/d and 7e/d, the abrupt declines in the concentrations of bioSi and uranium are identified (Fig. 3a and b). This is considered to be due to extreme insolation minima (65°N, June) during 5d and 7d (Fig. 3d, Karabanov et al., 1998; Prokopenko et al., 2002). Such drops in bioSi concentration in the Lake Baikal sediment are also observed in the BDP96-2 core collected from the underwater Academician ridge in Lake Baikal (Fig. 1, Prokopenko et al., 2001a, 2002). This geochemical evidence suggests a decrease in bioproductivity over the whole area of the lake. Karabanov et al. (1998) found that the 5d-sediment of Lake Baikal is composed of a glacial clay layer with low TOC and bioSi content, which suggests glacial advances in the mountains of the Sayan–Baikal region.

During the Holocene IRD cooling events, BSS06-G2 bioSi concentrations generally tend to show a slight decrease, while the uranium depicts an increase (Fig. 6), contrary to the glacial-interglacial patterns (Fig. 3). This tendency is consistent with geological and botanical evidence previously recorded in the Asian continental interior, which was influenced by westerly winds. Pollen analyses of the Ugii Nuur lake sediment in central Mongolia from the last 8.7 ka revealed that a temperature, indicated by *Chenopodiaceae* concentration, decreased, whereas the moisture index, mainly based on concentrations of *Pinus, Poaceae,* and *Cyperaceae,* increased during the IRD events (Wang et al., 2009). Holocene records of Gun Nuur lake sediment in northern Mongolia, obtained by analyses of organic matter, organic $\delta^{13}C$, and magnetic properties, suggested that an increase in vegetation cover and biomass in the watershed, representing the cold/wet condition, corresponded to the timing of the IRD events in the North Atlantic (Wang et al., 2004). These two Mongolian lakes are located in the watershed of Lake Baikal. Recently, a comprehensive attempt to reconstruct the climate during the Little Ice Age (corresponding to the last IRD event 0) was made in westerlies-dominated Central Asia (Chen et al., 2010). Chen et al. investigated the spatial and temporal patterns of effective moisture (precipitation) variations in the region over the last millennium, using 17 sediment records from different sites, and then demonstrated that the LIA climate conditions became humid over a wide area.

3.3 Synchrony in the response to climatic changes between bioSi production and uranium weathering intensity

The bioSi and uranium variations recorded in the Lake Baikal sediment follow the same patterns of global climate changes on both the glacial-interglacial (Fig. 3) and centennial-to-millennial-scales (Fig. 6). We attribute this to the sedimentary depositional systems of the Lake Baikal region which are capable of directly responding to climate changes in the form of diatom production and uranium weathering intensity.

There are two proposed hypotheses for the causes of the variation of diatom/bioSi concentration in Lake Baikal sediment: the first is the amount of nutrient (soluble Si, P, etc.) supplied from the watershed (Chebykin et al., 2002, 2004; Gavshin et al., 2001) and the second is the change in the regional temperature in response to solar insolation (Prokopenko et al., 2001a). As shown in Figs. 3 and 6, the bioSi-uranium relationships in the Lake Baikal sediment indicate that the bioSi does not always increase with the higher uranium. This implies that the deposition of these two components rarely influenced each other. If the diatom/bioSi production in the lake is dependent on nutrient supply from the watershed, the bioSi-uranium relationships must show a positive correlation since soluble uranium should flow into the lake together with soluble Si and P. Therefore, we consider the second hypothesis to be the cause of the diatom/bioSi.

A study supporting our interpretation was conducted using the bottom sediment of Lake Baikal. Prokopenko et al. (2006) investigated the temporal variations of the Holocene temperature and humidity in the Lake Baikal region using the GCM climate model simulation and pollen fossil analyses. The results show that the Baikal bioSi/diatom production is in phase with the annual and winter temperatures and is a strong counter-phase with the humidity over the Holocene period. We think that a direct influence of temperature change on Baikal diatom production would result in a correlation between the bioSi record and a proxy of global climate changes.

As shown in Figs. 3 and 6, in the Buguldeika saddle of Lake Baikal, the uranium curve bears a closer resemblance to the paleoclimate indices compared with the bioSi record. In Lake

Baikal, the dissolved and detrital uranium in the water column are derived mainly from uranium-bearing rocks in the Selenga drainage basin, and are transported to the lake via the Selenga River and its tributaries (Edgington et al., 1996, 1997). Most of this uranium is in a dissolved form. Uranium contained in the sediment is mainly preserved as an authigenic component from dissolved uranium adsorbed on the surface of suspended sediment loads in the river and lake water and diatom frustules produced in the lake, which accumulates on the lake bottom (Edgington et al., 1996, 1997; Goldberg et al., 2010; Sakaguchi et al., 2006). Therefore, there is a possibility that the uranium deposition may be affected by the supply of detritus materials and the production rate of diatoms.

On the other hand, chronological studies on the bottom sediment in Lake Baikal indicate that the depositional rates are nearly constant through the glacial-interglacial periods over the whole lake area: 14.8 cm/kyr for the last 264 kyr at the Buguldeika saddle (Colman et al., 1999) and 3.9 cm/kyr for the last 6.7 Ma at the underwater Academician Ridge (Kravchinsky et al., 2003). The observed sedimentary features of Lake Baikal suggest that the amount of materials on which the uranium adsorbed would vary little over the entire sequence, and the concentration of uranium in the bottom sediment would be only slightly dependent on the abundances of detritus materials and diatom frustules. Therefore, we believe that the uranium variation strongly influences the input of uranium into the lake, reflecting the weathering in the Selenga drainage basin which was associated with changes in moisture levels in the region.

4. Conclusion

Comparing the bioSi and uranium variations, the Lake Baikal records from the Buguldeika saddle and the paleoproxy records of global climate change showed only slight differences on glacial-interglacial and centennial-to-millennial scales – there are statistically no significant differences between these two records. Thus, we conclude that the bioSi and uranium records of Lake Baikal sediment follow the same degree of global climate changes on time scales covering centuries to tens of millennia.

5. Acknowledgement

The present study was supported by the Sasagawa Scientific Research grant No.23-601 from the Japan Science Society; the Sumitomo Foundation grant No.103359; awards from the 21st Century COE and the Global COE programs at Nagoya Univ. of the Ministry of Education, Culture, Sports, Science and Technology, Japan (Dynamics of the Sun-Earth-Life Interactive System, No. G-4; From Earth System Science to Basic and Clinical Environmental Studies, No. K-04).

6. References

BDP-93 Baikal Drilling Project Members (1997). Preliminary results of the first scientific Drilling on Lake Baikal, Buguldeika site, southeastern Siberia. *Quaternary International*, Vol.37, (June 1998), pp. 3-17, ISSN 1040-6182

Bond, G.; Kromer, B.; Beer, J.; Muscheler, R.; Evans, M. N.; Showers, W.; Hoffmann, S.; Lotti-Bond, R.; Hajdas, I. & Bonani, G. (2001). Persistent solar influence on North Atlantic climate during the Holocene. *Science*, Vol.294, No.5549, (December 2001), pp. 2130-2136, ISSN 0036-8075

Bond, G.; Showers, W.; Cheseby, M.; Lotti, R.; Almasi, P.; deMenocal, P.; Priore, P.; Cullen, H.; Hajdas, I. & Bonani, G.; (1997). A pervasive millennial-scale cycle in North Atlantic Holocene and Glacial climates. *Science*, Vol.278, No.5341, (November 1997), pp. 1257-1266, ISSN 0036-8075

Bush, A. B. G. (2004). Modelling of late Quaternary climate over Asia: a synthesis. *Boreas*, Vol.33, No.2, (May 2004), pp. 155-163, ISSN 0300-9483

Chebykin, E. P.; Edgington, D. N.; Grachev, M. A.; Zheleznyakova, T. O.; Vorobyova, S. S.; Kulikova, N. S.; Azarova, I. N.; Khlystov, O. M. & Goldberg, E. L. (2002). Abrupt increase in precipitation and weathering of soils in East Siberia coincident with the end of the last glaciation (15 cal kyr BP). *Earth and Planetary Science Letters*, Vol.200, Nos.1-2, (June 2002), pp. 167-175, ISSN 0012-821X

Chebykin, E. P.; Edgington, D. N.; Goldberg, E. L.; Phedorin, M. A.; Kulikova, N. S.; Zheleznyakova, T. O.; Vorobyova, S. S.; Khlystov, O. M.; Levina, O. V.; Ziborova, G. A. & Grachev, M. A. (2004). Uranium-series isotopes as proxies of Late Pleistocene climate and geochronometers in bottom sediments of Lake Baikal. *Geologiya i Geofizika (Russian Geology and Geophysics)*, Vol.45, No.5, (May 2004), pp. 539-556, ISSN 1068-7971

Chebykin, E. P.; Goldberg, E. L.; Kulikova, N. S.; Zhuchenko, N. A.; Stepanova, O. G. & Malopevnaya, Yu. A. (2007). A method for determination of the isotopic composition of authigenic uranium in Baikal bottom sediments. *Russian Geology and Geophysics*, Vol.48, No.6, (June 2007), pp. 468-477, ISSN 1068-7971

Chen, F. H.; Chen, J. H.; Holmes, J.; Boomer, I.; Austin, P.; Gates, J. B.; Wang, N. L.; Brooks, S. J. & Zhang, J. W. (2010). Moisture changes over the last millennium in arid central Asia: a review, synthesis and comparison with monsoon region. *Quaternary Science Reviews*, Vol.29, Nos.7-8, (April 2010), pp. 1055-1068, ISSN 0277-3791

Colman, S. M.; Jones, G. A.; Rubin, M.; King, J. W.; Peck, J. A. & Orem, W. H. (1996). AMS radiocarbon analyses from Lake Baikal, Siberia: Challenges of dating sediments from a large, oligotrophic lake. *Quaternary Science Reviews*, Vol.15, No.7, (January 1996), pp. 669-684, ISSN 0277-3791

Colman, S. M.; Peck, J. A.; Hatton, J.; Karabanov, E. B. & King, J. W. (1999) Biogenic silica records from the BDP93 drill site and adjacent areas of the Selenga Delta, Lake Baikal. *Journal of Paleolimnology*, Vol.21, No.1, (January 1999), pp. 9-17, ISSN 0921-2728

Colman, S. M.; Peck, J. A.; Karabanov, E. B.; Carter, S. J.; Bradbury, J. P.; King, J. W. & Williams, D. F. (1995). Continental climate response to orbital forcing from biogenic silica records in Lake Baikal. *Nature*, Vol.378, No.6559, (December 1995), pp. 769-771, ISSN 0023-0836

DeMaster, D. J. (1981). The supply and accumulation of silica in the marine environment. *Geochimica et Cosmochimica Acta*, Vol.45, No.10, (October 1981), pp. 1715-1732, ISSN 0016-7037

Edgington, D. N.; Robbins, J. A.; Colman, S. M.; Orlandini, K. A. & Gustin, M. -P. (1996). Uranium-series disequilibrium, sedimentation, diatom frustules, and paleoclimate change in Lake Baikal. *Earth and Planetary Science Letters*, Vol.142, Nos.1-2, (July 1996), pp. 29-42, ISSN 0012-821X

Edgington, D. N.; Robbins, J. A.; Colman, S. M.; Orlandini, K. A.; Gustin, M. -P.; Klump, J. V. & Granina, L. Z. (1997). Reply to the comment by R. Anderson on "Uranium-series disequilibrium, sedimentation, diatom frustules, and paleoclimate change in Lake Baikal". *Earth and Planetary Science Letters*, Vol.148, Nos.1-2, (April 1997), pp. 399-404, ISSN 0012-821X

EPICA community members (2004). Eight glacial cycles from an Antarctic ice core. *Nature*, Vol.429, No.6992, (June 2004), pp. 623-628, ISSN 0023-0836

Gavshin, V. M.; Bobrov, V. A. & Khlystov, O. M. (2001). Periodicity in diatom sedimentation and geochemistry of diatomaceous mud in Lake Baikal: global aspect. *Geologiya i Geofizika (Russian Geology and Geophysics)*, Vol.42, No.2, (February 2001), pp. 317-325, ISSN 1068-7971

Goldberg, E. L.; Chebykin, E. P.; Zhuchenko, N. A.; Vorobyeva, S. S.; Stepanova, O. G.; Khlystov, O. M.; Ivanov, E. V.; Weinberg, E. & Gvozdkov, A. N. (2010). Uranium isotopes as proxies of the environmental history of the Lake Baikal watershed (East Siberia) during the past 150 ka. *Palaeogeography Palaeoclimatology Palaeoecology*, Vol.294, Nos.1-2, (August 2010), pp. 16-29, ISSN 0031-0182

Goldberg, E. L.; Grachev, M. A.; Chebykin, E. P.; Phedorin, M. A.; Kalugin, I. A.; Khlystov, O. M. & Zolotarev, K. V. (2005). Scanning SRXF analysis and isotopes of uranium series from bottom sediments of Siberian lakes for high-resolution climate reconstructions. *Nuclear Instruments and Methods in Physics Research Section A: Accelerations, Spectrometers, Detectors and Associated Equipment*, Vol.543, No.1, (May 2005), pp. 250-254, ISSN 0168-9002

Imbrie, J.; Hays, J. D.; Martinson, D. G.; McIntyre, A.; Mix, A. C.; Morley, J. J.; Pisias, N. G.; Prell, W. L. & Shackleton, N. J. (1984). The orbital theory of Pleistocene climate: support from a revised chronology of the marine 6180 record, In: *Milankovitch and Climate, part 1*, Berger, A.; Imbrie, J.; Hays, J.; Kukula, G. & Saltzman, B., (Ed.), pp. 269-305, Plenum Reidel, ISBN 978-9027717917, Dordrecht, Netherlands

Karavanov, E. B.; Prokopenko A. A.; Williams, D. F. and Colman, S. M. (1998) Evidence from Lake Baikal for Siberian glaciation during Oxygen-Isotope substage 5d. *Quaternary Research*, Vol.50, No.1, (March 2001), pp. 46-55, ISSN 0033-5894

Karabanov, E.; Williams, D.; Kuzumin, M.; Sideleva, V.; Khursevich, G.; Prokopenko, A.; Solotchina, E.; Tkachenko, L. Fedenya, S.; Kerber, E.; Gvozdkov, A.; Khlustov, O.; Bezrukova, E.; Letunova, P. & Krapivina, S. (2004). Ecological collapse of Lake Baikal and Lake Hovsgol ecosystems during the Last Glacial and consequences for aquatic species diversity. *Palaeography, Palaeoclimatology, Palaeoecology*, Vol.209, Nos.1-4, (July 2004), pp. 227-243, ISSN 0031-0182

Kashiwaya, K.; Ochiai, S.; Sakai, H. & Kawai, T. (2001). Orbital-related long-term climate cycles revealed in a 12-Myr continental record from Lake Baikal. *Nature*, Vol.410, No.6824, (March 2001), pp. 71-74, ISSN 0023-0836

Koyama, M. & Matsushita, R. (1980). Use of neutron spectrum sensitive monitors for instrumental neutron activation analysis. *Bulletin of the Institute for Chemical Research, Kyoto University*, Vol.58, No.2, (August 1980), pp. 235-243, Retrieved from <http://hdl.handle.net/2433/76875>

Kravchinsky, V. A.; Krainov, M. A.; Evans, M. E.; Peck, J. A.; King, J. W.; Kuzmin, M. I.; Sakai, H.; Kawai, T. & Wiiliams, D. F. (2003). Magnetic record of Lake Baikal sediments: chronological and paleoclimatic implication for the last 6.7 Myr. *Palaeography, Palaeoclimatology, Palaeoecology*, Vol.195, Nos.3-4, (June 2003), pp. 281-298, ISSN 0031-0182

Laskar, J.; Joutel, F. and Boudin, F. (1993) Orbital, precessional, and insolation quantities for the Earth from -20 Myr to +10 Myr. *Astronomy and Astrophysics*, Vol.270, Nos.1-2, (March 1993), pp. 522-533, ISSN 0004-6361

Mackay, A. W. (2007). The paleoclimatology of Lake Baikal: A diatom synthesis and prospectus. *Earth-Science Review*, Vol.82, No.3-4, (June 2007), pp. 181-215, ISSN 0012-8252

Manabe, S. & Broccoli, A. J. (1985). The influence of continental ice sheets on the climate of ice age. *Journal of Geophysical Research*, Vol.90, No.D1, (February 1985), pp. 2167-2190, ISNN 0148-0227

Murakami, T.; Katsuta, N.; Yamamoto, K.; Takamatsu, N.; Takano, M.; Oda, T.; Matsumoto, G. I.; Horiuchi, K. & Kawai, T. (2010) A 27-kyr record of environmental change in central Asia inferred from the sediment record of Lake Hovsgol, northwest Mongolia. *Journal of Paleolimnology*, Vol.43, No.2, (February 2010), pp. 369-383, ISSN 0921-2728

Murakami, T.; Takamatsu, T.; Katsuta, N.; Takano, M.; Yamamoto, K.; Nakamura, T. and Kawai, T. (under review) Centennial- to millennial-scale climate shifts in continental interior Asia repeated between warm-dry and cool-wet conditions during the last three interglacial states: Evidence from uranium and biogenic silica in the sediment of Lake Baikal, southeast Siberia.

Petit, J. R.; Jouzel, J.; Raynaud, D.; Barkov, N. I.; Barnola, J. M.; Basile, I.; Bender, M.; Chappellaz, J.; Davis, J.; Delaygue, G.; Delmotte, M.; Kotlyakov, V. M.; Legrand, M.; Lipenkov, V.; Lorius, C.; Pépin, L.; Ritz, C.; Saltzman, E. & Stievenard, M. (1999). Climate and Atmospheric History of the Past 420,000 years from the Vostok Ice Core, Antarctica. *Nature*, Vol.399, No.6735, (June 1999), pp. 429-436, ISSN 0023-0836

Prokopenko, A. A.; Hinnov, L. A.; Williams, D. F. & Kuzmin, M. I. (2006). Orbital forcing of continental climate during the Pleistocene: a complete astronomically tuned climatic record from Lake Baikal, SE Siberia. *Quaternary Science Reviews*, Vol.25, Nos. 23-24, (December 2006), pp. 3431-3457, ISSN 0027-3791

Prokopenko, A. A.; Karabanov, E. B.; Williams, D. F.; Kuzmin, M. I.; Shackleton, N. J.; Crowhurst, S. J.; Peck, J. A.; Gvozdkov, A. N. & King, J. W. (2001a). Biogenic silica record of the Lake Baikal response to climatic forcing during the Brunhes. *Quaternary Research*, Vol.55, No.2, (March 2001), pp. 123-132, ISSN 0033-5894

Prokopenko, A. A.; Williams, D. F.; Karabanov, E. B. & Khursevich, G. K. (1999). Response of Lake Baikal ecosystem to climate forcing and pCO_2 change over the last glacial/interglacial transition. *Earth and Planetary Science Letters*, Vol.172, Nos.3-4, (October 1999), pp. 239-253, ISSN 0012-821X

Prokopenko, A. A.; Williams, D. F.; Karabanov, E. B. & Khursevich, G. K. (2001b). Continental response to Heinrich events and Bond cycles in sedimentray record of Lake Baikal, Siberia. *Global and Planetary Change*, Vol.28, No.1-4, (February 2001), pp. 217-226, ISSN 0921-8181

Prokopenko, A. A.; Williams, D. F.; Kuzmin, M. I.; Karabanov, E. B.; Khursevich, G. K. & Peck, J. A. (2002). Muted climate variations in continental Siberia during the mid-Pleistocene epoch. *Nature*, Vol.418, No.6893, (July 2002), pp. 65-68, ISSN 0023-0836

Reimer, P. J.; Baillie, M. G. L.; Bard, E.; Bayliss, A.; Beck, J. W.; Bertrand, C. J. H.; Blackwell, P. G.; Buck, C. E.; Burr, G. S.; Cutler, K. B.; Damon, P. E.; Edwards, R. L.; Fairbanks, R. G.; Friedrich, M.; McCormac, G.; Manning, S.; Ramsey, C. B.; Reimer, R. W.; Remmele, S.; Southon, J. R.; Stuiver, M.; Talamo, S.; Taylor, F. W.; van der Plicht, J. & Weyhenmeyer, C. E. (2004). IntCal04 terrestrial radiocarbon age calibration, 0–26 cal kyr BP. *Radiocarbon*, Vol.46, No.3, (December 2004), pp. 1029-1058, ISSN 0033-8222

Sakaguchi, A.; Yamamoto, M.; Sasaki, K. & Kashiwaya, K. (2006). Uranium and thorium isotope distribution in an offshore bottom sediment core of the Selenga Delta, Lake Baikal, Siberia. *Journal of Paleolimnology*, Vol.35, No.4, (May 2006), pp. 807-818, ISSN 0921-2728

Shichi, K.; Kawamuro, K.; Takahara, H.; Hase, Y.; Maki, T. & Miyoshi, N. (2007). Climate and vegetation changes around Lake Baikal during the last 350,000 years. *Palaeography, Palaeoclimatology, Palaeoecology*, Vol.248, Nos.3-4, (May 2007), pp. 357-375, ISSN 0031-0182

Stuiver, M.; Reimer, P. J.; Bard, E.; Beck, J. W.; Burr, G. S.; Hughen, K. A.; Kromer, B.; McCormac, G.; Van der Plicht, J. & Spurk, M. (1998) INTCAL98 radiocarbon age calibration, 24000-0 cal BP. *Radiocarbon*, Vol.40, No.3, (September 1998), pp. 1041-1083, ISSN 0033-8222

Tarasov, P.; Granoszeski, W.; Bezrukova, E.; Brewer, S.; Nita, M.; Abzaeva, A. & Oberhänsli, H. (2005). Quantitative reconstruction of the last interglacial vegetation and climate based on the pollen record from Lake Baikal, Russia. *Climate Dynamics*, Vol.25, No.6, (August 2005), pp. 625-637 ISSN 0930-7575

Tarasov, P.; Bezrukova, E.; Karabanov, E.; Nakagawa, T.; Wagner, M.; Kulagina, N.; Letunova, P.; Abzaeva, A.; Granoszewski, W. & Riedel, F. (2007). Vegetation and climate dynamics during the Holocene and Eemian interglacials derived from Lake Baikal pollen records. *Palaeogeography, Palaeoclimatology, Palaeoecology*, Vol.252, Nos.3-4, (September 2007), pp. 440-457, ISSN 0031-0182

Wang, W.; Feng, Z.; Lee, X.; Zhang, H.; Yuzhen, M. A.; Chenbang, A. N. & Guo, L. (2004). Holocene abrupt climatic shifts recorded in Gun Nuur lake core, northern Mongolia. *Chinese Science Bulletin*, Vol.49, No.5, (March 2004), pp. 520–526, ISSN 1001-6538

Wang, W.; Ma, Y.; Feng, Z.; Meng, H.; Sang, Y. & Zhai, X. (2009). Vegetation and climate changes during the last 8660 cal. a BP in central Mongolia, based on a high-resolution pollution record from Lake Ugii Nuur. *Chinese Science Bulletin*, Vol.54, No.9, (May 2009), pp. 1579-1589, ISSN 1001-6538

Watanabe, T.; Nakamura, T.; Nara, F. W.; Kakegawa, T.; Horiuchi, K.; Senda, R.; Oda, T.; Nishimura, M.; Matsumoto, G. I. & Kawai, T. (2009) High-time resolution AMS ^{14}C data sets for Lake Baikal and Lake Hovsgol sediment cores: Changes in radiocarbon age and sedimentation rates during the transition from the last glacial to the Holocene. *Quaternary International*, Vol.205, Nos.1-2, (August 2009), pp. 12-20, ISSN 1040-6182

Williams, D. F.; Peck, J.; Karabanov, E. B. Prokopenko, A. A.; Kravchinsky, V.; King, J. & Kuzmin, M. I. (1997). Lake Baikal record of continental climate response to orbital insolation during the past 5 million years. *Science*, Vol.278, No.5340, (November 1997), pp. 1114-1117, ISSN 0036-8075

8

Continental Erosion/Weathering Changes in Central Asia Recorded in the Holocene Sediment from Lake Hovsgol, Northwest Mongolia, by Synchrotron μ-XRF Mapping Analyses

Nagayoshi Katsuta[1], Takuma Murakami[2,3], Yuko Wada[2], Masao Takano[2], Masayuki Kunugi[4] and Takayoshi Kawai[2,5]
[1]*Faculty of Education, Gifu University*
[2]*Graduate School of Environmental Studies, Nagoya University*
[3]*Low Level Radioactivity Laboratory, Kanazawa University*
[4]*Environment Safety Center, Tokyo University of Science*
[5]*Association of International Research Initiatives for Environmental Studies*
Japan

1. Introduction

Lake Hovsgol (Fig. 1) is located in the southernmost part of the Baikal rift valley basins and occupies the second largest basin next to Lake Baikal. The lake lies 1645 m above sea level, and its surface area is 2,760 km^2 (136 km long, 20~40 km wide). It has a water volume of 380.7 km^3 and a maximum depth of 262.4 m (Goulden et al., 2006). The lake is surrounded by three types of vegetation regions: taiga-forest, steppe, and steppe-forest. The annual mean temperature is below zero (above zero during May to September), and the precipitation is 300~500 mm per year, most of which falls from April to October (Namkhaijantsan, 2006). The lake water contains Ca^{2+} at 797 μM. Its alkalinity is 2.60 (mEq/L), and it has a pH of 8.1 (Hayakawa et al., 2003). Geophysical observations reveal that Hovsgol's sediment is several kilometers thick (Fedotov et al., 2006), which suggests that the sedimentary sequences may document a long-term history of environmental changes in arid central Asia.

Recent studies on Lake Hovsgol cores indicate that the sediment chemistry records are important sources of information to understand environmental variations in the region and related climate changes. Oscillations in the climate proxy data acquired by stack of elements hosted in the carbonates and organic matter have been found to coincide with abrupt climate shifts in the Holocene and the last glacial/Holocene transition observed in the North Atlantic region (Fedotov et al., 2004a). Periodic variations in the 21 chemical elements in the bulk-sediment suggested that moisture change in central Asia occurred on glacial-interglacial scales, as well as with a period of ~8.7 kyr, through the last glacial/Holocene (Murakami et al., 2010). Phedorin et al. (2008) analyzed the past 1 Myr geochemical records of sediments from the two great Asian lakes (Hovsgol for Ti/Ca and Br; Baikal for diatoms and Br) with a synchrotron radiation (SR) induced μ-XRF line-scanning technique, and then found a 300-500 kyr long cycle, possibly associated with Earth's orbital eccentricity. Moreover, Phedorin et al. (2008) suggested that the variations with periods of 300-500 kyr influenced the evolution of terrestrial ecosystem.

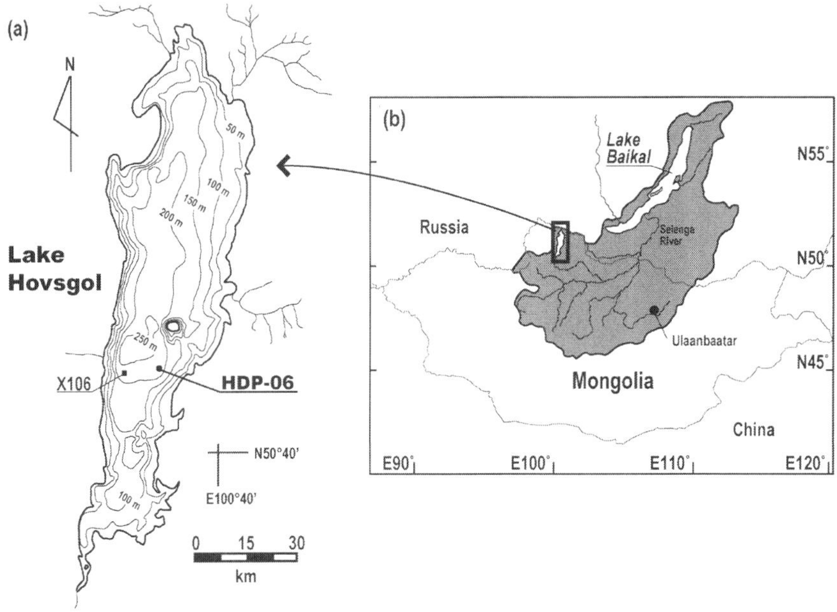

Fig. 1. Maps of (a) Lake Hovsgol in (b) northern Mongolia. (a) Bathymetric map of the lake showing drill sites for cores X106 and HDP-06. HDP-06 was used for analyses in the present study. The gray shading and lines in (b) indicate the Lake Baikal catchment area and the borders of China, Mongolia, and Russia, respectively.

In the present study, we investigated the distributions and their origins of 11 major and trace elements in the Holocene section of Hovsgol's sediment using the SR μ-XRF mappings. With respect to the Hovsgol's sediment, the two-dimensional distribution of the elements has not been studied until now. A visualization of distribution of the elements enables us to recognize detailed sedimentary or geochemical structures such as laminations, local elementally enriched grains and layers, and so on (Katsuta et al., 2007). Moreover, it makes possible to distinguish between primary structures and secondary structures deposited in the sediment. As a result of this study, we discovered an alternation of terrigenous and biological elements in the Hovsgol Holocene sections. In this chapter, we first describe methods for capturing the SR μ-XRF data of the sediment surfaces. After investigating distributions and features of the detected 11 elements, we assess the sources and natures of each element. We discuss the paleoenvironmental implications for the observed alternated pattern in the sedimentation, together with a record of the detritus input into Lake Hovsgol (Murakami et al. 2010).

2. Materials and analytical methods

2.1 Core HDP-06

In the present study, we used core HDP-06 which was collected at the base of the southeast slope of Lake Hovsgol, in northwest Mongolia (Fig. 1). The HDP-06 core drilling took place in March 2006 at 50°54′25″ N, 100°27′03″ E at a water depth of 235.5 m using an improvised push-coring type technology (Hovsgol Drilling Project Members, 2009). Core HDP-06 was about 26 m long and the sediment recovery was on the order of 80%. In the present study, we analyzed the upper section (about 18 cm) of HDP-06 core 1-1 (about 140 cm), which was undisturbed and fully recovered. The lithologic observation revealed that the core 1-1 bears sedimentary layers distinctive of Hovsgol's sediment, which is composed of diatomaceous clayey ooze in the Holocene section and calcareous clayey silt to silty clay in the last glacial section (Prokopenko et al., 2005).

2.2 Sample preparations

In order to acquire two-dimensional XRF images of the sediment surface by the SR μ-XRF techniques, we prepared thin sections of the resin-embedded samples. Subsamplings from an open core surface were carried out using aluminum U-channels (Fig. 2). After carefully retrieving the sediment slab from the U-channel, the slabs were impregnated with epoxy resin following the procedure of Tiljander et al. (2002). The U-channels were produced from a 0.2-mm-thick aluminum sheet using a sheet bender (Fig. 2a), which we produced ourselves.

A resin-embedded sediment block was shaped to a thickness of 0.3 mm for SR μ-XRF mapping. This was done to reduce the X-ray scattering from the inside of the sample. First, a pair of sediment slabs (Fig. 3a) was separated with a band saw because several sediment slabs were embedded together with the epoxy resin. Each sediment slab was cut into three pieces to produce approximately 5 cm × 5 cm samples at an angle of 45° to the bedding plane (Fig. 3b), in order to maintain the vertical continuity of the sediment. One face of each sample was polished and glued to a glass slide. After polishing the other face to achieve a thickness of about 0.3 mm, the thin section sample (Figs. 3c and d) was removed from the glass slide using a solvent.

2.3 SR μ-XRF mapping

The SR experiment was carried out at undulator beamline 37XU of SPring-8, Hyogo, Japan (Fig. 4a; Terada et al., 2004). A Si (111) double-crystal monochromator was used to acquire the incident X-ray beams at 37 keV. The X-ray beam size was adjusted using an XY slit and was 1 mm (V) × 0.5 mm (H). A thin-section sample was fixed on an acrylic holder (Fig. 3c). The sample holder was set on the X-Y computer-controlled step stage, which was rotated by 10° in the rectangular direction of the incident beam toward the detector (Fig. 3d). The stage had step sizes of 1 mm (V) and 0.5 mm (H). The XRF spectra from the sample were measured using a Ge solid-state detector coupled with a multichannel pulse-height analyzer (Fig. 4a). The measurement time was 25 s per step.

In Hovsgol's sediment, fluorescence X-rays of the 11 elements (Ti, Mn, Fe, Cu, Zn, As, Br, Rb, Zr, and Nb) were successfully detected (Figs. 4b and c). Based on the obtained XRF spectra, we determined the distribution map and profile of each element (Fig. 5). The XRF spectra were collected at each position on a sample. At each position, we computed the integrated numbers of X-ray photons with energy near each $K\alpha_1$ line (energy window within $K\alpha_1 \pm 0.10$ keV), and consequently produced XRF maps of the 11 elements. Moreover, the XRF profiles were acquired by averaging three vertical lines of the XRF maps, which were appropriately selected.

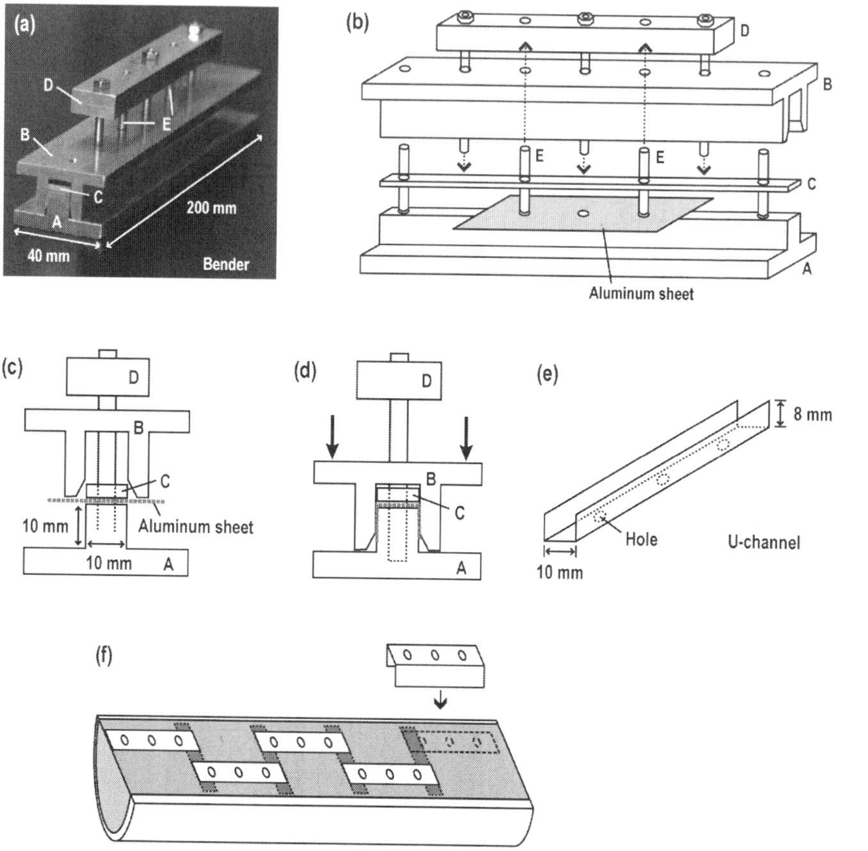

Fig. 2. Tools for subsampling open core surface. (a–d) Photograph and schematic sketches showing a sheet bender made of aluminum. (e) Schematic sketch showing an aluminum U-channel. An aluminum sheet (0.2 mm thick) was first cut into 120 mm × 36 mm rectangular pieces with a shearing machine, which in turn was set on a male die A with rods E (b). After pressing a female punch B and plank with rods D downward (c and d), the U-channel (e) was produced. Finally, the sample number and direction were inscribed onto the back of the U-channel using a scriber. The produced aluminum U-channels are pressed into halved cores, with adjacent U-channels overlapping each other as shown in (f). The sediment-filled U-channels were carefully extruded from the split cores using a metal spatula. The hole positions of an aluminum sheet (b) coincide with the positions of holes in B-D and rods E, and the holes have slightly large diameter compared with the rods. The insertion of the three rods in D and E into the corresponding holes in aluminum sheet (b) prevented the movement of the sheet during bending and simultaneously allowed the sheet to be bent at a right angle. Rods E were screwed into a male punch A like D, thereby allowing easy removal of the produced U-channel from the punch. The holes in U-channel (e) aided fluid flow in the later impregnation stage.

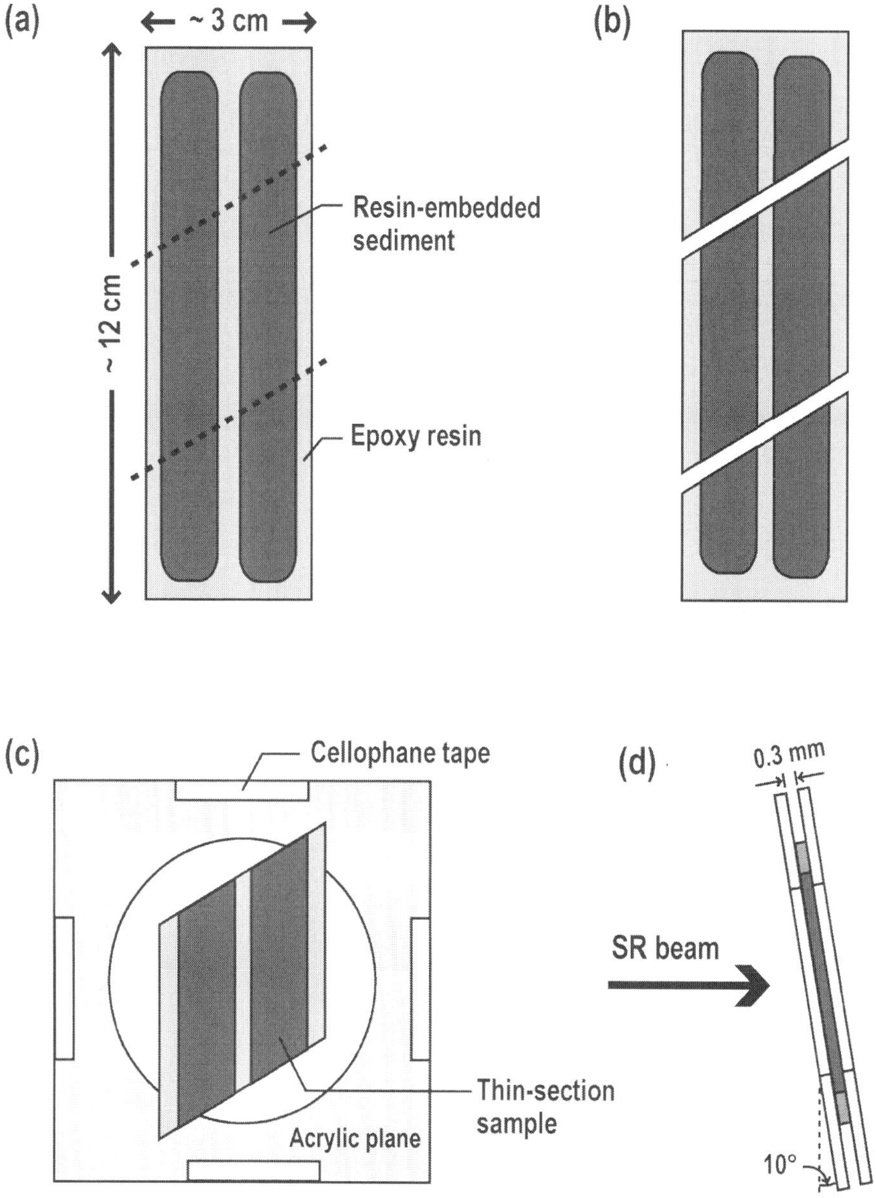

Fig. 3. Schematic diagrams of resin-embedded sediment sample and its experimental setup. (d) is indicated by red rectangle in Fig. 4a.

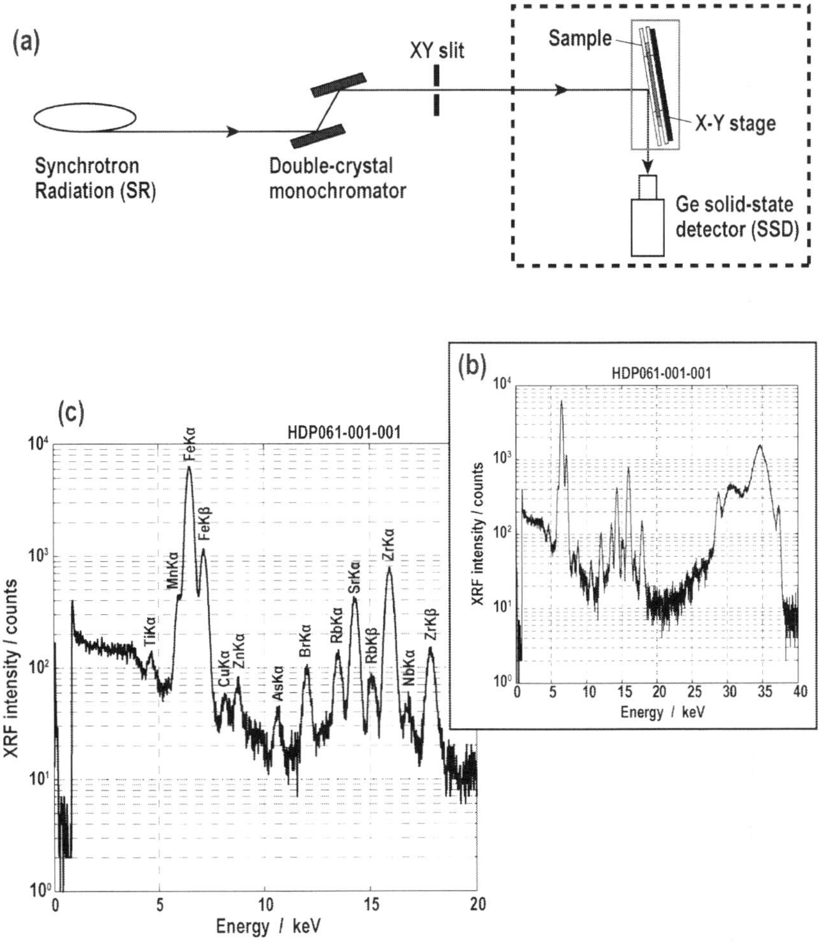

Fig. 4. (a) Schematic sketch showing experimental setup for SR μ-XRF mapping at BL37XU of SPring-8 and XRF spectra in ranges of (b) 0–40 keV and (c) 0–20 keV for upper section of HDP-06 core 1-1 from Lake Hovsgol. Sample of red rectangle in (a) corresponds to Fig. 3d.

3. Results and discussions

3.1 Distribution of elements in Lake Hovsgol sediment

Successive maps and profiles of the 11 elements in the upper section of HDP-06 core 1-1 are shown in Fig. 5. From a visual inspection of the distribution and features, the 11 elements were classified into three assemblages–group 1: Ti, Fe, Cu, Zn, As, Rb, Sr, Zr, and Nb; group 2: Br; and group 3: Mn, Fe, and As. Their distributions in the sediment are characterized as follows:

1. Groups 1 and 2 consist of several centimeter-scale layers that alternate with each other (Figs. 5a and c–k).
2. The manganese of group 3 is irregularly distributed over the entire section, occurring in thin layers and spots (Fig. 5b).
3. A portion of the arsenic from group 3 occurs in thin layers, together with Fe (Figs. 5c and f).

Based on the distributions and assemblages of the elements, group 1 is recognized to be elements composed of rock-forming minerals in the bedrock of the Hovsgol basin (Murakami et al., 2010). The group 1 is therefore terrigenous elements, which were supplied from the drainage basin by erosion and weathering processes. Bromine belonging to group 2 is a biophilic element whose abundance variation in the sediment reflects bioproduction in the lake (Phedorin et al., 2008). In Lake Hovsgol, the record of Br as alternative to diatom and bioSi, especially in the bottom sediment below 5.74 m (Fedotov et al., 2004b), is an important source of information to estimate biogenic production of the lake because the diatom frustules may be dissolved in the sediment high-pH pore water resulting from the presence of carbonates.

Manganese, Fe, and As of group 3 are elements sensitive to redox condition in the sediment. Existences of these three elements were indentified also in core X106 (Murakami et al. 2010). According to the study, the Mn in the core showed irregular distributions from the last glacial/Holocene transition to the Holocene section. The XRD analyses didn't indicate peaks of minerals containing Mn, thereby suggesting that the Mn in the sediment exists in an amorphous state. On the other hand, the coexistence of Fe and As identified in this study (red arrows in Figs. 5c and f) were identified in the last glacial/Holocene transition section of core X106. Because the section contains pyrite and dolomite as well as arsenic, Murakami et al. (2010) suggested the presence of sulfate-reducing bacteria.

These three elements of group 3 may have responded to redox changes and have subsequently migrated during the diagenetic process in the sediment. We therefore investigate paleoenvironmental implications for variations in the group 1 terrigenous elements and group 2 Br records of core HDP-06, as discussed below.

3.2 Continental erosion/weathering changes in central Asia during the Holocene period

Comparison of the terrigenous elements of group 1 (Ti, Fe, Cu, Zn, As, Rb, Sr, Zr, and Nb) with the Br of group 2 in the HDP-06 core 1-1 indicates a strong counterphase in the Holocene section (Fig. 5). At the present stage, the age of the section has not yet been determined. However, as shown in Figs. 6b and c, the content variation of Ti in the HDP-06 sediment is well correlative with that of dated X106 core (Murakami et al., 2010). Accordingly, by a correlation between these two profiles, we discuss hereafter nature of the temporal content variations of Ti and Br in HDP-06 core, together with paleoproxy records of X106 core.

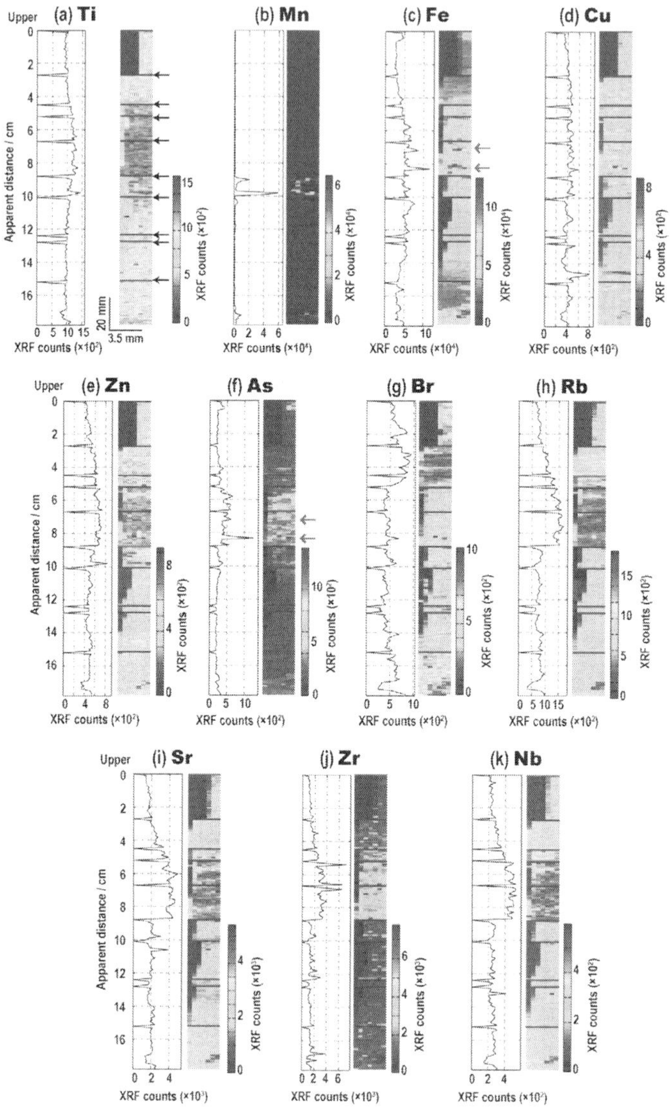

Fig. 5. Successive maps and profiles of major and trace elements in upper section of HDP-06 core 1-1: (a) Ti, (b) Mn, (c) Fe, (d) Cu, (e) Zn, (f) As, (g) Br, (h) Rb, (i) Sr, (j) Zr, and (k) Nb. The black arrows in (a) show the boundaries between the neighboring measurement areas, which were overlapped by approximately several millimeters. The vertical distance of each element profile hence represents the apparent core depth. The red arrows in (e) and (f) show the coexistence of Fe and As. The low XRF counts seen on the left side of each map were caused by the epoxy resin.

Figure 6 shows successive profiles for sediment chemistry of Lake Hovsgol, atmospheric CO_2 (Indermühle et al. 1999), and general circulation model (GCM)-predicted climatic parameters in central Asia (Bush, 2005). The second principal component (PC-2) score (Fig. 6d) was obtained by principal component analysis of 21 chemical components in bulk-sediment of X106 core by the ICP-MS (inductively coupled plasma mass spectrometry) analyses (Murakami et al., 2010). The variability of the PC-2 score was controlled by chemical elements from detrital materials, thereby indicating erosion/weathering intensity in the Hovsgol drainage basin. Temporal variations of annual mean temperature and precipitation minus evaporation (PME) in central Asia (Figs. 6e and f) are GCM outputs of simulation accounting to the combined effect of orbital and CO_2/H_2O (Fig. 6g) forcing.

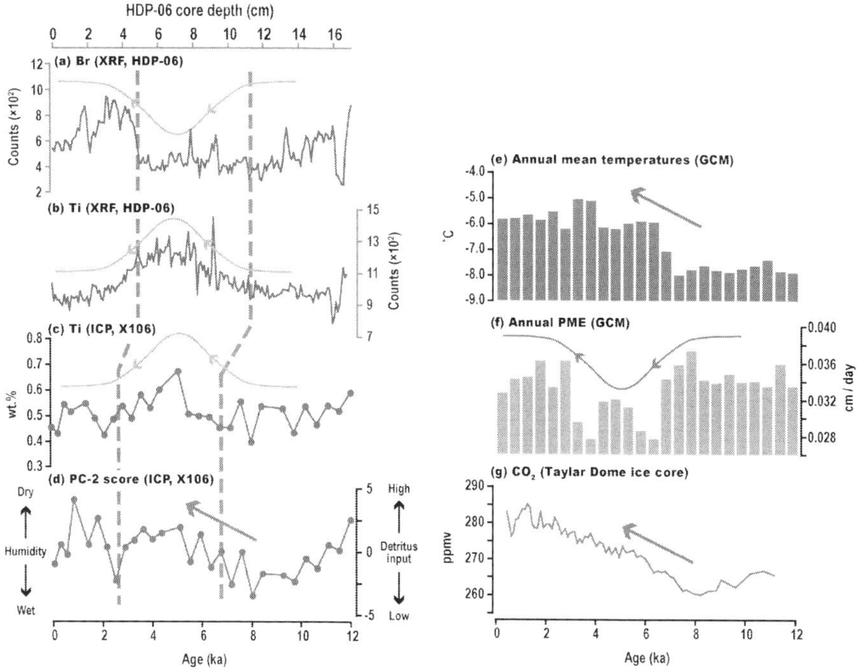

Fig. 6. Comparison of (a)-(d) paleoenvironmental proxy records from Lake Hovsgol, (e)-(f) GCM-simulated selected climatic parameters over central Asia (Bush, 2005), and (g) atmospheric CO_2 concentration record from Antarctica ice core at Talyor Dome (Indermühle et al., 1999) over the Holocene interval. (a) Br and (b) Ti profiles in Lake Hovsgol sediment of HDP-06 core 1-1 are captured by the SR μ-XRF analysis. (c) Ti and (d) PC-2 score in Lake Hovsgol core X106 are from Murakami et al. (2010). PME in (f) stands for precipitation minus evaporation. Arrows with curved or straight lines denoted in each plot emphasize significant environmental/climatic trend. The horizontal scale in (a)-(b) and that in (c)-(g) are shown by the core depth in the upper axis and by the age in lower axis, respectively. Vertical gray dashed lines in (a)-(d) indicates a visual correlation of Ti in cores between HDP-06 and X106.

The variations of Ti and Br contents in core HDP-06 are similar to that of PME over the Holocene period (Figs. 6a, b, and f). The relationship between Ti and PME shows an inverse correlation, whereas the relationship between Br and PME shows a positive correlation. In the mid-Holocene, the Ti intensity peaks at low PME, and the Br content increases in the early- and late-Holocene when the PME rises. On the other hand, the PC-2 score in core X106 (Fig. 6d), together with annual mean temperature in central Asia (Fig. 6e) and atmospheric CO_2 concentration (Fig. 6f), shows a gradual increase from about 8.0 ka to the present day.

These two-type variations observed in cores HDP-06 and X106 are considered to have resulted from changes in erosion/weathering intensity of central Asia (Asian continental interior) with moisture changes.

Evidences and suggestions supporting our hypothesis are provided by studies on the sediment from Lake Hovsgol. Prokopenko et al. (2007) showed that the early Holocene diatom/biogenic silica (bioSi) peaks correlated with the humidity maximum (Fig. 6f) reconstructed by the pollen fossil analyses and from predictions of GCMs (Bush, 2005). Based on these observations, Prokopenko et al. (2007) regarded that the early Holocene increase of diatom abundance in Lake Hovsgol was caused by increased nutrient supply with high precipitation and surface runoff. The early Holocene diatom/bioSi peaks correspond to the increased Br contents in core HDP-06 (Fig. 6a). Murakami et al. (2010) observed that low PC-2 detrital input occurred during the early Holocene (Fig. 4d). To explain the early Holocene decease of erosion/weathering intensity in the drainage basin with high humidity, Murakami et al. (2010) proposed that the detritus supply to Lake Hovsgol may have been controlled by the amount of vegetation cover: (1) vegetation cover in the catchment increased with high precipitation; (2) as a result, the nutrient supply to the lake enhanced, which in turn resulted in high productivity in the lake; (3) simultaneously, declined erosion through the drainage basin resulted in a reduced sediment supply into the lake.

The interpretation for the early Holocene detritus input by Murakami et al. (2010) can apply to the gradual increase from about 8.0 ka observed in the PC-2 score (Fig. 6d), the regional annual mean temperature (Fig. 6e), and atmospheric CO_2 concentration (Fig. 6g): (1) the annual mean temperature increases with the rise in atmospheric CO_2; (2) because of the exponential increase of the saturation vapor pressure with air temperature, the moisture decreases; (3) the resultant aridity of the continental interior has intensified the erosion/weathering processes.

4. Conclusions

We measured nondestructively 11 major and traces elements in the Holocene sediment of core HDP-06 core from Lake Hovsgol, using SR μ-XRF mapping techniques. A visual inspection of the acquired XRF images revealed that the 11 elements were classified into the three assemblages–group 1: Ti, Fe, Cu, Zn, As, Rb, Sr, Zr, and Nb, composed of rock-forming minerals; group 2: Br, recognized as a biophilic trace elements; and group 3: Mn, Fe, and As, sensitive to redox condition in the sediment.

Temporal variations in the first two groups, based on the age of core X106, are in phase with GCM-simulated PME in central Asia. This trend is remarkably different from the PC-2 score of core X106 indicating detritus input into the lake. The PC-2 score, together with atmospheric CO_2 and the regional annual mean temperature, shows a gradual increase from early Holocene to the present day. These two-type variations suggest that the continental erosion/weathering in central Asia occurred on two different processes and time-scales.

5. Acknowledgement

The authors are very grateful to the staff of Technical Center of Nagoya University. K. Suzuki provided helpful support in designing a sheet bender. S. Yogo made many helpful suggestions on sample preparation. A part of μ-XRF analysis was performed with the approval of SPring-8 (Proposal No. 2006B1091 and 2007A1563). We also thank to Y. Terada for μ-XRF mappings using synchrotron radiation. We are grateful to the HDP members for subsampling from the HDP-06 core. This study was supported by the Sumitomo Foundation grant No.103359; the Saijiro Endo Memorial Foundation; awards from Dynamics of the Sun-Earth-Life Interactive System, No. G-4 at Nagoya Univ., and Environmental Monitoring and Prediction of Long- and Short-Term Dynamics of Pan-Japan Sea Area, No. E-07 at Kanazawa Univ., of the 21st Century COE Program of the Ministry of Education, Culture, Sports, Science and Technology, Japan.

6. References

Bush, A. B. G. (2005). CO_2/H_2O and orbitally driven climate variability over central Asia through the Holocene. *Quaternary International,* Vol. 136, No. 1, pp. 15-23. ISSN 1040-6182

Fedotov, A. P., Chebykin, E. P., Semenov, M. Y., Vorobyova, S. S., Osipov, E. Y., Golobokova, L. P., Pogodaeva, T. V., Zheleznyakova, T. O., Grachev, M. A., Tomurhuu, D., Oyunchimeg, T., Narantsetseg, T., Tomurtogoo, O., Dolgikh, P. T., Arsenyuk, M. I. & De Batist, M. (2004a). Changes in the volume and salinity of Lake Khubsugul (Mongolia) in response to global climate changes in the upper Pleistocene and the Holocene. Palaeogeography, Palaeoclimatology, Palaeoecology, Vol. 209 Nos. 1-4, pp. 245–257, ISSN 0031-0182

Fedotov, A., Kazansky, A. Y., Tomurhuu, D., Matasova, G., Ziborova, G., Zheleznyakova, T., Vorobyova, S., Phedorin, M., Goldberg, E., Oyunchimeg, T., Narantsetseg, T., Vologina, E., Yuldashev, A., Kalugin, I., Tomurtogoo, O. & Grachev, M. (2004b). A 1-Myr record of paleoclimates from Lake Khubsugul, Mongolia. *EOS, Transactions, American Geophysical Union,* Vol. 85, No. 40, pp. 387, 390, ISSN 0096-3941

Fedotov, A., San'Kov, V., De Batist, M., Kazansky, A., Parfeevets, A., Miroshnitchenko, A. & Pouls, T. (2006). Chronology of the Baikal rift system. *EOS, Transactions, American Geophysical Union,* Vol. 87, No. 25, pp. 246, 250, ISSN 0096-3941

Goulden, C. E., Tumurtogoo, O., Karabanov, E. & Mongontsetseg, A. (2006). The geological history and geography of Lake Hövsgöl, In: *The Geology, Biodiversity and Ecology of Lake Hövsgöl (Mongolia),* Goulden, C. E., Sitnikova, T., Gelhaus, J. & Boldgiv, B., pp. 1–19, Leiden, The Netherlands, ISBN 9057821621

Hayakawa, K., Sekino, T., Yoshioka, T., Maruo, M. & Kumagai, M. (2003). Dissolved organic carbon and fluorescence in Lake Hovsgol: factors reducing humic content of the lake water. Limnology, Vol. 4, No. 1, pp. 25–33, ISSN 1439-8621

Hovsgol Drilling Project Members (2009). Sedimentary record from Lake Hovsgol, NW Mongolia: Results from the HDP-04 and HDP-06 drill cores. *Quaternary International,* Vol. 205, No. 1-2, pp. 21–37, ISSN 1040-6182

Indermühle, A., Stocker, T. F., Joos, F., Fischer, H., Smith, H. J., Wahlen, M., Deck, B., Mastroianni, D., Tschumi, J., Blunier, T., Meyer, R. & Stauffer, B. (1999). Holocene carbon-cycle dynamics based on CO_2 trapped in ice at Taylor Dome, Antarctica. Nature, Vol. 398, No. 6723, pp. 121-126, ISSN 0028-0836

Katsuta, N., Takano, M., Kawakami, S., Togami, S., Fukusawa, H., Kumazawa, M. & Yasuda, Y. (2007). Advanced micro-XRF method to separate sedimentary rhythms and event layers in sediments: its application to lacustrine sediment from Lake Suigetsu, Japan. *Journal of Paleolimnology,* Vol. 37, No. 2, pp. 259–271, ISSN 0921-2728

Murakami, T., Katsuta, N., Yamamoto, K., Takamatsu, N., Takano, M., Oda, T., Matsumoto, G. I., Horiuchi, K. & Kawai, T. (2010). A 27-kyr record of environmental change in central Asia inferred from the sediment record of Lake Hovsgol, northwest Mongolia. *Journal of Paleolimnology,* Vol. 43, No. 2, pp. 369–383, ISSN 0921-2728

Namkhaijantsan, G. (2006). Climate of the Hövsgöl Lake region, In: *The Geology, Biodiversity and Ecology of Lake Hövsgöl (Mongolia),* Goulden, C. E., Sitnikova, T., Gelhaus, J. & Boldgiv, B., pp. 63–76, Leiden, The Netherlands, ISBN 9057821621

Phedorin, M. A., Fedotov, A. P., Vorobieva, S. S. & Ziborova, G. A. (2008). Signature of long supercycles in the Pleistoncene history of Asian limic systems. *Journal of Paleolimnology,* Vol. 40, No. 1, pp. 445–452, ISSN 0921-2728

Prokopenko, A.A., Khursevich, G. K., Bezrukova, E. V., Kuzmin, M. I., Boes, X., Williams, D. F., Fedenya, S. A., Kulagina, N. V., Letunova, P. P. & Abzaeva, A. A. (2007). Paleoenvrionmental proxy records from Lake Hovsgol, Mongolia, and a synthesis of Holocene climate change in the Lake Baikal watershed. *Quaternary Research,* Vol. 68, No. 1, pp. 2-17, ISSN 0033-5894

Prokopenko, A. A., Kuzmin, M. I., Williams, D. F., Gelety, V. F., Kalmychkov, G. V., Gvozdkov, A. N. & Solotchin, P. A. (2005). Baisn-wide sedimentation changes and deglacial lake-level rise in the Hovsgol basin, NW Mongolia. *Quaternary International,* Vol. 136, No. 1, pp. 59–69, ISSN 1040-6182

Terada, Y., Goto, S., Takimoto, N., Takeshita, K., Yamazaki, H., Shimizu, Y., Takahashi, S., Ohashi, H., Furukawa, Y., Matsushita, T., Ohata, T., Ishizawa, Y., Uruga, T., Kitamura, H., Ishikawa, T. & Hayakawa, S. (2004). Construction and commissioning of BL37XU at SPring-8. *AIP Conference Proceedings,* Vol. 705, pp. 376-379, ISSN 0094-243X

Tiljander, M., Ojala, A., Saarinen, T. & Snowball, I. (2002). Documentation of the physical properties of annually laminated (varved) sediments at a sub-annual to decadal resolution for environmental interpretation. *Quaternary International,* Vol. 88, No. 1, pp. 5–12, ISSN 1040-6182

Part 3

Biological Responses to Environmental Change

Primary Succession in Glacier Forelands: How Small Animals Conquer New Land Around Melting Glaciers

Sigmund Hågvar
Department of Ecology and Natural Resource Management, Norwegian University of Life Sciences, Aas, Norway

1. Introduction

An easily observed effect of global warming is the gradual melting of glaciers in different parts of the world. Large areas of barren, pristine ground are left open for colonisation of various life forms (Fig. 1). From an ecological point of view, glacier forelands are interesting because they illustrate nature's ability to recover from severe disturbance. Since the successive development of communities starts without previous life forms, it is a primary succession. In contrast, a secondary succession starts with a species assemblage already present, for instance on a forest patch after clear-cutting. While the botanical succession in glacier forelands has been well studied, the parallel zoological succession is less described and understood. Which animal species are pioneers, what properties make them pioneers, how fast does species number increase, and how do plants and animals interact during succession? An ecological understanding of primary succession is not only of scientific interest, but also helps us to predict future ecosystems in areas freed from the ice cover.

2. Glacier forelands: Nature's ecological laboratory

In some glacier forelands, glaciologists have followed the varying position of the ice edge during long time, sometimes supported by old photographs. The age of certain characteristic moraines can, for instance, be well dated, and the age of sites between may be estimated. Several European glaciers had a maximum size at the end of the "Little Ice Age", which in Norway ended around A.D. 1750 with well-marked moraines. Forelands with dated sites up to 250 years age represent unique ecological laboratories for understanding nature's ability to conquer new land.

Ideally, a primary succession should be studied by following the gradual changes in flora and fauna in a fixed site over long time, from being newly freed from the ice cover, to having achieved a stable community structure. This is rarely possible, and the usual way is to substitute time with space, using plots with known age to estimate the future biological status of newly exposed land. The sequence of dated study plots in the foreland, illuding the succession on a given site over time, is called a chronosequence.

Fig. 1. Foreland at the Midtdalsbreen glacier snout, a part of the Hardangerjøkulen glacier in central south Norway. Photo: Sigmund Hågvar.

The ideal situation in a chronosequence is that the glacier has retreated at a constant speed, that climate conditions have been stable, and that the exposed ground has not been subject to reworking, for instance by glacier rivers. If temperature has been especially high during the last decades, the youngest sites may have developed faster than older sites did in their early phases of succession. Also, the source sites from which colonising organisms derive, can be influenced by climate change. A perfect chronosequence in all respects is hard to find, but certain forelands contain good historical information and well dated sites.

The botanical changes from a pioneer flora to a closed and stable plant community has been described in various glacier forelands, both in Norway (Matthews & Whittaker, 1987; Matthews, 1992; Vetaas, 1994, 1997), in the Austrian Alps (Moreau et al., 2005; Raffl, 1999; Raffl et al., 2006), and in Alaska (Chapin et al., 1994). Especially during the last decade, increased insight has also been given in the zoological succession along receding glaciers, from three different geographical areas in Europe: Svalbard, the Alps, and Norway. The invertebrate succession in two glacier forelands in Svalbard was described by Hodkinson et al. (2004). In Austria, early faunistic studies in glacier forelands by Janetschek (1949, 1958) and Franz (1969) were followed by Gereben (1994, 1995) on carabid beetles, and Paulus & Paulus (1997) on spiders. Recently, the foreland of the Austrian Rotmoos glacier has been under intense study, including invertebrate succession (Kaufmann, 2001, 2002; Kaufmann et al., 2002; Kaufmann & Raffl, 2002). In Italy, Zingerle (1999) studied spiders and harvestmen in the Dolomites, and Gobbi et al. (2006a,b, 2007) have described epigean arthropod succession in a glacier foreland in Central Italian Alps.

From southern Norway, three master/PhD theses based on pitfall trapping in glacier forelands focused mainly on surface active beetles and spiders (Alfredsen, 2010; Bråten & Flø, 2009; Vater, 2006). The other Norwegian faunistic studies in glacier forelands dealt with soil living mites (Acari) (Hågvar et al., 2009; Seniczak et al., 2006; Skubala & Gulvik, 2005) or springtails (Collembola) (Hågvar, 2010). Time has come to summarize and compare the invertebrate succession in these different geographical areas, looking for similarities, differences, and mechanisms.

Fig. 2. Midtdalsbreen glacier snout in Norway, August 2010: Behind the author, a 20 m broad belt of barren ground was freed from ice during this summer. Photo: Daniel Flø.

3. Life on barren ground: The pioneer animals

Several invertebrate groups are present on barren, vegetation-free ground close to the glacier boarder (Figs. 2-3). Typical representatives are springtails (Collembola) and mites (Acari), which are collectively named microarthropods, as well as beetles (Coleoptera), spiders (Araneae) and harvestmen (Opiliones). Since there is no organic layer, the pioneer invertebrates are surface active species, but they can find shelter in the crevices among stones, gravel and sand grains.

Fig. 3. Pitfall traps (with visible lids) on this barren, three year old moraine trapped many species of spiders, beetles, springtails and mites. Photo: Sigmund Hågvar.

3.1 Pioneer springtails (Collembola)

Table 1 lists pioneer springtails collected close to Midtdalsbreen glacier snout, which is a part of Hardangerjøkulen glacier in alpine south Norway. Pitfall catches from young ground illustrate the relative surface activity at ages 3, 36 and 47 years, while flotation a few meters from the ice edge (age 0 years) proved the presence of two species on freshly exposed ground (Fig.2). One of these, *Agrenia bidenticulata*, is characteristic for cold, moist habitats in arctic and alpine areas (Fjellberg, 2007). This specialized species disappeared already after 30-40 years, at which time two generalist species dominated the surface activity: *Lepidocyrtus lignorum* and *Isotoma viridis*. Table 1 also shows an intermediate phase after 3 years, where the surface activity was dominated by the large, sphere-formed species *Bourletiella hortensis,* and as much as eight species were already present. This case illustrates the great colonisation ability of springtails, how both specialists and generalists participate, and how the community structure may undergo rapid changes during the first few years. Although the Collembola fauna is different in forelands on Svalbard (Hodkinson et al., 2004) and in the Austrian Alps (Kaufmann et al., 2002), the colonisation pattern has certain features in common in these threes geographical sites: Springtails were among the earliest colonisers with a documented presence after only 2-4 years, and Isotomidae and Hypogastruridae were often pioneer families.

Age (year)	0	3	32-36	41-47
Sampling method	Flotation	Pitfall	Pitfall	Pitfall
Agrenia bidenticulata	84.6	24.7		
Desoria infuscata	15.4	1.5	6.5	
Bourletiella hortensis		59.9	1.1	
Isotoma viridis		5.2	28.3	21.3
Lepidocyrtus lignorum		0.4	50.0	65.3
Desoria olivacea		8.0	8.7	4.5
Desoria tolya		0.2	4.3	3.0
Ceratophysella scotica		0.1	1.1	6.1
NUMBER OF ANIMALS	26	1465	92	66

Table 1. Collembola sampled from young sites near the Midtdalsbreen glacier snout, Norway. Percentage dominance of various species are given. Pitfall catches illustrate surface activity. Flotation was used close to the ice boarder.

3.2 Pioneer mites (Acari)

Pitfall trapping near Midtdalsbreen glacier snout in Norway documented a considerable surface activity of small mites after 3 years, belonging to Actinedida (earlier called Prostigmata). Pitfalls and soil samples at 32-47 year age showed that other mite groups had then been added: Oribatida and Gamasina (Hågvar et al., 2009). Also on Svalbard and in the Alps, mites were recorded after only 2-4 years (Hodkinson et al., 2004; Kaufmann et al., 2002). Interestingly, the small, generalist oribatid species *Tectocepheus velatus* was a pioneer species both at Midtalsbreen, in another foreland at Jostedalsbreen glacier in south Norway (Skubala & Gulvik, 2005), as well as in two forelands on Svalbard (Hodkinson et al., 2004). Moreover, the species was found to be a characteristic pioneer on vegetation-free post-industrial dumps in Poland (Skubala, 2004). Also certain small species of the oribatid family Brachychthoniidae were early colonisers both in Norwegian forelands and in the industrial dumps. Maybe the pioneer community is more predictable for mites than for springtails in a primary succession.

3.3 Pioner spiders (Araneae)

Also spiders are among the first colonisers on barren ground, both on Svalbard, in Norway and in the Alps. Pioneer spiders often belong to the family Lycosidae (wolf spiders) or Linyphiidae (sheet web spiders). Wolf spiders are robust and agile night hunters with good eyesight and hunt without constructing a web. In the Italian Alps, the wolf spider *Pardosa saturatior* is in fact living among debris on the glacier surface (Gobbi et al., 2006a). In the foreland of the Austrian Rotmoos glacier this species is a typical pioneer together with *Pardosa nigra* (Kaufmann, 2001). In several South-Norwegian glacier forelands, the genus *Pardosa* was recognized as a pioneer by Vater (2006). The species *Pardosa trailli* (Fig. 4) was identified both in several young forelands in the Jotunheimen area by Vater (2006), and close to the Hardangerjøkulen glacier (Bråten & Flø, 2009).

The large group of sheet web spiders are small, delicate animals which weave horizontal, sheet like webs under which they are hanging. Above the sheets the spider inserts strands, into which insects fly and fall onto the web. In two Svalbard forelands, *Erigone arctica* was found after 16 years (Hodkinson et al., 2004). The same species is also a pioneer at the

Hardangerjøkulen glacier in South Norway, being recorded on a three year young moraine (Bråten & Flø, 2009). A related species, *Erigone tirolensis*, is a common pioneer at the Hardangerjøkulen glacier and in the Rootmostal foreland in Austria (Bråten & Flø, 2009; Kaufmann, 2001).

Fig. 4. The large and fast-running wolf spider *Pardosa trailli* is a pioneer species in Norwegian glacier forelands. Photo: Sigmund Hågvar.

3.4 Pioneer harvestmen (Opiliones)

Mitopus morio is a large and very active species which is common on newly exposed ground, less than 20 years old, in several South-Norwegian forelands studied by Vater (2006). Bråten & Flø (2009) found it abundantly on a three year old moraine close to Hardangerjøkulen glacier in south Norway. Together with *M. glacialis*, the species is also a pioneer in the Alps (Kaufmann, 2001).

3.5 Pioneer beetles (Coleoptera)

While beetles were absent in the forelands studied on Svalbard (Hodkinson et al., 2004), they belonged to the pioneers in all studies in Norway and the Alps. A characteristic picture is that species within the family Carabidae are present on barren ground close to the glacier. Moreover, certain genera, and even species, are common pioneers in these two geographical areas. Typical genera in this respect are *Nebria, Amara* and *Bembidion*. In South Norway,

Vater (2006) found the following carabid species in various foreland sites younger than 20 years: *Amara alpina, A. quenseli, Nebria nivalis,* and *Bembidion fellmanni.* The same species were pioneers close to Hardangerjøkulen glacier, except for another *Bembidion* species: *B. hastii* (Bråten & Flø, 2009, Fig. 5). In the Austrian Alps, Kaufmann (2001) recorded *Amara quenseli*, four *Nebria* species and a *Bembidion* species as early colonisers, and also Gobbi et al. (2006a) recorded *A. quenseli* as a pioneer species in the Italian Alps.

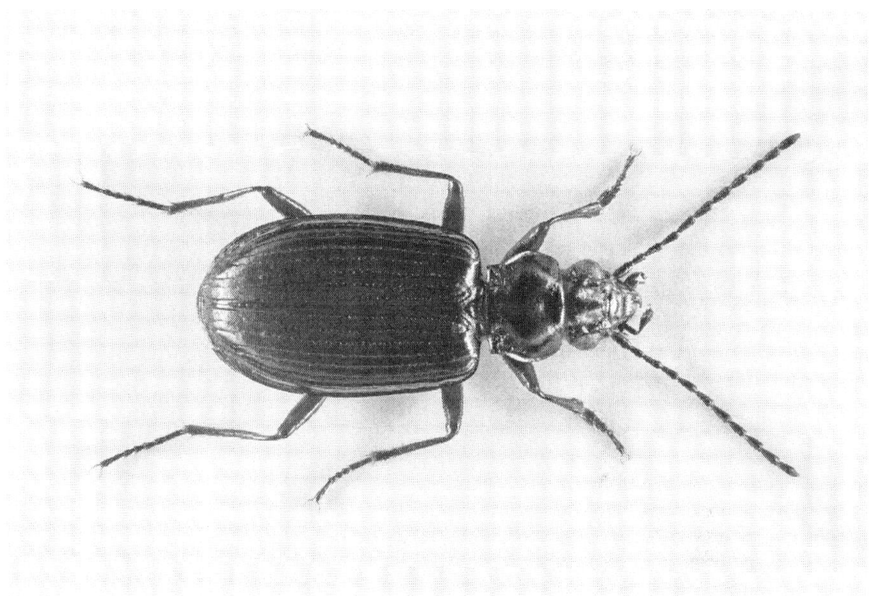

Fig. 5. *Bembidion hastii,* a pioneer predatious Carabidae beetle in Norwegian glacier foreland. Photo: Oddvar Hanssen.

3.6 Other pioneer groups

On a four year old barren moraine at Hardangerjøkulen glacier in South Norway, extraction of soil samples revealed the presence of Rotatoria and at least five different Nematoda genera. Tardigrada was found close to a single plant of *Poa alpina* (Christer Magnusson, pers. comm). In Austria, Kaufmann et al. (2002) found both Nematoda and Enchytraeidae in soil samples younger than 40 years. On Svalbard, Hodkinson et al. (2004) recorded larvae of terrestrial Chironomidae after 2 years and Enchytraeidae after 37 years.

4. Dispersion: How to get there?

Pioneer invertebrate species must be good dispersers, but our knowledge in this field is limited. The easiest dispersion would be by air, either by active flight, by passive wind transport, or a combination. On Svalbard, areal dispersal of invertebrates over the foreland of Midtre Lovénbreen glacier was studied by Coulson et al. (2003). Large numbers of Diptera, Hymenoptera and Araneae were caught in water and sticky traps, some only 15 m from the glacier snout on 2 year old ground. Sticky traps were placed either just above ground level, or at a height of 1 m. It was concluded that spiders caught 1 m above ground must have been aerially dispersed. The actual spider family, Linyphiidae, is known for their ability to fly by wire, called "ballooning". By raising the abdomen and gradually releasing a thread in the breeze, the spider is finally lifted upwards and can be blown very far away. Holm (1958) suggested that many spider species on Svalbard had originally arrived from Greenland as aerial plankton. The airborne Diptera and Hymenoptera in the glacier foreland represented a food source for the spiders. Another interesting observation was that more than 95 % of the animals caught in sticky traps were taken close to the ground, and very few at 1 m height. The vast majority of animals were dispersing at, or below, 0.25 m. Furthermore, animals were trapped from all directions, despite some prevailing wind directions during the study. It was concluded that these arctic insects appear to make flights of short duration and remain close to the ground where wind velocities are considerably reduced and air temperatures elevated. This behaviour enables them to perform directional flight largely independent of wind direction (Coulson et al., 2003).

Although several springtails and mites are early colonisers on Svalbard, Coulson et al. (2003) did not catch these groups in the sticky traps. Later, Mangnussen (2010) achieved some springtails and mites in water traps on Svalbard outside glaier forelands. His traps had a sticky rim to avoid crawling into the trap. Interestingly, the airborne transport of springtails seemed to occur at low wind speeds and in periods with high air humidity, indicating a high surface activity during such conditions. Since springtails can jump, they may be taken further by air currents. In Alaska, wind-blown springtails and mites have been collected in suspended plankton nets (Gressitt and Yoshimoto 1974). Elsewhere, they have even been taken as aerial plankton at altitudes of 1,500 m (Glick, 1939; Riley et al., 1995), so local dispersion by wind seems likely.

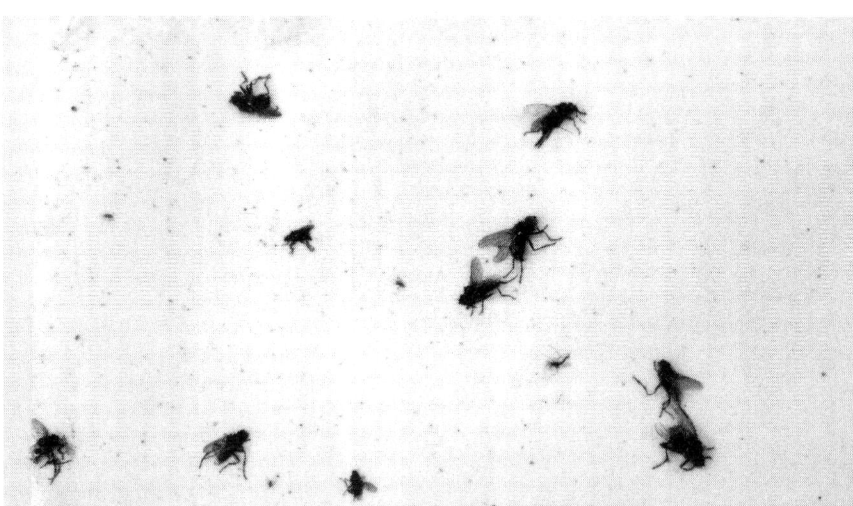

Fig. 6. Sticky trap illustrating activity of various Diptera on a 3 year old moraine at Midtdalsbreen glacier snout, Norway. Photo: Sigmund Hågvar.

Most beetles can fly, but not all. Not surprisingly, typical pioneer beetles are generally fully winged, while non-flying species are late colonisers in glacier foreland (Gobbi et al., 2007). However, dispersing by foot may be efficient in certain very active invertebrates, as the large Opiliones of the genus *Mitopus,* mentioned above.

The combined knowledge today indicates that pioneer communities in glacier forelands are rather predictable, and that dispersal may not seriously restrict community development (e.g. Hodkinson et al., 2004; Kaufmann, 2001; Vater, 2006). However, this does not mean that pioneer species have to be ecologically similar.

5. Ecological similarities and differences between pioneer invertebrates

5.1 Specialists or generalists?

Pioneer invertebrates in European glacier forelands comprise both specialists and generalists. Even specialists represent a heterogeneous group, depending on their speciality. Some are "cold-loving", represented by the springtail *Agrenia bidenticulata* (Hågvar, 2010) and certain carabid beetle species of the genus *Nebria,* for instance *Nebria nivalis* (Bråten & Flø, 2009; Gobbi et al., 2007; Kaufmann, 2001; Vater, 2006). Such cold-adapted species may increase their distribution if the area of pioneer ground increases due to an increased melting rate, but may eventually disappear locally if the glacier or snow field melts away. A second group of specialists are those preferring open, barren ground. Some of these, both among microarthropods, beetles and spiders, have an alpine and/or arctic distribution. However, some also occur in lowland areas on various sandy, gravely or stony habitats, for instance carabid species of the genus *Bembidion* (Bråten & Flø, 2009) or the springtail species *Bourletiella hortensis* (Hågvar, 2010).

Ecological generalists from several taxonomic groups are found in pioneer communities. These species tolerate a wide range of habitats, both in the lowland and in mountains. Examples from European glacier forelands are the carabid beetle *Amara quenseli,* the harvestman *Mitopus morio* , the springtail *Isotoma viridis,* and the oribatid mite *Tectocepheus velatus* (Bråten & Flø, 2009; Hågvar, 2010; Hågvar et al., 2009; Hodkinson et al., 2004; Kaufmann, 2002; Vater, 2009). An interesting point is that is rather predictable which "generalists" are present among the pioneers, in the same way as the specialists are predictable. Clearly, only a few "generalists" can extend their ecological niche far enough to thrive on pioneer ground close to a glacier - including the ability to arrive there. Later successional stages may contain several other "generalist species", but they do not have this extra flexibility.

5.2 Parthenogenetic or bisexual?

Some springtails and mites are parthenogenetic, which means that one single individual can start a local population. This ability is an obvious advantage for a pioneer species if dispersion is a limiting factor. In a glacier foreland in south Norway, Hågvar et al. (2009) found that the two characteristic pioneer mites were parthenogenetic. However, parthenogenetic mites were found along the whole foreland gradient, including some slow-dispersing species. Among springtails in the same foreland, the pioneer species were mainly bisexual (Hågvar, 2010). Therefore, among microarthropods, parthenogenesis is not more typical among pioneer species than among later colonisers. This may indicate efficient dispersal of individuals.

5.3 Short or long life cycle?

Pioneer species with a short life cycle might have an advantage compared to species with a long life cycle in establishing a high and permanent population. Most of the typical pioneer springtail species in alpine south Norway have a one-year life cycle, which is relatively "fast" under these conditions (Fjellberg, 1974; Hågvar, 2010). However, the pioneer oribatid mite *Tectocepheus velatus*, is assumed to use two or more years to fulfil the life cycle in the same area (Solhøy, 1975). This species was represented mainly with juveniles in the pioneer site at Hardangervidda, indicating local reproduction (Hågvar, 2010). For this species, a slow development does not seem to be a handicap in colonisation and establishment.

5.4 Resident survivors or continuously colonising?

High densities of pioneer microarthropods could be due to continuous transport by air. Theoretically, the pioneer ground might be an ecological sink, receiving animals which continuously die. However, filled guts in sampled microarthropods indicate feeding activity on the pioneer ground. Whether pioneer ground may to a large degree be a sink for ballooning spiders, is an open question.

5.5 Saprophagous super-pioneers?

In a glacier foreland in the Austrian Alps, Bardgett et al., (2007) found that pioneer, heterotrophic microbial communities to a large degree used ancient carbon released by the glacier as an energy source. Only after more than 50 years of organic matter accumulation did the soil microbial community change to one supported primarily by modern carbon, most likely from recent plant production. This means that also pioneer microarthropods feeding on fungi and bacteria could use ancient carbon, allowing microarthropods to establish resident populations immediately after the ground is laid free of ice. Inblown organic material will also gradually add substrate for saprophagous food chains. Microbial-feeding animals like microarthropods, rotatoria, tardigrada, nematoda and enchytraeidae may be the first animals which establish viable and resident populations independently of resources from outside. If so, they are the super-pioneers among animals.

6. The predator first- hypothesis

According to conventional ecological textbooks, a primary succession should start with the establishment of plants. These would offer life conditions for herbivore animals, which finally allow the presence of predators. Hodkinson et al. (2002) showed that in practice, newly exposed substrates, as a fresh glacial moraine or a cooled volcanic lava flow, are to a large degree inhabited by various predator invertebrates. In other words, the autotrophs are preceded by a largely unrecognized heterotrophic phase. Summing up literature documenting aerial transport and deposition of invertebrates, they assumed that pioneer predators were fed by a fallout of invertebrates onto land and water surfaces (Figs. 6-7). In adition, a fallout of detritus would favour scavenging detritivores. It was suggested that these heterotrophic communities conserve nutrients, particularly nitrogen, and facilitate the establishment of green plants. In a glacier foreland on Svalbard, Hodkinson et al. (2001) showed that spider densities were highly correlated with allochthonous inputs of potential prey items, predominantly chironomid midges. Coulson et al. (2003) further documented aerial transport of invertebrates in the same foreland.

Pioneer foreland communities containing macroarthropod predators have been documented both on Svalbard, in Norway, and in the Alps (Bråten & Flø, 2009; Gobbi et al., 2006a,b, 2007; Hodkinson et al., 2004; Kaufmann, 2001; Kaufmann & Raffl, 2002; Vater, 2006). While spiders represent the pioneer predators on Svalbard, a mixture of carabid beetles, various spiders and one or two harvestman species are typical on the European mainland.

Fig. 7. Air-borne insects sampled on the surface of the Hardangerjøkulen glacier, south Norway. These specimens have been a part of the air plankton, but low temperatures above the glacier have made them fall down. Photo: Marte Lilleeng.

6.1 The predator first-hypothesis challenged

The predator first-hypothesis is at first sight an ecological paradox, but can be explained if the predators are fed by airborne food as for instance chironomid midges. But how stable is the airborn input of suitable and sufficient food to the pioneer ground? As already pointed at by Hodkinson et al. (2002), detritivores such as Collembola can also be eaten by predators such as spiders and carabid beetles. But to what degree is this occurring, and how important are resident Collembola or mite species as a stable food source? Gut content analyses are needed to answer these questions, preferably by recognizing prey items via their specific DNA. Perhaps the input of predators is very high, for instance of ballooning spiders, and that predators to a large degree eat other predators? Are pioneer sites in practice large sinks, where the majority of even predators do not survive? And which of the pioneer beetles, spiders and harvestmen do really reproduce on the barren ground?
Recent studies in the foreland of Midtalsbreen glacier snout, south Norway, indicate that chlorophyll-based food chains may start very early. Interestingly, the key organisms in this respect are mosses. On a large moraine which was freed from ice in 2005, twenty pitfall

traps were operated during the snow-free season 2008. Besides a pioneer fauna of beetles, spiders and harvestmen (Bråten and Flø, 2009), mites and springtails, the traps contained inblown fragments of various mosses. These fragments, among them so-called bulbils (Fig. 8) are able to develop into moss colonies. However, because these diaspores are tiny and end up between stones and gravel, they are not visible by eye in the field. By studying the gut content of springtails in the traps, it was revealed that most individuals of the large, sphere-formed species *Bourletiella hortensis* had eaten leaves and/or rhizoids of mosses (Figs. 9-10). This species can be very active and was observed to make jumps up to 10 cm length on the moraine, so it can obviously locate the inblown moss fragments. If the predatory beetles, spiders and harvestmen can eat this species, chlorophyll-based food chains may start very early.

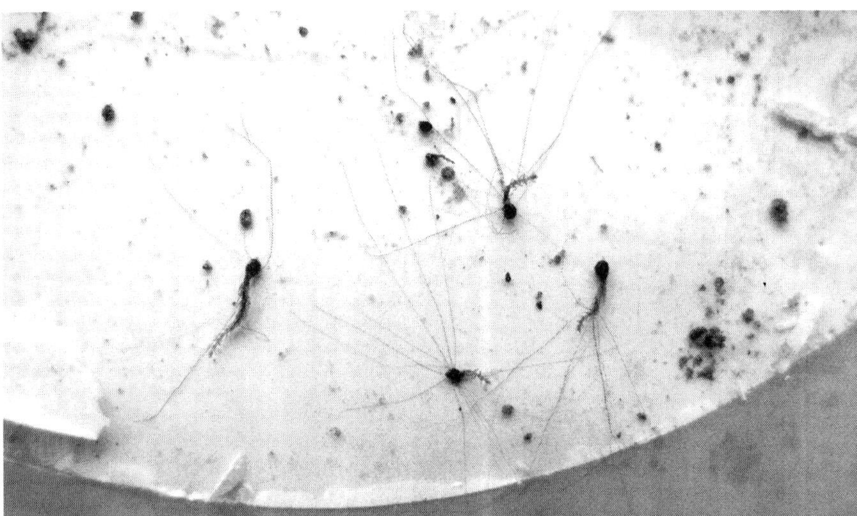

Fig. 8. Certain mosses can easily be wind-dispersed by so-called bulbils. This picture shows how individual moss plants, including rhizoids, develop from bulbils placed on moist plaster of Paris. Photo: Sigmund Hågvar.

The first moss patches may also serve as a habitat for certain moss-living macroinvertebrates. After four years, in 2009, dry extraction of a small moss patch on the moraine revealed two larvae of terrestrial Chironomidae, as well as larvae and a pupa of the beetle *Simplocaria metallica* (Byrrhidae). The pupa soon hatched in the laboratory. Few insects are moss-eaters, but the family Byrrhidae is an exeption (Figs. 11-12). The importance of mosses for early faunal succession in glacier forelands should be closer studied. Maybe pioneer mosses represent a "driver" which facilitates the colonisation of certain invertebrates.

It should also be noted that some pioneer carabid beetles in the genus *Amara* are considered to be omnivorous, for instance *Amara quenseli* and *Amara alpina* (e.g. Lindroth, 1986) and these might for instance feed on inblown seeds.

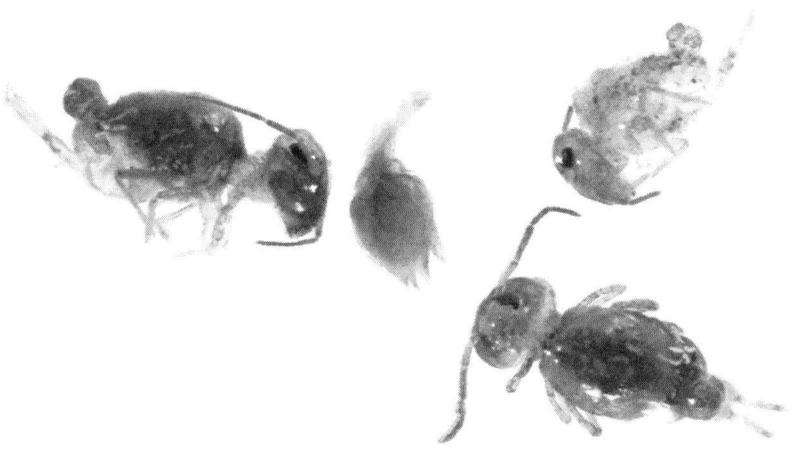

Fig. 9. Three specimens of the moss-eating springtail *Bourletiella hortensis*. An inblown moss fragment of 1 mm length is in the middle. Photo: Marte Lilleeng.

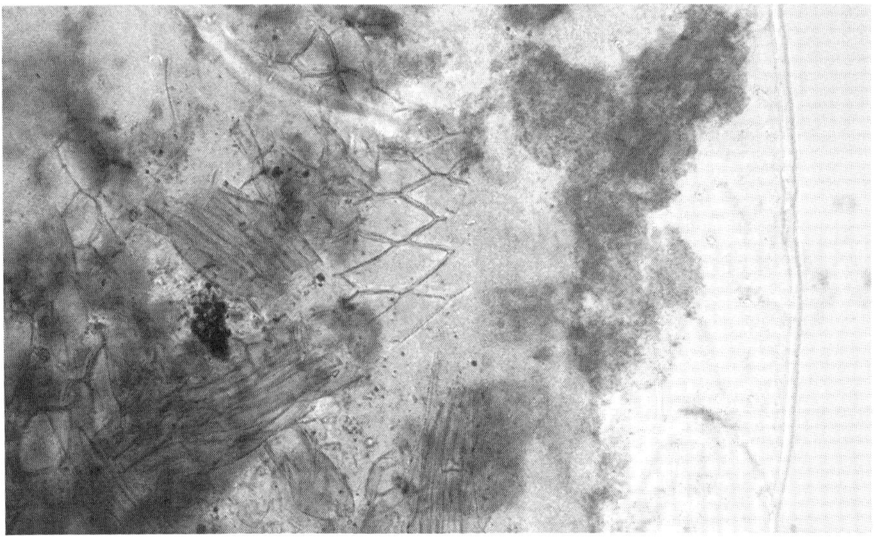

Fig. 10. Gut content of the springtail *Bourletiella hortensis* showing brown moss fragments. Photo: Marte Lilleeng.

7. A new look at pioneer communities?

The predator first-hypothesis is valuable by pointing to the fact that many predators are present rather immediately, before any visible primary production. Their food requirements can probably be fulfilled by aerial transport of invertebrates. But animal life on a young moraine is more complicated than that. The super-pioneers among animals are microflora-feeding groups belonging to the decomposer food chain, and some of these may serve as food for predators. Furthermore, certain moss-eating microarthropods, beetles and Chironomidae are present after few years, being part of chlorophyll-based food chains.

Fig. 11. Newly hatched adult and larva of the moss-eating beetle *Simplocaria metallica*. Photo: Marte Lilleeng

Fig. 12. Pioneer moss patch after four years. Midtdalsbreen glacier snout, south Norway. Photo: Sigmund Hågvar

Maybe also cyanobacteria with chlorophyll are present very early. Some of the typical pioneer beetles are omnivorous and may eat inblown seeds. Finally, who eats who is still an open question, as well as whether pioneer ground is a sink or a reproduction ground. The ecology of pioneer communities may be more complicated than earlier thought

Fig. 13. This rim of pioneer mosses along a large stone after four years is due to inblown moss fragments which have aggregated along the stone. Midtdalsbreen glacier snout, south Norway. Photo: Sigmund Hågvar.

8. Succession patterns after the pioneer stage

The succession from a pioneer stage to a mature, stable community goes through various phases which are more or less predictable. Not surprisingly, both the botanical and the zoological succession can be related to three more or less interrelated factors: Time since glaciation, distance to glacier and vegetation cover (e.g. Bråten & Flø, 2009; Gobbi, 2007; Hågvar, 2010; Hågvar et al., 2009; Hodkinsen et al., 2004; Kaufmann, 2001; Matthews, 1992; Vater, 2006). In the zoological succession, an obvious element is that herbivores like Chrysomelidae and Curculionidae have to wait for their host plant to be established. Furthermore, certain microarthropods and saprophagous beetles depend on a certain thickness of the soil organic layer (Bråten & Flø, 2009; Hågvar, 2010; Hågvar et al., 2009).

Different taxonomic groups may show different succession patterns in the same foreland. For instance, at the Midtdalsbreen glacier foreland in south Norway, springtails colonised faster than oribatid mites. After 70 years, 84 % of the springtail species in the chronosequence were present, but only 57 % of the oribatid mites (Hågvar, 2010; Hågvar et al., 2009). Beetles followed a pattern similar to springtails, while spiders colonised more gradually, similar to oribatid mites (Bråten & Flø, 2009). The general rate of succession can also differ between geographical sites. In the foreland of the Rotmoos glacier in Austria, most beetle and spider species were present after only 40-50 years (Kaufmann, 2001). This is a faster colonisation rate than observed in alpine south Norway (Bråten & Flø, 2009; Vater, 2009). The difference is probably due to a milder climate in the Austrian site, since certain taxa absent in the Norwegian sites were present there, e.g. Lumbricidae, Formicidae, and Diplopoda.

In an extensive study of eight different glacier forelands in south Norway, Vater (2006) demonstrated how altitude and local climate influenced colonisation rate, even within a small geographical area. In the forested sub-alpine zone the colonisation rate of macroarthropods was high, while the succession was very slow in a high alpine foreland. A "geoecological model" was proposed by Vater (2006) to explain three distinctive pathways of succession, representing the subalpine, the low/mid-alpine, and the high alpine zone, respectively. Certain characteristic species could, however, be pioneers at very different altitudes.

Although the colonisation rate may vary considerably between sites due to climatic differences, the sequence between different taxa or ecological groups may show striking similarities. We see that both within south Norway, and in comparison with the Alps (Bråten & Flø, 2009; Kaufmann, 2001; Kaufmann & Raffl, 2002; Vater, 2009). Among beetles, surface active predators within the family Carabidae are typical pioneers, while the species-rich family Staphylinidae dominates later. This family contains many small species which are favoured by the development of an organic soil layer. Most herbivorous species are also relatively late colonisers, except for the moss-eating genus *Simplocaria* (Byrrhidae), which can inhabit pioneer moss patches. An interesting aspect is that herbivore beetles do not necessarily colonise promptly when the food plant is established. In the Midtdalsbreen foreland, *Chrysomela collaris* (Chrysomelidae) was found after about 80 years, while the food plant *Salix herbacea* belonged to the pioneer species. The explanation may partly be that the beetle is a slow disperser, partly that a certain cover of the food plant is needed.

From the Alps, Kaufmann (2001) concluded that faunal colonisation and succession in alpine glacier forelands, to a large extent, followed predictable and deterministic assembly rules and that stochastic effects were of minor importance. Studies in Norway and Svalbard support this general picture. However, Kaufmann (2001) also stressed that favourable sun and light conditions may facilitate successional progress in local patches.

A general question in succession studies is the turnover rate of species. To answer this, most species have to be identified. Based on a limited taxonomical resolution, Vater (2006) concluded that most macroinvertebrates remained after colonisation. However, in the Alps, several pioneer species of spiders and beetles were absent in later successional stages (Gobbi, 2006b, 2007; Kaufmann, 2001). Bråten & Flø (2009) found that most spiders remained after colonisation, while beetles showed a certain turnover of species. Within mites and springtails, most species remain after colonisation, although their abundance and relative dominance may vary throughout the chronosequence (Hågvar, 2010; Hågvar et al., 2009).

9. Climate change and succession pattern

A gradually warmer climate will make it more difficult to use dated sites as a substite for time. Kaufmann (2002) concluded that an increase of 0.6°C in summer temperatures approximately doubled the speed of initial colonisation, whereas later successional stages were less sensitive to climate change. It is also possible that the surrounding source habitats may be influenced and increase their dispersion of species into the foreland.

Since the first effects of climate change are likely to be observed in terrestrial habitats at northern latitudes (IPCC, 2007), Norwegian studies may be especially relevant. Alpine areas of southern Norway have both had a marked temperature increase during the last 2-3 decades, and been subject to deposition of long-distance transported atmospheric nitrogen (Hole & Engardt, 2008; Ytrehus et al., 2008). Fertilization effects are most probable in nutrient-deficient ecosystems, such as alpine habitats with poorly developed soils. In a nutrient poor alpine *Dryas* heath in south Norway, experimental plots were artificially heated and/or fertilized to study the combined above-ground (plants) and below-ground (soil animals) effects (Hågvar & Klanderud, 2009). Nutrient addition and nutrient addition combined with warming resulted in several effects below ground on microarthropods as previously shown above ground on plants: Increased biomass, high dominance of a few rapid-growing species, contrasting responses of closely related species, and a reduction in species numbers. An earthworm (*Dendrobaena octaedra*) which was very rare in control plots, seemed to be favoured by the changes. These short-term responses (after 4 years) may have profound long-term effects in this alpine ecosystem.

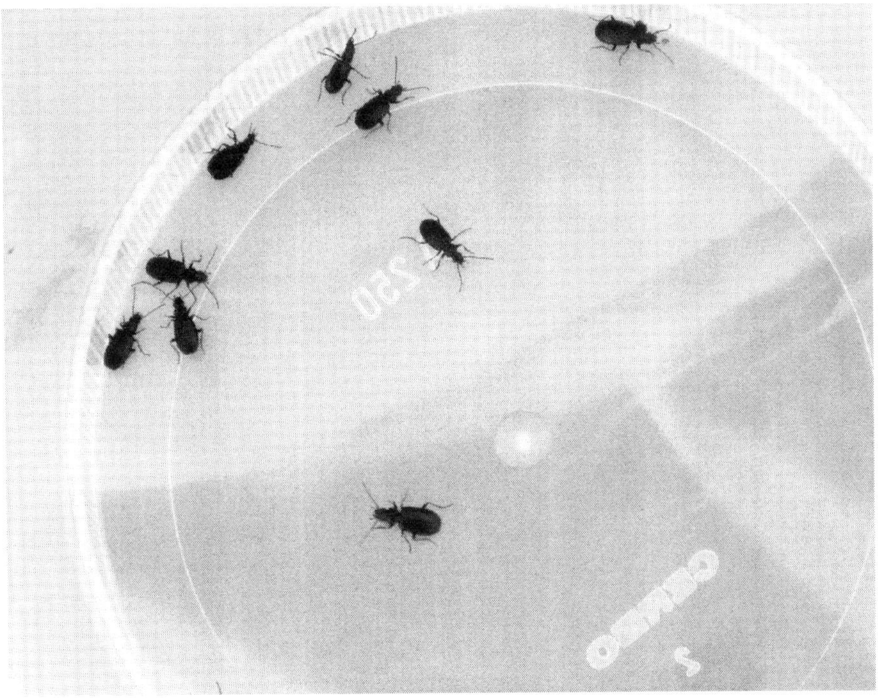

Fig. 14. Pitfall trap on a three year old moraine with several specimens of the predatious and very active Carabidae species *Bembidion hastii*. But what is the density of the species, what does it eat, and does it reproduce here? Photo: Sigmund Hågvar.

10. Future research

Although we are beginning to understand several trends and mechanisms in the primary succession of glacier forelands, more field studies with a high taxonomic resolution, and from different geographical areas, are needed. It is a special challenge to explain the ecological mechanisms working in pioneer communities on barren ground. What is the importance of pioneer microflora as food for pioneer microarthropods, and how important are resident microarthropods as food for pioneer beetles and spiders? Gut content analyses based on DNA primers of potential prey items would be highly welcome. And how important are pioneer mosses as a driver in succession?

Another improvement would be to add more quantitative samplings methods. Pitfall trapping favours fast-moving surface species (Fig. 14). Species with low densities may give considerable catches if they are very active, while species with higher densities may be lacking in the pitfall traps if they are rather immobile (for instance web-building spiders or moss-eating beetles).

Long-term monitoring of selected plots can illustrate effects of changed climate on succession. Plots which are already well studied should be re-studied at intervals. A better understanding of primary succession makes it easier to forecast what future ecosystems may be like in areas freed from ice. It can also increase our general insight into ecology, maybe by removing the "predator first-paradox" as a paradox.

11. Conclusion

Due to global warming, glaciers are receding in many parts of the world, leaving considerable areas of barren ground. While the botanical succession in such glacier forelands have been well studied, the parallel zoological succession is less described and understood. Glacier forelands illustrate nature's ability to recover from severe disturbance, and it is of considerable ecological interest to understand the succession process. Succession studies also help us to predict future ecosystems in deglaciated terrain.

This chapter summarizes and compares zoological studies in glacier forelands within three main areas in Europa: Svalbard, south Norway, and the Alps. A common technique is to study sites with known age in different distances from the ice. The sequence of dated study plots is called a chronosequence, and the various plots act as a substitute for following the same plot over time. Not surprisingly, time, distance and vegetation cover use to be highly correlated factors in a glacier foreland.

Several invertebrates are present before any vegetation is visible. Typical representatives are springtails (Collembola), mites (Acari), beetles (Coleoptera), spiders (Araneae), and harvestmen (Opiliones). The actual species are surface active animals, but they find shelter in the crevices among stones, gravel and sand grains. Springtails and mites are saprophagous, while species from the other groups are mainly predators. It has been called an ecological paradox that predators preceed both plant-eaters and plants. However, the pioneer ground receives airborne insects (mainly Diptera), on which the predators can feed. This fertilizes the ground and contributes to the gradual establishment of plants. However, chlorophyll-based food chains may start surprisingly early, for instance based on pioneer mosses on which certain springtails and beetles can feed.

In the Alps, most arthropod species colonise during a period of 40-50 years, while the colonisation is slower in Norway. High Arctic forelands on Svalbard have a poor fauna, but springtails, mites and certain spiders are early colonisers even there. Certain invertebrate taxa are typical pioneers in all three geographical areas, or common to Norway and the Alps. It is also concluded that the main pattern of the zoological succession is rather predictable. This indicates that dispersion may not be a serious problem. Herbivorous invertebrates are often relatively late colonisers.

Some pioneers are highly specialised, cold-tolerant species. These may go locally extinct if the glacier melts away. Other are open ground-specialists, and may live also in open habitats in the lowland. Several are generalists, with an extra flexibility to inhabit the harsh conditions close to a glacier. Pioneers may be parthenogenetic or bisexual, or have a short or long life cycle. Although pioneer species form an ecologically heterogeneous group, the pioneer community is often rather predictable.

Some of the remaining questions are: Is dispersal such an easy task? What do the various pioneer species eat? Is the pioneer ground an ecological sink, continuously fed from outside? How do plants and animals interact through succession? More field studies with a high taxonomic resolution, and in various geographical areas, are welcomed. Climate change may generally speed up the succession rate around melting glaciers.

12. References

Alfredsen, A. N. (2010). Primary succession, habitat preferences and species assemblages of carabid beetles in front of the retreating glacier Midtdalsbreen, Finse, southern Norway. *Master thesis,* University of Bergen, 83 pp.

Bardgett, R. D.; Richter, A.; Bol, R.; Garnett, M. H.; Bäumler, R.; Xu, X.; Lopez-Capel, E.; Manning, D. A.; Hobbs, P. J.; Hartley, I. R.; & Wanek, W. (2007). Heterotrophic microbial communities use ancient carbon following glacial retreat. *Biological Letters* Oct 22: 3 (5): 487-490.

Bråten, A. T. & Flø, D. (2009). Primary succession of arthropods (Coleoptera and Araneae) on a newly exposed glacier foreland at Finse, southern Norway. *Master thesis,* Norwegian University of Life Sciences, 85 pp.

Chapin, F. S.; Walker, L. R.; Fastie, C. L.; & Sharman, L. C. (1994). Mechanisms of primary succession following deglaciation at Glacier Bay, Alaska. *Ecological Monographs,* 64, 149-175.

Coulson, S. J.; Hodkinson, I. D. & Webb, N. R. (2003). Aerial dispersal of invertebrates over a High Arctic glacier foreland. *Polar Biology,* 26, 530-537.

Fjellberg, A. (1974). A study of the Collembola fauna at Stigstuv, Hardangervidda. Abundance, biomass and species diversity. *Master Thesis,* University of Bergen, Norway, 141 pp. (In Norwegian.)

Fjellberg, A. (2007). The Collembola of Fennoscandia and Denmark. Part II: Entomobryomorpha and Symphypleona. *Fauna Entomologica Scandinavica,* 42, 1-266.

Franz, H. (1969). Besiedlung der jüngst vom Eise freigegebenen Gletschervorfelder und ihrer Böden durch wirbellose Tiere. Neue Forschungen im Umkreis der Glocknergruppe. *Wissenschaftliche Alpenvereinshefte,* 21, 291-298.

Gereben, B. A. (1994). Habitat-binding and coexistence of carabid beetles in a glacier retreat zone in the Zillertal Alps. In: *Carabid beetles: ecology and evolution.* Desender, K., Dufrene, M., Loreau, M., Luff, M. L. and Maelfait, J.-P. (eds.), pp. 139-144. Dordrecht: Kluwer.

Gereben, B. A. (1995). Co-occurrence and microhabitat distribution of six Nebria species (Coleoptera: Carabidae) in an alpine glacier retreat zone in the Alps, Austria. *Arctic and Alpine Research*, 27, 371-379.

Glick, P. A. (1939). The distribution of insects, spiders and mites within the air. *Technical Bulletines U.S. Department of Agriculture*, 673, 1-151.

Gobbi, M.; De Bernardi, F.; Pelfini, M.; Rossaro, B. & Brandmayr, P. (2006a). Epigean arthropod succession along a 154-year glacier foreland chronosequence in the Forni Valley (Central Italian Alps). *Arctic, Antarctic, and Alpine Research*, 38, 357-362.

Gobbi, M.; Fontaneto, D. & De Bernardi, F. (2006b). Influence of climate changes on animal communities in space and time: the case of spider assemblages along an alpine glacier foreland. *Global Change Biology*, 12, 1985-1992.

Gobbi, M.; Rossaro, B.; Vater, A.; De Bernardi, F.; Pelfini, M. & Brandmayr, P. (2007). Environmental features influencing Carabid beetle (Coleoptera) assemblages along a recently deglaciated area in the Alpine region. *Ecological Entomology*, 32, 682-689.

Gressitt, J. L. & Yoshimoto C. M. (1974). Insect dispersal studies in northern Alaska. *Pacific Insects*, 16, 11-30.

Hågvar, S.; Solhøy, T. & Mong, C. (2009). Primary succession of soil mites (Acari) in a Norwegian glacier foreland, with emphasis on Oribatid species. *Arctic, Antarctic and Alpine Research*, 41, 219-227.

Hågvar, S. (2010). Primary succession of springtails (Collembola) in a Norwegian glacier foreland. *Arctic, Antarctic and Alpine Research*, 42, 422-429.

Hågvar, S. & Klanderud, K. (2009). Effect of simulated environmental change on alpine soil arthropods. *Global Change Biology*, 15, 2972-2980.

Hodkinson, I. D.; Coulson, S. J.; Harrison, J. & Webb, N. R. (2001). What a wonderful web they weave: spiders, nutrient capture and early ecosystem development in the high Arctic - some counter-intuitive ideas on community assembly. *Oikos*, 95, 349-352.

Hodkinson, I. D.; Webb, N. R. & Coulson, S. J. (2002). Primary community assembly on land - the missing stages: why are the heterotrophic organisms always there first? *Journal of Ecology*, 90, 569-577.

Hodkinson, I. D.; Coulson, S. J. & Webb, N. R. (2004). Invertebrate community assembly along proglacial chronosequences in the high Arctic. *Journal of Animal Ecology*, 73, 556-568.

Hole, L. & Engardt, M. (2008). Climate change impact on atmospheric nitrogen deposition in northwestern Europe: a model study. *Ambio*, 37, 9-17.

Holm, Å. (1958). The spiders of the Isfjord region of Spitsbergen. *Zoologiska Bidrag Från Uppsala*, 33, 29-67.

IPCC (2007). *Fourth Assessment Report of the Intergovernmental Panel on Climate Change*. Cambridge University Press, Cambridge, UK.

Janetschek, H. (1949). Tierische Successionen auf hochalpinem Neuland. *Berichte des naturwissenschaftlich-medizinischen Vereins Innsbruck*, 48/49, 1-215.

Janetschek, H. (1958). Über die tierische Wiederbesiedlung im Hornkees-Vorfeld (Zillertaler Alpen). *Schlern-Schriften*, 188, 209-246.

Kaufmann, R. (2001). Invertebrate succession on an Alpine glacier foreland. *Ecology*, 82, 2261-2278.

Kaufmann, R. (2002). Glacier foreland colonisation: Distinguishing between short-term and long-term effects of climate change. *Oecologia*, 130, 470-475.

Kaufmann, R.; Fuchs, M. & Gosterxeier, N. (2002). The soil fauna of an alpine glacier foreland: Colonization and succession. *Arctic, Antarctic and Alpine Research,* 34, 242-250.

Kaufmann, R. & Raffl, C. (2002). Diversity in primary succession: The chronosequence of a glacier foreland. *In* Körner, C. and Spehn, E. (eds.), *Global Mountain Biodiversity: A Global Assessment.* London: Parthenon, pp. 177-190.

Lindroth, C. H. (1986). The Carabidae (Coleoptera) of Fennoscandia and Denmark. Part 2. *Fauna Entomologica Scandinavica,* 15 (2), 227-498.

Magnussen, T. (2010). Aerial dispersal of invertebrates on Svalbard and the influence of weather. *Master thesis,* University of Oslo, 40 pp.

Matthews, J. A. (1992). *The Ecology of Recently-deglaciated Terrain: A Geoecological Approach to Glacier Forelands and Primary Succession.* Cambridge: Cambridge University Press, 386 pp.

Matthews, J. A. & Whittaker, R. J. (1987). Vegetation succession on the Storbreen glacier foreland, Jotunheimen, Norway: a review. *Arctic and Alpine Research,* 19, 385-395.

Moreau, M.& Laffly, D.; Joly, D. & Brossard, T. (2005). Analysis of plant colonization on an arctic moraine since the end of the Little Ice Age using remotely sensed data and a Bayesian approach. *Remote Sensing of Environment,* 99, 244-253.

Paulus, U. & Paulus, H. F. (1997). Die Zönologie von Spinnen auf dem Gletschervorfeld des Hornkees in den Zillertaler Alpen in Tirol (Österreich) (Arachnida, Araneae). *Berichte des naturwissenschaftlich-medizinischen Vereins Innsbruck,* 80, 227-267.

Raffl, C. (1999). Vegetationsgradienten und Sukzessionsmuster in einem Gletschervorfeld in den Zentralalpen (Ötztaler Alpen, Tirol). *Diploma Thesis,* University of Innsbruck. 102 pp.

Raffl, C.; Mallaun, M.; Mayer, R. & Erschbamer, B. (2006). Vegetation succession pattern and diversity changes in a glacier valley, central Alps, Austria. *Arctic, Antarctic, and Alpine Research,* 38, 421-428.

Riley, J. R.; Reynolds, D. R.; Mukhopadhyay, S.; Ghosh, M. R. & Sarkar, T. K. (1995). Long-distance migration of aphids and other small insects in northeast India. *European Journal of Entomology,* 92, 639-653.

Seniczak, A.; Solhøy, T. & Seniczak, S. (2006). Oribatid mites (Acari: Oribatida) in the glacier foreland at Hardangerjøkulen (Norway). *Biological Letters,* 43, 231-235.

Skubala, P. (2004). *Colonization and development of oribatid mite communities (Acari: Oribatida) on post-industrial dumps.* Katowice: Wydawnictwo Uniwersytetu Slaskiego, 208 pp.

Skubala, P. & Gulvik, M. (2005). Pioneer oribatid mite communities (Acari, Oribatida) in newly exposed natural (glacier foreland) and anthropogenic (post-industrial dump) habitats. *Polish Journal of Ecology,* 53, 105-111.

Solhøy, T. (1975). Dynamics of oribatei populations on Hardangervidda. In: *Fennoscandian Tundra Ecosystems. Part 2. Animals and Systems Analysis.* Wielgolaski, F. E. (ed), pp. 60-65. Springer-Verlag, Berlin.

Vetaas, O. R. (1994). Primary succession of plant assemblages on a glacier foreland - Bødalsbreen, southern Norway. *Journal of Biogeography,* 21, 297-308.

Vetaas, O. R. (1997). Relationships between floristic gradients in a primary succession. *Journal of Vegetation Science,* 8, 665-676.

Vater, A. E. (2006). Invertebrate and arachnid succession on selected glacier forelands in southern Norway. *PhD. thesis,* University of Wales, 472 pp.

Ytrehus, B.; Bretten, T.; Bergsjø, B. & Isaksen, K. (2008). Fatal Pneumonia Epizootic in Musk ox (*Ovibos moschatus*) in a period of extraordinary weather conditions. *EcoHealth*, 5, 213-223.

Zingerle, V. (1999). Spider and harvestman communities along a glaciation transect in the Italian Dolomites. *Journal of Arachnology*, 27, 222-228.

10

Excess Supply of Nutrients, Fungal Community, and Plant Litter Decomposition: A Case Study of Avian-Derived Excreta Deposition in Conifer Plantations[1]

Takashi Osono
Center for Ecological Research, Kyoto University
Japan

1. Introduction

1.1 Excess supply of nutrients and terrestrial ecosystems

Human activities have greatly accelerated emissions of both carbon dioxide and biologically reactive nutrients such as nitrogen (N) to the atmosphere (Canfield et al., 2010), which cause environmental changes affecting ecosystem processes and biodiversity in forests. Excess supply of N of anthropogenic origin to forest soils, such as combustion of fossil fuels, production of N fertilizers, and cultivation of N-fixing legumes, is an example of such environmental changes often leading to a decrease of the rate of carbon dioxide evolution and decomposition (Fog, 1988; Berg and Matzner, 1997) and a concomitant increase in the amount of soil carbon stock (deVries et al., 2006; Zak et al., 2008). These changes are primarily attributable to the reduced activity of fungal ligninolytic enzymes that play crucial roles in the turnover of soil organic carbon and are known to be sensitive to N deposition (Sinsabaugh, 2010). However, such changes in the enzymatic activity are not consistently associated with changes in the abundance and diversity of fungi that are responsible for the activity (Waldrop and Zak, 2006; Blackwood et al., 2007; Hassett et al., 2009). This discrepancy merits further studies to examine the response of ecological and functional properties of fungal communities to excess supply of N and its consequences on the dynamics of carbon and N in forest soils.

The transfer of nutrients by waterbirds from aquatic to terrestrial ecosystems provides similar situations to the anthropogenic supply of nutrients because birds feed on fish in the aquatic zone and deposit their waste rich in nutrients to the terrestrial parts of their habitats. Such allochthonous input of N and other nutrients to terrestrial ecosystems can lead locally to substantial enrichment of soils and plants and alter food webs, nutrient cycling, and

---

[1]This manuscript should be cited as follows: Osono, T. (2011). Excess supply of nutrients, fungal community, and plant litter decomposition: a case study of avian-derived excreta deposition in conifer plantations, In: *Environmental Change*, S.S. Young and S.E. Silvern, (Ed.), 000-000, InTech, ISBN979-953-307-109-0, Rijeka, Croatia

ecosystem processes in bird colonies (Mizutani and Wada, 1988; Anderson and Polis, 1999). In contrast, much less concern has been directed toward the diversity and activity of saprobic fungi in forest soils affected by excess supply of avian-derived N and the consequences for carbon sequestration in forest soils.

1.2 Cormorant populations in lakeside forests in Japan

The great cormorant, *Phalacrocorax carbo* L., is a colonial piscivorous bird that is distributed almost all over the world (Johnsgard, 1993). In Japan, the cormorants breed and roost in trees in riparian woods and feed on fishes in lakes, rives, and coastal areas (Ishida et al., 2003). The population of cormorants increased rapidly after the 1980s as the number of new colonies increased (Kameda et al., 2003). For example, there were no breeding records of cormorants between World War II and 1982 within the watershed of Lake Biwa, currently one of the main habitats of cormorants in Japan, whereas the population size increased rapidly in the 1990s to reach more than 17,000 during the breeding season from January to August in 2003 (Kameda et al., 2006). The increased populations have caused serious conflicts with fisheries and forests in their habitats (Kameda et al., 2003).

Isaki Peninsula (35°12'N, 136°5'E, 57 ha), located on the southeast side of Lake Biwa (Fig. 1) and covered with plantations of Japanese cypress (*Chamaecyparis obtusa* Sieb. et Zucc.), was selected for the present study. The mean annual temperature is 15.1°C and annual precipitation is 1,474.5 mm at the Hikone Weather station about 20 km from the Isaki Peninsula. After cormorant nests were first discovered at Isaki Peninsula in 1988, the area of the colony expanded from 1.3 ha in 1992 to 19.3 ha in 1999 and the number of nests from 30 to 40 in 1989 to 5,300 in 1999 (Fig. 1) to become one of the major habitats of the cormorants in the watershed of Lake Biwa (Fujiwara and Takayanagi, 1999). Five study sites were chosen on Isaki Peninsula, Sites C, T, P, A, D, which had the same vegetation composition but were in different stages of breeding colony establishment (Table 1). A study plot (50 × 50 m) was established at each site and used to study the effects of cormorant colonization on soils and vegetation.

1.3 Responses of forest ecosystems to cormorant colonization

During the breeding season, the input of bird excreta at Site P was estimated at 2.2 t/ha/month (Kameda et al., 2000). Because the excreta are rich in N (11.1% w/w on average) and other nutrients such as P and Ca, the excreta input was estimated to be the equivalent of 0.24 t/ha/month of excreta-derived N, which corresponds to about 10,000 times the ordinary input by precipitation (Fig. 2) (Kameda et al., 2000). In addition, litterfall input at Site P during the breeding season was estimated at 2.6 t/ha/month, which was 7 to 22 times greater than that at Site C (Fig. 2) (Hobara et al., 2001). This increase of litterfall at Site P was due to damage of the overstory by the cormorants. *Chamaecyparis obtusa* was one of the most heavily damaged tree species at forest stands colonized by the cormorants (Ishida, 1996b). A part of forest stands intensively colonized by the cormorants declined due to high mortality of *C. obtusa* (Site D; Fig. 2) (Fujiwara and Takayanagi, 2001).

The forest decline was also partly and indirectly attributable to changes in soil properties caused by excess supply of excreta-derived nutrients. A dramatic increase in inorganic N pools, a decrease in carbon to N ratio, and an increase in nitrification rate were observed in forest floor materials and in soils at Sites P and A (Ishida, 1996a; Hobara et al., 2001), indicative of N saturation at the study sites exposed to bird colonization (Aber et al., 1998). Excreta-derived N was incorporated into not only soils but also aboveground tissues of plants, as indicated by natural ^{15}N abundance as a natural tracer (Kameda et al., 2006). Because cormorants are piscivorous birds and one of the top predators in aquatic food webs, $\delta^{15}N$ of their tissues and excreta is markedly higher (i.e., 13 to 17‰) than those of N from precipitation and N fixation (-1 to 1‰). The data of $\delta^{15}N$ in soils and plants were used to construct 'N stable isotope map' of Isaki Peninsula (Fig. 1) showing the spatial patterns of cormorant effects (Kameda et al., 2006).

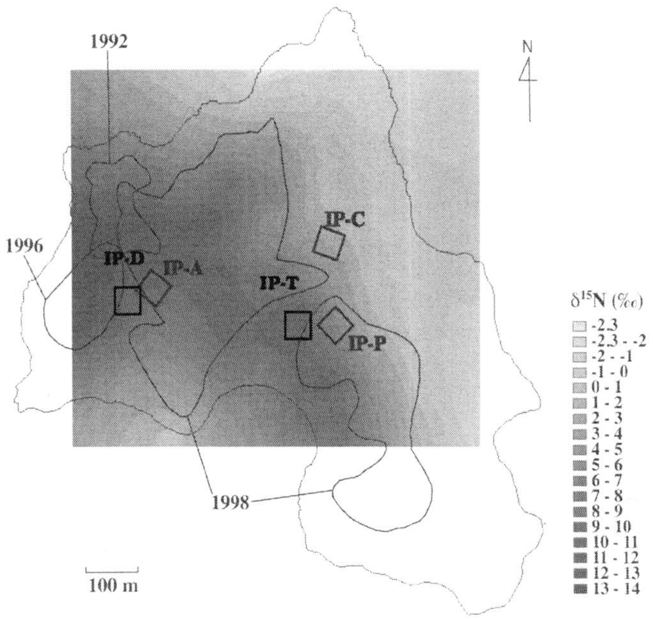

Fig. 1. Study sites, cormorant colony boundaries and the year of colony establishment, and nitrogen stable isotope map of Isaki Peninsula (IP) at Lake Biwa, Japan. The nitrogen stable isotope map shows the intensity and duration of cormorant colonization (Kameda et al., 2006). See Table 1 for the description of study sites.

Site	Colonization	Description
C	No colonization	Never colonized by cormorants (control)
T	Spring 1999	Temporarily colonized for 3 months before cormorants were expelled by hunters; no cormorants thereafter
P	1997-2003	Presently colonized; cormorants abundant
A	1996-1999	Abandoned after 3 years of colonization; no cormorants
D	1992-1996	Declined after 4 years of intensive colonization; no cormorants

Table 1. Study sites and descriptions of breeding colony of cormorants at Isaki Peninsula.

Fig. 2. Surface of the forest floor covered with dead twigs fallen at Site A (left), leaves of understory vegetation covered with excreta deposited at Site P (middle), and dead trees of *Chamaecyparis obtusa* in a declined forest stand at Site D (right).

1.4 Purposes

In this chapter I summarize a series of published papers reporting the effects of excess supply of N as avian excreta on fungal communities and plant litter decomposition in conifer plantations colonized by cormorants (Osono et al., 2002, 2006a, 2006b, unpublished data; Katsumata, 2004) to present a comprehensive picture of their relationships and to predict long-term patterns in the accumulation of dead plant tissues and excreta-derived nutrients on the forest floor. The following hypotheses are addressed. (i) The excess supply of nutrients affected the abundance, diversity, and species composition of saprobic fungal communities, as well as their nutrition and activity (Sections 2, 3, and 4). (ii) Such changes in fungal diversity and activity in turn affected the decomposition processes of dead plant tissues, such as needles, twigs, and stems (Section 5). (iii) Dead plant tissues abundantly supplied to the forest floor serve as reservoirs of excreta-derived N (Section 6). The studies explicitly demonstrate that the changes in fungal communities and decomposition of dead plant tissues had consequences regarding the long-term patterns of accumulation of carbon and N in soils of forest stands colonized by cormorants.

2. Excreta deposition and fungal communities

It is usually difficult to study both the biomass and the species composition of fungal assemblages simultaneously with any single method (Osono, 2007). Thus, fungal biomass and species composition were studied separately. Firstly, dead needles and twigs of *C. obtusa* were collected from the forest floor, and the length of hyphae in the tissues was examined with a direct observation method as a measure of fungal biomass and compared among forest stands with different histories of cormorant colonization (Osono et al., 2002). Twigs were defined as woody tissues with a diameter less than 0.5 cm.

2.1 Fungal biomass in dead needles and twigs

The total hyphal length was generally longer in needles than in twigs and was in the order: Sites C > P > A (Fig. 3), suggesting that the biomass of fungi was reduced in forest stands supplemented with excreta. The length of clamp-bearing hyphae, belonging to the Basidiomycota (Fig. 4), accounted for 10 to 11% of the total hyphal length at Site C but was reduced markedly at Sites P and A (Fig. 3).

The reduced fungal biomass at Sites P and A was possibly attributable to the inhibitory effects on fungal growth of excreta rich in ammonia, uric acid, and salts (see Section 4.1) and to the decreased availability of carbon compounds owing to condensation of N-rich compounds (Osono et al., 2002). Söderström et al. (1983) also reported a decrease in microbial biomass after N fertilization in coniferous forest soils. The lower length of clamp-bearing hyphae (i.e., biomass of basidiomycetous fungi) at Sites P and A than at Site C might also have been due to a biochemical suppression of lignin-degrading enzymes of some fungi in the Basidiomycota caused by excess excreta deposition (Keyser et al., 1978; Fenn et al., 1981). This may have reduced competitiveness relative to that of other non-ligninolytic fungi and hence hyphal growth of basidiomycetes at Sites P and A.

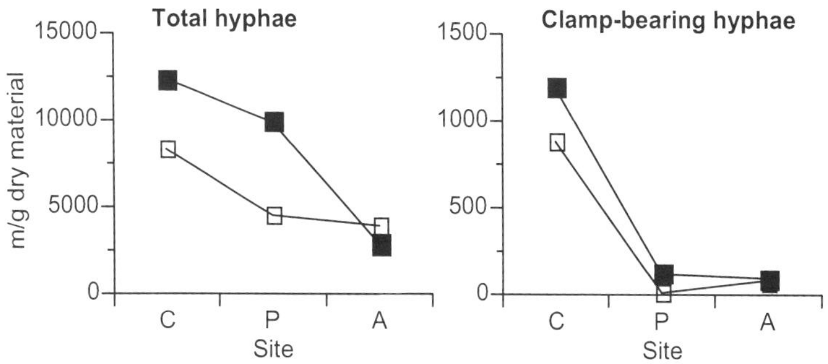

Fig. 3. Total hyphal lengths and lengths of clamp-bearing hyphae in dead needles and twigs of *Chamaecyparis obtusa* examined with an agar film method. ◆ needles; □ twigs. Sites are as in Table 1. Data after Osono et al. (2002).

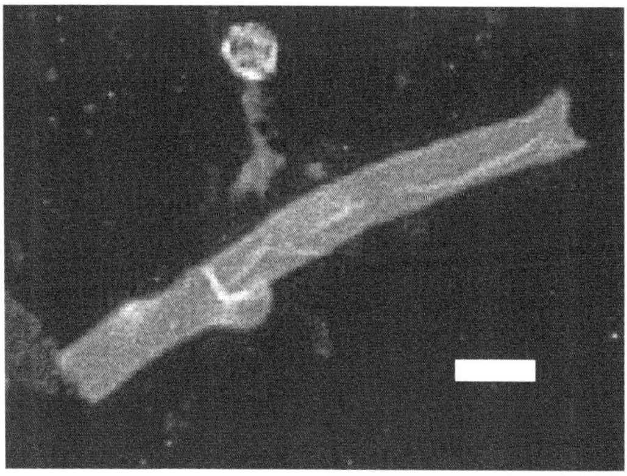

Fig. 4. A hypha with a clamp connection (arrow) observed under a microscope. Bar = 5 μm.

2.2 Diversity and species composition of fungi

Secondly, species richness, diversity, and equitability of fungal assemblages associated with the dead needles and twigs were examined with a culture-dependent, surface disinfection method (Fig. 5). A total of 231 isolates of 70 fungal species were isolated from dead needles and twigs at Sites C, P, and A. Species richness (i.e., the number of species isolated) in needles was higher at Site A than at Sites C and P, but the species richness in twigs was similar among the sites. Diversity index was higher in twigs than in needles and was higher at Site A than at Sites C and P. Equitability was higher in twigs than in needles and in the order: Sites A > P > C in both needles and twigs.

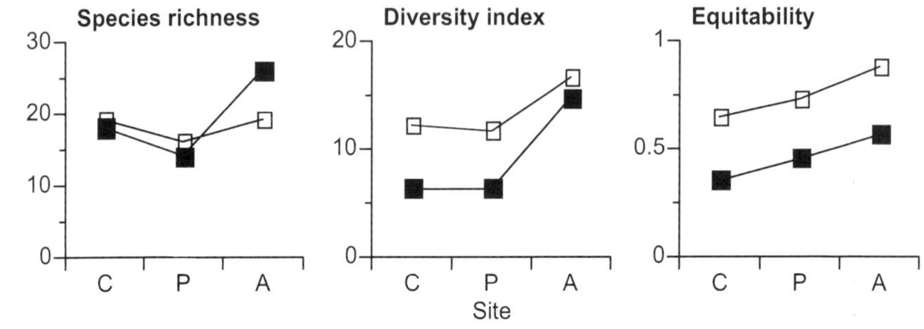

Fig. 5. Diversity of fungal assemblages in dead needles and twigs of *Chamaecyparis obtusa*. ◆ needles; □ twigs. Sites are as in Table 1. Species richness (S) equals to the total number of species. Simpson's diversity index (D) and equitability (E) were calculated as: $D = 1/\sum Pi^2$, $E = D/S$, where P*i* was the relative frequency of the *i*th species in each fungal assemblage (Osono et al., 2002).

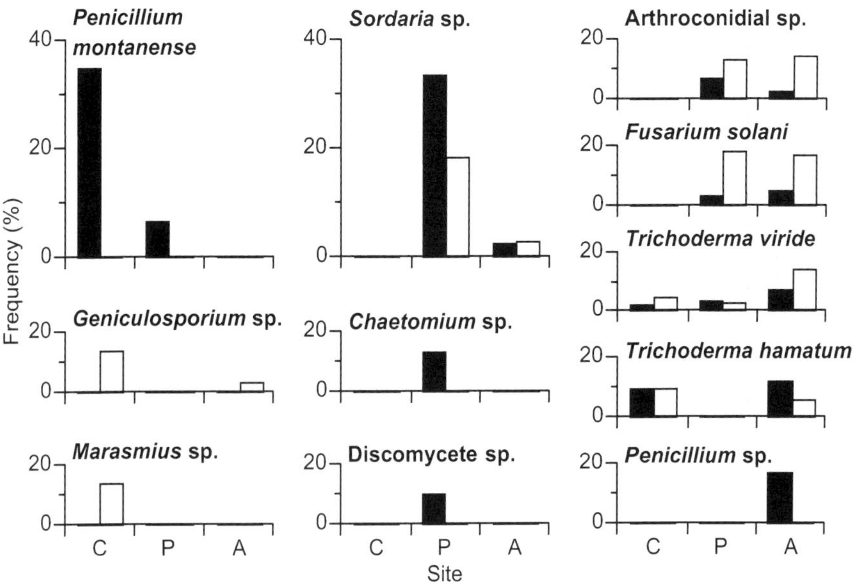

Fig. 6. Relative frequency (%) of major fungal species in dead needles and twigs of *Chamaecyparis obtusa* (Osono et al., 2002). Black bar, needles; open bar, twigs. Sites are as in Table 1.

A few studies have examined the effects of bird colonization on soil fungal assemblages. Ninomiya et al. (1993) and Schoenlein-Crusius et al. (1996) observed no difference in fungal diversity between soil affected by the presence of birds and control soil, which contrasted with the results of the present study. Osono et al. (2002) summarized previous studies on the effects of ornithologenic and anthropogenic eutrophication on the diversity of soil saprobic fungal assemblages and found that the response was variable depending on the study. The inconsistency of the eutrophication effect on fungal diversity suggests that factors other than nutrient addition may also affect the diversity, such as the amount and/or form of nutrients added, time after fertilization, physical and chemical properties of soils, and different methodologies used for fungal isolation.

Clear differences were found for the patterns of occurrence of 11 major fungal species among the sites (Fig. 6). *Penicillium montanense*, *Geniculosporium* sp., and *Marasmius* sp. dominated at Site C were decreased at Sites P and A. Koide and Osono (2003) reported a similar result that an udentified species of *Geniculosporium* was isolated from leaf litter of *Camellia japonica* at Site C but not at Site A. This contrasted to *Sordaria* sp., *Chaetomium* sp., Discomycete sp., an unidentified arthroconidial species, and *Fusarium solani*, which showed marked increases at Site P in both needles and twigs. Arthroconidial sp. and *F. solani* also occurred frequently at Site A, as did *Trichoderma viride*, *T. hamatum*, and *Penicillium* sp. The absence of a ligninolytic basidiomycete *Marasmius* sp. from twigs at Sites P and A was consistent with the decrease in clamp-bearing hyphae (Fig. 3) and may have been due to enzymatic suppression by excessive inorganic-N or N-rich compounds in these sites as discussed above. *Sordaria* sp. is considered to be a coprophilous species associated with bird excreta.

In summary, the abundance of basidiomycetes (Fig. 3) and the relative frequency of ligninolytic *Marasmius* sp. (Fig. 6) were reduced at presently colonized (Site P) and abandoned forest stands (Site A), possibly due to excess supply of nutrients in excreta, such as N. To verify this possibility, effects of excreta addition on fungal growth and decomposition was examined under pure culture conditions in Section 4.

3. Utilization of excreta-derived nutrients by fungi

Utilization of cormorant-derived N by fungi was demonstrated by investigating the natural ^{15}N abundance in fruit bodies of litter- and wood-decomposing fungi collected in the study sites. ^{15}N enrichments in plant tissues, forest floor materials, and mineral soils due to excreta deposition were demonstrated in the cormorant colonies at Isaki Peninsula (Section 1.3; Fig. 1), which was associated with such processes as trophic enrichment through aquatic food webs and ammonia volatilization from soils (Kameda et al., 2006). Using natural ^{15}N abundance as a natural tracer thus makes it possible to test whether fungi utilized excreta-derived N in the colonized forests.

The ^{15}N values of fruiting bodies at Site C were 0.1 to 1.5‰ on average and at similar levels to that in precipitation at the vicinity of the study sites (Fig. 7) and were within the range for saprobic fungi previously reported (e.g., Kohzu et al., 1999; Trudell et al., 2004). $\delta^{15}N$ was significantly (generalized linear model, χ^2=39.0, P<0.001) different among Sites C, P, and A and was significantly (χ^2=15.4, P<0.001) higher in litter- than in wood-decomposing fungi (Fig. 7). Mean $\delta^{15}N$ values of fruiting bodies were in the order: Sites A > P > C for both litter- and wood-decomposing fungi (Fig. 7). $\delta^{15}N$ of dead needles, forest floor materials, and woody debris were also higher at Sites P and A than at Site C, and fruiting bodies of fungi were generally enriched in ^{15}N relative to their substrata collected at the same sites. Fruiting bodies of litter-decomposing fungi at Sites P and A and those of wood-decomposing fungi at Site A had similar or higher $\delta^{15}N$ values than that in excreta (Fig. 7).

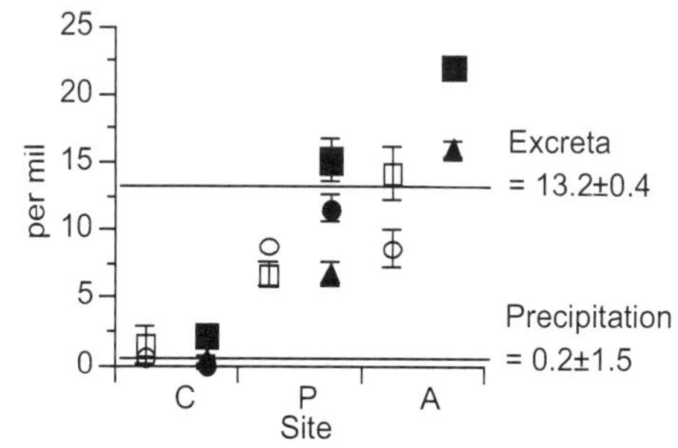

Fig. 7. Nitrogen stable isotope ratios of fruiting bodies of litter- (◆) and wood-decomposing fungi (□) (T. Osono, unpublished). Nitrogen stable isotope ratios of substrates for fungi are also shown: ℓ dead needles of *Chamaecyparis obtusa*; ‗forest floor materials; ○ woody debris. Values indicate means ± standard errors. Sites are as in Table 1. Horizontal lines indicate $\delta^{15}N$ values for excreta (means ± standard errors, n=12) and for precipitation (n=5) (Kameda et al., 2006). A total of 44 samples of fungal fruiting bodies representing 24 taxa were qualitatively collected from February 2000 to April 2003 and used for the analysis.

These results showed the effects of ^{15}N-enriched excreta deposition on fruiting bodies of litter- and wood-decomposing fungi at the forest stands colonized by the cormorants. Previous studies have been successful in using N stable isotope ratios to demonstrate the transfer of animal-derived N to biotic components in terrestrial ecosystems, such as seabird rookeries (Mizutani and Wada, 1988; Wainright et al., 1998) and bear habitats where salmons are transferred from coastal waters to riparian forests (Wilkinson et al., 2005; Nagasaka et al., 2006). The uptake of excreta-derived N can alter metabolic activity of fungal mycelia, which is investigated in the next section.

4. Reduction of fungal growth and decomposition by excreta

The results of Sections 2 and 3 suggest possible effects of excreta on fungal growth and decomposition of plant tissues. These effects were verified with pure culture tests of fungal growth and decomposition on an agar medium supplemented with excreta in comparison with those on a control medium without excreta (Osono et al., 2006b).

In September 2000, water collectors with 15-cm diameter funnels on the top were installed on the forest floor within each of Sites C and P to collect throughfall (i.e., excess water shed from wet leaves onto the ground surface). The water samples from Site C contained throughfall (rainfall plus leaf leachates), whereas that from Site P contained the throughfall plus excreta of the cormorants. The water sample from Site P had higher pH and electrical conductivity and higher contents of total carbon, total N, and NH_4-N than that from Site C (Osono et al., 2006b). Throughfall from Sites C and P was mixed with 2% agar (w/v) and sterilized to prepare agar media that were denoted here as media C and P, respectively.

4.1 Excreta addition reduced fungal growth

Linear growth rates of 22 fungal isolates (12 basidiomycetes, 11 ascomycetes, and 1 zygomycete) were compared between media C and P. Nineteen of the 22 isolates were collected in the study sites, and another three isolates in the Basidiomycota were obtained from a culture collection. The mean value of linear growth rates on medium P was significantly lower than that on medium C (Fig. 8), indicating that excreta of the cormorants generally suppressed the mycelial extension of fungi. When taxonomic groups of fungi were examined separately, the linear growth rates for the Basidiomycota were significantly (paired t-test, $n=12$, $P<0.05$) lower on medium P than on medium C, whereas the difference was not significant for the Ascomycota (paired t-test, $n=11$, $P>0.05$). These results are consistent with the field measurements showing that hyphal lengths in needle and twigs were shorter at Site P, where the forest floor suffered excreta deposition, than at Site C and that the reduction was obvious for clamp-bearing hyphae that belong to the Basidiomycota (Fig. 3). Possible inhibitory factors responsible for the decrease of fungal growth include the toxicity of ammonia and uric acid and the higher pH and salt concentration in excreta, as discussed above.

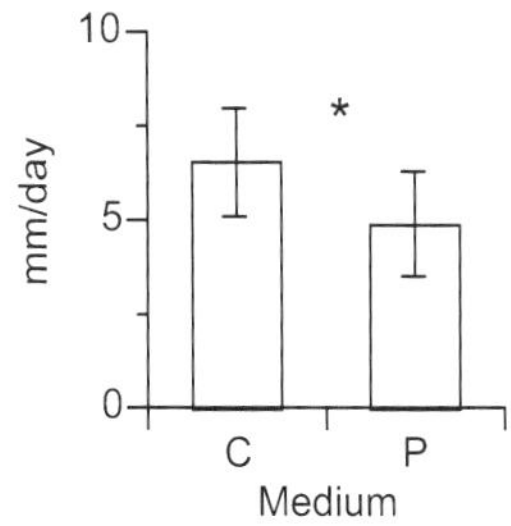

Fig. 8. Linear growth rate of fungal colony on media C and P at 20°C under pure culture conditions. Medium P contained excreta. The original data are in Osono et al. (2006b). Values indicate means ± standard errors for 22 fungal species tested. Results of paired t-tests are shown. * $P<0.05$.

4.2 Excreta addition retarded fungal decomposition of needles

Another pure culture decomposition test was carried out for 13 (eight basidiomycetes and five ascomycetes) of the 22 fungal isolates to evaluate the effect of excreta addition on decomposition (Osono et al., 2006b). Dead needles of *C. obtusa* collected at Site C were used as a substratum. The mean value of mass loss of needles on medium P was significantly lower than that on medium C (Fig. 9), indicating that excreta of the cormorants generally reduced the fungal decomposition. This reduction in decomposition was due to the suppression of decomposition of acid-unhydrolyzable residue (AUR) in needles, as the mass loss of AUR was significantly lower on medium P than on medium C (Fig. 9). In contrast, the mass loss of total carbohydrates was not significantly different between the media C and P (Fig. 9). The mass loss of N was significantly lower on medium P than on medium C (Fig. 9), indicating more accumulation of N in needles when fungi were incubated on medium P. $\delta^{15}N$ of needles decomposed by fungi on medium P (1.21±0.15‰, mean ± standard error, $n=13$) was significantly (paired t-test, $P<0.001$, $n=13$) higher than that on medium C

(0.51±0.06‰), suggesting that N in excreta was translocated into needles during the fungal decomposition on medium P.

When taxonomic groups of fungi were examined separately, the mean values of mass loss of AUR were significantly lower on medium P than on medium C for the Basidiomycota (paired t-test, n=8, P<0.05), whereas the difference was not significant for the Ascomycota (paired t-test, n=5, P>0.05), suggesting that AUR decomposition by basidiomycetes is more sensitive to excreta than that by ascomycetes. The AUR fraction, which has been commonly denoted as Klason lignin, contains lignin, tannin, and cutin (Preston et al., 1997) as well as humic substances produced secondarily during fungal decomposition (Fukasawa et al., 2009). The AUR fraction thus represents recalcitrant components in plant tissues and often limits decomposition and nutrient dynamics (Osono and Takeda, 2004). Because a high concentration of inorganic N can cause biochemical suppression of lignin-degrading enzymes responsible for AUR decomposition (Keyser et al., 1978; Fenn et al., 1981; Osono and Takeda, 2001), excreta rich in N are probably responsible for the observed sensitivity of ligninolytic basidiomycetes to excreta on medium P.

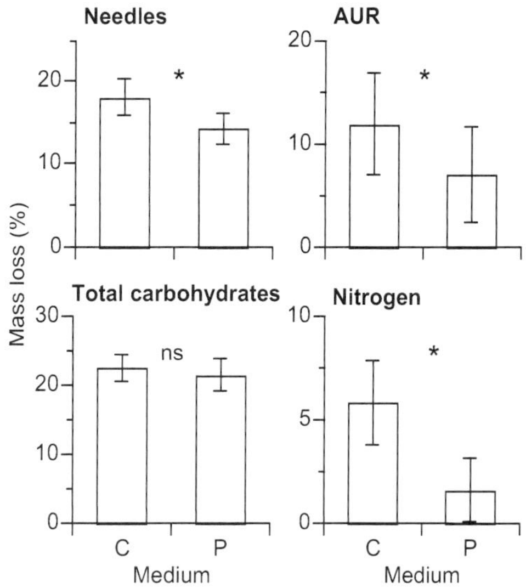

Fig. 9. Mass loss (% original mass) of dead needles of *Chamaecyparis obtusa* and of acid-unhydrolyzable residue (AUR), total carbohydrates, and nitrogen in the needles on medium C and P. Medium P contained excreta. The original data are in Osono et al. (2006b). The needles were sterilized with ethylene oxide gas, inoculated with fungal isolates, and incubated at 20°C for 12 weeks in the dark. Values indicate means ± standard errors for 13 fungal species tested. Results of paired t-tests are shown. * P<0.05, ns non significant.

In summary, the pure culture tests demonstrated that cormorant excreta negatively affected fungal growth and decomposition of needles and that ligninolytic basidiomycetes are more sensitive to excreta than ascomycetes. The reduced growth and decomposition by ligninolytic basidiomycetes due to excreta can alter the decomposition processes of dead plant tissues in the field, because these fungi are primary agents removing recalcitrant compounds from the tissues and mobilizing nutrients (Osono, 2007). Consequently, it is hypothesized that the reduction in biomass (Fig. 3) and activity (Figs. 8 and 9) of ligninolytic basidiomycetes due to excreta addition would result in the reduction of long-term decomposition rates, the accumulation of recalcitrant compounds in decomposing plant tissues, and the concomitant immobilization of nutrients in the tissues. These hypotheses are examined in detail in the next section.

5. Excreta deposition and decomposition of dead plant tissues in the field

A litterbag experiment (Fig. 10) was performed to follow the two-year decomposition of needles and twigs of *C. obtusa* on the forest floor and to compare them between Sites C and P to estimate the possible effects of excreta on the decomposition (Osono et al., 2006a). In another field survey, mass and N content of coarse woody debris (CWD: logs, snags, and stumps with diameter equal to or greater than 10 cm) were examined in the study sites to estimate the decomposition processes in cormorant-colonized forests.

Fig. 10. Litterbags to study long-term decomposition of dead plant tissues in the field. In the study of Osono et al. (2006a), needles and twigs collected at Site C were enclosed in polypropylene shade cloth (10 × 10 cm, mesh size of approx. 2 mm) and incubated on the forest floor at Sites C and P for two years. The litterbags were retrieved at 3- (the first year) or 6-month (the second year) intervals to analyze remaining mass and contents of organic chemical constituents and nutrients.

5.1 Rate of mass loss of needles and twigs and recalcitrant compounds

Over the two-year period, the mass loss was slower at Site P than at Site C and faster in needles than in twigs (Fig. 11). AUR mass loss in needles and twigs showed similar trends to mass loss of whole tissues and was slower at Site P than at Site C (Fig. 11). In contrast, mass loss of total carbohydrates in needles and twigs showed similar patterns between Sites C and P (data not shown; Osono et al., 2006a). These results support the hypotheses that the excreta deposition can lead to a reduction in decomposition rates and the accumulation of recalcitrant compounds in the decomposing plant tissues. The reduced AUR decomposition at Site P was primarily attributable to the reduced biomass and activity of ligninolytic basidiomycetes due to excess supply of excreta-derived N, as discussed above.

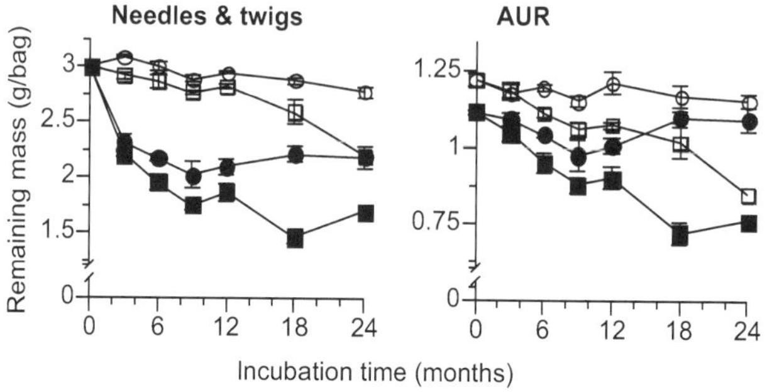

Fig. 11. Changes in remaining mass of needles and twigs of *Chamaecyparis obtusa* (left) and of acid-unhydrolyzable residue (AUR) in needles and twigs (right) at Sites C and P examined for two years in the field (Osono et al., 2006a). ◆ needles at Site C; ● needles at Site P; □ twigs at Site C; ○ twigs at Site P. Sites are as in Table 1. Values indicate means ± standard errors (n=3).

5.2 Immobilization of excreta-derived nitrogen

The mass of N in needles at Site P increased rapidly during the first 3 months and was relatively constant thereafter, whereas that at Site C decreased during decomposition (Fig. 12). The mass of N in twigs also increased at Site P, whereas such an increase was not detected at Site C (Fig. 12). The net increase, i.e. net immobilization, of N at Site P indicates the incorporation of external N into decomposing plant tissues. $\delta^{15}N$ values of the plant tissues at Site P increased rapidly during the first 3 months to reach the value of cormorant's excreta (13.2 ± 0.4‰, mean ± standard error, n=12; Kameda et al., 2006), whereas such an increase was not detected at Site C (Fig. 12). This stable isotope tracer successfully demonstrated that this exogenous N incorporated into the decomposing plant tissues was derived from excreta.

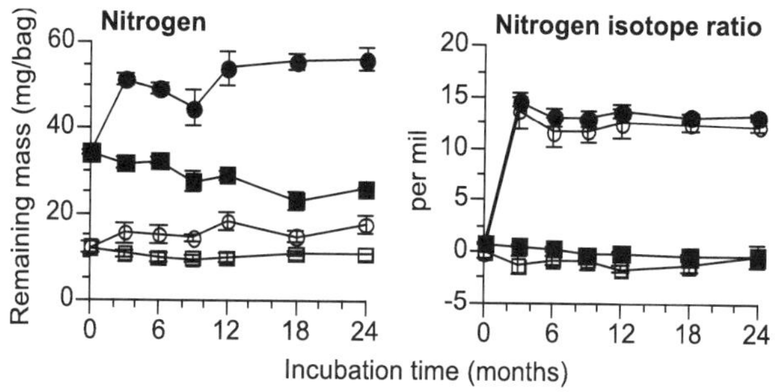

Fig. 12. Changes in remaining mass of nitrogen and nitrogen stable isotope ratio ($\delta^{15}N$, ‰) in needles and twigs of *Chamaecyparis obtusa* at Sites C and P in the field (Osono et al. 2006a). Symbols are as in Fig. 11. Values indicate means ± standard errors (n=3).

Causal relationships can be expected among the increased N immobilization, the AUR accumulation, and the reduced mass loss of whole tissues. The secondary formation of nitrogenous recalcitrant substances can be stimulated during plant litter decomposition under N-rich conditions (Berg, 1986, 1988). The $\delta^{15}N$ values (10.5 to 12.3‰) of AUR in needles and twigs at Site P, compared to those at Site C (-1.0 to 1.1‰), clearly indicated that excreta-derived N was incorporated into AUR during decomposition (Osono et al., 2006a). The formation of nitrogenous recalcitrant compounds registered as AUR resulted in the reduced net loss of mass of AUR, which in turn retarded the loss of mass of whole tissues.

5.3 Decomposition of coarse woody debris

Coarse woody debris (CWD) serves as a major pool and source of carbon and nutrients in forest ecosystems because of its long turnover time (Harmon et al., 1986). In Isaki peninsula, the mass of CWD ranged from 15.5 to 42.0 t/ha at Sites P, A, and D (Fig. 13). These values were 2 to 5.5 times that at Site C (7.7 t/ha, Fig. 13) and generally larger than CWD mass in most undisturbed coniferous forests (Harmon et al., 1986). The greater CWD mass in the colonized forests was due to the increased mortality of stems as snags in the colonized forest stands (Fujiwara and Takayanagi, 2001; see Section 1.3) which accounted for 68 to 87% of total CWD mass at Sites P, A, and D (Fig. 13). Most snags persisted as standing-dead for 10 years after the bird colonization at Site D, but gradually shifted from decay class I to II during the period (Fig. 13).

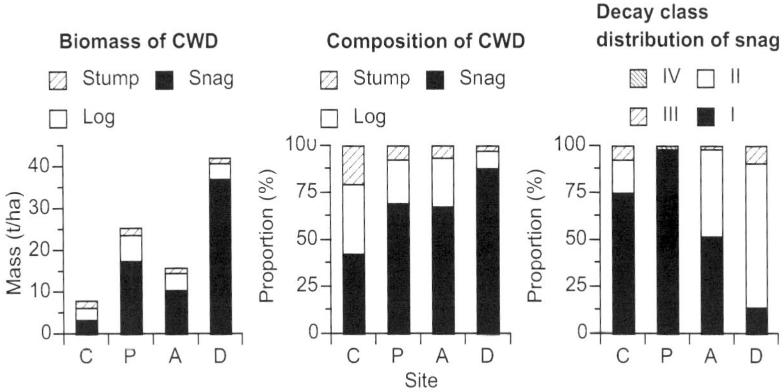

Fig. 13. Mass and composition of coarse woody debris (CWD) and decay class distribution of snags at Sites C, P, A, and D at Isaki Peninsula (Katsumata, 2004). Sites are as in Table 1. CWD (diameter equal to or greater than 10 cm) were investigated in belt transects (4 m width, a total length of 2,030 m, 0.07 to 0.30 ha for each site) in 2003. CWD were recorded for each of three categories (log, snag, stump) and each of five decay classes [decay class I (recently dead and minimally decomposed) to V (strongly decomposed)] according to visual criteria for coniferous CWD (Sollins, 1982). No snag was classified into decay class V in the study of Katsumata (2004).

The nitrogen content of CWD of *C. obtusa* was generally low regardless of the category (log, snag, or stump) and the decay class (I to V), mostly ranging from 0.8 to 1.5 mg/g (Fig. 14). The exceptions were logs in decay class IV at Sites P and A that had higher N content (mean values of 6.6 and 5.8 mg/g, respectively) (Fig. 14). However, the differences in N contents in CWD among the categories or the decay classes were not statistically significant (generalized linear model, $P>0.05$) because of a large variation in N content between CWD samples. Measurements of N isotope ratio in log samples of decay class IV and V indicated that $\delta^{15}N$ was 0.6‰ for a log at Site C, whereas it ranged from 4.2 to 14.8‰ (mean = 8.6‰, n=10) for logs at Sites P, A, and D (Fig. 7), suggesting that excreta-derived N can be incorporated into logs during decomposition and that some logs served as a reservoir of excreta-derived N on the forest floor.

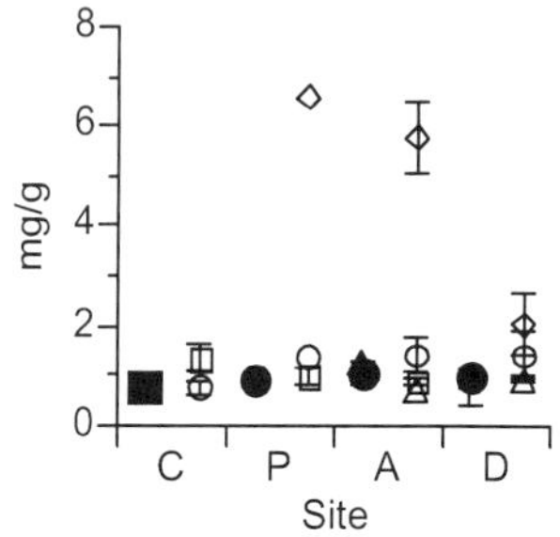

Fig. 14. Nitrogen content (mg/g) in coarse woody debris (CWD) of *Chamaecyparis obtusa*. ◆ snag, decay class I; ℓ snag, decay class II; ‗ snag, decay class III; □ log, decay class I; ○ log, decay class II; △ log, decay class III; ◇ log, decay class IV. Sites are as in Table 1. Values indicate means ± standard errors.

6. Predicting the dynamics of dead plant tissues and excreta-derived nitrogen in colonized forest

The previous sections demonstrated that the excess supply of N as excreta altered the patterns of decomposition of dead plant tissues due to the changes in the ecological (i.e., abundance, diversity, and species composition) and physiological (growth and ligninolytic activity) properties of saprobic fungi. The impact of birds on forest stands, however, is not limited to the supply of a large amount of excreta to the forest floor and the concomitant changes in biological properties in soils. Cormorants break needles and twigs for nesting material and frequently drop these on the forest floor. Such behavior results in a high volume of litterfall in the colonized forests (Section 1.3), which can lead to tree mortality and forest decline. The fallen needles and twigs abundantly supplied to the forest floor are expected to serve as large reservoirs of carbon and excreta-derived nutrients (Section 5). Moreover, the amount of litterfall and excreta deposition is expected to depend on the density of bird colonization, which varies in time and space (Fujiwara and Takayanagi, 2001).

In order to predict the impact of bird colonization on nutrient dynamics in soils, therefore, it is necessary to quantitatively relate the density of bird colonization to the amount of litterfall and to the amount of dead plant tissues and nutrients in the decomposing tissues. In this section, empirical linear models are constructed to describe the relationship between the number of cormorant nests (as an index of bird density) and (i) litterfall amount (denoted as Nest number-litterfall amount or NNLA model), (ii) amount of dead plant tissues remaining after a given period of decomposition (Nest number-residual mass or NNRM model), and (iii) amount of nutrients accumulated in dead plant tissues after a given period of decomposition (Nest number-residual nutrient or NNRN model).

6.1 Nest number-litterfall amount (NNLA) models

Litterfall amount was measured for needles, twigs, and coarse woody debris (CWD) at the study sites and linearly related to the number of cormorant nests (Fig. 15). Here, CWD is sometimes equivalently referred to as stems when mentioning them as the living compartment of forest stands. The regression equations and coefficients of determination are:

$$\text{Needle: } LF_{NDL} = 0.0226 \times NE + 1.180 \ (n=6, R^2=0.94, P<0.01) \quad (1)$$

$$\text{Twig: } LF_{TWG} = 0.0104 \times NE - 0.164 \ (n=6, R^2=0.74, P<0.05) \quad (2)$$

$$\text{CWD: } LF_{CWD} = 0.0122 \times NE - 0.096 \ (n=4, R^2=0.71, P=0.16) \quad (3)$$

where LF_{NDL}, LF_{TWG}, and LF_{CWD} are litterfall amount (t/ha/year) of needles, twigs, and CWD, respectively, and NE is the number of cormorant nests (/ha/year). The coefficient of determination for CWD was not statistically significant at the 5% level because of the large variation of data and the low number of study sites examined.

These regression equations provide useful implications about the relationship between forest decline and nest number. According to previous literature, the biomass of needles, twigs, and stems in plantations of *C. obtusa* varied with location and stand age, ranging from 10 to 20 (13 on average, n=22) t/ha for needles, 10 to 30 (15 on average) t/ha for twigs, and 50 to 200 (107 on average) t/ha for stems (Kawahara, 1974). Substituting the biomass data into equations 1, 2, and 3 yields the number of cormorant nests at which all biomass within a forest stand can be transformed into litterfall. Those are: 390 to 830 (520 on average) nests/ha/year for needles, 980 to 2900 (1460 on average) nests/ha/year for twigs, and 4100 to 16400 (8780 on average) nests /ha/year for stems.

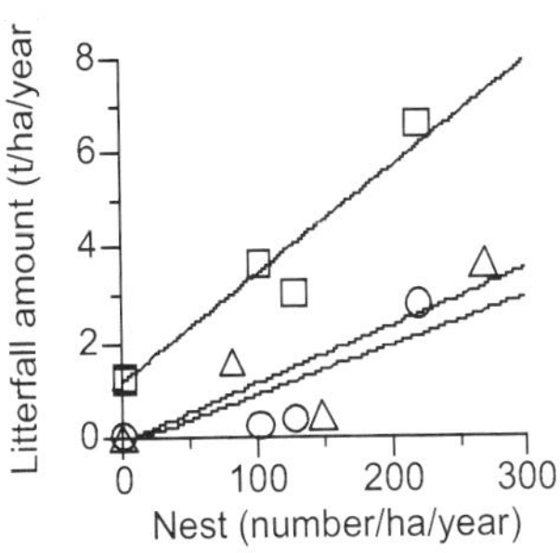

Fig. 15. Litterfall amount as related to the number of cormorant nests. □ needles; ○ twigs; △ coarse woody debris (CWD). Regression lines are for equations 1, 2, and 3, respectively. For needles and twigs, the litterfall amount and the nest number were measured simultaneously during the breeding season of cormorants in 1999 and 2000 at Sites T, P, and D (Fujiwara, 2001). For CWD, the litterfall amount was calculated by measuring diameters at breast height of trees that died (as standing-dead) during the periods of bird colonization starting from the years of colony establishment (1992 to 1997) to 2000 at Sites P, A, and D, as well as Site C (Fujiwara, 2001) and by converting these data to mass according to an allometric equation of Toda et al. (1991). The nest numbers at these sites during the periods of bird colonization were from Fujiwara (2001).

This calculation suggests that all needles in a forest stand can fall when the annual number of nests reaches 400 to 800, or when the cumulative number of nests over some years reaches those values. Because the needle is a photosynthetic organ and because *C. obtusa* is known to lack the ability to sprout (i.e. re-grow) after mechanical loss, 400 to 800 nests per ha will be a critical level at which the forest stand cannot maintain primary production and will start to decline. This prediction is in agreement with the observation at Site D, where the cormorants colonized intensively at least for 4 years, from 1992 to 1996, and then declined (Fujiwara, 2001). The number of cormorant nests in 1992 was 269 /ha at Site D (Fujiwara, 2001), which corresponds to a litterfall rate of needles of 7.3 t/ha/year according to equation 1. This estimated litterfall rate would be high enough to result in forest decline at Site D if similar colonization density of cormorants was maintained for 4 years.

6.2 Nest number-residual mass (NNRM) model

The exponential equation of Olson (1963) is used to describe the changes in remaining mass of needles, twigs, and CWD with respect to the period of decomposition. Data of the 2-year decomposition experiment at Site P (Fig. 11; Osono et al., 2006a) were used to estimate decomposition constants (*k*, /year) for needles and twigs (Equations 4 and 5). Katsumata (2004) showed that more than 68% of CWD was present as snags in the study sites and that most of the snags persisted as standing-dead for more than 10 years but gradually shifted from decay class I to II (Fig. 13). Thus, a total of 32 snags, including those in decay class I at Sites C (no bird colonization; i.e., 0 year after colonization) and P (3 years) and in decay class I and II at Sites A (6 years) and D (10 years), were sampled to measure mass per volume. The mass per volume data of CWD were used to construct the pattern of changes in remaining mass per volume of snags and to estimate a decomposition constant for CWD by means of a chronosequence approach (Equation 6). The exponential equations are expressed as:

$$\text{Needles: } MR_{NDL,t} = LF_{NDL} \times \exp^{-0.27t} \text{ (n=7, } R^2\text{=0.28)} \quad (4)$$

$$\text{Twigs: } MR_{TWG,t} = LF_{TWG} \times \exp^{-0.03t} \text{ (n=7, } R^2\text{=0.75)} \quad (5)$$

$$\text{CWD: } MR_{CWD,t} = LF_{CWD} \times \exp^{-0.02t} \text{ (n=6, } R^2\text{=0.75)} \quad (6)$$

where $MR_{NDL,t}$, $MR_{TWG,t}$, and $MR_{CWD,t}$ are the remaining mass (t/ha) of needles, twigs, and CWD, respectively, at time *t* and *t* is the time in years. The decomposition constants (*k*) are 0.27, 0.03, and 0.02 /year, respectively. The coefficient of determination for needles was low because of asymptotic pattern of changes in remaining mass over the study period (Fig. 11). Substituting equations 1-3 into exponential equations 4-6 yields the equations describing the relationship between the nest number and the remaining mass of needles, twigs, and CWD, respectively, at a given decomposition time *t* (NNRM model):

$$\text{Needles: } MR_{NDL,t} = (0.0226 \times NE + 1.180) \times \exp^{-0.27t} \quad (7)$$

$$\text{Twigs: } MR_{TWG,t} = (0.0104 \times NE - 0.164) \times \exp^{-0.03t} \quad (8)$$

$$\text{CWD: } MR_{CWD,t} = (0.0122 \times NE - 0.096) \times \exp^{-0.02t} \quad (9)$$

6.3 Nitrogen immobilization and Nest number-residual nutrient (NNRN) model

Excreta-derived N was immobilized in decomposing needles and twigs (Section 5.2) and in CWD (Section 5.3). Osono et al. (2006a) estimated the potentials of these plant tissues to immobilize excreta-derived N (denoted here as the immobilization potential, which means the maximum amount of exogenous N immobilized per initial material) and the duration of this immobilization phase according to the method described in Mellilo and Aber (1984). Firstly, the linear relationships between the percent remaining mass of plant tissues and N content in the remaining materials were described as regression equations (Fig. 16):

$$\text{Needles: } MR_{NDL}/LF_{NDL} \times 100 = -20 \times NIT_{NDL} + 120 \ (n=7, R^2=0.76, P<0.01) \quad (10)$$

$$\text{Twigs: } MR_{TWG}/LF_{TWG} \times 100 = -23 \times NIT_{TIG} + 110 \ (n=7, R^2=0.32, P=0.18) \quad (11)$$

$$\text{CWD: } MR_{CWD}/LF_{CWD} \times 100 = -174 \times NIT_{CWD} + 105 \ (n=6, R^2=0.13, P=0.48) \quad (12)$$

where NIT_{NDL}, NIT_{TWG}, and NIT_{CWD} are N content (%, w/w) of needles, twigs, and CWD, respectively.

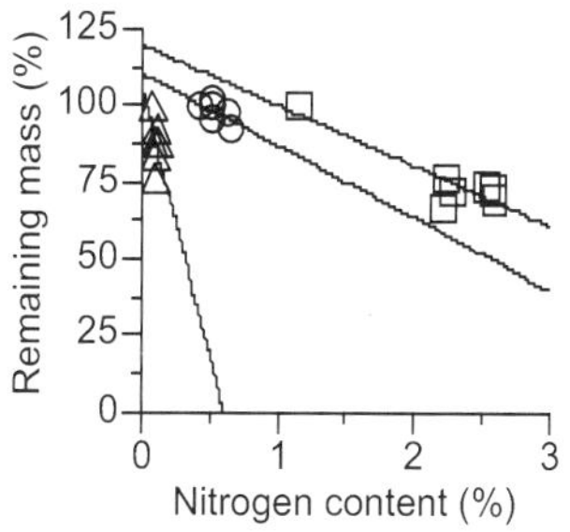

Fig. 16. Linear relationships between the percent remaining mass of decomposing plant tissues (% of the original mass: $MR_t/LF \times 100$) and the N content (%, w/w) in the remaining tissues for needles, twigs, and CWD. Symbols are as in Fig. 15.

Substituting the intercepts and slopes of equations 10-12 and the decomposition constants of equations 4-6 into the equations of Mellilo and Aber (1984), the immobilization potential was calculated to be 6.6, 8.6, and 0.8 mg N/g initial material and the duration of the immobilization phase to be 1.6, 19.9, and 32.2 years for needles, twigs, and CWD, respectively.

Using these values, Osono et al. (2006a) estimated the N immobilization potential of litterfall to be 10.3 and 7.2 kg/ha/month for needles and twigs, respectively. These values accounted for 4.1% and 3.0% of total N input as excreta during the breeding season at Site P (i.e., 240 kg/ha/month; Kameda et al., 2006). This tentative calculation thus suggests that the increased litterfall at Site P due to breeding activity of the cormorants has a potential to immobilize only a total of 7% of total excreta-derived N deposited on the forest floor during 2 (needles) and 20 (twigs) years of decomposition. The major fate of excreta-derived N thus can be leaching into deeper soil layers (Hobara et al., 2005) and volatilization as ammonia gas into the atmosphere.

Finally, substituting equations 10-12 into equations 7-9 yields equations describing the relationship between the nest number and N mass (kg/ha) in the remaining needles, twigs, and CWD, respectively, at a given decomposition time *t* (NNRN model):

Needles:

$$NIT_{NDL,t} = [(0.0226 \times NE + 1.180) \times \exp^{-0.27t}] \times [(100 \times \exp^{-0.27t} - 120)/(-20)] \times 10 \quad (13)$$

Twigs:

$$NIT_{NDL,t} = [(0.\ 0104 \times NE + 0.164) \times \exp^{-0.03t}] \times [(100 \times \exp^{-0.03t} - 110)/(-23)] \times 10 \quad (14)$$

CWD:

$$NIT_{NDL,t} = [(0.\ 0122 \times NE + 0.096) \times \exp^{-0.02t}] \times [(100 \times \exp^{-0.02t} - 105)/(-174)] \times 10 \quad (15)$$

6.4 Dynamics of dead plant tissues and excreta-derived nitrogen in colonized forest

Using the empirical models in equations 7-9 and 13-15, long-term patterns of the remaining mass of dead plant tissues and in the N mass during decomposition were estimated for forest stands colonized by cormorants (Fig. 17). The models show the different roles of plant tissues as components of the forest floor and reservoirs of excreta-derived N.

At the time of litterfall (i.e., 0 year), needles, twigs, and CWD account for approx. 50%, 25%, and 25% of total litterfall, respectively (Fig. 17). However, needles almost disappear before 20 years of decomposition because the mass loss for them is much faster than that for twigs and CWD. After 20 years twigs and CWD constitute the dominant components of the detritus pool in the forest stand. Although not shown in Fig. 17, CWD becomes two times more important quantitatively than twigs at ~60 years of decomposition.

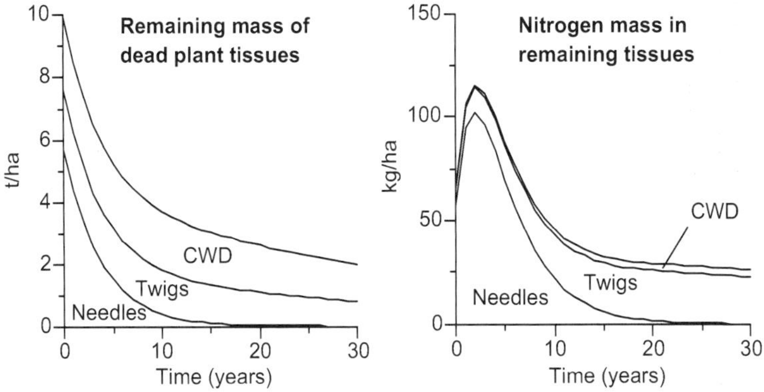

Fig. 17. Estimated long-term changes in the remaining mass of dead plant tissues and the nitrogen mass in the remaining tissues during decomposition in forest stand colonized by cormorants at 200 nests/ha, the mean value at Isaki Peninsula in 2000 (Fujiwara, 2001). The models start with litterfall in a single year, and the figures show decomposition for 30 years.

Needles account for 87% of N in dead plant tissues at the time of litterfall because the initial N contents are 3 and 15 times higher than in twigs and CWD, respectively (Fig. 17). During the first two years of decomposition, N mass in needles increased 1.8 times compared to that of the initial mass due to the net immobilization of excreta-derived N (Section 5.2). The N mass reached its maximum value at 3 years of decomposition but decreased thereafter due to the net release from needles. The N mass in needles becomes smaller exponentially, and it almost disappears before 20 years of decomposition. This suggests that needles serve as a temporary reservoir and then as a source of N thereafter up to 20 years of decomposition. In contrast, twigs immobilize N slowly for 20 years to become the dominant reservoir of N thereafter. The model predicts that CWD (snags) plays a negligible role in N retention. Note, however, that a part of CWD can be a reservoir of N when it falls down to become logs (Section 5.3), a process not incorporated in the model.

It should also be noted that some equations in the models have low coefficients of determination. This especially holds true for the litterfall amount of CWD (Eq. 3), the changes in remaining mass of needles (Eq. 4), and the dynamics of N in twigs and CWD (Eqs. 11 and 12). When the decomposition of needles is assumed to follow an asymptotic function, for example, needles become a more important, longer-term reservoir of dead plant tissues and N in the detritus pool. Obviously, longer-term studies of tree mortality and decomposition of needles, twigs, and CWD will be necessary to construct more accurate empirical models. Still, the present models provide useful insights into the effects of the density of cormorant colonization on the amount of litterfall and into the differential roles of dead plant tissues as reservoirs of carbon and excreta-derived N.

7. Conclusion

The series of studies demonstrated that excess supply of excreta-derived N changed the community structure, nutrition, and substrate utilization of saprobic fungi, which by altering the decomposition processes led to carbon sequestration, accumulation of excreta-derived N, and thus a slow turnover of carbon and N in forest soils affected by the cormorant (Fig. 18). Most of the previous studies examined the effects of excess supply of nutrients on fungal communities, microbial activities, decomposition processes, or soil carbon accumulation separately, and the interactions and possible causal relationships between these biological and ecosystem processes have rarely been explored. This case study of cormorant-derived excreta deposition in conifer plantations at Isaki Peninsula thus can provide useful implications for the understanding of biological mechanisms underlying the N-induced sequestration of soil carbon in forest ecosystems supplemented with excess nutrients.

The past century is the first time since the evolution of modern N cycle linked to microbial processes with robust natural feedbacks and controls that human activities may have produced the largest impact on global N cycle (Canfield et al., 2010). Disrupted N cycles due to excess supply of N of anthropogenic origin and the concomitant buildup of N and carbon sequestration in forest soils are one of major global issues because of its potential influence on the evolution of carbon dioxide and feedback to global warming (Nadelhoffer et al., 1999; de Vries et al., 2006). The present study highlights potential importance of fungi and their indispensable roles linking N deposition with carbon sequestration in soils. Future research directions include the dynamics of phosphorus (Conley et al., 2009), another major nutrient that is abundantly contained in excreta (Hobara et al., 2005), and which limits primary production more frequently than N and has different effects on soil processes and fungal activity.

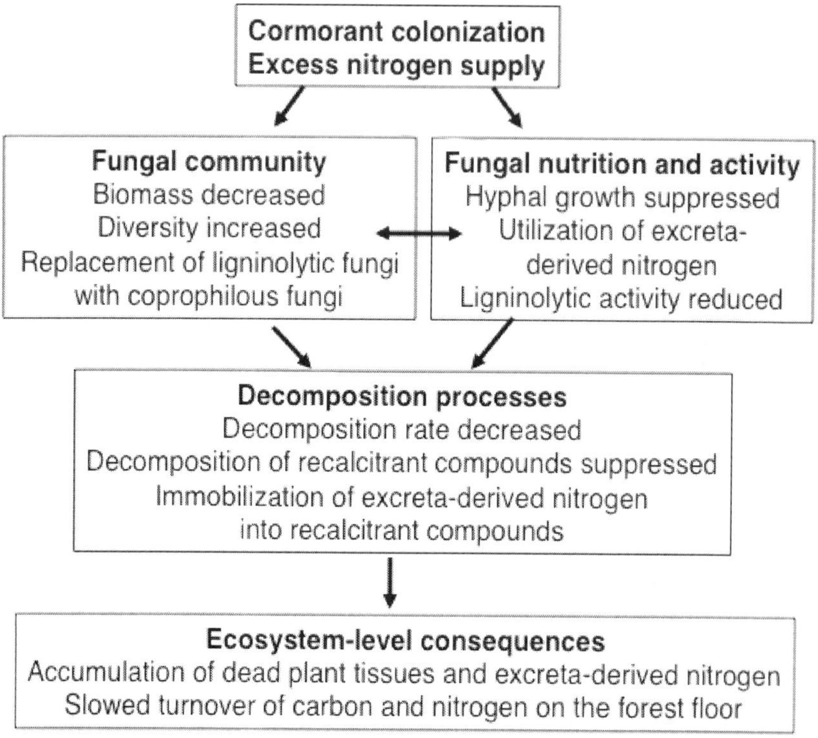

Fig. 18. Schematic diagram showing the possible effects of excess supply of excreta-derived nitrogen on fungal community and activity and decomposition processes and its ecosystem consequences.

8. Acknowledgments

I thank Drs. Kayoko Kameda, Satoru Hobara, Keisuke Koba, Akira Ishida, Yu Fukasawa, and Hiroshi Takeda, Mrs. Satomi Fujiwara, and Mr. Shingo Katsumata for collaborations in the research projects; Drs. Seiji Tokumasu and Susumu Iwamoto for helpful identification of fungi; and Dr. Elizabeth Nakajima for proofreading of the manuscript. Dr. Kayoko Kameda kindly provided original data of Fig. 1. This study was supported by Global COE Program A06 of Kyoto University.

9. References

Aber, J.; McDowell, W.; Nadelhoffer, K.; Magill, A.; Berntson, G.; Kamakea, M.; McNulty, S.; Currie, W.; Rustad, L.; Fernandez, I. (1998). Nitrogen saturation in temperate forest ecosystems, hypotheses revisited. *Bioscience*, 48, 921-934

Anderson, W.B.; Polis, G.A. (1999). Nutrient fluxes from water to land: seabirds affect plant nutrient status on Gulf of California islands. *Oecologia*, 118, 324-332

Berg, B. (1986). Nutrient release from litter and humus in coniferous forest soils - a mini review. *Scandinavian Journal of Forest Research*, 1, 359-369

Berg, B. (1988). Dynamics of nitrogen (^{15}N) in decomposing Scots pine (*Pinus sylvestris*) needle litter. Long-term decomposition in a Scots pine forest. VI. *Canadian Journal of Botany*, 66, 1539-1546

Berg, B.; Matzner, E. (1997). Effect of N deposition on decomposition of plant litter and soil organic matter in forest systems. *Environmental Review*, 5, 1-25

Berg, B.; McClaugherty, C. (2003). *Plant Litter, Decomposition, Humus Formation, Carbon Sequestration*. Springer Verlag, Berlin

Blackwood, C.B.; Waldrop, M.P.; Zak, D.R.; Sinsabaugh, R.L. (2007). Molecular analysis of fungal communities and laccase genes in decomposing litter reveals differences among forest types but no impact of nitrogen deposition. *Environmental Microbiology*, 9, 1306-1316

Canfield, D.E.; Glazer, A.N.; Falkowski, P.G. (2010). The evolution and future of Earth's nitrogen cycle. *Science*, 330, 192-196

Conley, D.J.; Paerl, H.W.; Howarth, R.W.; Boesch, D.F.; Seitzinger, S.P.; Havens, K.E.; Lancelot, C.; Likens, G.E. (2009). Controlling eutrophication: nitrogen and phosphorus. *Science*, 323, 1014-1015

De Vries, W.; Reinds, G.J.; Gundersen, P.; Sterba, H. (2006). The impact of nitrogen deposition on carbon sequestration in European forests and forest soils. *Global Change Biology*, 12, 1151-1173

Fenn, P.; Choi, S.; Kirk, T.K. (1981). Ligninolytic activity of *Phanerochaete chrysosporium*: physiology of suppression by NH_4^+ and L-glutamate. *Archives in Microbiology*, 130, 66-71

Fog, K. (1988). The effect of added nitrogen on the rate of decomposition of organic matter. *Biological Review*, 63, 433-462

Fujiwara, S. (2001). *Forest decline in cormorant colonies*. Master thesis, Kyoto University (in Japanese)

Fujiwara, S.; Takayanagi, A. (2001). The influence of the common cormorant (*Phalacrocorax carbo* Kuroda) on forest decline. *Applied Forest Science*, 10, 85-90 (in Japanese with English abstract)

Fukasawa, Y.; Osono, T.; Takeda, H. (2009). Effects of attack of saprobic fungi on twig litter decomposition by endophytic fungi. *Ecological Research*, 24, 1067-1073

Harmon, M.E.; Franklin, J.F.; Swanson, F.J.; Sollins, P.; Gregory, S.V.; Lattin, J.D.; Anderson, N.H.; Cline, S.P.; Aumen, N.G.; Sedell, J.R.; Lienkaemper, G.W.; Cromack, Jr., K.; Cummins, K.W. (1986). Ecology of coarse woody debris in temperate ecosystems. *Advances in Ecological Research*, 15, 133-302

Hassett, J.E.; Zak, D.R.; Blackwood, C.B.; Pregitzer, K.S. (2009). Are basidiomycete laccase gene abundance and composition related to reduced ligninolytic activity under elevated atmospheric NO_3^- deposition in a northern hardwood forest? *Microbial Ecology*, 57, 728-739

Hobara, S.; Osono, T.; Koba, K.; Tokuchi, N.; Fujiwara, S.; Kameda, K. (2001). Forest floor quality and N transformations in a temperate forest, affected by avian-derived N deposition. *Water, Air, and Soil Pollution*, 130, 679-684

Hobara, S.; Koba, K.; Osono, T.; Tokuchi, N.; Ishida, A.; Kameda, K. (2005). Nitrogen and phosphorus enrichment and balance in forests colonized by cormorants: implications of the influence of soil adsorption. *Plant and Soil*, 268, 89-101

Ishida, A. (1996a). Changes in soil properties in the colonies of the common cormorant, *Phalacrocorax carbo*. *Journal of Forest Research*, 1, 31-35

Ishida, A. (1996b). Effects of the common cormorant, *Phalacrocorax carbo*, on evergreen forests in two nest sites at Lake Biwa, Japan. *Ecological Research*, 11, 193-200

Ishida, A.; Narusue, M.; Kameda, K. (2003). Management of Great Cormorant *Phalacrocorax carbo hanedae* colonies in Japan. *Vogelwelt*, 124, 331-337

Johnsgard, P.A. (1993). *Cormorants, Darters, and Pelicans of the World*. Smithosonian Institution Press, London

Kameda, K.; Koba, K.; Yoshimizu, C.; Fujiwara, S.; Hobara, S.; Koyama, L.; Tokuchi, N.; Takayanagi, A. (2000). Nutrient flux from aquatic to terrestrial ecosystem mediated by the Great Cormorant. *Sylvia*, 36, 54-55

Kameda, K.; Ishida, A.; Narusue, M. (2003). Population increase of the Great Cormorant *Phalacrocorax carbo hanedae* in Japan: conflicts with fisheries and trees and future perspectives. *Vogelwelt*, 124, 27-33

Kameda, K.; Koba, K.; Hobara, S.; Osono, T.; Terai, M. (2006). Mechanism of long-term effects of cormorant-derived nitrogen in a lakeside forest. *Hydrobiologia*, 567, 69-86

Katsumata, S. (2004). *Biomass and nutrient contents of coarse woody debris in forest stands colonized by the cormorants*. Bachelor thesis, Kyoto University (in Japanese)

Kawahara, T. (1974). Stock and circulation of nutrients within ecosystem, In: *Chamaecyparis obsuta forests: its ecology and natural regeneration*, T. Shidei, (Ed.), 131-209, Chikyusha, Japan (in Japanese)

Keyser, P.; Kirk, T.K.; Zeikus, J.G. (1978). Ligninolytic enzyme system of *Phanerochaete chrysosporium*: synthesized in the absence of lignin in responses to nitrogen starvation. *Journal of Bacteriology*, 135, 790-797

Kohzu, A.; Yoshioka, T.; Ando, T.; Takahashi, M.; Koba, K.; Wada, E. (1999). Natural ^{13}C and ^{15}N abundance of field-collected fungi and their ecological implications. *New Phytologist*, 144, 323-330

Koide, K.; Osono, T. (2003). Chemical composition and mycobiota of bleached portion of *Camellia japonica* leaf litter at two stands with different nitrogen status. *Japanese Journal of Forestry Society*, 85, 359-363 (in Japanese with English abstract)

Mellilo, J.M.; Aber, J.D. (1984). Nutrient immobilization in decaying litter: an example of carbon-nutrient interactions, In: *Trends in Ecological Research for the 1980s*, J.H. Cooley, F.B. Gooley (Eds.), 193-215, Plenum Press, New York

Mizutani, H.; Wada, E. (1988). Nitrogen and carbon isotope ratios in seabird rookeries and their ecological implications. *Ecology*, 69, 340-349

Nadelhoffer, K.J.; Emmett, B.A.; Gundersen, P.; Kjønaas, O.J.; Koopmans, C.J.; Schleppi, P.; Tietema, A.; Wright, R.F. (1999). Nitrogen deposition makes a minor contribution to carbon sequestration in temperate forests. *Nature*, 398, 145-148

Nagasaka, A.; Nagasaka, Y.; Ito, K.; Mano, T.; Yamanaka, M.; Katayama, A.; Sato, Y.; Grankin, A.L.; Zdorikov, A.I.; Boronov, G.A. (2006). Contributions of salmon-derived nitrogen to riparian vegetation in the northwest Pacific region. *Journal of Forest Research*, 11, 377-382

Ninomiya, A.; Antunes, M.F.R.; Schoenlein-Crusius, I.H. (1993). Fungi from soil affected by birds in the Parque Estadual Das Fontes Do Ipiranga, São Paulo State, Brazil. *Revista de Microbiologia*, 24, 49-53

Olson, J. (1963). Energy storage and the balance of produces and decomposers in ecological systems. *Ecology*, 44, 322-331

Osono, T. (2007). Ecology of ligninolytic fungi associated with leaf litter decomposition. *Ecolgical Research*, 22, 955-974

Osono, T.; Takeda, H. (2001). Effects of organic chemical quality and mineral nitrogen addition on lignin and holocellulose decomposition of beech leaf litter by *Xylaria* sp. *European Journal of Soil Biology*, 37, 17-23

Osono, T.; Hobara, S.; Fujiwara, S.; Koba, K.; Kameda, K. (2002). Abundance, diversity, and species composition of fungal communities in a temperate forest affected by excreta of the Great Cormorant Phalacrocorax carbo. *Soil Biology & Biochemistry*, 34, 1537-1547

Osono, T.; Takeda, H. (2004). Accumulation and release of nitrogen and phosphorus in relation to lignin decomposition in leaf litter of 14 tree species. *Ecological Research*, 19, 593-602

Osono, T.; Hobara, S.; Koba, K.; Kameda, K.; Takeda, H. (2006a). Immobilization of avian excreta-derived nutrients and reduced lignin decomposition in needle and twig litter in a temperate coniferous forest. *Soil Biology & Biochemistry*, 38, 517-525

Osono, T.; Hobara, S.; Koba, K.; Kameda, K. (2006b). Reduction of fungal growth and lignin decomposition in needle litter by avian excreta. *Soil Biology & Biochemistry*, 38, 1623-1630

Preston, C.M.; Trofymow, J.A.; Sayer, B.G.; Nie, J. (1997). ^{13}C nuclear magnetic resonance spectroscopy with cross-polarization and magic-angle spinning investigation of the proximate-analysis fractions used to assess litter quality in decomposition studies. *Canadian Journal of Botany*, 75, 1601-1613

Schoenlein-Crusius, I.H.; Trufem, S.F.B.; Malatinsky, S.M.M.; Ninomiya, A.; Antunes, M.F.R. (1996). Mucorales (Zygomycotina) from soil affected by excrement of birds in the Parque Estadual das Fontes do Ipiranga, São Paulo, Brazil. *Revista Brasileira de Botanica*, 19, 7-10

Sinsabaugh, R.L. (2010) Phenol oxidase, peroxidase and organic matter dynamics of soil. *Soil Biology & Biochemistry*, 42, 391-404

Söderström, B.; Bååth, E.; Lundgren, B. (1983). Decrease in soil microbial activity and biomass owing to nitrogen amendments. *Canadian Journal of Microbiology* 29, 1500-1506

Sollins, P. (1982). Input and decay of coarse woody debris in coniferous stands in western Oregon and Washington. *Canadian Journal of Forest Research*, 12, 18-28

Toda, H.; Haibara, K.; Arai, M. (1991) Nutrient circulation in a small watershed under an established sugi (*Cryptomeria japonica*) and hinoki (*Chamaecyparis obtusa*) stand. *Bulletin of University Forest of Tokyo University of Agriculture and Technology*, 28, 1-22 (in Japanese with English abstract)

Trudell, S.A.; Rygiewicz, P.T.; Edmonds, R.L. (2004). Patterns of nitrogen and carbon stable isotope ratios in macrofungi, plants and soils in two old-growth conifer forests. *New Phytologist*, 164, 317-335

Wainright, S.C.; Haney, J.C.; Kerr, C.; Golovkin, A.N.; Flint, M.V. (1998). Utilization of nitrogen derived from seabird guano by terrestrial and marine plants at St. Paul, Pribilof Islands, Bering Sea, Alaska. *Marine Biology*, 131, 63-71

Waldrop, M.P.; Zak, D.R. (2006). Response of oxidative enzyme activities to nitrogen deposition affects soil concentrations of dissolved organic carbon. *Ecosystems*, 9, 921-933

Wilkinson, C.E.; Hocking, M.D.; Reimchen, T.E. (2005). Uptake of salmon-derived nitrogen by mosses and liverworts in coastal British Columbia. *Oikos*, 108, 85-98

Zak, D.R.; Holmes, W.E.; Burton, A.J.; Pregitzer, K.S.; Talhelm, A.F. (2008). Simulated atmospheric NO_3^- deposition increases soil organic matter by slowing decomposition. *Ecological Applications*, 18, 2016-2027

11

Effect of Environmental Change on Secondary Metabolite Production in Lichen-Forming Fungi

Christopher Deduke, Brinda Timsina and Michele D. Piercey-Normore
Department of Biological Sciences, University of Manitoba, Winnipeg, MB, Canada

1. Introduction

The production and regulation of secondary metabolites in non-lichenized fungi, mainly ascomycetes, has been reviewed by a number of authors with an emerging understanding of the biosynthesis and the pathways involved in regulation (Keller et al., 2005; Yu & Keller 2005; and others). However, lichenized fungi make up almost half of all known ascomycetes (Kirk et al., 2001) and are known to produce over 800 secondary metabolites, most of which are unique to lichenized fungi. Many of these compounds have bioactive properties (Huneck, 1999) and some studies have shown or suggested that secondary metabolite production is influenced by changes in culture conditions, which might be regarded as environmental changes. Intense investigation of the changes in production of these unique bioactive secondary metabolites from lichen fungi have been hampered by problems associated with isolating and growing cultures of lichen fungi. Lichens have been studied for more than two centuries as morphological entities but experimental lichenology has remained a nearly unexplored scientific field for many decades because of the slow growing nature of lichens. Thomas (1939 in Stocker-Worgotter, 2001) reported the first successful resynthesis of *Cladonia pyxidata*. Since the 1970's, one major goal of experimental lichenology has been the improvement and optimization of culture conditions of lichen fungi. Culture techniques for lichen fungi have improved in recent years allowing for further research on these challenging organisms. Therefore, with greater access to cultures of lichen symbionts and progression of knowledge of non-lichenized fungi, studies are just beginning to accumulate on genes involved in production of secondary metabolites from lichen fungi; and the effects of the environment on the expression of these genes by observations in ecological studies, and through experimentation by manipulating culture conditions.

Fungal secondary metabolism is covered by extensive body of literature (see Bennett & Ciegler, 1983). Secondary metabolism is not required for survival and its products are dispensible whereas primary metabolism is essential for survival with anabolic and catabolic activities to maintain life. Secondary metabolites are chemically diverse but are produced from a few key intermediates of primary metabolism, and are generally categorized by the intermediates from which they are produced. Bennett and Ciegler (1983) summarize six categories of secondary metabolites derived from different primary intermediates. Although fungal secondary metabolites are extensive, they are generally produced by one of just a few major pathways (Moore, 1998). The mevalonic acid pathway produces terpenes, sterols, and carotenoids. The malonate (or acetyl-polymalonyl; Elix & Stocker-Worgotter, 2008) pathway produces polyketides. Other metabolites are produced by the shikimate-chorismate (or shikimic acid; Elix & Stocker-Worgotter, 2008) pathway. This chapter focuses mainly on the polyketides produced by the acetyl-polymalonyl pathway. Polyketides constitute structurally diverse molecules produced by the successive condensation of small carboxylic acids, typically co-enzyme A activated malonate by a mechanism similar to fatty acid biosynthesis (Hopwood & Sherman, 1990). The diversity of polyketide structure produced from this pathway reflects the diversity of their biological activities.

For more thorough reviews of the structure and regulation of secondary metabolites in fungi the reader is referred to (Hopwood, 1997; Drew & Demain, 1977; Katz & Donadio, 1993; Hutchinson & Fujii, 1995; Keller & Hohn, 1997; Bennett & Ciegler, 1983). Reviews and inventories of lichen metabolites are summarized by Culberson C. F. & Elix (1989), Elix & Stocker-Worgotter (2008); Culberson C. F. (1969); Culberson C. F. (1970); Culberson C. F., et al. (1977b); Stocker-Worgotter (2008); and a recent classification of lichen substances (Culberson C. F. & Elix, 1989). The adaptive significance of secondary metabolites produced by lichen fungi has been speculated (Lawrey 1977) and numerous functional studies (reviewed in Huneck, 1999), but few studies have linked adaptation of lichen substances with environmental change.

This chapter provides a synopsis of secondary metabolite production in fungi with a focus on lichenized fungi. The synopsis includes a review of the effect of environmental parameters and fungal development on production and regulation of secondary metabolites by focusing on three genera of lichen-forming fungi but not exclusive to these taxa. The chapter also describes an original ecological study of secondary metabolite production for two species of lichen fungi along a geographic gradient. It concludes by summarizing these findings in light of the significance of secondary metabolic changes in terms of ecological and anthropogenic prospects.

2. Overview of secondary metabolite production and regulation

Current changes in climate have prompted a number of studies to predict future changes in species distributions, ecosystem changes, effect on rare species, and effects on invasive species. For example, *Lecanora populicola* has been predicted to expand its distribution with continued environmental change (Ellis et al., 2007). However, integral to the distribution of this species is the host tree, *Populus* sp., on which the lichen grows. *Biatora helvolla* was shown to follow the distribution of its host spruce tree after the glaciation in the Alps (Printzen et al., 1999). In another study, lichen diversity was influenced more by forest structure than by climate (Moning et al., 2009) and a recommendation was to allow native broad leaved trees to grow to maturity promoting rare lichen species. One of the more significant changes with global warming is the depletion of the atmospheric ozone layer allowing more UV-B radiation exposure to living organisms (Kerr & McElroy, 1993). Since many lichen compounds absorb UV-B (Solhaug et al., 2003) an increase in production of those compounds is expected with continued climate change. Bjerke et al. (2003) showed an increase in the levels of two tridepsides (methyl gyrophorate and gyrophoric acid) and unidentified trace metabolites after a five-year study of two arctic lichen species of *Peltigera*.

But there were no changes in thallus dimensions or nitrogen fixation activity. A shift in secondary metabolism to allow survival in a particular habitat may promote changes in species and therefore functional attributes of phenotype. One of the functional changes of lichen-fungi dealt with in this chapter is that of secondary metabolite production. To some extent fungal secondary metabolites reflect taxonomy, but some studies have suggested that secondary metabolites may also be influenced by environmental change. Environmental changes influence many cellular activities and also serve as triggers for a change in mode of reproduction, influencing the entire biology of the species.

Since most species have diagnostic compounds that are consistently produced because of genetic inheritance and species adaptation to particular niches, chemical diversity can be correlated with taxonomy. The chemical correlation with taxonomy is referred to as chemotaxonomy (reviewed by Hawksworth, 1976; Frisvad et al., 2008). Knowledge of species taxonomic diversity is a first clue to understanding the polyketide diversity in any habitat. *Ramalina americana* was split into two different species (*R. culbersoniorum* and *R. americana*) based on secondary metabolite and nucleotide sequence divergence (LaGreca, 1999). The *Cladonia chlorophaea* complex contains at least five chemospecies, which are named and determined by the secondary metabolite produced (Culberson C. F. et al., 1977a). Other examples exist to show variability among individuals within the same geographic area. Secondary metabolites may also vary even within chemospecies. For example, the diagnostic metabolite for *C. grayi* is grayanic acid, and for *C. merochlorophaea* is merochlorophaeic acid. However, these species may or may not produce fumarprotocetraric acid, a polyketide that is considered to be an accessory compound since it is not consistently produced among individuals within a species. One suggestion for the quantity of accessory compounds to vary is changes in the environment (Culberson C. F. et al., 1977a) affecting regulatory pathways that depend on fungal developmental and environmental cues.

2.1 Exploring diversity of secondary metabolites within three genera of lichen-forming fungi

Since lichens are named according to their fungal partner (Kirk et al., 2001), 13,500 known species of lichenized fungi are somewhat scattered throughout the ascomycete families and reflect one of several ecological groups of fungi. Other ecological groups of fungi include mycorrhizal fungi, plant and animal pathogenic fungi, and saprobic fungi. These ecological groups may be considered artificial groups that reflect changes in feeding habits because of environmental plasticity that are present in most taxonomic groups. The lichenized fungi are currently classified among three classes of ascomycetes, Sordariomycetes, Lecanoromycetes, and Eurotiomycetes, and approximately 20 species of basidiomycetes. The majority of lichen-forming fungi belong to the Lecanoromycetes (Tehler & Wedin, 2008). Three genera within the Lecanoromycetes include *Cladonia*, a large ground-dwelling genus; *Ramalina*, epiphytes on rocks and trees; and *Xanthoparmelia*, an almost exclusive rock-dwelling genus. The substratum on which fungi grow allows for a diversity of nutrients to be available to the fungus (Brodo, 1973). The three genera grow on different substrata, have large thalli, have broad global distributions, and therefore provide a good contrast for examining secondary metabolite diversity.

The genus *Cladonia* is a large genus within the family Cladoniaceae comprised of more than 400 species (Ahti, 2000) and contains more than 60 described secondary metabolites with 30 of those being major metabolites in high concentration (Ahti, 2000) and the remaining being minor accessory compounds in lower concentration. Secondary metabolites produced by members of the genus have been extensively studied with some variability in polyketide diversity (Huovinen & Ahti, 1986a, 1986b, 1988; Huovinen et al., 1989a, 1989b, 1990). Members of the genus are mostly ground dwelling on soil or moss and sometimes occur on thin soil over rock. Other species are found on decaying wood or tree bases. All species have a primary crustose or squamulose thallus in direct contact with the substratum and a vertical fruticose thallus (podetium) often culminating in the sexually produced fruit body (apothecium) at its apex (Fig. 1A). The fungi in this genus associate with Eukaryotic unicellular green algae in the genus *Asterochloris*.

The genus *Ramalina* is comprised of 46 species in North America (Esslinger, 2011) and is often considered to be highly variable in its polyketide production. The genus is characterized by producing B-orcinol depsides and depsidones. Usnic acid is the most common cortical compound in the genus. The *R. farinacea* complex produces a variety of metabolites that are all biosequentially related (Culberson W. L., 1966) with similar variability in the *R. siliquosa* complex (Culberson C. F. et al., 1992, 1993). Members of the *R. americana* species complex alone contain more than 55 metabolites (Culberson C. F. et al., 1990, 2000). Culberson C. F. et al. (1990) described comprehensively the biogenetic relationships and geographic correlations of the chemical variation within *R. americana*. While some species within the genus grow on rocks or cliffs, other species prefer the bark of trees, and some of the generalists may be found on both rock and tree bark. The genus contains fruticose species that are attached to their substratum by a single or several holdfasts giving the thallus a tufted or sometimes pendant appearance (Fig. 1B). The degree of contact between substratum and thallus is less than that for either *Cladonia* or *Xanthoparmelia*. Species of *Ramalina* associate with eukaryotic unicellular green algae in the genus *Trebouxia*.

Xanthoparmelia is a large genus distributed globally with more than 406 species (Hale 1990) but in present times is thought to exceed 800 species (Crespo et al., 2007). It is also polyketide diverse containing more than 38 major compounds and 53 accessory compounds (Hale, 1990). Salazinic, stictic, fumarprotocetraric, and norstictic acids are the most common medullary metabolites and usnic acid is the main cortical compound in the genus. Species in this genus are large foliose lichens that grow on non-calcareous rock and sometimes on mineral soils as the substratum. The thallus is attached to the substratum by large numbers of rhizines, which are clusters of fungal hyphae that extend from the underside of the thallus and penetrate the substratum (Fig. 1C). With many rhizines on each thallus the degree of contact with the substratum is greater than that with *Ramalina* but less than that with *Cladonia*. *Xanthoparmelia* species associate with eukaryotic unicellular green algae in the genus *Trebouxia*.

The heteromerous thallus in each of the three genera contains highly organized layers of tissue and each layer has a specific function (see Fig. 1 inserts; Budel & Scheidegger, 2008). Because of the cylindrical nature of the thallus, fruticose lichens have outer, middle, and sometimes inner layers of thallus tissue extending upright (podetium; *Cladonia*) or outward (pendant or tufted; *Ramalina*) from the substratum, whereas foliose thalli have upper, middle and lower layers of tissue because of the flattened, leaf-like nature of the thallus against the substratum (*Xanthoparmelia*). The outer/upper layer may be comprised of a cortex (except some *Cladonia* spp.) with thick walled conglutinated fungal hyphae densely adhered to one another. This layer sometimes contains pigments or other secondary metabolites that have a number of hypothesized protective functions. The middle layer of tissue is comprised of the medulla, which is a layer of loosely woven fungal hypae often with air spaces. Secondary metabolites that confer an external hydrophobic property, and a continuous or patchy layer of algal cells are present in the upper or outer layer of the medulla. The lower or inner layer of tissue varies tremendously depending on the taxonomy and habitat of members of the genus. The genus *Cladonia* contains an inner hollow tube with a margin of conglutinated fungal hyphae similar to a cortex. This hollow tube is diagnostic of the genus and it provides the upright podetial thallus with increased support to successfully release fungal spores into the air current for effective dispersal. The inner layers of the primary squamulose thallus are comprised of medullary hyphae. The inner layer of *Ramalina* is a continuation of medullary hyphae with no differentiated inner tissue. The lower layer of *Xanthoparmelia* species consists of a thin lower cortex to which rhizines are attached for anchorage on rock substrata.

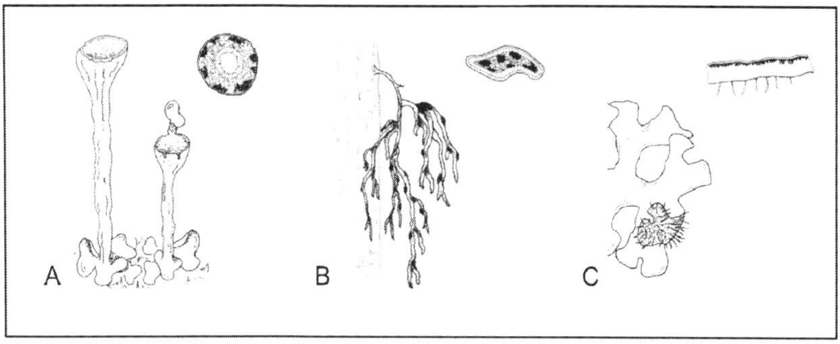

Fig. 1. Illustration of lichen growth forms for A. upright fruticose podetium and leafy squamules of *Cladonia* sp., B. pendant fruticose thallus of *Ramalina* sp. showing the single holdfast attachment to a tree, and C. foliose thallus of *Xanthoparmelia* sp. with an overturned lobe showing rhizines on the underside of the lobe. Inserts show thallus cross sections for each growth form (see text for details).

2.2 Regulation and production of secondary metabolites based on current knowledge of fungi

Spatial scale plays a role in interpretation of secondary metabolite production and in determination of the function of metabolites within the thallus. Concentrations of usnic acid can vary on a microscopic scale, within a thallus, by containing higher amounts in some regions of the thallus than other regions (Bjerke et al., 2005). In some species, production of a compound may not be evenly distributed, but appear to be randomly produced in specific parts of the thallus medulla. Usnic acid production was concentrated in the apothecium, pycnidium, and on the outer layer of hyphae around the algal cells of some lichens (Culberson C. F. et al., 1993; Liao et al., 2010). It is known that the cortex produces an array of compounds that are not produced by the medullary hyphae (Elix & Stocker-Worgotter, 2008). Specific functions have been studied and assigned to the compounds produced more commonly by specific tissues (see section 3.1).

Secondary compound production also varies among individuals within the geographic distribution of a single species. The concentrations of secondary compounds such as usnic acid can vary greatly in Arctic populations of *Flavocetraria nivalis* (Bjerke et al., 2004). Intraspecifically, the chemospecies of some lichens have been observed to sort geographically (Hale, 1956; Culberson C. F. et al., 1977a; McCune, 1987; Culberson C. F. et al., 1990). Other studies have shown that these geographic patterns are not consistent (Culberson W. L. et al., 1977). Quantitative variation may be present within genetically identical species that produce biosequentially related secondary metabolites (Culberson W. L. & Culberson C. F., 1967; Culberson W. L. et al., 1977b). Various chemotypes of *Cladonia acuminata* are reported (Piercey Normore, 2003, 2007) as well as a number of other species with chemotypes. The presence of fumarprotocetraric acid may vary even within the same location for members of the species *Cladonia arbuscula* (Piercey-Normore, 2006, 2007) and *Arctoparmelia centrifuga* (Clayden, 1992). *Cladonia uncialis* will produce squamatic acid when it is growing in coastal North America but squamatic acid is not present in specimens growing in continental North America (pers. observations). *Ramalina siliquosa* produces bands of six chemical races on the rocky coast of Wales at different distances from the oceanic spray (Culberson W. L. & Culberson C. F., 1967). Other groups of lichens also show similar habitat specific correlations such as *Cladonia chlorophaea* complex and *Parmelia bolliana* (Culberson W. L., 1970). The production of some secondary compounds, such as rhizocarpic acid, have been shown to correspond with increases in altitude (Rubio et al., 2002). However, the absence of an altitudinal correlation with usnic acid is also reported (Bjerke et al., 2004). The genus *Thamnolia* is comprised of a single species world-wide with two chemical variants, *T. vermicularis* and *T. vermicularis* var. *subuliformis*. *T. vermicularis* contains thamnolic acid and is predominant in the Antarctic. It slowly decreases in frequency across the equator in alpine habitats to the Arctic. *T. vermicularis* var. *subuliformis* contains baeomycesic and squamatic acids and has the opposite trend. It is more predominant in the Arctic and decreases in frequency toward the Antarctic region. The varieties are identical in appearance but are distinguished by their secondary chemistry. With environment and geographic distribution playing such an important role in the production of secondary compounds, one might expect secondary compound production to correspond with variability of lichen phenotype.

Fungal secondary metabolites such as polyketides are produced by large multidomain enzymes, called polyketide synthases (PKS). In fungi, polyketide synthesis is catalysed by iterative Type I PKS, which are structurally and mechanistically similar to fatty acid synthases. PKSs are multidomain proteins that catalyse multiple carboxylic acid condensations (Keller et al., 2005). The fungal PKSs consist of a linear succession of domains of ketosynthetase (KS), acyl transferase (AT), dehydratase (DH), enoyl reductase (ER), ketoreductase (KR), acyl carrier protein (ACP) and thioesterase (TE) (reviewed in Keller et al., 2005). The simplest fungal PKS includes the KS, AT and ACP domains, which are the minimal set of domains required for carboxylic acid condensation (Hopwood, 1997). Some fungal PKSs include KR, DH and ER domains in addition to the minimal domains, which catalyse the reduction of carbonyl groups after each cycle of condensation (Proctor et al., 2007). Fungal polyketides usually undergo modifications (reductions, oxygenations, esterifications, etc.) after they are formed. This modification is catalysed by enzymes in addition to the PKS (Proctor et al., 2007). The genes encoding the PKS and modifying enzymes are often located adjacent to each other in gene clusters. The genes in a cluster are co-regulated with transcription of all the genes being repressed or activated simultaneously

(Keller & Shwab, 2008). The polyketides produced are reduced to different degrees by the reducing domains, which are further modified by enzymes resulting in a highly diverse collection of molecules in both structure and function.

Studies of genetic regulation of fungal secondary metabolism are at an early stage (Fox & Howlett, 2008) and in lichen fungi there are few publications directly on gene expression (Brunauer et al., 2009; Chooi et al., 2008). Secondary metabolism has been studied separately with a focus on metabolite variation within and between species (Culberson W. L., 1969; Hawksworth, 1976), evolutionary hypotheses proposed for biosynthetic pathway evolution (Culberson W. L. & Culberson C. F., 1970), and phylogenomic analysis of polyketide synthase genes (Schmitt & Lumbsch, 2009; Kroken et al., 2003). The increasing number of phylogenomic analyses show that a single fungal genome may contain more than one PKS gene and each species of fungi can produce more than one polyketide or polyketide family (Proctor et al., 2007; Boustie & Grube, 2005). Each gene paralog may encode a particular polyketide product. Multiple paralogs of PKS genes have been detected (Table 1) in members of the lichen families Parmeliaceae (Opanowicz et al., 2006) and the Cladoniaceae (Armaleo et al., 2011). Six paralogs of the KS domain of PKS genes have been detected so far in the Parmeliaceae and a high number of paralogous PKS genes are expected to be present in the genomes of the Parmeliaceae because they are rich in diverse phenolic compounds. *Cladonia grayi* has been shown to contain up to 12 paralogs even though it is known to produce only two polyketides.

Paralogs may have arisen by gene duplication, mobile elements, gene fusion, or other mechanisms reviewed by Long et al. (2003). Alternative explanations for multiple, apparently non-functional, genes include horizontal gene transfer from bacteria to fungi (Schmitt & Lumbsch, 2009), horizontal gene transfer between different fungi (Khaldi et al., 2008), or adaptions triggering gains and losses through evolution (Blanco et al., 2006). Numbers of paralogs reported for lichen fungi in Table 1 are low and appear to correspond with the number of polyketides. However, these numbers are expected to be higher than reported because of recent knowledge of the numbers of paralogs present from genome sequencing projects in *Aspergillus* (Gilsenan et al., 2009), *Cladonia grayi* (Armaleo et al., 2011), and more than 200 projects in progress or completed for other ascomycetes (http://www.ncbi.nlm.nih.gov/genomes/leuks.cgi). It has been reported that the number of secondary metabolite genes far exceeds the number of known compounds in an organism (Sanchez et al., 2008). For example in *Aspergillus nidulans* as many as 27 polyketide synthase genes have been identified whereas only seven secondary metabolites are known for this species and 16 paralogs are reported for *C. grayi* when only two polyketides are known to be produced by this species. Genome sequencing has also revealed unique gene clusters among various organisms, probably because an organism may have evolved to produce different secondary metabolites to best suit its biological and ecological requirements (Sanchez et al., 2008). The primer series used in Table 1 (for this study) amplified two paralogs in *Flavocetraria cucullata* and a single gene in *Alectoria ochroleuca* (Table 1). An earlier study by Opanowicz et al., (2006) reported three paralogs in both *Flavocetraria cucullata* and two paralogs in *Alectoria ochroleuca*. Variation in the number of paralogs may exist within and between populations, but more likely in this study variation may exist because of the limitation of primers available, where a larger number of paralogs might be expected to be present in all genomes.

Species	No. compounds reported	Source for no. of compounds	No. putative PKS paralogs reported	Source for no. of paralogs
Alectoria ochroleuca	2	Culberson C F. (1970)	2	Opanowicz et al. (2006)
Alectoria ochroleuca	2	This study	1	This study
Aspergillus fumigatus	Unknown	Not applicable	14	Nierman et al. (2005)
Aspergillus nidulins	7	Sanchez et al. (2008)	27	Sanchez et al. (2008)
Aspergillus terreus	Unknown	Not applicable	30	Nierman et al. (2005)
Cetraria islandica	3	Culberson C F. (1970)	3	Opanowicz et al. (2006)
Cetraria islandica	3	This study	3	This study
Cladonia grayi	2	Culberson C F. (1970)	12	Armaleo et al. (2011)
Flavocetraria cucullata	3	Culberson C F. (1970)	3	Opanowicz et al. (2006)
Flavocetraria cucullata	2	This study	2	This study
Flavocetraria nivalis	1	Culberson C F. (1970)	1	Opanowicz et al. (2006)
Flavocetraria nivalis	1	This study	1	This study
Fusarium graminearum	4	Hoffmeister & Keller (2007)	15	Hoffmeister & Keller (2007)
Gibberella moniliformis	Unknown	Not applicable	15	Schmitt et al. (2008)
Hypogymnia physodes	4	Culberson C F. (1970)	1	Opanowicz et al. (2006)
Neurospora crassa	Unknown	Not applicable	7	Galagan et al. (2003)
Ramalina intermedia	4	Bowler & Rundel (1974)	3	This study
Ramalina farinacea	7	Worgotter et al. (2004)	3	This study
Tukermannopsis chlorophylla	2	Culberson C F. (1970)	1	Opanowicz et al. (2006)
Tukermannopsis chlorophylla	1	This study	1	This study
Usnea filipendula	2	This study	1	This study
Xanthoparmelia conspersa	8	Culberson et al. (1981)	2	Opanowicz et al. (2006)
Xanthoria elegans	3	This study	1	This study
Xanthoria elegans	3	This study	1	Brunauer et al. (2009)

Table 1. Diversity of secondary metabolites and PKS paralogs expected for lichenized fungi and comparison with selected non-lichenized fungi from this study and summarized from the literature.

2.3 Hypothesized roles of secondary metabolite production

A fungus undergoes maximum growth when all required nutrients are available in optimal quantities and proportions. If one nutrient becomes altered, then primary metabolism is affected and fungal growth is slowed. The intermediates of primary metabolism that are no longer needed in the quantity in which they are produced, may be shifted to another pathway. It is thought that the intermediates may be used in the secondary metabolic pathways (Moore, 1998) serving as an alternative sink for the extra products of primary metabolism while allowing nutrient uptake mechanisms to continue to operate. The continued operation of primary metabolism allows continued growth but without the close integration of processes results in non-specific secondary end products maintaining effective growth (Bu'Lock, 1961 in Moore, 1998). This leaves the impression that secondary metabolism has no specific role or advantage in the fungus. However, secondary metabolism may give the fungus a selective advantage. It has been reported in many publications that secondary metabolites have a variety of functions (see below).

Secondary metabolism is often triggered at a stage of fungal growth and development when one or more nutrients become limiting and growth slows down (Moore, 1998). It is thought that when mycelial growth slows, carbohydrates are not used in growth processes and they become constant. As these carbohydrates are metabolized, secondary metabolites are produced and accumulate. The production of secondary metabolites may not serve specific functions but they may confer a selective advantage with multiple inadvertent ecological functions. Secondary metabolites may serve mainly as products of an unbalanced primary metabolism resulting from slowed growth, including metabolites that are no longer needed for growth.

Lichens and their natural products have been used for centuries in traditional medicines and are still of considerable interest as alternative treatments (Miao et al., 2001). Most natural products in lichens are small aromatic polyketides synthesized by the fungal partner in the symbiosis (Elix & Stocker-Worgotter, 2008). Polyketides are produced by a wide range of bacteria, fungi, and many plants. The finding of polyketides in forest soils, where they are exposed to harsh environmental conditions with other competing organisms, has led to the suggestion that those polyketides with antagonistic properties may structure the microbial communities in the soil (Kellner & Zak, 2009). Polyketide-producing organisms that do not live in soil may derive benefit from these compounds, which allow them to survive in discrete ecological niches by reacting to environmental variables such as light or drought, or protecting themselves from predators and parasites (Huneck, 1999). Secondary metabolites have also been hypothesized to play a role in herbivory defence, antibiotics, or as metal chelators for nutrient acquisition (Gauslaa, 2005; Lawrey, 1986, Huneck, 1999). Recently it was hypothesized that polyketides play a role in protection against oxidative stress in fungi (Luo et al., 2009; Reverberi et al., 2010) and that some metabolites such as fumarprotocetraric acid, perlatolic, and thamnolic acids contribute to the ability of lichens to tolerate acid rain events and consequences (Hauck, 2008; Hauck et al., 2009).

One explanation for high levels and numbers of secondary metabolites in lichen fungi is the slow growth of the lichen. It is known that lysergic acids are produced in the slow growing over-wintering structures (ergot) of the non-lichenized fungus *Claviceps purpurea*. The ergot in *C. purpurea* represents the slow growing overwintering stage of the fungus following the fast growing mycelial stage during the summer season where infection of the host occurs. However, lichens have no fast growing stage in comparison with *C. purpurea* and there appears to be no limitation to production of polyketides. The detoxification of primary metabolites is another hypothesis that has been proposed to explain the production of secondary metabolites. If growth of the fungus slows down, but metabolism is still very active, toxic products of primary metabolism may accumulate. The transformation of these into secondary metabolites may be one method to prevent toxic accumulation of byproducts. This hypothesis may be integrated within the first hypothesis on slow growth rates to explain the production of secondary compounds by fungi.

Regardless of the reason for secondary metabolite production (biproduct, detoxification of primary metabolism, or leftover products after growth slows) they often elicit a function that is advantageous to survival of the lichen within its ecological niche. The advantage(s) may in part be understood by the location of the compounds within the thallus such as atranorin and usnic acid occurring more frequently in the cortical hyphae than the medullary hyphae and having a function related to photoprotection. These chemical characters are thought to be adaptive features because of their perceived ecological role. The presence or absence of polyketides has also been shown to be gained and lost multiple times in the evolution of the Parmeliaceae (Blanco et al., 2006). If the compounds allow adaptations of lichens to their habitats and are expressed when triggered by a combination of ecological conditions (Armaleo et al., 2008), the repeated gain and loss through evolution is a result of environmentally induced expression rather than the evolutionary gain and loss of functional genes.

3. Observations on how specific environmental parameters influence changes in secondary metabolite production

Production and regulation of secondary metabolites in fungi is complex with numerous environmental and developmental stimulants (Fox & Howlett, 2008) that may directly influence polyketide synthase transcription or may influence one another indirectly initiating complex signal transduction cascades. This multifaceted system makes it difficult to separate the effects of environmental parameters, developmental stages, and other factors, from one another. This section attempts to separate and describe studies involving these parameters and their effects on PKS gene expression, but concludes by integrating the significance of all parameters together.

3.1 Effects of abiotic parameters: Temperature, light, pH, and humidity or drought

Studies are beginning to accumulate that have linked environmental and culture conditions such as dehydration or aerial hyphal growth with production of secondary metabolites (Culberson C. F. & Armaleo, 1992) or exposure to strong light and drought (Stocker-Worgotter, 2001). Culberson C. F. & Armaleo (1992) showed that grayanic acid was not produced by cultured *Cladonia grayi* until aerial hyphae began to develop in the cultures. Stocker-Worgotter (2001) showed that baeomysesic and squamatic acids were not produced by *Thamnolia vermicularis* var. *subuliformis* until the culture media began to dehydrate and they were exposed to high light conditions under relatively low temperatures (15C). These conditions reflect the conditions in the natural habitat of *Thamnolia* spp. where thalli typically grow in polar or alpine habitats exposed to cooler temperatures, under high light conditions, and dehydrating winds, that affect thallus evaporation and water content (Larson, 1979). These observations suggest that environmental parameters may trigger the production of certain compounds in some species. Numerous studies have shown a correlation between light levels and production of usnic acid (Armaleo et al., 2008; McEvoy et al., 2007a; Rundel, 1969; Bjerke et al., 2002; McEvoy et al., 2006) or other compounds (Armaleo et al., 2008; Bjerke et al., 2002; McEvoy et al., 2007b) within thalli. The amount of atranorin in the cortex of *Parmotrema hypotropum* was shown to correlate positively with the amount of yearly light reaching the thallus (Armaleo et al., 2008). In the same study norstictic acid on the medullary hyphae showed a negative correlation with yearly light levels. The authors suggested that the higher quantites of medullary compound with lower light levels may be an adaptive link between the need for production of these hydrophobic compounds when water potential increases within the thallus (from low light levels) to allow efficient carbon dioxide diffusion to the algae. As light levels decrease the water potential in the thallus increases and therefore the need for hydrophobic compounds also increases. Based on the difference in polyketide production between the medulla and the cortex with different environmental triggers for different metabolites, Armaleo et al. (2008) proposed that two different pathways with two different sets of genes were responsible for production of these compounds. This is a plausible explanation since a larger number of paralogs are present compared with the number of polyketides actually produced by many species (Table 1). On the other hand, other studies did not report a relationship between light and polyketide production (Fahselt, 1981; Hamada, 1991; Bjerke et al., 2004).

Growth media and available nutrients may influence the secondary metabolites produced by lichen fungi. The presence of gene clusters for production of a potentially larger variety of polyketides than is produced within each species, is supported by the work of Brunauer et al. (2007). Cultured lichen fungi have been shown to produce secondary metabolites that are not present in the naturally collected lichen. The authors offered two explanations for this 1) the lichen fungus may adapt to the conditions in the artificial media triggering induction of an alternate pathway, and 2) enzyme activity may be shifted by availability of certain trace elements, carbohydrates, or unusual pH of the medium. These external factors may affect expression of genes involved in regulation of secondary metabolities or on the genes directly involved in metabolite production. For example, the transcription factor, *VeA* (velvet family of proteins) is regulated by light levels and has been reported to repress penicillin biosynthesis (Sprote & Brakhage, 2007). The velvet complex subunits coordinate cell development and secondary metabolism in fungi (Bayram & Braus, 2011). These proteins are reported to be conserved among several species of fungi including *Aspergillus* spp., *Neurospora crassa, Acremonium chrysogenum,* and *Fusarium verticilloides* (Bayram et al., 2008; Dreyer et al., 2007; Kumar et al., 2010).

The effect of pH on gene expression in fungi is reviewed by Penalva & Arst (2002). Regulation of gene expression by pH, is thought to be mediated by a transcription factor (pacC). Higher pH, resulting in alkaline conditions that mimic PacC mutations, causes an increased production of penicillin in *Aspergillus nidulins* and in *Penicillium chrysogenum.* Carbon source also influences penicillin production where some sources will repress the effects of an alkaline pH on penicillin production (Suarez & Penalva, 1996). On the other hand, acidic growth conditions promote production of aflatoxins in *Aspergillus parasiticus* and *A. nidulins* (Keller et al., 1997). If pH regulation is an important determinant in plant pathogenicity (Penalva & Arst, 2002) and in sclerotial development in *Schlerotinia sclerotiorum* (Rollins et al., 2001), then it might also be expected to influence the controlled parasitic interaction (Ahmadjian & Jacobs, 1981) between lichen fungi and algae and the production of polyketides in fungi linking observations on environmental parameters and developmental changes in culture. For example, Stocker-Worgotter (2001) showed that species within the genera *Umbilicaria* and *Lassalia* produce their diagnostic secondary metabolites only when grown on an acidic medium (potato-dextrose-agar). Species of *Umbilicaria* and *Lassalia* (*U. mammulata, L. papulosa*) typically grow on acidic granite rocks and have not been reported on any other substratum, suggesting that the pH of the substratum may also influence PKS gene expression in these species. However, other factors specific to the rock habitat may also influence PKS gene expression such as mineral composition of the rock or the presence of other organisms. The significance of the substratum to lichen fungi is reviewed by Brodo (1973). The bark of different tree species and the diversity of rock types can have different pHs, nutrients, and water holding capacity making them suitable for some species but not for other species. Lichens growing under other conditions have also shown changes in production of secondary metabolites. The quantity of depsides was highest in *Ramalina siliquosa* cultures when the pH was 6.5 and incubation temperature was 15C (Hamada, 1989). Hamada (1982) also showed that the depsidone, salazinic acid, was highest in *R. siliquosa* when the annual mean temperature was approximately 17C.

Microorganisms capable of growing over a wide range of pH have gene expression under control of the pH of their growth medium (Penalva & Arst, 2002). It has been found that the signals generated in response to environmental conditions are relayed through proteins including CreA for carbon, AreA for nitrogen and PacC for pH signaling. These proteins may have positive or negative effects on metabolite production. With regard to two *Cladonia* species, *C. pocillum* and *C. pyxidata*, it has been suggested that pH is the driving environmental factor responsible for the morphological difference between the two species (Gilbert, 1977; Kotelko & Piercey-Normore, 2010). Secondary metabolite production varies among members of the *Cladonia chlorophaea* complex, which have been found to share virtually identical morphologies but different secondary metabolites (Culberson C. F. et al., 1988; Culberson W. L., 1986). *Cladonia grayi* and *C. merochlorophaea* grow at lower pH than *C. chlorophaea sensu stricto* or other members of the complex. If pH is regulating production of polyketides that are diagnostic among these chemospecies, then the species complex represents the range of versatility the species has acquired to adapt to changing environmental conditions.

3.2 Carbon source may influence the secondary metabolite pathway

The lichen association involves a fungal partner and an autotrophic partner, a green alga or cyanobacterium. The carbon source provided by the photobiont has been shown to have an impact on the secondary metabolism of the mycobiont. The more common of these green algal photobionts are in the genera *Trebouxia, Myrmecia* and *Coccomyxa.* These algae are thought to produce the sugar ribitol, and *Trentepohlia* produces erythritol (Honegger, 2009). This sugar alcohol is transferred to the mycobiont where it is metabolized into mannitol. This is an irreversible reaction where mannitol becomes unavailable to the fungal partner. Secondary compounds produced by *Xanthoria elegans* were strongly induced by the presence of mannitol with negligible effects by ribitol (Brunauer et al., 2007). An early study of nutritional implications in *Pseudevernia furfuracea* examined the production of polyketides after applying different carbon sources to natural thalli incubated in a moist water-filled chamber (Garcia-Junecda et al., 1987). Production of atranorin is not enhanced by glucose but it is enhanced by remobilization of storage carbohydrates to produce acetate as the starting intermediate. Production of lecanoric acid is enhanced by glucose and may be a result of the catabolism of mannitol or glucose. The production of atranorin was favoured when catabolism of mannitol or glucose was repressed by a synthetic inhibitor. Hamada et al. (1996) measured the yield of secondary metabolites from nine species of lichen fungi and compared media supplemented with 0.4% and 10% sucrose. All species showed an increase in metabolite production in the 10% sucrose media. It follows that if ecological conditions are varied (as in the microenvironment of a lichen thallus) and/or algal physiology is varied (Hoyo et al., 2011), then a combination of different polyketides may be produced within a single thallus by the availability of different types of starting units.

It has been reported that the availability and type of carbon and nitrogen source affect polyketide production (Keller et al., 2002). As the sole carbon source, sugars like glucose, sucrose or sorbitol, have been found to support high aflatoxin production along with increased fungal growth and sporulation. On the other hand, peptones and more complex sugars such as galactose, xylose, lactose and mannose do not support aflatoxin production. Studies on *Aspergillus* species have shown different effects of nitrogen sources in growth medium on aflatoxin and sterigmatocystin production (Keller et al., 2002). Keller et al. (1997) reported an increased amount of sterigmatocystin and aflatoxin production in ammonia-based medium and a decreased amount in nitrate-based medium.

The ability of lichens to adapt to changes in light levels, depends on the stability of thylakoid membranes, which protect them from attack by reactive oxygen species (Berkelmans & van Oppen, 2006). Therefore, the choice of algal partner would depend largely on the habitat conditions in which the developing lichen thallus is found. If the choice of alga depends on habitat conditions, and different algae produce different starting units, then the polyketide production would also depend on the habitat conditions and the alga. For lichen thalli that are thought to contain multiple algae simultaneously (Piercey-Normore, 2006; Hoyo et al., 2011), the predominant alga would provide the majority of starting carbohydrates, with a specific combination of carbohydrates available for different biosynthetic pathways.

3.3 Environmental cues affecting secondary metabolite production

The development of non-lichenized fungi and secondary metabolite production appears to be coordinated (reviewed in Schwab & Keller, 2008; Bennett & Ciegler, 1983). Morphogenesis of the macrolichens (fruticose and foliose) is highly complex compared with crustose lichens and the vegetative phase of many non-lichenized fungi. The macrolichen thallus is comprised of differentiated "tissues" arranged in layers (see section 2) that often produce different metabolites (see Honegger (2008) for a review of morphogenesis in lichens). Thallus development in lichens has been examined using microscopy (Honegger, 1990; 1993; Joneson & Lutzoni, 2009) and recently a study has described a number of genes that correlate with symbiont recognition and early thallus development (Joneson et al., 2011). Observations of cultures of lichen-forming fungi have suggested that thallus development may be involved in production of secondary metabolites. For example, a major compound umbilicaric acid produced by *Umbilicaria mammulata* was produced by cultures of *U. mammulata* only after lobe-like structures were formed in dehydrating medium. Similarly, cultures of *Cladonia crinita* produced its major substance, fumarprotocetraric acid and its satellite substances only after podetial structures were formed (Stocker-Worgotter, 2001). Species of *Ramalina* produced secondary metabolites only after layers of mycelia became differentiated (Stocker-Worgotter, 2001). As further research is conducted on development in lichens it is expected that more links between development and production of secondary metabolites will become evident.

Regulation of fungal secondary metabolism to some extent is thought to depend on the chromosomal organization of biosynthetic genes. A global transcription factor, which is encoded by genes that are unlinked with biosynthetic gene clusters, may also control the production of secondary metabolism (Fox & Howlett, 2008). Genes encoding global transcription factors regulate multiple physiological processes and are thought to respond to pH, temperature, and nutrients. Signal cascades that regulate fungal morphogenesis are necessary for fungi to sense environmental change and adapt to those changes. These signaling cascades have been studied more intensely with reference to fungi that are human pathogens (Shapiro et al., 2011). Environmental cues may iniatiate a shift between morphological growth forms that is necessary for survival of the fungus but causes disease in the host. Studies on mycotoxin production and regulation of the genes responsible for mycotoxin production in species of *Aspergillus* have shown that the gene, *veA*, regulates production of three aflatrem biosynthetic genes and another toxin in *A. flavus* (Duran et al., 2007). *veA* (velvet A) has also been shown to regulate penicillin production in *A. nidulans* (Kato et al., 2003). The same gene, *veA*, has also been reported to be involved with regulation of aflatoxin production in *A. parasiticus*, suggesting that the regulatory mechanism may be conserved among species of *Aspergillus* (Duran et al., 2007). Another gene, *laeA*, has also been shown to regulate expression of biosynthetic gene clusters in species of *Aspergillus* (Bok & Keller, 2004; Keller et al., 2005; Fox & Howlett, 2008). In addition, it has been shown that laeA negatively affects the regulation of veA (Kale et al., 2008). The loss of laeA results in gene silencing (Bok et al., 2006b; Perrin et al., 2007).

4. Variation in secondary metabolite production may change along the geographic distribution of a species – An empirical study

4.1 Background to the study

The most widely studied secondary metabolite produced by lichen-forming fungi is usnic acid, a cortical compound that absorbs UV light. Seasonal and geographic variation has been shown to occur in populations of the usnic acid producing lichens *Flavocetraria nivalis* and *Nephroma arcticum* in Arctic and Antarctic regions (Bjerke et al., 2004, 2005; McEvoy et al., 2007). These are regions that are highly exposed to strong UV light, desiccating winds, and harsh temperature changes. Other secondary metabolites examined on large geographic scales include alectoronic acid, a-collatolic acid, and atranorin produced by *Tephromela atra*, a crustose lichen that grows on tree bark. That study showed a significant variation between localities (Hesbacher et al., 1996) but no relationship with tissue age, grazing, or reproductive strategy. In a study on the *Cladonia chlorophaea* complex the levels of fumarprotocetraric acid increased from coastal North Carolina to the Appalacian mountains in the interior of the state (Culberson C. F. et al., 1977a). The authors interpreted this geographical gradient of higher levels of fumarprotocetraric acid in mountain populations, as providing protection against harsher environmental conditions in the mountains than in the coastal area. If environment influences secondary metabolite production, then changes should be observed along a gradient of environmental conditions over a species distribution.

Although Hesbacher et al. (1996) showed that thallus age has no affect on secondary compound concentrations for atranorin and alectoronic acid, Golojuch & Lawrey (1988) showed that concentrations of vulpinic and pinastric acids are higher in younger lichens. Bjerke et al. (2002) showed that the most exposed sections of the thallus (such as the tips of *C. mitis*) accumulate greater concentrations of secondary compounds than less exposed sections of the thallus. However, it is not known if the metabolites are actively produced in the exposed and younger tips, or if the metabolites are lost in the older parts of the thallus as the thallus ages and the fungal tissue degrades, giving the appearance that the tips have more metabolites. High concentrations of secondary metabolites were reported in the sexual and asexual reproductive bodies rather than the somatic (vegetative) lichen tissue (Liao et al., 2010; Culberson C. F. et al., 1993). Geographic and intrathalline variation suggest a functional role for these metabolites that has been described in a theory called optimal defence theory (ODT). The theory states that plants and fungi will allocate secondary compounds where they are most beneficial to the organism (Hyvärinen et al., 2000), implying an active production of secondary metabolites, which is contrary to the current theories of secondary metabolite production (see section 2.3). The inconsistency in findings to explain geographic trends and the intrathalline variation in secondary metabolite production may be addressed by increasing sample size and geographic distance to capture the population variation and prevent saturation of larger scale geographic variation. Relationships between metabolite production and geographic location should be evident in a north - south direction because of differences in climate. It would also be expected that the production of intrathalline metabolites would be coordinated because of their hypothesized function regarding environmental changes.

The objectives of this study were 1) to test the relationship between the quantity of secondary metabolite produced and geographic location over latitudinal range, and 2) to test the relationship between metabolites produced within a thallus to determine whether production of one compound is dependent on production of another compound.

Fig. 2. Shield lichens inhabit exposed rock of the Precambrian shield in North America showing A. *Arctoparmelia centrifuga*, a yellow-green foliose thallus with concentric rings of growth, and B. *Xanthoparmelia viridulombrina*, yellow-green foliose thallus with brown apothecia (arrow) and wide lobes. Photo of *A. centrifuga* by T. Booth.

4.2 Methods

4.2.1 Species and sampling strategy

Two species were chosen for this experiment, *Arctoparmelia centrifuga* and *Xanthoparmelia viriduloumbrina* (Fig. 2). Both lichen species are saxicolous, foliose lichens that grow on the Precambrian shield in North America belonging to the family, Parmeliceae (Ascomycotina). Originally part of the *Xanthoparmelia* genus, *Arctoparmelia* was reclassified as a separate genus (Hale, 1986) and currently both genera are in the Parmeliaceae. *Arctoparmelia centrifuga* is a yellow-green foliose lichen that grows in concentric rings (Fig. 2A). The center of the ringed pattern discolours with age, the source of its specific epithet ('retreat from centre'). The thallus lacks a lower cortex, appearing white underneath, and is found growing on exposed rock. The major compounds produced by *A. centrifuga* include atranorin, usnic acid, alectoronic acid, and an unidentified aliphatic acid (Culberson C. F., 1969). *Xanthoparmelia viriduloumbrina* is a yellow-green foliose lichen with straplike lobes. The underside is brown, with brown rhizines. Maculae, which are absent from this species (Lendemer, 2005), are discolourations on the thallus surface caused by the absence of the photobiont beneath the cortex. The lichen grows on exposed rocks and a morphologically similar species *X. stenophylla* has a pH tolerance ranging between 4.1 and 7.0 (Hauck & Jürgens, 2008). The secondary compounds produced by *X. viriduloumbrina* include usnic acid, salazinic acid, consalazinic acid and an accessory compound, lobaric acid (Hale, 1990). Both species, X. *viriduloumbrina* and *A. centrifuga*, reproduce sexually and the algal partner is *Trebouxia*.

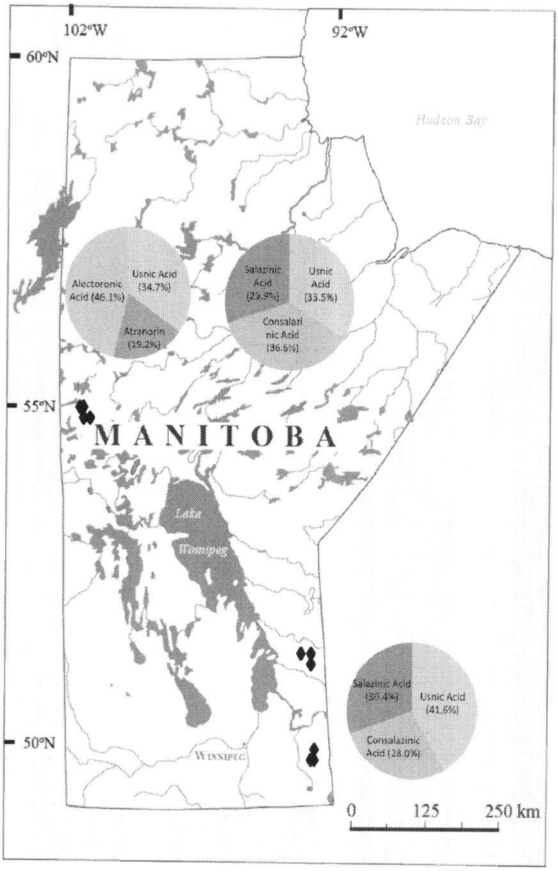

Fig. 3. Map of Manitoba, Canada, showing latitude (left) and longitude (top), location of collection sites (black diamonds), and proportion of secondary metabolites from *X. viridulombrina* (usnic, salazinic and consalazinic acids) in northern and southern sites, and proportion of secondary metabolites from *A. centrifuga* (usnic, alectoronic acids, and atranorin) in northern sites. (Map was provided by R. Lastra).

Sampling for both species occurred along a northwest-southeast transect covering a distance of approximately 700km along the Precambrian shield in the province of Manitoba (Fig. 3). The Precambrian Shield extends northwest-southeast along on the eastern shore of Lake Winnipeg. Twenty-nine transects measuring 40m in length and evenly spaced 1m x 1m quadrats were placed every 10m for sample collection. Vouchers were collected and deposited in the University of Manitoba Herbarium (WIN-C). Ninety-five samples of *A. centrifuga* were collected and 109 samples of *Xanthoparmelia viriduloumbrina* were collected in the summer of 2010.

4.2.2 Quantitative Thin Layer Chromatography

Portions of young thallus lobes weighing 5mg DW (Mettler PM460 DeltaRange) were placed in 1.5 mL Eppendorf tubes. Extraction of secondary compounds was done following Culberson C. F. (1972) with 3.3mL acetone washes and three incubations for 5, 5, and 10 minutes. Acetone extracted samples were processed using thin layer chromatography (TLC; Orange et al., 2001; Culberson C. F., 1972, 1974). The protocol was standardized by placing 46uL on each spot of the silica-coated glass TLC plate (Fisher Scientific, Ottawa, Ontario, Canada) and placed in solvent A (toluene 185 mL: dioxane 45 mL: glacial acetic acid 5mL) for migration of the solvent to the top of the plate. After drying, pictures were taken of each plate for short-wave (254 nm) and long-wave (365 nm) ultraviolet light. These photos were used to quantify the secondary compound. The plates were then sprayed with 10% sulphuric acid and baked in an 80C oven until colours developed (10 minutes). Secondary metabolites were determined by comparison with known characteristics (Culberson C. F., unpub; Orange et al., 2001), by using a standard for Rf comparison, and an usnic acid commercial standard (ChromaDex, Santa Ana, CA).
Secondary compounds were quantified using Digimizer (Version 4.0.0. MedCalc Software, Mariakerke, Belgium, 2005-2011). Photos of TLC plates taken under short and long wave UV light were used. Three compounds for each species were quantified. Two measures were used to arrive at compound quantity (in pixels). The first was the area of the spot. The second measure was brightness or average intensity under UV light. This was the average pixel value on a scale between 0 (black) and 1 (white). The purpose of the brightness quantity was to account for the thickness of the silica plate. At 250 μm thick, greater saturation of the extract could occur in an area on the plate. The two values of spot area and brightness where multiplied together to get a total pixel value for the individual compound. Usnic acid, atranorin, salazinic acid and consalazinic acid were all quantified under short-wave ultraviolet light and were analyzed by inverting the quenched spots on the plate to allow the pixel area to be determined. Pixels in the dark quenched spots cannot be determined. Alectoronic acid was quantified by its fluorescence under long-wave ultraviolet light (365nm). No inversion was necessary because brightness values were already positive.

4.2.3 Data analysis

Univariate statistics were done using JMP® (Version 8.0.1 SAS Institute Inc., Cary, NC, 2009). Quantities of secondary compounds were log transformed and plotted against the independent variable, latitude for northern sites, southern sites, and all sites for X. *viriduloumbrina;* and for northern sites only for *A. centrifuga*. Spearman's correlation was used to measure the relationship between compound quantities and latitude. Four correlations were calculated , one for *A. centrifuga* and three for *X. viriduloumbrina*. Pairwise regression analyses between compounds for each species were done. P values were recorded for the significant relationships. Pie charts were created to show the proportion of secondary compounds in northern and southern sites for each species based on the average log transformed pixel quantity for each secondary compound.

4.3 Results

Xanthoparmelia viriduloumbrina was collectected in all locations of both northern and southern sites. *A. centrifuga* was collected only in northern sites because the species was absent from the southern sites. *Xanthoparmelia viriduloumbrina* consistently produced three compounds (usnic, consalazinic and salazinic acids) and occasionally one accessory compound, lobaric acid with up to two unknown compounds. *A. centrifuga* consistently produced three compounds (usnic and alectoronic acids and atranorin) and up to four unknown compounds (Fig. 4).

The proportion of secondary compounds within *X. viriduloumbrina* was relatively similar between the three collection sites (Fig. 2). The pie-charts showed the cortical compound usnic acid was the most abundant compound overall and within the southern site. The medullary compound consalazinic acid had the highest proportion in the northern site. Alectoronic acid was the largest proportion of the three compounds for *A. centrifuga*.

Secondary metabolites produced by *Xanthoparmelia viriduloumbrina* showed four significant correlations with latitude. Spearman's correlations were conducted for *Xanthoparmelia viriduloumbrina* on each secondary compound, usnic acid, consalazinic acid and salazinic acid, across the entire study area (n=109; 5 degrees latitude). Salazinic acid decreased significantly from the southern to the northern collection sites (Spearman's rho = -0.3330 and p = 0.0004) (Fig. 5A). There were no significant trends for usnic acid (p = 0.1321) or consalazinic acid (p = 0.5720) for all collections sites.

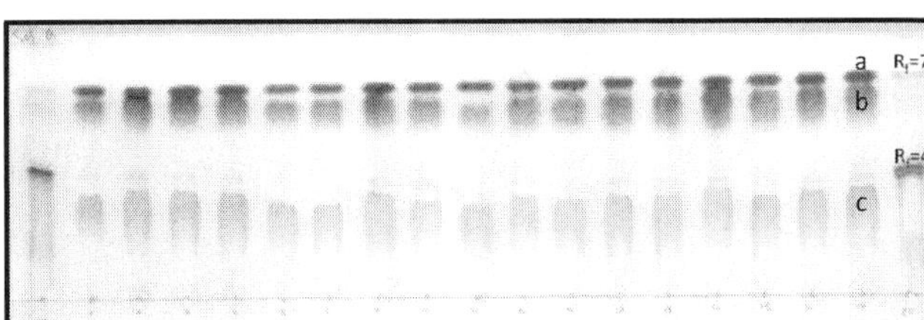

Fig. 4. Image of a developed TLC plate showing 17 polyketide profiles for *Arctoparmelia centrifuga*. Each profile contains a yellow-brown spot at Rf class of 7 (a) determined to be atranorin, a blue-green spot at Rf class of 6 (b) determined to be usnic acid, and a peach spot at Rf class of 3 (c) determined to be alectoronic acid. Profiles shown on the far left and right are the reference profiles for Rf classes 4 and 7.

Similarly, Spearman's correlation analyses in the northern collection sites (n=35; 2 degrees latitude) produced two significant results. Salazinic and consalazinic acids increased significantly in *Xanthoparmelia viriduloumbrina* from southern to northern sites even within a 2 degree latitude (salazinic acid; Spearman's rho = 0.7124 and p = 0.0001) (consalazinic acid; Spearman's rho = 0.3523 and p = 0.0379) (Fig. 5B and C). Usnic acid produced no significant trend (p = 0.3364). Spearman's correlations were also conducted for *Xanthoparmelia viriduloumbrina* in the southern collection sites (n=74; 2 degrees latitude) where salazinic acid decreased significantly from southern to northern sites (Spearman's rho = -0.3371 and p = 0.0033) (Fig. 5D). Usnic acid and consalazinic acid showed no significant correlation (rho = 0.2627; p = 0.1770 respectively) in the southern sites. However, metabolites produced by *A. centrifuga* showed no significant correlations with latitude. Analyses with A. centrifuga could only be conducted for northern sites because the species was absent from the southern sites.

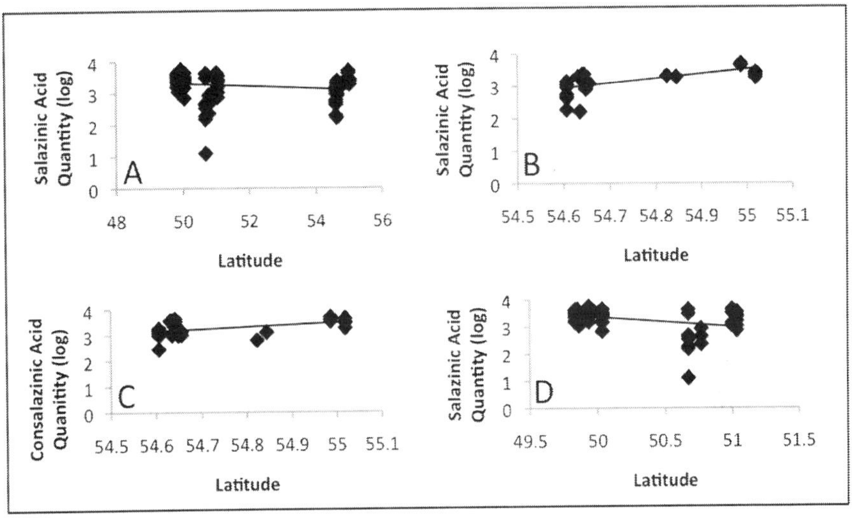

Fig. 5. Relationship between log transformed quantified secondary metabolites produced by *Xanthoparmelia viriduloumbrina* and latitude for A. salazinic acid from specimens collected from all sites; B. salazinic acid from specimens collected only in northern sites; C. consalazinic acid from specimens collected only in northern sites; and D. salazinic acid from specimens collected only in southern sites.

Pairwise regression analyses were conducted between the three metabolites produced by each species to determine whether the production of one compound is related to the production of another compound. Within *Xanthoparmelia viriduloumbrina* regressions between secondary compounds were significant between all three combinations. The relationship between usnic acid and consalazinic acid, between consalazinic acid and salazinic acid, and between usnic acid and salazinic acid were all significant at p=0.0001 (Fig. 6A, B, and C). The regression analyses between secondary compounds produced by *Arctoparmelia centrifuga* showed one significant relationship. Changes in the quantity of usnic acid and atranorin were significant at p=0.0001 (Fig. 6D). Other combinations showed no significant relationship.

4.4 Discussion

4.4.1 Shield lichens adapt to different habitats

The significant decrease in the quantity of salazinic acid from southern to northern latitudes (Fig. 5) are great enough to suggest that *X. viriduloumbrina* is responding to environmental changes. Hamada (1982) reported that dark rock colours, higher temperatures, and southern exposures result in larger quantities of salazinic acid in thalli of *R. siliquosa*. The average mean temperature in the northern sites for 2006 was 1.7°C lower than that in the southern sites (National Climate Data and Information Archieve, 2011). If the overall difference in salazinic acid across all sites reflects a large scale response to temperature, then the significant increase in levels of salazinic acid within the northern sites, suggests a response to more localized environmental parameters as the mean annual temperature would not differ as significantly as it would across all sites, in such a small area. Salazinic acid has also been shown to change with other environmental parameters. The production of salazinic acid is dependent on osmotic pressure and may increase with increased sucrose and low nitrogen levels (Hamanda & Miyagawa, 1995; Behera & Makhija, 2001). The increased production of salazinic acid in low nitrogen and high sucrose culture conditions with *Bulbothrix setschwanensis* (Behera & Makhija, 2001) supports the finding that salazinic acid is produced only in cultures with the algal partner of *B. setschwanensis* present (Behera et al., 2000). The quantity of salazinic acid decreased initially under ozone stress and then increased in what was thought to be stress induced defence (MacGillvray & Helleur, 2001). One explanation is that the compound has antioxidant properties (Amo de Paz et al., 2010) having potential use in treatment of Alzheimer's and Parkinson's diseases (Amo de Paz et al., 2010), and a modified structure of the molecule may be cytotoxic to some tumor cells (Micheletti et al., 2009). The similar trend in consalazinic acid could be explained by the increasing quantity of salazinic acid. The relationship between consalazinic acid and salazinic acid has been known for a long time since they are quite similar chemically and consalazinic acid is considered a co-metabolite of salazinic acid (O'Donovan et al., 1980).

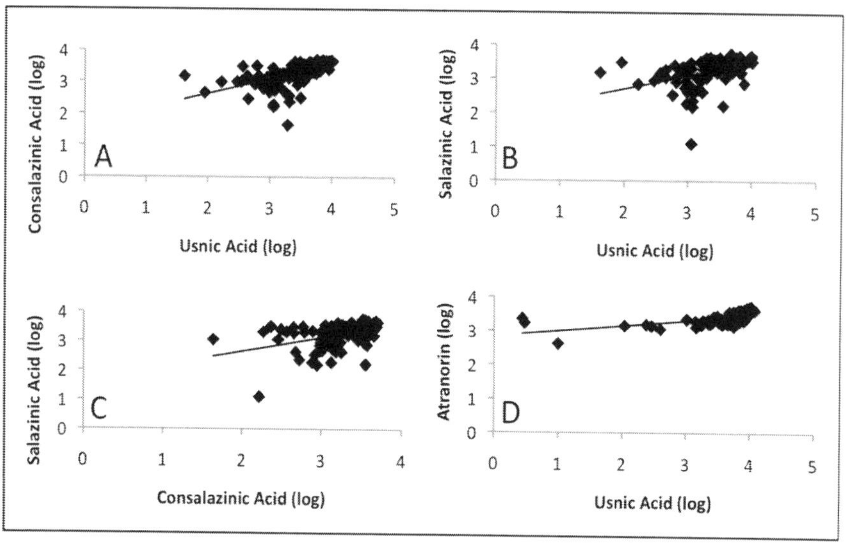

Fig. 6. Pairwise regression analysis of log transformed quantities of secondary metabolites produced by each species showing significant linear relationships between A. consalazinic and usnic acid in X. *viriduloumbrina* (y=1.69+0.46x); B. salazinic and usnic acids in *X. viriduloumbrina* (y=1.91+0.39x) C. salazinic and consalazinic acids in *X. viriduloumbrina* (y=1.69+0.47x); and D. artranorin and usnic acid in *A. centrifuga* (y=2.84+0.15x). All other comparisons were not significant.

4.4.2 Absence of expected relationships suggest localized adaptation

The absence of a relationship between the cortical secondary metabolites and geographic location was unexpected since the literature contains numerous examples of changes in usnic acid or atranorin with light levels. However, the major photoprotective function that has been assigned to usnic acid and atranorin was not accounted for in this study. The 5 degree latitude difference in this study resulted in a temperature and daylength difference. But the change in UV light levels was not likely to be sufficient to produce changes in cortical compounds as was evident for McEvoy et al. (2006) and Bjerke et al. (2002), where increased light gradients were measured from forested locations to exposed alpine locations. In this study the habitat was relatively constant with open jack pine bedrock of the Precambrian shield regardless of whether the location was in the northern or southern regions. The literature on usnic acid is large and includes environmental science as well as medical applications (Cocchietto et al., 2002; Ingólfsdóttir, 2002) suggesting that the functions of usnic acid are numerous and diverse.

Similarly, the bioactive function assigned to the medullary metabolite, alectoronic acid, is not related to habitat. Alectoronic acid concentration was highest in heavily grazed thalli and lowest in thalli with the lowest level of grazing damage by snails (Hesbacher et al., 1996) but the differences were not significant. These differences were however, correlated with geographic distance within 10 km. Alectoronic acid is also known to have antimicrobial properties (Gollapudi et al., 1994) suggesting that levels of alectoronic acid may change in response to the presence of other living organisms or damage they inflict on the lichen thallus. Changes in production of alectoronic acid are not dependent on thallus age and like many secondary compounds, will exhibit intrathalline variation (Hesbacher et al., 1996). Localized production of usnic and alectoronic acids may occur depending on light levels or microbial/herbivore activity that was not measured in this study.

4.4.3 Environmental change influences production of metabolites in a coordinated fashion

Since the proportion of metabolites for each of the northern and southern regions was similar (Fig. 3), some of them showed a significant relationship with one another (Fig. 6). Environmental changes may be coordinating the production of the metabolites. The coordinated production of usnic acid with salazinic acid is consistent with the results of Valencia-Islas et al. (2007) and Amo de Paz et al. (2010), who show that usnic acid and salazinic acid share similar effects due to air pollution and antioxidant behavior. The significant relationship between usnic acid and consalazinic acid is also expected. If consalazinic is a co-metabolite of salazinic acid (O'Donovan et al., 1980), and usnic increases significantly with salazinic (Fig. 6B), then it follows that consalazinic would also increase with usnic.

The coordinated production of two cortical compounds, usnic acid and atranorin, is also a significant relationship. These metabilites are not biogenically related and therefore the coordinated production cannot be explained as pathway intermediates. However, the extensive literature describing their photoprotecive properties and pollution sensitivities suggest that similar environmental features may influence both metabolites. Valencias-Islas et al., (2007) reported that concentrations of atranorin were greater than those of usnic acid, which were greater than those of salazinic acid. Salazinic acid increased at the expense of chloratranorin and atranorin suggesting the same starting carbohydrates were used for production of both compounds; hence, the pathways were in competition for the starting carbohydrates. The relationship between salazinic and consalazinic acids could be explained by the biogenic relationship. However, the relationship between usnic acid and atranorin, produced from different pathways, do not have a biogenic relationship but may be explained by environmental changes.

5. Significance of secondary metabolite production with respect to on-going climate change

A number of environmental predictions of future global climate conditions are predicted in the fourth assessment of the United Nations Intergovernmental Panel on Climate Change (2007). The outlook included an increase in average temperature; an increase in intensity and length of droughts; an increase in global water vapour, evaporation and precipitation rates which will cause increasing tropical precipitation and decreasing subtropic precipitation; an increase in sea levels from glacial melt; and anthropogenic carbon dioxide production will further increase atmospheric carbon dioxide levels (Meehl et al., 2007). Most of these changes will have implications on the future adaptability and secondary metabolite production of lichen species. These secondary metabolites protect against increasing environmental stresses such as light exposure, water potential changes, microbial and herbivore interactions, and other changes associated with changes in environmental conditions.

Increases in temperature may require the increase of secondary metabolites such as salazinic acid to mitigate the effects of higher temperatures on lichen biology. The relationship between temperature and production of salazinic acid is thought to be related to the effect of hydrophobic properties of the metabolite. The metabolite, being produced by medullary hyphae, would ensure a hydrophobic environment to optimize carbon dioxide transfer to the algal cells. A higher temperature increases the water potential of the thallus and more need for hydrophobic conditions to allow optimal carbon dioxide exchange between air spaces and algal cells. However, a higher thallus temperature may also promote the initiation of transferring one algal partner for another partner. Depending on the taxonomic extent of different algal partners this may invoke different carbohydrate starting units or trigger a different biosynthetic pathway for secondary metabolite production. The predicted increases in average annual temperature in northern geographic areas may also promote temperate species of lichens to move further north into previously uninhabitable environments. Simultaneously, this may cause a more northerly movement of lichens that are adapted to or can tolerate cooler environments. The effects on epiphytic lichens will also be significant based on the availability of host tree species and how well the host trees adapt to climate change. Cool temperature plant species that do not adapt well to warmer temperatures may become fewer in number in northern regions. Fewer plant species may reduce the availability of suitable habitat for lichens specialized to growing on the bark of specific tree species. Species of lichens that are generalists, colonizing a number of different tree species or other substrata, will be better adapted to environmental changes than specialist species, because previously lost tree hosts may be replaced by succeeding species of plant host.

Droughts will further affect the plant community. Plants that are not drought resistant may become fewer in number and replaced by drought resistant species. Extreme drought may cause further loss of plants and increase soil erosion. Such a situation would create the opportunity for terricolous lichen expansion but perhaps on a scale too slow to prevent significant losses. Under the scenario of increased degree and frequency of drought, it might be expected that there will be an increased production of mineral chelating compounds and hydrophilic compounds; or institution of physiological mechanisms to retain water within the thallus. These physiological changes might be expected because rain would become less reliable as a source of water and nutrients.

Increasing carbon dioxide and atmospheric nitrogen levels may negatively affect lichen species overall. Being poikilohydric organisms, their passive absorption of air, water and substrate nutrients will be impacted by increased acidity due to pollution. Past research has shown that ozone and carbon dioxide kill the photobiont, which ultimately kills the lichen. Some secondary metabolites have the ability to mitigate these effects and some lichens are better adapted to polluted environments than others. Increases in pollution will entail increases in secondary compound quantities that neutralize the negative effects of acidity with the lichen. Usnic acid is a compound found within lichens inhabiting acidic environments. Higher acidity from pollution will negatively affect these species because of usnic acid's limited ability to control acidity. However, basic substrates have the ability to buffer against acidification, which is the result of most types of pollution. This could mean that those lichens will be better able to adapt to increased acid levels than usnic acid containing lichens. On the other hand, lichens growing on basic substrata could be at risk from acidification of limestone causing deterioration of the substratum or a change in the pH to a pH that is intolerable by the lichen.

Pollution is also thought to be responsible for the increased levels of ultraviolet light caused by the loss of atmospheric ozone. Cortical compounds and other compounds within the thallus that offer protection to the sexual and asexual reproductive structures and photobionts, may ensure that those lichen species will have some protection from increase ultraviolet light. Species lacking those photoprotective compounds may endure degradation of photobionts and an increased frequency of mutations due to ultraviolet light exposure. Environmental stress may stimulate the production of cortical compounds in species that normally do not produce them; in species that do not produce them frequently; and in increased quantities for the species that already produce them.

If biochemical diversity decreases in response to climate change (Hauck, 2011), fewer secondary metabolites will be available for herbivore defense and, therefore, more grazing on lichen thalli will occur. Metabolites that would normally be lost to the soil, where they have an effect on growth of plants and microbes, may become reduced in type and concentration of metabolite. The lower concentration of the metabolites in the soil will have a reduced effect on growth of plants and microbes. This reduced impact will allow more microbes and plants to grow among mats of lichens and perhaps outcompete lichen growth sooner than would be expected. With fewer compounds there might also be less protection from ultraviolet light and a diminished ability for lichens to adapt to environmental changes that require secondary metabolites. However, fungi are plastic and may adapt in other ways or produce an array of different types of compounds with similar effects. This scenario of the production of other ecologically valuable metabolites may be plausible since so many gene paralogs have been reported (Table 1) that have no known associated function.

6. Acknowledgments

The authors thank T. Booth and K. Fontaine (University of Manitoba) for field assistance; J. Roth and R. Lastra (University of Manitoba) for statistical advice; NSERC and Faculty of Science (University of Manitoba) for funding; and Manitoba Conservation for collecting permit number WPB-25028.

7. References

Ahmadjian, V. & Jacobs, J. B. (1981) Relationship between fungus and alga in the lichen *Cladonia cristatella* Tuck. Nature 289:169–172

Amo de Paz, G.; Raggio, J.; Gómez-Serranillos, M.P.; Palomino, O.M.; González-Burgos, Carretero, M.E. & Crespo, A. (2010) HPLC isolation of antioxidant constituents from *Xanthoparmelia* spp. Journal of Pharmaceutical and Biomedical Analysis 53:165-171.

Armaleo, D.; Zhang, Y. & Cheung, S. (2008) Light might regulate divergently depside and depsidone accumulation in the lichen *Parmotrema hypotropum* by affecting thallus temperature and water potential. Mycologia 100:565-576.

Armaleo, D.; Sun, X. & Culberson, C. F. (2011). Insights from the first putative biosynthetic gene cluster for a lichen depside and depsidone. Mycologia, 103:741-754.

Ahti, T. (2000) Cladoniaceae. Flora Neotropica, 78, Organization for Flora Neotropica and New York Botanical Garden, Bronx. 362 pp.

Ahti, T. & Hawksworth, D. L. (2005) *Xanthoparmelia stenophylla*, the correct name for *X. somloënsis*, one of the most widespread usnic acid containing species of the genus. Lichenologist 37:363-366.

Bayram, O. & Braus, G. H. (2011) Coordination of secondary metabolism and development in fungi: the velvet family of regulatory proteins. FEMS Microbiol. Rev. doi: 10.1111/j.1574-6976.2011.00285.x.

Bayram, O.; Krappmann, S.; Ni, M.; Bok, J. W.; Helmstaedt, K.; Valerius, O.; Braus-Stromeyer, S.; Kwon, N.-J.; Keller, N. P.; Yu, J.-H. & Braus, G. H. (2008) VelB/VeA/LaeA complex coordinates light signal with fungal development and secondary metabolism. Science 320:1504-1506.

Behera B.C.; Makhija U. & Adawadker B. 2000. Tissue cultures of *Bulbothrix setschwanesis* (lichenized ascomycete) *in vito*. Current Science 78:781-783.

Behera, B.C. & Makhija, U. (2001) Effects of varios culture conditions on growth and production of salazinic acid in *Bulbothrix setschwanesis* (lichenized ascomycete) *in vitro*. Current Science 80:1424-1427.

Bennett, J. W. & Ciegler, A. (1983) Secondary metabolism and differentiation in fungi. Marcel Dekker, Inc. New York. 478 pp.

Berkelmans, R & van Oppen, M. J. H. (2006) The role of zooxanthellae in the thermal tolerance of corals: a 'nugget of hope' for coral reefs in an era of climate change. Proceedings of the Royal Society B 273:2305–2312.

Bjerke, J. W.; Lerfall, K. & Elvebakk, A. (2002) Effects of ultraviolet radiation and PAR on the content of usnic and divaricatic acids in two arctic-alpine lichens. Photochem Photobiol Sci 1:678–685.

Bjerke, J. W.; Zielke, M. & Solheim, B. (2003) Long-term impacts of simulated climatic change on secondary metabolism, thallus structure, and nitrogen fixation activity in two cyanolichens from the Arctic. New Phytologist.159:361-367.

Bjerke, J. W.; Elvebakk, A.; Dominiguez, E. & Dahlback, A. (2005) Seasonal trends in usnic acid concentrations of Arctic, alpine and Patagonian populations of the lichen *Flavocetraria nivalis*. Phytochemistry 66:337–344.

Bjerke, J. W.; Joly, D.; Nilsen, L. & Brossard, T. (2004) Spatial trends in usnic acid concentrations of the lichen *Flavocetraria nivalis* along local climatic gradients in the Arctic (Kongsfjorden, Svalbard). Polar Biol 27:409–417.

Blanco, O.; Crespo, A.; Ree, R. H. & Lumbsch, H. T. (2006) Major clades of Parmelioid lichens (Parmeliaceae, Ascomycota) and the evolution of their morphological and chemical diversity. Molecular Phylogenetics and Evolution 39:52-69.

Bok, J. W. & Keller , N. P. (2004) LaeA, a Regulator of Secondary Metabolism in *Aspergillus* spp. Eukaryotic Cell. 3:527–535

Bok, J. W.; Hoffmeister, D.; Maggio-Hall, L. A.; Murillo, R.; Glasner, J. D. & Keller, N. P. (2006) Genomic mining for *Aspergillus* natural products. Chem. Biol. 13:31–37.

Boustie, J. & Grube, M. (2005) Lichens – a promising source of bioactive secondary metabolites. Plant Genetic Resources: Characterization and Utilization 3:273-287.

Bowler, P. A. & Rundel, P.W. (1974) The *Ramalina intermedia* complex in Norh America. Bryologist. 77:617-623.

Brodo, I. M. (1973) Substrate ecology. *In:* V. Ahmadjian & M. E. Hale (eds.): The Lichens. Academic Press, New York and London, pp. 401-441.

Brunauer, G.; Hager, A.; Grube, M., Turk, R., & Stocker-Worgotter, E. (2007) Alterations in secondary metabolism of aposymbiotically grown mycobionts of *Xanthoria elegans* and cultured resynthesis stages. Plant Physiology and Biochemistry. 45:146-151.

Brunaeur, G.; Muggia, L.; Stocker-Worgotter, E. & Grube, M. (2009) A transcribed polyketide synthase gene from *Xanthoria elegans*. Mycol. Res. 113:82-92.

Budel B. & Scheidegger, C. (2008) Thallus morphology and anatomy pp. 40-68. In Lichen Biology (Nash III, T. H.) Cambridge University Press. UK.

Chooi, Y. H., Stalker, D. M., Davis, M. A., Fujii, I., Elix, J. A., Louwhoff, S. H. & Lawrie, A. C. (2008) Cloning and sequence characterization of a non-reducing polyketide synthase gene from the lichen *Xanthoparmelia semiviridis*. Mycological Research 112:147-61.

Clayden, S. R. (1992) Chemical divergence of eastern North American and European populations of *Arctoparmelia centrifuga* and their sympatric usnic acid deficient chemotypes. The Bryologist 95: 1 - 4.

Cocchietto, M.; Skert, N.; Nimis, P.L. & Sava G. (2002) A review of usnic acid, an interesting natural compound. Naturwissenschaften 89:137-146.

Crespo, A.; Lumbsch, H. T.; Mattsson, J-E.; Blanco, O.; Divakar, P. K.; Articus, K.; Wiklund, E.; Bawingan, P. A. & Wedin, M. (2007) Testing Morphology-based hypotheses of phylogentic relationships in Parmeliaceae (Ascomycota) using three ribosomal markers and the nuclear RPBI gene. Molecular Phylogenetics and Evolution 44: 812-824.

Culberson, C. F. (1969) Chemical and Botanical Guide to Lichen Products. University of North Carolina Press, Chapel Hill. 628 pp.

Culberson, C. F. (1970) Supplement to "chemical and Botanical Guide to Lichen Products". Bryologist 73:177-377.

Culberson, C. F. (1972) Improved conditions and new data for the identification of lichen products by a standardized thin-layer chromatographic method. Journal of Chromatography 72:113 - 125.

Culberson, C. F. (1974) Conditions for the use of Merck silica gel 60 F254 plates in the standardized thin-layer chromatographic technique for lichen products. Journal of Chromatography B. 97:107–108.

Culberson, C.F.; Culberson, W.L. & Johnson, A. (1981) A standardized TLC Analysis of β-Orcinol Depsidones. Bryologist 84:16-29.

Culberson, C. F.; Culberson, W. L. & Johnson, A. (1988) Gene flow in lichens. American Journal of Botany 75:1135-1139.

Culberson, C. F. & Armaleo, D. (1992) Induction of a complete secondary-product pathway in a cultured lichen fungus. Experimental Mycology 16:52-63.

Culberson, C. F. & Elix, J. A. (1989) Lichen substances. In Methods in Plant Biochemistry Vol 1: Plant phenolics. Ed. P. M. Dey and J. B. Harbourne, pp 509-535. London, Academic Press.

Culberson, C. F.; Culberson, W. L. & Arwood, D. A. (1977a) Physiography and fumarprotocetraric acid production in the *Cladonia chlorophaea* group in North Carolina. Bryologist 80:71-75.

Culberson, C. F.; Culberson, W. L. & Johnson, A. (1977b) Second Supplement to "chemical and Botanical Guide to Lichen Products". American Bryological and Lichenologial Society, Missouri Botanical Garden, St. Louis. 400 pp.

Culberson, C. F.; Culberson, W. L. & Johnson, A. (1988) Gene flow in lichens. American Journal of Botany 75:1135-1139.

Culberson, C. F.; Culberson, W. L., & Johnson, A. (1990) The *Ramalina americana* complex (Ascomycotina, Ramalinaceae): Chemical and geographical correlations. Bryologist 93:167-186.

Culberson, C. F.; Culberson, W. L. & Johnson, A. (1992) Characteristic lichen products in cultures of chemotypes of the *Ramalina siliquosa* complex. Mycologia 84:705-714.

Culberson, C.F.; Culberson, W. L. & Johnson, A. (1993) Occurrence and histological distribution of usnic acid in the *Ramalina siliquosa* species complex. Bryologist 96:181–184.

Culberson, C. F.; LaGreca, S.; Johnson, A. & Culberson, W. L. (2000) Trivaric acid, a new tridepside in the *Ramalina americana* chemotype complex (lichenized Ascomycota, Ramalinaceae). Bryologist 102:595-601.

Culberson, W. L. (1966) Chimie et taxonomie des lichens du groupe *Ramalina farinacea* en Europe. Revue de Bryologie et Lichenologie. 34:841-851.

Culberson, W. L. (1969) The use of chemistry in the systematics of the lichens. Taxon 18:152-166.

Culberson, W. L. (1970) Chemosystematics and ecology of lichen-forming fungi. Ann. Rev. Ecol and Systematics 1:153-170.

Culberson, W. L. (1986) Chemistry and sibling speciation in the lichen-forming fungi: ecological and biological considerations. Bryologist 89:123-131.

Culberson, W. L. & Culberson, C. F. (1967) Habitat selection by chemically differentiated races of lichens. Science. 158:1195-1197.

Culberson, W. L. & Culberson, C. F. (1970) A phylogenetic view of chemical evolution in the lichens. Bryologist 73:1-31.

Culberson, W.L.; Culberson, C. & Johnson, A. (1977b) Mycologia 69:604–614.

Drew, S. W. & Demain, A. L. (1977) Effect of primary metabolites on secondary metabolism. Annu. Rev. Microbiol. 31:343-356.

Dreyer, J.; Eichhorn, H.; Friedlin, E.; Kürnsteiner, H. & Kück, U. (2007) A Homologue of the *Aspergillus velvet* gene regulates both Cephalosporin C Biosynthesis and hyphal fragmentation in *Acremonium chrysogenum.* Applied and Environmental Microbiology. 73:3412-3422.

Duran, R.M.; Cary, J. W. & Calvo, A. M. (2007) Production of cyclopiazonic acid, aflatrem, and aflatoxin by *Aspergillus flavus* is regulated by veA, a gene necessary for sclerotial formation. Appl Microbiol Biotechnol 73:1158-1168.

Elix, J. A.; Ferguson, B. A. & Sargent, M.V. (1974) The structure of alectoronic acid and related lichen metabolites. Australian Journal of Chemistry 27:2403–2411.

Elix, A. & Stocker-Wörgötter, E. (2008) Biochemistry and secondary metabolites. *In:* T. H. Nash, III: *Lichen Biology. Second Edition*. Cambridge University Press, Cambridge. viii+486 pages, pp. 104-133.

Ellis, C. J.; Coppins, B. J.; Dawson, T. P. & Seaward M. R. D. (2007) Response of British lichens to climate change scenarios: trends and uncertainties in the projected impact for contrasting biogeographic groups. Biological Conservation 140:217-235.

Esslinger, T. L. (2011) A cumulative checklist for the lichen-forming fungi, lichenicolous and allied fungi of the continental United States and Canada. North Dakota State University. Online. <http://www.ndsu.edu/pubweb/~esslinge/chcklst/chcklst7.htm>. (First posted 1 December 1997, Most Recent Version (#17) 16 May 2011), Fargo, North Dakota. Accessed June 24, 2011.

Fahselt, D. (1981) Lichen products of *Cladonia stellaris* and *C. rangiferina* maintained under artificial conditions. Lichenologist 13:87–91.

Frisvad, J. C., Anderson, B. & Thrane, U. (2008) The use of secondary metabolite profiling in chemotaxonomy of filamentous fungi. Mycological Research 112:231-240

Fox, E. M. & Howlett, B. J. (2008) Secondary metabolism: regulation and role in fungal biology. Current Opinion in Microbiology 11:481-7.

Gallopudi, S. R.; Telikepalli, H.; Jampani, H. B.; Mirhom, Y. W.; Drake, S. D.; Bhattiprolu, K. R.; Velde, D. V. & Mitscher, L. A. (1994) Alectosarmentin, a new antimicrobial dibenzofuranoid lactol from the lichen, *Alectoria sarmentosa.* Journal of Natural Products 57:934-938.

Galagan, J.E.; Calvo, S. E.; Borkovich, K.; Selker, E.; Read, N.; Jaffe, D.; FitzHugh5, W.; Ma, L.; Smirnov, S.; Purcell, S.; Rehman, B.; Elkins, T.; Engels, R.; Wang, S.; Nielsen, C.; Butler, J.; Endrizzi, M.; Qui, D.; Ianakiev, P.; Bell-Pedersen, D.; Nelson, M.; Werner Washburne, M.;. Selitrennikoff, C.; Kinsey, J.; Braun, E.; Zelter, A.; Schulte, U.; Kothe, G.; Jedd, G.; Mewes, W.; Staben, C.; Marcotte, E.; Greenberg, D.; Roy, A.; Foley, K.; Naylor, J.; Stange-Thomann, N.; Barrett, R.; Gnerre, S.; Kamal, M.; Kamvysselis, M.; Mauceli, E.; Bielke, C.; Rudd, S.; Frishman, D.; Krystofova, S.; Rasmussen, C.; Metzenberg, R.; Perkins, D.; Kroken, S.; Cogoni, C.; Macino, G.; Catcheside, D.; Li, W.; Pratt, R.; Osmani, S.; DeSouza, C.; Glass, L.; Orbach, M.; Berglund, J.; Voelker, R.; Yarden, O., Plamann, M.;, Seiler, S.; Dunlap, J.; Radford, A.; Aramayo, R.; Natvig, D.; Alex, L.; Mannhaupt, G.; Ebbole, D.; Freitag, M.; Paulsen, I.; Sachs, M.; Lander, E.; Nusbaum, C. & Birren, B. (2003) The genome sequence of the filamentous fungus *Neurospora crassa*. Nature. 859-868.

Garcia-Junceda, E.; Gonzalez, A. & Vicente, C. (1987) Photosynthetical and Nutritional Implications in the Accumulation of Phenols in the Lichen *Pseudevernia furfuracea*. Biochemical Systematics and Ecology. 15:289-296,

Gauslaa, Y. (2005) Lichen palatability depends on investments in herbivore defence. Oecologia. 143: 94-105.

Gilbert OL (1977) Phenotypic plasticity in *Cladonia pocillum*. Lichenologist 9:172-173.

Gilsenan, J. E. M.; Atherton, G.; Bartholomew, J.; Giles, P. F.; Attwood, T. K.; Denning, D. W. & Bowyer, P. (2009) *Aspergillus* genomes and the *Aspergillus* cloud. Nucleic Acids Research. 37:D509-D514.

Golojuch, S. T. & Lawrey, J. D. (1988) Quantitative variation in vulpinic and pinastric acids produced by *Tuckermannopsis pinastri* (lichen-forming Ascomycotina, Parmeliaceae). American Journal of Botany 75:1871-1875.

Hale, M. E. Jr. (1956) Chemical strains of the lichen *Parmelia furfuracea*. American Journal of Botany 43:456–459.

Hale, M. E. Jr. (1986) *Arctoparmelia*, a new genus in the Parmeliaceae (Ascomycotina). Mycotaxon 25:251-254.

Hale, M. E. Jr. (1990) A synopsis of the lichen genus *Xanthoparmelia* (Vainio) Hale (Ascomycotina, Parmeliaceae). Smithsonian Contributions to Botany 74:1-250.

Hamada, N. (1982) The effect of temperature on the content of the medullary depsidone salazinic acid in *Ramalina siliquosa* (lichens). Canadian Journal of Botany 60:383-385.

Hamada, N. (1989) The effect of various culture conditions on depside production by an isolated lichen mycobiont. Bryologist 92:310-313.

Hamada, N. (1991) Environmental factors affecting the content of usnic acid in the lichen mycobiont of *Ramalina siliquosa*. Bryologist 94:57-59.

Hamada, N. & Miyagawa, H. (1995) Secondary metabolites from isolated lichen mycobionts cultured under different osmotic conditions. Lichenologist 27:201-205.

Hamada, N.; Miyagawa, H.; Miyagawa, H. & Inoue, M. (1996) Lichen substances in mycobionts of crustose lichens cultured on media with extra sucrose. Bryologist 99:71-74.

Hauck, M. (2008) Susceptibility to acidic precipitation contributes to the decline of the terricolous lichens *Cetraria aculeata* and *Cetraria islandica* in central Europe. Environmental pollution. 152:731-735.

Hauck, M. (2011) Eutrophication threatens the biochemical diversity in lichens. Lichenologist 43:147-154.

Hauck, M. & Jürgens, S. R. (2008) Usnic acid controls the acidity tolerance of lichens. Environmental Pollution 156:115-122.

Hauck, M.; Jurgens, R-S.; Huneck. S. & Leuschner, C. (2009) High acidity tolerance in lichens with fumarprotocetraric, perlatoloc or thamnolic acids is correlated with low pK_{a1} values of these lichen substances. Environmental pollution. 157:2776-2780.

Hawksworth, D. L. (1976) Lichen chemotaxonomy. *In:* D. H. Brown, D. L. Hawksworth & R. H. Bailey (eds.): *Lichenology: Progress and Problems*. Academic Press, London, pp. 139-184

Hesbacher, S.; Froberg, L.; Baur, A.; Baur, B. & Proksch, P. (1996) Chemical variation within and between individuals of the lichenized Ascomycete *Tephromela atra*. Biochemical Systematics and Ecology. 8:603-609.

Honegger, R. (1990) Mycobiont-photobiont interactions in adult thalli in axenically resynthesized pre-thallus stages of *Xanthoria parietina* (Teloschistales, lichenized ascomycetes). *In:* H. M. Jahns (ed.): *Contributions to Lichenology in Honour of A. Henssen*. Bibliotheca Lichenologica. No. 38. J. Cramer, Berlin-Stuttgart, pp. 191-208.

Honegger, R. (2008) Mycobionts. *In:* T. H. Nash, III: *Lichen Biology. Second Edition*. Cambridge University Press, Cambridge. 486 pages, pp. 27-39.

Honegger, R. (2009) Lichen forming fungi and their photobionts. In: The Mycota V, Plant Relationships, second ed. H. Deising (ed.) Springer-Verlag, Berlin.

Honegger, R. (1993) Tansley Review No. 60. Developmental biology of lichens. New Phytologist 125:659-677.

Hoffmeister, D. & Keller, N. P. (2007) Natural products of filamentous fungi: enzymes, genes, and their regulation. Nat. Prod. Rep. 24:393-416.

Hopwood, D. A. (1997) Genetic contributions to understanding polyketide synthases. Chem. Rev. 97:2465-2497.

Hopwood, D. A. & Sherman, D. H. (1990) Molecular genetics of polyketides and its comparison to fatty acid biosynthesis. Annu Rev Genetics 24:37-66.

Hoyo, A.; lvarez, R. A.; del Campo, E. M.; Gasulla, F.; Barreno, E. & Casano, L. M. (2011) Oxidative stress induces distinct physiological responses in the two *Trebouxia* phycobionts of the lichen *Ramalina farinacea*. Annals of Botany. 107:109-118.

Huneck, S. (1999) The significance of lichens and their metabolites. Die Naturwissenschaften 86: 559-570.

Huovinen, K. & Ahti, T. (1986a) The composition and content of aromatic lichen substances in the genus *Cladina*. Ann. Bot. Fenn. 23:93-106.

Huovinen, K. & Ahti, T. (1986b) The composition and content of aromatic lichen substances in *Cladonia* section *Unciales*. Ann. Bot. Fenn. 23:173-188.

Huovinen, K. & Ahti, T. (1988) The composition and content of aromatic lichen substances in *Cladonia* section *Perviae*. Ann. Bot. Fenn. 25:371-383.

Huovinen, K.; Ahti, T. & Stenroos, S. (1989a) The composition and content of aromatic lichen substances in *Cladonia* section *Cocciferae*. Ann. Bot. Fenn. 26:133-148.

Huovinen, K.; Ahti, T. & Stenroos, S. (1989b) The composition and content of aromatic lichen substances in *Cladonia* section *Helopodium* and subsection *Foliosae*. Ann. Bot. Fenn. 26:297-306.

Huovinen, K.; Ahti, T. & Stenroos, S. (1990) The composition and content of aromatic lichen substances in *Cladonia* section *Cladonia* and group *Furcatae*. Biblioth. Lichenol. 38:209-241.

Hutchinson, C. R. & Fujii, I. (1995) Polyketide synthase gene manitoulation: A structure-function approach in engineering novel antibiotics. Annu. Rev. Microbiology 49:201-238.

Hyvärinen, M.; Koopmann R.; Hormi O. & Tuomi J. (2000) Phenols in reproductive and somatic structures of lichens: a case of optimal defence? Oikos 91:371-375.

Ingólfsdóttir, K. (2002) Usnic acid. Phytochemistry 61:729-736.

JMP®, Version 7. SAS Institute Inc., Cary, NC, 1989-2007.

Joneson, S. & Lutzoni, F. (2009) Compatibility and thigmotropism in the lichen symbiosis: a reappraisal. Symbiosis 47:109-115.

Joneson, S.; Armaleo, D. & Lutzoni, F. (2011) Fungal and algal gene expression in early developmental stages of lichen-symbiosis. Mycologia 103:291-306

Kale, S. P.; Milde, L.; Trapp, M. K.; Frisvad, J. C & Keller, N. P. (2008) Requirement of LaeA for secondary metabolism and sclerotial production in *Aspergillus flavus*. Fungal Genetics and Biology 45:1422-1429.

Kato, N.; Brooks, W. & Calvo, A.M. (2003) The expression of sterigmatocystin and penicillin genes in *Aspergillus nidulans* is controlled by veA, a gene required for sexual development. Eukaryotic Cell 2:1178–1186.

Katz, L. & Donadio, S. (1993) Polyketide synthesis: Prospects for hybrid antibiotics. Annu. Rev. Microbiol. 47: 875-912.

Keller, N. P. & Hohn, T. M. (1997) Metabolic pathway gene clusters in filamentous fungi. Fungal Genetics and Biology 21:17-29.

Keller, N. P.; Nesbitt, C.; Sarr, B.; Phillips, T. D. & Burow, G. B. (1997) pH regulation of sterigmatocystin and aflatoxin biosynthesis in *Aspergillus* spp. Phytopathology 87:643–648.

Keller, N. P.; Calvo, A.; Wilson, R. & Bok, J. (2002) Relationship between secondary metabolism and fungal development. Mol Microbiol Reviews. 66:447-459.

Keller, N. P.; Turner, G. & Bennett, J. W. (2005) Fungal secondary metabolism - from biochemistry to genomics. Nat Rev Microbiol. 3:937-47.

Kellner, H. & Zak, D. R. (2009) Detection of expressed fungal type 1 polyketide synthase genes in a forest soil. Soil Biology and Biochemistry. 41:1344-1347.

Kerr, J. B. & McElroy, C. T. (1993) Evidence for large upward trends of ultraviolet-B radiation linked to ozone depletion. Science 262:1032-1034.

Khaldi, N.; Collemare, J.; Lebrun, M.-H. & Wolfe, K. H. (2008) Evidence for horizontal transfer of a secondary metabolite gene cluster between fungi. Genome Biology 9:R18.

Kirk, P. M.; Cannon, P. F.; David, J. C. & Stalpers, J. A. (2001) Ainsworth and Bisby's Dictionary of Fungi, CAB International, Oxon, UK

Kotelko, R. & Piercey-Normore, M. D. (2010) *Cladonia pyxidata* and *C. pocillum*; genetic evidence to regard them as conspecific. Mycologia 102:534-545. [

Kroken, S.; Glass, N. L.; Taylor, J. W.; Yoder, O. C. & Turgeon, B. G. (2003) Phylogenomic analysis of type 1 polyketide synthase genes in pathogenic and saprobic ascomycetes. PNAS. 100:15670-15675.

Kumar, L.; Breakspear, A.; Kistler, C.; Ma, L.-J. & Xie, X. (2010) Systematic discovery of regulatory motifs in Fusarium graminearum by comparing four Fusarium genomes. BMC Genomics. 11:208

LaGreca, S. (1999) A phylogenetic evaluation of the *Ramalina americana* chemotype complex (lichenized Ascomycota, Ramalinaceae) based on rDNA ITS sequence data. Bryologist 102:602-618.

Larson, D. W. (1979) Lichen water relations under drying conditions. New Phytol 82:713-731.

Lawrey, J. D. (1977) Adaptive significance of O-methylated lichen depsides and depsidones. Lichenologist 9:137-142.

Lawrey, J. D. (1986) Biological role of lichen substances. Bryologist 89:111-122.

Lendemer, J. C. (2005) *Xanthoparmelia viriduloumbrina*, a neglected species from eastern North America. Mycotaxon 92:441–442.

Liao, C. R. J.; Piercey-Normore, M. D.; Sorenson, J. L. & Gough, K. M. (2010) In situ imaging of usnic acid in selected *Cladonia* spp. by vibrational spectroscopy. The Analyst. 135:3242-3248.

Long, M.; Betran, E.; Thornton, K. & Wang, W. (2003) The origin of new genes: glimpses from the young and old. Nature Reviews Genetics 4:865-875.

Luo, H.; Yamamoto, Y.; Kim, J. A.; Jung, J. S.; Koh, Y. J. & Hur, J.-S. (2009) Lecanoric acid, a secondary lichen substance with antioxidant properties from *Umbilicaria antarctica* in maritime Antarctica (King George Island). Polar Biology 32:1033-1040.

McCune, B. (1987) Distribution of chemotypes of *Rhizoplaca* in North America. Bryologist 90:6–14.

McEvoy, M.; Nybakken, L.; Solhaug, K. A. & Gauslaa, Y. (2006) UV triggers the synthesis of the widely distributed secondary lichen compound usnic acid. Mycol Progress 5:221–229.

McEvoy, M.; Solhaug, K. A. & Gauslaa, Y. (2007a) Solar radiation screening in usnic acid-containing cortices of the lichen *Nephroma arcticum*. Symbiosis 43:143-150.

McEvoy, M.; Gauslaa, Y. & Solhaug, K. A. (2007b) Changes in pools of depsidones and melanins, and their function, during growth and acclimation under contrasting natural light in the lichen *Lobaria pulmonaria*. New Phytologist 175:271-282.

MacGillvray, T. & Helleur, R. (2001) Analysis of lichens under environmental stress using TMAH thermochemolysis-gas chromatography. Journal of Analytical and Applied Pyrolosis 58-59: 465-480.

Meehl, G. A.; Stocker, T. F.; Collins, W. D.; Friedlingstein, P.; Gaye, A. T.; Gregory, J. M.; Kitoh, A.; Knutti, R.; Murphy, J. M.; Noda, A.; Raper, S. C. B.; Watterson, I. G.; Weaver, A. J. & Zhao, Z.-C. (2007) Global Climate Projections. In: Climate Change 2007: The Physical Science Basis.Contribution of Working Group I to the Fourth Assessment Report of the Intergovernmental Panel on Climate Change [Solomon, S.; Qin, D.; Manning, M.; Chen, Z.; Marquis, M.; Averyt, K. B.; Tignor M. & Miller H. L. (eds.)]. Cambridge University Press, Cambridge, United Kingdom and New York, NY, USA.

Miao, V.; Coeffet-LeGal, M. F.; Brown, D.; Sinnemann, S.; Donaldson, G. & Davies, J. (2001) Genetic approaches to harvesting lichen products. Trends in Biotechnology. 19:349-355.

Moning, C.; Werth, S.; Dziock, F.; Bassler, C.; Bradtka, J.; Hothorn, T. & Muller, J. (2009) Lichen diversity in temperate montane forests is influenced by forest structure more than climate. Forest Ecology and Management. 258:745-751.

Micheletti, A. C.; Beatriz, A.; de Lima, D. P.; Honda, N. K.; do Ó Pessoa, C.; de Moraes, M. O.; Lotufo, L. V.; Magalhães, H. I. F.; and Carvalho, N. C. P. (2009) Chemical constituents of *Parmeotrema lichexanthonicum* Elisario & Adler – isolation, structure modification, and elevation of antibiotic and cytotoxic activities. Química Nova 32: 12-20

Moore, D. (1998) Fungal morphogenesis. Cambridge University Press, UK. 469 pp.

National Climate Data and Information Archive. 2011. Environment Canada. Online <http://climate.weatheroffice.gc.ca/climateData/canada_e.html>. Pinawa: ID 503B1ER; Flin Flon: ID 5050920. Accessed July 7, 2011.

Nierman, W. C.; Pain, A.; Anderson, M. J.; Wortman, J. R.; Kim, H. S.; Arroyo, J.; Berriman, M.; Abe, K.; Archer, D. B.; Bermejo, C.; Bennett, J.; Bowyer, P.; Chen, D.; Collins, M.; Coulsen, R.; Davie, R.; Dyer, P. S.; Farman, M.; Fedorova, N.; Feldblyum, T. V.; Fischer, R.; Fosker, N.; Fraser, A.; García, J. L.; García, M. J.; Goble, A.; Goldman, G. H.; Gomi, K.; Griffith-Jones, S.; Gwilliam, R.; Haas, B.; Haas, H.; Harris, D.; Horiuchi, H.; Huang, J.; Humphray, S.; Jiménez, J.; Keller, N.; Khouri, H.; Kitamoto, K.; Kobayashi, T.; Konzack, S.; Kulkarni, R.; Kumagai, T.; Lafton, A.; Latge, J. P.; Li, W.; Lord, A.; Lu, C.; Majoros, W. H.; May, G. S.; Miller, B. L.; Mohamoud, Y.; Molina, M.; Monod, M.; Mouyna, I.; Mulligan, S.; Murphy, L.; O'Neil, S.; Paulsen, I.; Peñalva, M. A.; Pertea, M.; Price, C.; Pritchard, B. L.; Quail, M. A.; Rabbinowitsch, E.; Rawlins, N.; Rajandream, M. A.; Reichard, U.; Renauld, H.; Robson, G. D.; Córdoba, S. R.; Rodríguez-Peña, J. M.; Ronning, C. M.; Rutter, S.; Salzberg, S. L.; Sanchez, M.; Sánchez-Ferrero, J. C.; Saunders, D.; Seeger, K.; Squares, R.; Squares, S.; Takeuchi, M.; Tekaia, F.; Turner, G.; Vazquez de Aldana, C. R.; Weidman, J.; White, O.; Woodward, J.; Yu, J. H.; Fraser, C.; Galagan, J. E.; Asai, K.; Machida, M.; Hall, N.; Barrell, B. & Denning, D. W. (2005) Genome sequence of the pathogenic and allergenic filamentous fungus *Aspergillus fumigatus*. Nature, 438:1151-1156.

O'Donovan, D. G.; Roberts, G. & Keogh M. F. (1980) Structure of the B-orcinal depsidones connorstictic and consalazinic acids. Phytochemistry 19: 2497-2499.

Opanowicz, M.; Blaha, J. & Grube, M. (2006) Detection of paralogous polyketide synthase genes in Parmeliaceae by specific primers. Lichenologist 38:47-54.

Orange, A.; James, P, W. & White, F. J. (2001) Microchemical Methods for the Identification of Lichens. British Lichen Society. 101 pp.

Paul, A.; Hauck, M. & Leuschner, C. (2009) Iron and phosphate uptake in epiphytic and saxicolous lichens differing in their pH requirements. Environmental and Experimental Biology 67:133-138.

Penalva, M. & Arst, H. N. (2002) Regulation of Gene Expression by Ambient pH in Filamentous Fungi and Yeasts. Microbiology and Molecular Biology Reviews. 66:426-446.

Perrin, R. M.; Federova, N. D.; Bok, J. W.; Cramer, R. A.; Wortman, J. R.; Kim, H. S.; Nierman, W. C. & Keller, N. P. (2007) Transcriptional regulation of chemical diversity in *Aspergillus fumigatus* by LaeA. PLoS Pathog. 3, e50.

Piercey-Normore, M. D. (2003) A field survey of the genus *Cladonia* (Ascomycotina) in Manitoba, Canada. Mycotaxon 86:233-247.

Piercey-Normore, M. D. (2006) Lichens from the Hudson Bay Lowlands: diversity in the southeastern peatlands of Wapusk National Park, Manitoba. Canadian Journal of Botany 84:1781-1793.

Piercey-Normore, M. D. (2007) The genus *Cladonia* in Manitoba: exploring taxonomic trends with secondary metabolites. Mycotaxon 101:189-199.

Printzen, C.; Lumbsch, H. T.; Schmitt, I. & Feige, G. B. (1999) A study on the genetic variability of *Biatora helvola* using RAPD markers. Lichenologist 31:491-499.

Proctor, R. H.; Butchko, R. A. E.; Brown, D. W. & Moretti A. (2007) Functional characterization, sequence comparisons and distribution of a polyketide synthase gene required for perithecial pigmentation in some *Fusarium* species. Food additives and Contaminant. 24:1076-1087

Reverberi, M.; Ricelli, A.; Zjalic, S.; Fabbri, A. A. & Fanelli, C. (2010) Natural functions of mycotoxins and control of their biosynthesis in fungi. Appl Microbiol Biotechnol 87:899-911.

Rollins, J. A. & Dickman, M. B. (2001) pH signaling in *Sclerotinia sclerotiorum*: identification of a *pacC/RIM1* homolog. Appl. Environ. Microbiol. 67:75–81.

Rubio, C.; Fernández, E.; Hidalgo, M. & Quilhot, W. (2002) Bol. Soc. Chilena Química 47 <http://www.scielo.cl/scielo.php?pid=S0366-16442002000100012&script=sci_arttext> Mar 2, 2011.

Rundel, P. W. (1969) Clinal variation in the production of usnic acid in *Cladonia subtenuis* along light gradients. Bryologist 72:40–44.

Schmitt, I. & Lumbsch, H. T. (2009) Ancient horizontal gene transfer from bacteria enhances biosynthetic capabilities of fungi. PLoS ONE 4:e4437.

Schmitt, I.; Kautz, S. & Luumbsch, H. T. (2008) 6-MSAS-like polyketide synthase genes occur in lichenized ascomycetes. Mycol. Res. 112:289-296.

Schwab, E. K. & Keller, N. P. (2008) Regulation of secondary metabolite production in filamentous ascomycetes Mycol. Res. 112:225-230.

Shapiro, R. S.; Robbins, N. & Cowen, L. E. (2011) Regulatory Circuitry Governing Fungal Development, Drug Resistance, and Disease. Microbiology and Molecular Biology Reviews, 75:213-267.

Solhaug, K. A.; Gauslaa, Y.; Nybakken, L. & Bilger, W. (2003) UV-induction of sun-screening pigments in lichens. New Phytologist 158:91-100.

Sprote, P. & Brakhage, A. A. (2007) The light-dependant regulator velvet A of *Aspergillus nidulins* acts as a repressor of the penicillin biosynthesis. Arch Microbiol 188:69-79.

Stocker-Worgotter, E. (2001) Experimental studies of the lichen symbiosis: DNA-analyses, differentiation and secondary chemistry of selected mycobionts, artificial resynthesis of two- and tripartite symbioses. Symbiosis 30:207-227.

Stocker-Worgotter, E. (2008) Metabolic diversity of lichen-forming ascomycetous fungi: culturing, polyketide and shikimate production, and PKS genes. Natural Product Reports. 25:188-200.

Stocker-Worgotter E.; Elix, J. A. & Grube, M. (2004) Secondary chemistry of lichen-forming fungi: Chemosyndromic variation and DNA-analyses of cultures and chemotypes in the *Ramalina farinacea* complex. Bryologist 107:152-162.

Sua´rez, T. & Pen˜alva, M. A. (1996) Characterization of a *Penicillium chrysogenum* gene encoding a PacC transcription factor and its binding sites in the divergent *pcbAB-pcbC* promoter of the penicillin biosynthetic cluster. Mol. Microbiol. 20:529–540.

Tehler, A. & Wedin, M. (2008) Systematics of lichenized fungi. *In:* T. H. Nash, III: Lichen Biology. Second Edition. Cambridge University Press, Cambridge. 486 pages, pp. 336-352.

Valencia-Islas N.; Zambrano, A. & Rojas, J. L. (2007) Ozone reactivity and free radical scavenging behavior of phenolic secondary metabolites in lichens exposed to chronic oxidant air pollution from Mexico City. Journal of Chemical Ecology 33:1619-1634.

Sanchez, J. F.; Chiang, Y.-M. & Wang, C. C. C. (2008) Diversity of polyketide synthases found in the *Aspergillus* and *Streptomyces* genomes. Molecular Pharmaceutics. 5:226-233.

Yu, J. H. & Keller, N. (2005) Regulation of secondary metabolism in filamentous fungi. Annual Review of Phytopathology 43: 437-458

Part 4

Land Use and Land Cover Change

12

Investigating Soils, Vegetation and Land Use in a Lunette Dune-Pan Environment: The Case of Sekoma Lunette Dune-Pan Complex, Botswana

S. Mosweu[1], J.R. Atlhopheng[1] and M.P. Setshogo[2]
[1]*Department of Environmental Sciences*
[2]*Department of Biological Sciences*
University of Botswana, Gaborone, Botswana

1. Introduction

The association between vegetation and environmental factors has been a subject of ecological studies over time (e.g. Monier & Amer, 2003; McDonald et al., 1996). Some of these studies have addressed facilitative and competitive interactions between woody and herbaceous plants (Maestre et al., 2003), whilst others focused on the feedbacks in the dynamics of plant communities (Schwinning et al., 2005). On the other hand, there is considerable empirical research work on pans and their associated landforms (e.g. Lancaster, 1986; Goudie & Thomas, 1986; Cooke et al., 1993). Common land forms associated with pans like lunette dunes have particularly received significant attention from researchers (e.g. Lancaster, 1978; Goudie & Thomas, 1986; Holmgren & Shaw, 1996). Most of the afore-mentioned studies have mainly focused on the morphology, sedimentology and the origin of lunette dunes and pans. In addition, they have considered the significance of lunette dunes in palaeo-environmental reconstruction (Holmgren & Shaw, 1996; Lancaster, 1989; Marker & Holmes, 1995).

Livestock production dominated by cattle rearing plays a pivotal economic role in the Kalahari area (van de Maas et al., 1994; Chanda et al., 2003; Mosweu et al., 2010). The most limiting factor in livestock production in Kalahari over the years has been the availability of surface water and fodder resources. Consequently, lunette dune-pan environments continue to play a central role as sources of both water (Figure 1) and fodder resources for livestock in the area. As a result, lunette dune-pan environments exist in the Kalahari as unique interspersed micro-ecosystems that are significantly intertwined with the livelihoods of rural communities of the area (Chanda et al., 2003; Mosweu & Areola, 2008).

Although some research work has been conducted on the lunette dunes, pans, vegetation-environment relationships, and land use in the Kalahari environment (e.g. Chanda et al., 2003; Privette et al., 2004; Shugart et al., 2004; Wang et al., 2007; Mosweu, 2008), paucity still exists in researches that consider lunette dunes, pans and their environs as unique micro-ecosystems of significance to rural communities inhabiting semi-arid and arid regions. This scenario prevails in spite of the fact that the state of lunette dune-pans as micro-ecosystems remains vital in the sustainability of the livelihoods of the Kalahari rural communities and other communities residing in semi-arid and arid regions elsewhere. It is on this basis that the main aim of this chapter was to examine the interrelationships among the soil, vegetation, topography and land use in a lunette dune-pan environment with a view to elucidate their interactions and the consequent environmental changes thereof. Thus, the specific objectives of this study were to investigate the following in a lunette dune-pan environment:

- Soil physical and chemical characteristics;
- Woody vegetation properties;
- Land use attributes; and
- The correlations amongst soil, vegetation and land use characteristics.

Fig. 1. Hand-dug well located in Sekoma pan.

2. The study site

A lunette dune-pan complex located in the Sekoma village (Figure 2) in the Kalahari region of Botswana was chosen as a case study area. The state of the environment, current land use practices and geographical position of the Sekoma lunette dune-pan system present an ideal environment for the investigation of environmental changes and ecosystem dynamics particularly in lunette dune-pan micro-ecosystems. The geographical location of the study site along the Kalahari Transect (KT) 'megatransect' which has been established by the International Geosphere-Biosphere Programme (IGBP) for the study of both regional and universal environmental changes (e.g. Shugart et al., 2004; Wang et al., 2007) positions this study within an international context of studies focusing on environmental changes.
The lithology of the area is characterized by the dolomite Precambrian aquifer system (Geological Survey Department, 1995). The general structure of vegetation in the area is shrub savanna and the vegetation is classified as southern Kalahari bush savanna (Department of Surveys and Mapping, 2001). The mean annual rainfall in the area is about 400 mm (Bhalotra, 1985) and the rainfall season is characterized by erratic rainfall patterns. The lunette dune-pan complex is situated between the former (Sekoma West) and current (Sekoma) locations of the village (Figure 2).

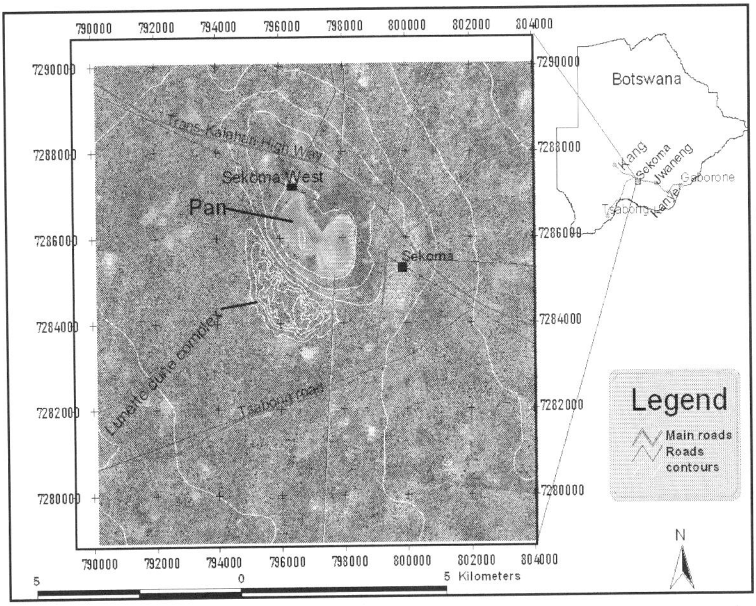

Fig. 2. Study site location (Author using base maps from Department of Surveys and Mapping, Gaborone, Botswana).

3. Research methods

3.1 Sample collection

Stratified transect sampling was used in this study. This is a systematic sampling method in which sampling points were arranged linearly and continuously. Transects were established from the pan fringes across selected lunette dunes (Tshube, Leremela and Kebuang) to the end of the slip face slope of each dune (Figure 3). Sampling was carried out at the pan fringes, wind ward slope, dune crest and slip face slope which were referred to as sampling points 1, 2, 3, and 4 respectively (Figure 3). At the slopes, sampling was carried out at the approximate mid-point of the slopes. A similar method was used with success in other studies including salt-marshes, inter-tidal zones, study of pattern and succession on dunes, altitudinal gradients, from dry to wet heath and across gradients of trampling intensity (Goldsmith & Harrison, 1976).

Three quadrats of 20 m^2 separated by 10 m were located at sampling points marked 1, 2, 3 and 4 (Figure 3) along the transects. The quadrats were identified as indicated in Figure 3 (Tshube 1-12; Leremela 13-24; Kebuang 25-36; 'S' denotes site). Vegetation and soil sampling was conducted in each of the quadrats. Soil samples were collected in the center of each quadrat using an auger that had a sample collection chamber length of 20 cm and a volume of ca. 23.75 mL. Therefore, about 23.75 mL per sample volume were collected. It was observed in the preliminary study that a soil profile established in the dunes did not show soil horizons. Therefore, soil samples were collected at predetermined sampling depths (SDs) of 0-20 cm, 40-60 cm, 80-100 cm, 130-150 cm and 180-200 cm. Methods used in this study to investigate vegetation and soil are summarized in Tables 1 and 2.

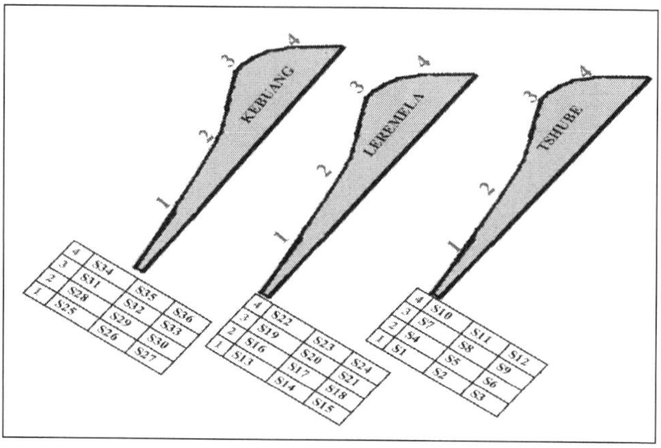

Fig. 3. Detailed layout of quadrats for sampling.

Parameter	*Method of Analysis*	*References*
Species Cover	Crown-Diameter method	Muller-Dombois &Ellenberg, 1974; Krebs (1989)
Species density	Simple counts	Krebs (1989)
Species composition	list of plant species within a particular quadrat	Bonham(1989);Krebs (1989)
Species distribution	Spatial range of species	Bonham(1989);Krebs (1989)

Table 1. Methods of vegetation study.

Parameter	*Method of Analysis*	*Analytical Instrument*	*References*
Available Phosphorus	Olsen's	UV- Visible Spectrophotometer	ISRIC, 1993
Particle size (sand, silt-clay)	Sieve	Retsch shaker and sieve	Buurmanet al. (1996)
Electrical conductivity& pH	1:2 (soil:water) ratio	InoLabcond 730 WTW series electrical conductivity meter & HANNA pH 210 pH meter	Sonnevelt & van den Ende (1971); Janzen (1993) Soon & Warren (1993).
Soil Organic Carbon	Walkley-Black wet oxidation	Various apparatus	Van Reeuwijk (1993)
Effective Cation Exchange Capacity (ECEC) and Exchangeable cations (Ca, Mg, Na, K), Al, Fe &Mn	Barium Chloride ($BaCl_2$)	Atomic Absorption Spectrophotometer (AAS)	Hendershot & Duquette (1986)

Table 2. Summary of methods of soil study.

3.2 Mini-social survey

A non-probability sampling procedure known as purposive sampling (Rea & Parker, 2005) was employed in a mini-social survey to gather data on the perceptions of the communities about the spatial and temporal environmental changes that they had witnessed in the lunette-dune pan environment over the years. The method facilitated the use of professional assessment, instead of randomness, in choosing the respondents (Rea & Parker, 2005). The survey was therefore, restricted only to key informants who were considered to be endowed with indigenous knowledge within the Sekoma community. Consequently, two focused group discussions, one constituted by the chief and village elders and the other by the Village Development Committee (VDC) members were conducted in the village. Open-ended questions were posed to the groups to facilitate freedom of expression. In the questionnaire, the most predictable answers had been pre-stated for data capturing convenience, but were not read out to respondents to minimize the researcher's influence on the respondent's view. Recording of the responses was conducted during the interview process. In addition, notes were made on the relevant additional information provided by the respondents.

4. Results

4.1 Pedo-geomorphological characteristics of the lunette dunes

To explore the distribution pattern of selected soil resources in the lunette dune-pan environment, correlation analysis (Table 3) was used to establish the relationships between soil variables at different sampling depths (SD) and the distance from the pan fringes. It was observed that only sodium indicated a significant negative correlation ($r = 0.991$, $P = 0.009$) at SD 0–20 cm in the Tshube lunette dune at α = 0.01. Aluminium and organic carbon also exhibited negative correlations ($r = -0.980$, $P = 0.020$ and $r = -0.958$, $P = 0.042$ respectively) with distance at α = 0.05. At SD 40-60 cm, sodium ($r = -0.958$, $P = 0.042$) and EC ($r = - 0.985$, $P = 0.015$) showed negative significant relationships with distance at α = 0.05. It was observed that at SD 80-100 cm all soil variables indicated negative relationships with distance except sand fraction, but the relationships were not significant at both α = 0.01 and 0.05. ECEC was the only soil variable that showed significant and negative relation with distance ($r = -0.998$, $P = 0.002$) at SD 130-150 cm and α = 0.01. Furthermore, all other soil variables indicated negative relationships with distance except sand fraction, phosphorus and pH. Magnesium ($r = -1$, $P = 0.011$), manganese ($r = -0.999$, $P = 0.033$) and phosphorus ($r = 0.999$, $P = 0.029$) were the only soil variables that exhibited significant relationships with distance at α = 0.05 in relation to SD 180-200 cm (Table 3) in the Tshube lunette dune.

In Leremela lunette dune, none of the selected soil variables showed a significant relationship with distance from the pan fringes (SP1) to the slip face slope (SP 4) at SD 0-20 cm and α = 0.01 and 0.05 (Table 3). However, all variables displayed negative relationships with distance except sand fraction, aluminium and manganese. Magnesium ($r = 0.984$, $P = 0.016$) was the only soil variable that indicated positive significant relationship with distance at α = 0.05 in relation to the SD 40-60 cm. With the exception of sand fraction, manganese and phosphorus, all other soil variables were negatively related to distance at SD 40-60 cm. From SD 60-200 cm, soil variables and distance were not significantly related at α = 0.01 and 0.05.

All selected soil variables did not show significant relationships with distance at SD 0-20 cm in Kebuang lunette dune. Furthermore, all soil variables were negatively related to distance except sand fraction and aluminium at SD 0-20 cm. Potassium (*r* = *-0.984, P* = *0.016*) and EC (*r* = *-0.964, P* = *0.036*) were the only soil variables that showed negative significant relationships with distance at $\alpha = 0.05$ with respect to SD 40-60 cm. At SD 80-100 cm, the relationships between all soil variables and distance were not significant at both $\alpha = 0.01$ and 0.05. In addition, all soil variables were negatively related to distance except sand fraction and aluminium. Only calcium (*r* = *-1, P* = *0.013*) displayed a perfect negative relationship with distance at $\alpha = 0.05$ and SD 130-150 cm sampling depth. At $\alpha = 0.01$ and 0.05 significant levels, all selected soil variables were not significantly related to distance at SD 180-200 cm in Kebuang lunette dune (Table 3). It was also observed that all the relationships were negative except for sand fraction and pH.

4.2 Plant species distribution patterns and community composition

In Detrended Correspondence Analysis (DCA) diagram, each site point lies at the centroid of the points of the species that occurs at the sampling site (Hill, 1979). Therefore, Figure 4 mirrors the approximate plant species distribution patterns and plant community composition in the lunette dune-pan environment. On the basis of Figure 4, Inferences were made about the species that were likely to be found at a particular sampling site. Sites that were close to the point of the species were likely to exhibit high density of that particular species, and the density of a species was expected to decrease with the increase in distance from its location.

Two main plant communities were identified in the lunette dune-pan environment. The first one was dominated by *Acacia mellifera* and the other by *Grewia flava* (Figure 4). *A. mellifera* community was dominant particularly at the sampling points that were located on the slip face of the lunette dunes, and between the lunette dune-pan complex and the settlement area. *G. flava* community was predominated the wind ward slope.

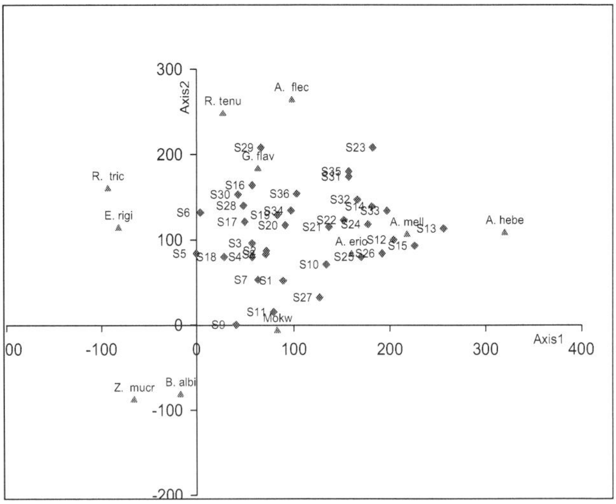

Fig. 4. Plant species distribution in the lunette dune-pan environment (*scale* = *1; multiplier* = *100*).

	SD (cm)		Sand	SC	K	Mg	Na	Al	Mn	Fe	Ca	ECEC	P	%OC	pH	EC
Tshube	0-20	*r*	0.93	-0.93	-0.91	0.73	-0.99**	-0.98*	-0.38	-0.91	0.65	-0.41	-0.63	-0.96*	-0.70	-0.87
		P	0.07	0.07	0.10	0.27	0.01	0.02	0.62	0.10	0.35	0.60	0.37	0.04	0.30	0.13
	40-60	*r*	0.77	-0.77	-0.92	-0.88	-0.96*	-0.10	-0.36	-0.77	0.80	0.58	0.81	-0.38	-0.23	-0.99*
		P	0.23	0.23	0.084	0.12	0.04	0.90	0.64	0.23	0.19	0.42	0.19	0.62	0.77	0.02
	80-100	*r*	0.92	-0.92	-0.75	-0.54	-0.86	-0.86	-0.56	-0.84	-0.20	-0.93	-0.20	0.40	-0.66	-0.73
		P	0.08	0.08	0.26	0.46	0.14	0.14	0.45	0.16	0.81	0.07	0.80	0.60	0.34	0.27
	130-150	*r*	0.94	-0.94	-0.74	-0.95	-0.78	-0.93	-0.79	-0.83	-0.17	-1.00**	0.67	-0.44	0.88	-0.83
		P	0.06	0.06	0.26	0.05	0.22	0.07	0.21	0.17	0.83	0	0.33	0.56	0.12	0.17
	180-200	*r*	0.98	-0.98	-0.70	-1.00*	-0.99	0.35	-1.00*	0.33	0.37	-0.96	1.00*	0.15	0.54	-0.82
		P	0.14	0.14	0.50	0.01	0.07	0.77	0.03	0.79	0.76	0.19	0.03	0.91	0.64	0.39
Leremela	0-20	*r*	0.88	-0.88	-0.83	-0.75	-0.65	0.28	0.05	-0.62	-0.61	-0.58	-0.26	-0.64	-0.67	-0.87
		P	0.12	0.12	0.17	0.25	0.35	0.72	0.95	0.38	0.39	0.42	0.74	0.36	0.34	0.13
	40-60	*r*	0.89	-0.89	-0.87	0.98(*)	-0.47	-0.66	0.69	-0.75	-0.65	-0.82	0.78	-0.86	-0.82	-0.79
		P	0.12	0.12	0.13	0.02	0.53	0.34	0.31	0.25	0.36	0.18	0.22	0.14	0.18	0.21
	80-100	*r*	0.79	-0.79	-0.99	0.23	-1.00	0.28	0.24	-0.89	-0.92	-0.91	0.94	-0.35	-0.72	-0.86
		P	0.42	0.42	0.09	0.85	0.06	0.82	0.85	0.30	0.26	0.27	0.22	0.77	0.49	0.35
	130-150	*r*	0.61	-0.61	-0.97	0.79	-0.90	-0.09	0.98	0.87	-0.99	-0.97	0.38	0.63	-0.86	-0.83
		P	0.58	0.58	0.15	0.42	0.28	0.94	0.12	0.33	0.08	0.14	0.76	0.57	0.35	0.38
	180-200	*r*	0.95	-0.95	-0.91	0.92	-0.83	0.69	0.86	-0.91	-0.81	-0.83	0.69	0.70	-0.67	-0.98
		P	0.20	0.20	0.28	0.25	0.37	0.51	0.34	0.28	0.40	0.38	0.52	0.51	0.53	0.14
Kebuang	0-20	*r*	0.94	-0.94	-0.94	-0.93	-0.84	0.66	-0.85	-0.24	-0.84	-0.46	-0.72	-0.77	-0.42	-0.92
		P	0.06	0.06	0.06	0.08	0.16	0.34	0.15	0.76	0.16	0.54	0.28	0.23	0.58	0.08
	40-60	*r*	0.84	-0.84	-0.98*	-0.49	0.64	0.82	-0.77	-0.20	-0.71	-0.70	-0.23	-0.58	0.54	-0.96*
		P	0.16	0.16	0.02	0.51	0.36	0.18	0.23	0.80	0.29	0.30	0.77	0.42	0.46	0.04
	80-100	*r*	0.84	-0.84	-0.87	-0.80	-0.85	0.02	-0.99	-0.43	-0.88	-0.78	-0.66	-0.07	-0.15	-0.77
		P	0.37	0.37	0.33	0.41	0.35	0.99	0.08	0.72	0.32	0.43	0.54	0.96	0.91	0.45
	130-150	*r*	1.00	-1.00	-0.93	-0.95	-0.03	0.39	-0.99	-0.80	-1.00*	-0.98	-0.93	0.81	0.85	0.96
		P	0.05	0.05	0.23	0.20	0.98	0.74	0.07	0.41	0.01	0.15	0.24	0.41	0.35	0.18
	180-200	*r*	0.85	-0.85	-0.33	-0.94	-0.94	-0.17	-0.98	-0.66	-0.92	-0.87	-0.13	-0.90	0.97	-0.16
		P	0.35	0.35	0.79	0.22	0.22	0.89	0.14	0.55	0.26	0.33	0.92	0.28	0.16	0.90

** Correlation is significant at the 0.05 level (2-tailed).*
***Correlation is significant at the 0.01 level (2-tailed). SD: Sampling Depth*

Table 3. Correlation analysis of sampling depth and distance from the pan fringes to SP4.

4.3 Vegetation - Environment relationships

Canonical Correspondence Analysis (CCA) diagrams show the interrelationships between vegetation and selected soil variables that were observed in the lunette dune-pan environment (Figures 5-9). The diagrams display points that represent species and sampling sites, and arrows that symbolize soil variables. The species and sampling points mutually portray the dominant patterns in community composition to the extent that these could be elucidated by the selected soil variables (ter Braak, 1988). The species points and the arrows of the soil variables mutually depict the plant species distribution along each of the soil variable (ter Braak, 1988). It is worth noting that only the direction and the relative length of the arrows convey essential information (ter Braak, 1995). The length of an arrow representing a soil variable was considered to be equal to the rate of change in the score as inferred from Figures 5-9, hence a measure of how much the plant species distribution differ along that soil variable (Gauch, 1982). As a result, important soil variables were represented by longer arrows (Figures 5-9). The species points that are positioned on the edge or very close to the edge of a CCA diagram are often considered to be rare species (ter Braak, 1995), and such species are usually considered to be very insignificant in CCA. Consequently, plant species of that nature were excluded in the analysis.

Similar researches that have been carried out in the past on vegetation-soil associations focused on the 0-20 cm top soil layer (e.g. Moleele & Perkins, 1998; Moleele, 1999; Smet & Ward, 2006). However, it is widely acknowledged that different plant species exhibit various rooting systems as well as different responses to various environmental factors along environmental gradients (Gauch, 1982). For instance, potassium, phosphorus and sodium gradients may not necessarily be the same at SD 0-20 cm and 180-200 cm, and some plant species may not be able to access essential soil resources from the depth of 200 cm and beyond. To this end, an attempt was made to examine the effect of changes in soil properties due to soil depth variation on the vegetation-soil interrelationships. The assessment was premised on the inferences from Figures 5-9. Worth highlighting is the scale of the diagrams (Figures 5-9); 1 unit in the plot corresponds to 1 unit for the sites, to 1 unit for the species and to 10 units for the soil variables.

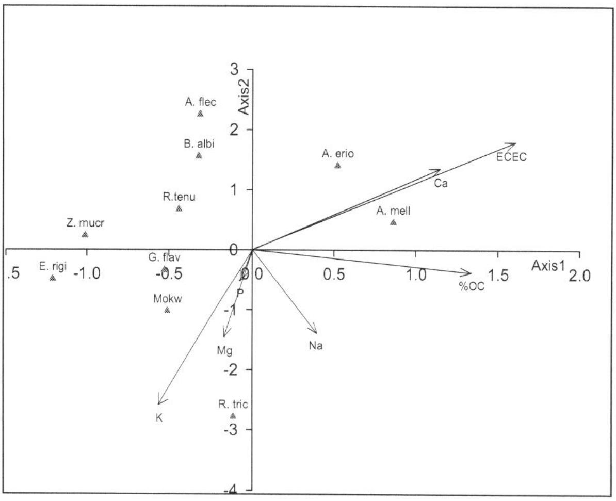

Fig. 5. Relationships between soil properties (SD 0-20 cm) and plant species distribution.

The CANOCO programme excluded pH, electrical conductivity, aluminium, and manganese from the CCA for SD 0-20 cm because they exhibited negligible variance (Figure 5). Silt-clay was also eliminated from the analysis because the programme detected collinearity when fitting the variables. *Acacia hebeclada, Gardenia volkensii*, S13, S14, S15 and sand fraction were not displayed in the diagram because they introduced polarity in the data points. This suggested that the distribution of *A. hebeclada* and *G. volkensii* were not associated with the distribution of the selected soil variable. Furthermore, it was evident that S13, S14, and S15 were not inhabited by the common plant species in the lunette dune environment. It was observed that *A. mellifera* was closely associated with sites that had high level of calcium, ECEC and organic carbon. These were sampling sites that were mostly located on the slip face of the lunette dunes. The ranking of selected soil variables at SD 0-20 cm on the basis of their significance was ECEC, organic carbon, calcium, potassium, sodium, magnesium and phosphorus in descending order of significance for plants existing in the lunette dune-pan environment (Figure 5).
Shallow solum (depth >20cm) that existed at S2, S13, S14, S26, and S27 resulted in the exclusion of the sampling sites from the CCA for SD 40-60 cm (Figure 6). S15, S23, S33, *G. volkensii, A. hebeclada*, sand, silt-clay were rejected because they polarized data points. EC, pH, aluminium and manganese indicated negligible variance and were therefore removed from the CCA. It was observed that *A. mellifera* still dominated sampling sites that were relatively fertile at SD 40-60 cm, and the distribution of other species were insignificantly influenced by the distribution of selected soil variables at SD 40-60 cm. Figure 6 suggests that up to the depth of 60 cm, *A. mellifera* had a competitive edge over *G. flava* with respect to soil nutrients. The significant soil variables at SD 40-60 cm were potassium, organic carbon, ECEC, and phosphorus (Figure 6).
In the CCA for SD 80-100 cm (Figure 7), EC, pH, aluminium and manganese were excluded from the analysis due to their negligible variance. S1, S2, S13, S14, S15, S25, S26, and S27 were not included in the CCA because the solum at the sampling sites was shallow (depth >60cm). Sand and silt-clay were also excluded from the diagram as they indicated collinearity. It is also worth noting that *G. volkensii, A. hebeclada* and *Acacia erioloba* were not displayed because they caused polarity of data points. It was observed that the density of *A. mellifera* was positively related to phosphorus and potassium in the soil and negatively related to other variables at SD 80-100 cm. On the other hand, the density of *G. flava, Ehretia rigida* and *Rhigozum trichotomum* were closely linked to the distribution of ECEC, calcium, and organic carbon in the soil. This may suggest that the competitive capacity of *A. mellifera* for soil nutrients diminished with increase in soil depth. The most significant soil variables at SD 80-100 cm were iron, phosphorus, ECEC, sodium and calcium (Figure 7).
The challenge of shallow solum associated with the pan fringes continued to cause exclusion of some sampling points in the CCA. Consequently, S1, S2, S5, S13, S14, S15, S25, S26, and S27 were excluded for SD 130-150 cm (Figure 8). *G. volkensii* and *A. hebeclada* were also eliminated from the analysis due to data polarity. Sand and silt-clay were disregarded for collinearity. EC, pH, aluminium and manganese were excluded due to negligible variance. It was noted that the number of soil variables that had positive relationships with the distribution of *A. mellifera* continued to decrease with an increase in soil depth. Only phosphorus showed positive relationship with *A. mellifera* distribution in the lunette dune-pan environment. Contrarily, *G. flava* distribution had positive relationships with ECEC, sodium and calcium in the soil. This may indicate that *G. flava* gained a competitive

advantage over *A. mellifera* for soil nutrients as soil depth increased. Significant soil variables that had impacts on plant species distribution in the lunette dune environment at SD 130-150 cm were iron, sodium, calcium and ECEC.

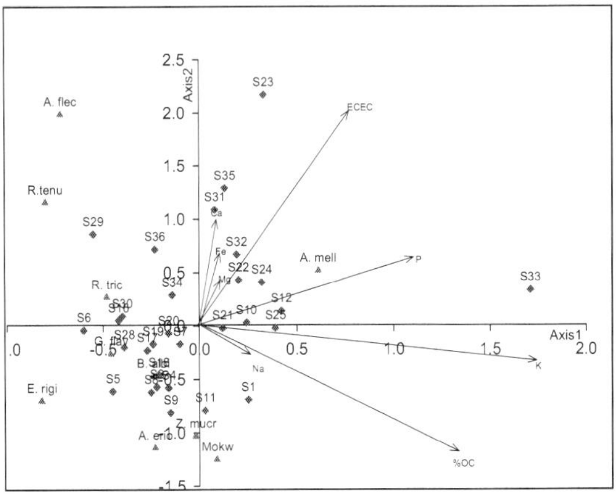

Fig. 6. Relationships between soil properties (SD 40-60 cm) and plant species distribution.

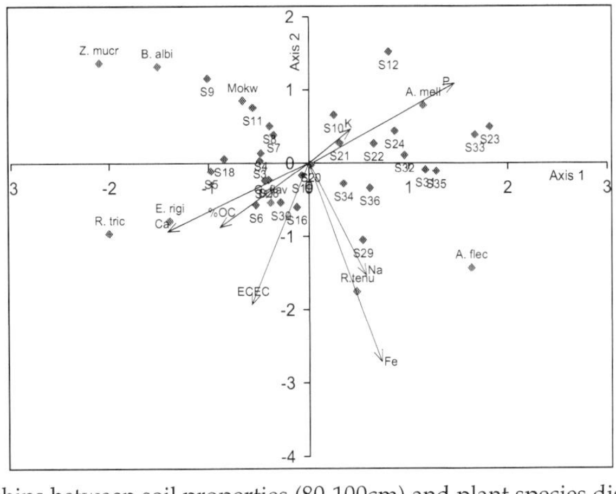

Fig. 7. Relationships between soil properties (80-100cm) and plant species distribution.

Due to the shallowness of the solum, S1, S2, S3, S4, S5, S13, S14, S15, S25, S26, and S27 were not included in the CCA for SD 180-200 cm (Figure 8). Furthermore, *G. volkensii*, *A. hebeclada*, EC, pH, aluminium, manganese, sand and silt-clay were excluded in the CCA owing to negligible variance. It was observed that the number of soil variables that had positive relationships with the distribution of *A. mellifera* still continued to decrease with an increase in soil depth. In addition, similar to SD 130-150 cm, only phosphorus indicated positive relationship with *A. mellifera* distribution in the lunette dune-pan environment. In contrast, *G. flava* distribution had positive relationships with ECEC, sodium, calcium, potassium and iron in the soil. This therefore gave impetus to the observation that *G. flava* gained a competitive advantage over *A. mellifera* for soil nutrients as soil depth increased. Phosphorus, sodium, potassium, iron, calcium and ECEC were identified as significant soil variables in relation to plant species distribution in the lunette dune-pan environment.

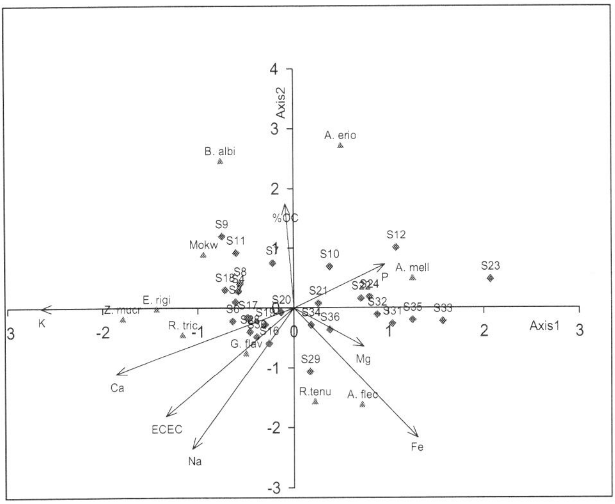

Fig. 8. Relationships between soil properties (130-150cm depth) and plant species distribution.

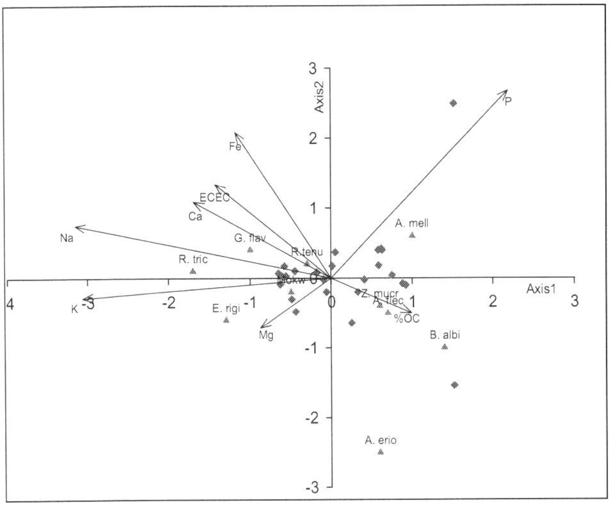

Fig. 9. Relationships between soil properties (180-200cm depth) and plant species distribution.

Axis	DCA	CCA 0-20cm		CCA 40-60cm		CCA 80-100cm		CCA 130-150cm		CCA 180-200cm	
	λ	λ	*r*	λ	*r*	λ	*r*	λ	*r*	λ	*r*
1	0.42	0.26	0.79	0.26	0.87	0.28	0.87	0.27	0.85	0.20	0.81
2	0.26	0.15	0.75	0.19	0.78	0.17	0.83	0.16	0.81	0.19	0.92
3	0.16	0.10	0.65	0.13	0.79	0.12	0.80	0.08	0.73	0.14	0.78
4	0.08	0.08	0.64	0.08	0.67	0.10	0.78	0.08	0.71	0.11	0.82

Table 4. Eigen values of the first four axes and the species-environment correlations.

The eigen values (λ) of the DCA and CCA were determined to further assess the degree to which the selected soil variables could explain plant species distribution in the lunette dune-pan environment (Table 5). The eigen value is usually referred to as the "per centage variance accounted for" (ter Braak, 1988). It always ranges from one (1) to zero (0), and the higher the value the more important the ordination axis. Furthermore, eigen values of ca. 0.3 and higher are usually common in ecological applications (ter Braak, 1988). However, an ordination diagram that explains only a low per centage of the total variance in the species data may still be informative (ter Braak, 1988). Eigen values are usually in the form of a decreasing order with values for axes 1 and 2 being larger than those of axes 3 and 4 as is the case in Table 5 which shows the species-environment correlations (*r*) and the eigen values for the first four axes. It was observed that some eigen values were lower than 0.3 (Table 5). This suggested limitations on the use of data on selected soil variables to explain variation in plant species distribution. This was not out of the ordinary as it is widely acknowledged that plant species distribution in any ecosystem is a function of numerous environmental factors, and that it is practically impossible for any scientific research to exhaustively and concurrently incorporate all environmental factors of potential significance into a particular study. Therefore, the selected soil variables were considered sufficient to comprehensively shed light on the patterns of plant species distribution in the lunette dune-pan environment in Sekoma.

4.4 Social survey

It was established that the village of Sekoma did not originate where it was currently located. The village originated in the western side (Sekoma West) of the lunette dunes and a considerable portion of the community decided to migrate to the eastern side (Sekoma) of the lunette dunes between the years 1924-1927. However, some few members of the community decided to remain in Sekoma West and they still inhabited the area at the period of this research. They indicated that there was nothing major that caused the migration. However, observations indicated that some changes in their environment instigated the migration. For instance, observation of abandoned old hand-dug wells located in the western side of the pan suggested a possible exhaustion of underground water resources at that site. The migration implied a shift in land use pressure from one side of the lunette dune-pan complex to the other. During the discussions, it became apparent that over the years the local community had amassed a wealth of indigenous knowledge with regard to the changes in their environment. The following is an account of the perceptions of the local community pertaining to the lunette dune-pan environment:

- The community perceived the existence of the lunette dunes in their environs as a natural phenomenon.

- They noticed an increase in dune size and height with simultaneous shrinkage of the pan which occurred gradually over the years.
- They recognized that the lunette dunes were not a single hammock of sand, but a dune field of distinct sand dunes. Consequently, they identified the main lunette dunes as Tae, Kebuang, Boisi, Leremela and Tshube from east to west. They were also aware of the perpetual development of some minor dunes in the area.
- Excessive wind erosion was identified as the main agent of soil transfer from the environment onto the lunette dunes. This coupled with the trees which had grown on the lunette dunes trapping the aeolian soil particles, were acknowledged as the main drivers linked with the continuous development of lunette dunes.
- They noticed a rapid increase in the height and size of the lunette dunes between the years 1985-1987 which they attributed to the severe drought that occurred in the area during that period. They pointed out that due to the drought, vegetation was devastated leaving large areas of bare land, creating conducive conditions for rapid soil erosion in the area.
- They associated the drought with bush encroachment or thickening which was evident within the lunette dune-pan environment.
- As evidence to the climatic changes that they observed in the area over a long period of time, they cited a decline in the amount of rainfall that the Sekoma area received over the years. Furthermore, they indicated that in the past, the annual rainfall was sufficient to fill the pan and that the pan was able to hold surface water for longer periods. They realized that this was no longer the case. They attributed the changes to loss of surface water holding capacity due to pan sedimentation. Pan sedimentation was associated with excessive soil erosion that continued to occur over the years causing the pan shrink. As a result, the community relied heavily on hand-dug wells located within the pan as the main source of water for livestock.
- The community indicated that the lunette dunes did not contribute significantly to productivity in pastoral farming in their area due to shortage of fodder resources in the lunette dune-pan environment.
- Finally, they pointed out that the lunette dune-pan environment was subjected to increasing land use pressure due to the increase in the population of the community and livestock in the area. However, they indicated that the situation had been aggravated by some farmers from other villages that had relocated close to Sekoma due to shortage of fodder and water in their areas.

5. Discussion

5.1 Pedo-geomorphology of the lunette dune complex

Soils in the Kalahari area are sandy grains constituted mainly by quartz and small amounts of zircon, garnet, feldspar, ilmenite and tourmaline (Wang et al., 2007; Leistner, 1967). Analysis of soil properties of the Sekoma lunette dune-pan environment did not indicate otherwise as the lunette dunes were more than 95% sandy up to the depth of 200 cm. A soil profile established in one of the lunette dunes indicated no signs of soil horizons up to the depth of 200cm. This showed dominance of sand fraction in the soil texture of the lunette dunes. Goudie and Wells (1995) and Lancaster (1978) pointed out that the deflation of sediment directly from the pan floor during dry climatic condition periods resulted in the formation of the lunette dunes in the Kalahari. Paradoxically, sandy soils were dominant in the lunette dunes compared to the fine textured soil associated with the pan floor. However, Lawson (1998) mentioned that presently sediment deflation from the pan floor was limited in Kalahari. Therefore, the observed soil texture suggested that the Sekoma pan had contributed insignificant amount of sediments to the development of the lunette dunes in the recent years. The dominance of the sand fraction also implied that the sandy environs of the surrounding area had recently contributed significantly to the sedimentation of the lunette dunes compared to the pan floor. This may in turn point to the spatial and temporal environmental changes that had occurred in the area with particular reference to changes in land use, and climatic conditions including, *inter alia,* direction of wind flow, rainfall patterns, increase in livestock population and occurrences of veldt fires.

Correlation analyses indicated that most of the relationships between soil and geomorphological variables were not statistically significant in the three selected lunette dunes. This suggested that the geomorphological properties, particularly the dune slope did not have an influence in the distribution of selected soil variables in the lunette dune-pan environment. Furthermore, lack of distinct patterns in the distribution trends of the selected soil variables in the lunette dune-pan environment pointed to the existence of considerable spatial heterogeneity in the soil resources distribution in the environment. Similar findings in relation to soil resources distribution in arid zones elsewhere have been cited (e.g. Wang et al., 2007; Wezel et al., 2000). Heterogeneity in soil resources in arid regions has often been attributed to the existence of resources islands that normally form under shrub canopies (Wang et al., 2007; Wezel et al., 2000). The islands usually represent micro-sites of favourable conditions for plant growth (Wang et al., 2007; Dhillion 1999).

5.2 Vegetation of the lunette dune complex

Two main plant communities that inhabited the Sekoma lunette dune-pan complex were dominated by *G. flava* and *A. mellifera*. The *G. flava* community occupied the wind ward slopes in all the sampled dunes, but also existed at the crest in the Leremela lunette dune. The *A. mellifera* community inhabited the slip face slope in all the sampled dunes, but also existed at the pan fringes in Leremela and Kebuang and at the crest in Kebuang lunette dunes. The density of *A. mellifera* was higher close to the village as compared to further afield. Leremela and Kebuang lunette dunes were the closest lunette dunes to the settlement area of Sekoma. Furthermore, the hand-dug wells used for livestock watering were located closer to Kebuang lunette dune as compared to the other two lunette dunes. Consequent to this was the evidence of pronounced land use pressure footprints on Kebuang lunette dune. Bush encroachment species predominated by *A. mellifera* was one of the prominent land use pressure footprints in the lunette dune-pan environment. Hence, land use was identified as one of the significant factors that influenced environmental changes, particularly the distribution of plant species and community composition, in the Sekoma lunette dune-pan environment.

The dominance of *A. mellifera* in the lunette dune-pan environment was indicative of the competitive capability of *A. mellifera* in areas that were subjected to intense land use pressure. The abundance of *A. mellifera* under conditions similar to that of the study site has been linked to the species morphological features which enhance its establishment and survival when subjected to harsh environmental conditions (Moleele, 1999). For instance, in spite of the high nutritive value associated with the species, its thorny nature makes it less susceptible to browsing by livestock (Tolsma et al., 1987). Similar studies conducted elsewhere have indicated that species that were more resistant to browsing were normally found in abundance closer to the 'foci-point', which could either be a water source or settlement area (Perkins & Thomas, 1993; Moleele & Perkins, 1998; Moleele, 1999).

Fig. 10. High livestock density in the Sekoma pan with lunette dunes on the background.

Herbaceous cover in the area was non-existent during the time of sampling (Figure 11). It may be argued that this could be linked to the sampling period as it was conducted at the beginning of the rainy season. However, the Kalahari communities have intrinsic inclination towards keeping cattle over small stock. On the other hand, the physiological constraints of cattle limit their movements from their water source (Moleele & Perkins, 1998; Moleele, 1999). Consequently, cattle spent most of their time within the lunette dune-pan environment (Figure 10) close to their water sources. In light of this, the intensity of land use, particularly pastoral farming, was identified as the primary contributing factor to lack of herbaceous cover in the lunette dune-pan environment. In fact, similar researches conducted elsewhere (e.g. Skarpe, 1986; Ringrose et al., 1996; Moleele & Perkins, 1998; Moleele, 1999) have indicated that the development of bare land patches is often caused by overgrazing and trampling due to high livestock density. This condition facilitated the predominance of species like *A. mellifera* and *G. flava* which have innate ability to adapt to hostile environmental conditions through their competitive edge over others (Skarpe, 1990; Moleele & Perkins, 1998; Moleele, 1999) leading to bush encroachment or thickening in the lunette dune-pan environment.

Browse resources contribute significantly to livestock feed in environments where grazing resources are limited (Scholte, 1992; Moleele, 1999). Hence, the scarcity of grazing resources in the lunette dune-pan environment compelled livestock to heavily depend on browse resources. Scholte (1972) and Moleele (1999) indicated that the establishment and survival of woody species is determined by their survival mechanisms against browsing pressure. In view of this, plant species that had the capacity to withstand browsing pressure (*A. mellifera* and *G. flava*) became dominant in the lunette dune-pan environment over the years as land use pressure increased. Therefore, the phenomenon of environmental changes characterized by the development of imprints of selectivity of livestock on browse resources was inevitable in the lunette dune-pan complex.

Fig. 11. Common bare ground condition in the lunette dune-pan environment.

5.3 Local community perceptions on environmental changes

The social survey provided evidence of a wealth of indigenous knowledge that had been accumulated through informal observations and experiences by the local community. The community perceived wind as the main agent transporting soil particles from the pan and the environs onto the lunette dunes. The perceptions also indicated that the lunette dunes and the plants that grew thereon served as barriers that trapped the aeolian soil particles and lead to continuous process of dune development. The perceptions had considerable overlap with findings from empirical research (e.g. Lancaster, 1978b).

The community perceived the lunette dune-pan environment as an important water source pertinent to their pastoral farming activities. However, it was evident that potential developments in the area of pastoral farming were bedevilled by lack of grazing resources which was a major concern for the community. It was indicated that lack of grazing resources in the area was mainly caused by environmental changes that were characterized by an increase in the livestock population and a decline in the annual rainfall. Therefore, livestock grazing was perceived to be insignificant in the lunette dune-pan environment, hence the lunette dunes were considered insignificant in relation to fodder provision in Sekoma. However, field observations indicated that in spite of the changes in the environment, the lunette dune complex continued to contribute substantially in fodder provision over the years mainly through browsing resources that they sustained.

6. Conclusion

Changes in land use patterns as well as its intensity had affected the lunette dune-pan complex and continue to cause significant spatial and temporal environmental changes in the Sekoma area. The general changes in the climatic factors over the years had influenced changes in the land use patterns, and also contributed to environmental changes observed in the area. The predominance of bush encroachment species, particularly *A. mellifera* was evidence of the precedence of land use intensity over other drivers of environmental changes. The establishment of a sustainable environmental management strategy that could mitigate against the impacts of major drivers of environmental changes in the area was therefore necessary. The fact that the Sekoma community exhibited a wealth of indigenous knowledge in relation to the environmental changes taking place in the lunette dune-pan complex was desirable from the sustainable environmental management perspective. The findings of this study, concomitant with the indigenous technical knowledge of the Sekoma community could therefore form the basis upon which sustainable environmental management planning for the Sekoma lunette dune-pan complex could be established to facilitate natural resources and ecosystem conservation. Furthermore, attention of scientists who conduct their research works in arid environments has been drawn to the need for special consideration of lunette dune-pan complexes that normally exist as interspersed micro-ecosystems in arid environments. More studies are therefore essential to further elucidate environmental changes and ecosystem dynamics of lunette dune-pan micro-ecosystems in arid and semi-arid zones globally. This is particularly important in view of the empirical research observations (e.g., Chanda et al., 2003; Mosweu, 2008; Mosweu & Areola, 2008) which indicated that the livelihoods of most communities living in arid and semi-arid zones revolve around the sustainability of lunette dune-pan micro-ecosystems.

7. References

Bhalotra, Y.P.R. 1985. Rainfall maps of Botswana. Gaborone. Department of Meteorological Services.

Bonham, C.D. 1989. Measurements for Terrestrial Vegetation.Wiley, New York.

Chanda, R., Totolo, O., Moleele, N., Setshogo, M., & Mosweu, S. 2003. Prospects for subsistence livelihood and environmental sustainability along the Kalahari Transect: The case of Matsheng in Botswana's Kalahari rangelands. *Journal of Arid Environments 54, 425-445.*

Cooke, R., Warren, A., & Goudie, A. 1993.Desert geomorphology. UCL Press, London

Department of Surveys and Mapping .2001. Botswana National PC Atlas 1.0C. Gaborone, Botswana.

Dhillion, S.S. 1999. Environmental heterogeneity, animal disturbances, micro site characteristics, and seedling establishment in a *Quercushavardii* community. *Restoration Ecology 7, 399-406.*

Gauch, H.G. 1982a. Multivariate analysis in community ecology.Cambridge University Press. England.

Geological Survey Department. (1995). Groundwater pollution vulnerability map of Republic of Botswana. Gaborone, Botswana: Government Printers.

Goldsmith, F.B. & Harrison, C.M. 1976.Description and analysis of vegetation. In: S.B Chapman. (Ed.). Methods in plant ecology.Backwell Scientific Publications, Oxford, UK.Pg 85-155.

Goudie, A.S. & Thomas, D.S.G. 1986.Lunette dunes in Southern Africa.*Journal of Arid Environments 10, 1-12.*

Goudie, A.S., & Wells, G.L.1995. The nature, distribution and formation of pans in arid zones.*Earth Science Reviews 38, 1-69.*

Hendershot, W.H. & Duguetts, M.1986.A simple barium chloride method for determining cation exchange capacity and exchangeable cations.*Soil Sci. Soc. Am. J. 50, 605-608.*

Hill, M.O. 1979. TWINSPAN: A Fortran Program for Arranging Multivariate Data in an Ordered Two-Way Table by Classification of the Individual and Attributes. Cornell University, NY.

Holmgren, K. & Shaw, P. 1996. Paleoenvironmental reconstruction from near-surface pan sediments: An example from Lebatse pan, southeast Kalahari, Botswana. *Geogr. Ann., 79A (1-2), 83-93.*

International Soil Reference and Information Centre. 1993. Procedures for soil analysis. 4th Ed. Wageningen, Netherlands.

Janzen, H.H. 1993. Soluble salts. In: Cater, M.R. (Ed.). Soil Sampling and Methods of Analysis.Canadian Society of Soil Science, Lewis Publishers, London.

Krebs, C.J. 1989. Ecological methodology. Harper Collins, New York.

Lancaster, I. N. 1978. The pans of Southern Kalahari, Botswana.*Geographical Journal 144, 81-98.*

Lancaster, I. N. 1986. Grain-size characteristics of linear dunes in the southwest Kalahari.*Journal of Sedimentology and Petrology 56, 395-400.*

Lancaster, I. N. 1989. Late Quaternary Palaeoenvironments in the southwestern Kalahari. *Palaeogepgraphy, Palaeoclimatology, Palaeoecology 70, 367-376.*

Lawson, M. 1998. Environmental change in South Africa: A luminescence-based chronology of late Quaternary lunette dune development. Unpublished PhD Thesis, University of Sheffield. In: Carr, A.S., Thomas, D.S.G. & Bateman, M.D. 2006. Climatic and sea level controls on late Quaternary eolian activity on the Agulhas Plain, South Africa. Quaternary Research 65, 252-263.

Leistner, O.A. 1967. The plant ecology of the southern Kalahari.Botanical survey memoir No 38. The Government Printer, Pretoria.

Maestre, F. T., Bautista, S., & Cortina. J. 2003. Positive, negative and net effects in grass-shrub interactions in Mediterranean semiarid grasslands.*Ecology 84, 3186-3197.*

Marker, M.E. & Holmes, P.J. 1995. Lunette dunes in the northeast Cape, South Africa, as geomorphic indicators of palaeoenvironmental change. *Catena 24, 259-273.*

McDonald, D.J., Crowling, R.M., & Boucher, C. 1996. Vegetation-environment relationships on a species-rich coastal mountain range in the fynbos biome (South Africa). *Vegetatio 123, 165-182.*

Moleele, N.M. & Perkins, J.S. 1998. Encroaching woody plant species and boreholes: is cattle density the main driving factor in the Olifants Drift communal grazing lands, South-eastern Botswana? *Journal of Arid Environments 40, 245-253.*

Moleele, N.M. 1999. A review of browse: A neglected food resource for cattle in Botswana. PhD Thesis, Stockholm University, Sweden.

Monier, M.A El-G., & Amer, W.M. 2003. Soil-vegetation relationships in a coastal desert plain of Southern Sinai, Egypt. *Journal of Arid Environments 55, 607-628.*

Mosweu, S. 2008. Soil resources distribution, woody plant properties and land use in a lunette dune-pan system in Kalahari, Botswana. *Scientific Research and Essay, 3(9) 242-256.*

Mosweu, S. & Areola, O. 2008. The impacts of pan quarrying on livestock watering in a semi-arid region: Case study of Kang pan in the Kalahari, Botswana. *Botswana Journal of Agriculture and Applied Sciences, 5(1) 36-44.*

Mosweu, S., Atlhopheng, J.R. & Setshogo, M.P. 2010. Variation in the distribution of soil attributes along the lunette dune slopes. *Botswana Journal of Agriculture and Applied Sciences, 6(1) 34-47.*

Muller-Dombois, D., & Heinz, E. 1974.Aims and Methods of Vegetation Ecology. John Wiley and Sons, New York.

Perkins, J.S., & Thomas, D.S.G. 1993. Environmental responses and sensitivity to permanent cattle ranching, semi-arid western central Botswana. In: D.S.G. Thomas and R.J. Allison. (Eds.). Landscape sensitivity. John Wiley and Sons, Chichester.

Ringrose, S., Vanderpost, C. & Matheson, W. 1996. The use of remotely sensed data and GIS data to determine causes of vegetation cover change in Southern Botswana. *Applied Geography 16, 225-242.*

Scholte, P.T. 1992. Leaf litter and Acacia pods as feed for livestock during the dry season in Acacia-Commiphorabushland, Kenya.*Journal of Arid Environments 22, 271-276.*

Schwinning, S., Starr, B.I., & Ehleringer, J.R. 2005. Summer and winter drought in cold desert ecosystem (Colorado Plateau), Part II: Effects on plant carbon assimilation and growth. *Journal of Arid Environments 61, 61-78.*

Skarpe, C.1986. Plant community structure in relation to grazing and environmental changes along a North South transect in Western Kalahari. *Vegetatio 68, 3-18.*

Skarpe, C.1990. Structure of woody vegetation in disturbed and undisturbed arid savanna, Botswana.*Vegetatio 87, 11-18.*

Smet.M., & Ward, D. 2006. Soil quality gradients around water-points under different management systems in a semi-arid savanna, South Africa.*Journal of Arid Environments 64, 251-269.*

Sonnevelt, C., & van den Ende, J. 1971. Soil analysis by means of a 1:2 volume extract. *Plant Soil 35, 505-506.*

Soon, Y.K., & Warren, C.J. 1993. Soil Solution. In: M.R Cater. (Ed.). Soil Sampling and Methods of Analysis.Canadian Society of Soil Science.Lewis Publishers, London.

Ter Braak, C.J.F. 1988. Program CANOCO Manual. Ministry of Agriculture and fisheries, Agricultural Mathematics Group (DLO), Wageningen, The Netherlands

Rea, L.M. & Parker, R.A. 2005. *Designing and Conducting Survey Research: A comprehensive guide.* 3rd ed. San Francisco: Jossey-Bass.

Ter Braak, C.J.F. 1995. Ordination. In: R.H.G. Jongman, C.J.F. ter Braak, & O.F.R. van Tongeren. (Eds.). Data analysis in Community and landscape ecology. Cambridge University Press, Great Britain.

Tolsma, D.J., Ernst, W.H., &Verway, R.A. 1987. Nutrients in soil and vegetation around two artificial water points in eastern Botswana. *Journal of Applied Ecology 24, 991-1000.*

Van Reeuwijk, L. P., (1993), Procedures for Soil Analysis. International Soil Reference and Information Centre Technical Paper No.1.

Wang, L., D'Odorico, P., Ringrose, S., Coetzee, S., & Macko, S.A. 2007. Biogeochemistry of Kalahari sands. *Journal of Arid Environments, 71(3): 259-279*

Wenzel, A., Rajot, J.L.,& Herbrig, C. 2000. Influence of shrubs on soil characteristics and their function in the Sahalian agro-ecosystems in semi-arid Niger. *Journal of arid environments 44, 383-398.*

13

Late Quaternary Environmental Changes and Human Interference in Africa

Wolfgang Römer
Department of Geography, RWTH (University) Aachen, Germany

1. Introduction

The African continent has been subjected to several changes in its environmental conditions in the past. These changes have affected vegetation patterns, soil development and earth- surface processes. Repeated change has caused the development of a complex pattern of inherited features in the present-day landscape that regulate its susceptibility towards modern change in environmental conditions. Since the late Pleistocene and early Holocene, human interference in ecodynamics has increased dramatically. Humans have been altering the environment since they first controlled fire and invented agriculture. However, the exponential growth of population in the last 100 years has brought with it an accelerated rate of landscape degradation. The superimposition of anthropogenous sources of interference and climatic factors has often changed the type and intensity of earth-surface processes. This results in an imbalance and often triggers an array of self-reinforcing processes. These processes operate on different spatial scales and in different time frames and are discussed in chapter 2.

In many areas of Africa, intensified use of land has induced serious soil erosion. Particularly in the semi-arid tropical and subtropical zones of Africa, soil-erosion processes are supported by the variable nature of rainfalls, the strong seasonal contrasts in the availability of moisture and the poor vegetation cover and soils and sediments, which are characterised by a high level of erodibility. Chapter 3 provides an attempt to summarise some of the processes and impacts which are associated with soil erosion in Africa. Extreme events played and play an important role in the African morphodynamic system and may pose a threat to humans. The spatial and temporal distribution of extreme events and factors which determine the magnitude, frequency and the impact of such events are discussed in chapter 4. The increasing demand for arable land has resulted in the enlargement of those areas affected by biomass burning. Chapter 5 provides an overview of the impact of savanna fires on the vegetation and the emission of greenhouse gases into the atmosphere.

The objective of this paper is to present a synthesis of the recent research on the influence of human interference on earth-surface processes and the differing reaction paths in the African landscapes.

2. Environmental changes in Africa

2.1 Long-term environmental change

In the course of the Cenozoic period, the African continent experienced several phases characterised by very different environmental conditions. On a time scale of 10^6 to 10^8 years, these changes were associated with the break-up and fragmentation of the ancient continent Pangaea into individual continents in the Mesozoic. The plate tectonic motion of the continents initiated a number of different processes, including the drift of continental masses into polar areas, the uplift of the Tibetan Plateau, the formation of new mountain belts and the establishment of a new ocean circulation system. These changes resulted in a climatic system that was completely different from that of the Cretaceous and early Cenozoic periods (Haq, 1981; Seibold and Berger, 1995; Goudie, 1999; Skinner and Porter, 2000). In the Neogene, the new position of the continents supported the growth of the Antarctic ice sheet and the drop in sea-surface temperatures. This induced a trend towards drier climatic conditions. In the Quaternary the changes culminated in relatively rapid climatic fluctuations. The progression of different climates brought with it changes in the vegetation cover and in the denudation rates. Further processes which influenced the Cenozoic evolution of Africa are slow epirogenic crustal movements, which were responsible for the development of large basins and swells (Summerfield, 1999, Römer, 2004). Periods of local uplift produced elevated continental margins, and intense rift processes promoted intense volcanism and block faulting ((Petters, 1991; Summerfield, 1999). The cumulative effects of these processes ranged from the development of new drainage systems, the rejuvenation of old erosion surfaces, to the development of uplifted and highly dissected plateaux along the continental margins, and the relatively young volcanic areas and uplifted block-faulted mountain zones along the rift valleys in eastern Africa. The different landscapes tend to respond in different ways and at different rates of environmental change. The distinctive response of the landscapes and geomorphic forms, however, depends not only on the lithological and structural conditions in the different geotectonic domains. The strength, propagation and prolongation of the response is also modulated by the different coupling strength between hillslopes, major river systems and oceans (Wirthmann, 2000; Römer, 2012).

Africa encompasses rain forest, savanna, desert and Mediterranean environments. The present pattern of bioclimatic zones is the result of the climatic changes that occurred after the Pleistocene. However, large areas of Africa are covered with sediments and soils that are derived from the Pleistocene period. The sediments and soils are an integral part of the present ecosystem and exert influences on the rate at which earth-surface processes progress, the physico-chemical processes in the soils and the distribution of the vegetation.

2.2 Climatic change in the last millennium

Even within a timescale of 10^2 to 10^3 years, the environmental conditions in Africa are highly variable. In southern Africa, the climate was as warm or warmer from 900 to 1300 (medieval warm period) than at present, but became colder than present from 1300 to 1810 (Tyson and Partridge, 2000). The transitions from the medieval warm period to the period of the "Little Ice Age" (1300 to 1850), and from the end of the "Little Ice Age" to the recent period are well documented in southern Africa. Historical reports, studies of lake levels and dendroclimatological analyses show that it is likely that some of the climatic changes at the end of the "Little Ice Age" occurred synchronously in the northern and the southern hemisphere. According to Nicholson (1999, p. 69), a trend towards increasing dryness is indicated in the droughts that occurred from the 1780's until the 1830's for the northern and the southern hemisphere. However, in the Sahelian zone, the droughts appear to have lasted for two decades whilst in southern Africa they lasted only for a few consecutive years (Nicholson, 1999, p. 80). In southern Africa, the wet to dry conditions are related to tropical easterlies and westerly disturbances. The strengthening of the tropical easterlies is associated with warm, wet spells and high rainfall amounts in the summer rainfall regions of southern Africa. Conversely, dry spells result from the more frequent westerly disturbances (Tyson and Partridge, 2000). According to this model, atmospheric adjustments resulting in cooler conditions imply a decrease in rainfalls over summer rainfall areas of southern Africa whilst a warming caused by adjustments in the tropical circulation is associated with a general increase in rainfalls. In the Sahelian zone, the process of aridification during the transition from the eighteenth to the nineteenth century appears to have been related to a later northward advance of the ITCZ. This implies a shorter rainy season in the Sahel but wetter conditions on the coast of Guinea (Nicholson, 1999). Based on reports of prevailing winds and hydrological records, an earlier advance of the ITCZ seems to explain the wetter conditions in the Sahel and West Africa during the seventeenth and eighteenth century (Nicholson, 1999, p.69). However, even these wet periods were interrupted by severe droughts, which lasted one or two decades (Nicholson, 1999). Some pronounced events appear to coincide with global climatic trends in the atmospheric circulation pattern that are associated with the "Little Ice Age". The severe droughts of the 1820's and 1830's in Africa correspond with the last intensification of the "Little Ice Age" and may be related to a global period of anomalous circulation (Nicholson, 1999). The cooler and drier conditions in Southern Africa during the "Little Ice Age" are clearly documented from several tree ring analyses (Tyson and Partridge, 2000).

More recent changes are associated with links between sea surface temperatures and rainfall amounts over tropical Southern Africa. Studies of Richard et al. (2001) imply a higher likelihood between El Niño events, higher water temperatures in the Indian Ocean and reduced rainfall amounts over tropical Southern Africa since 1970. A similar direction of change is expected by Landman and Mason (1999) for Namibia and South Africa. According to their investigation, wetter conditions in north-eastern Southern Africa and northern Namibia tend to be associated more often with warm events in the Indian Ocean, whilst prior to the late 1970's, there was a stronger correspondence between warm events in the Indian Ocean and dry conditions over Namibia and South Africa.

2.3 Environmental change and human interference

In Africa, the superimposition of environmental changes and human interference is not a recent phenomenon. In the savannas, humans appear to have used fire for over 1.5 million years (Gowlett et al., 1981, Goldammer, 1993). Colluvial deposits containing artefacts of the late Pleistocene and early Holocene periods point towards an increase of erosion resulting from deterioration of the vegetation cover (Lewis, 2008) caused by human interference. Extensive woodland clearance is also documented from Tanzania, where charcoal production for iron smelting in the last 900 years has led to an increase in soil erosion (Schmidt, 1997, Eriksson et al., 2000). Agriculture and fires that were ignited by humans appear to have played a key role in the development of the vegetation pattern and in the composition of the plant communities in most parts of the savanna areas.

The highly variable environmental conditions in Africa have, at all times, increased the pressure to expand the utilisation of land into areas which do not support large populations on a subsistence basis. The semi-arid tropics of Africa, in particular, have been subjected on several occasions to marked environmental degradation (Seuffert, 1987, Mensching, 1990). As the amount of summer rainfall in these areas determines the quantity of forage and crop yield, rainfall amounts also determine the economic base of the human population. An example of the complex interaction of the various processes is the Sahel zone, where the rapid growth of the population over the last century coincides with an enlargement of areas for agriculture (Seuffert, 1987). Varying rainfall amounts and consecutive years with low rainfall amounts in conjunction with intensified cultivation methods, vegetation clearance and increased livestock husbandry resulted in a degradation of many areas. In the Sahel zone, the carrying capacity of the land was exceeded and the southward expansion into more humid areas caused further environmental degradation (Mensching, 1990).

3. Soil erosion

3.1 Factors contributing to soil erosion

Land degradation resulting from inappropriate cultivation practices, high grazing intensities and clearance of the vegetation is often associated with an increase in erosion by water and wind. In the tropical and subtropical areas of Africa, the effects of soil erosion appear to be correlated with a decline of the productivity of the cultivated land and of the per capita ratio of cultivated area (Beckedahl, 2002). Soil erosion may induce irreversible damage to arable land, and tends to produce ecologically unstable landscapes and socioeconomic problems (Scoones et al., 1996). Human interference contributes to soil erosion in a direct and indirect way by modifying the topography for buildings, clearance of vegetation for pasture and cultivation, or by compacting the soil by the use of machines. Although the extrapolation of anthropogenously induced soil loss over longer periods remains a challenge, most authors agree that land degradation by soil erosion has affected large areas of Africa (Reading, et al., 1995; Scoones et al., 1996; Valentin et al., 2004; Bork, 2006; Dahlke and Bork, 2006; Nyssen et al., 2004). Beckedahl (2002) states that about 85% of the area of Africa north of the equator is potentially endangered by soil erosion and that the area of the arable land has decreased from 0.3 ha/person (1986) to about 0.23 ha/person (2000). A further reduction of the arable area per person to 0.15 ha is predicted for the year 2050 (Beckedahl, 2002, p.18).

The process of soil erosion is a function of a number of interrelated factors. These include the climatic conditions, the soil, the relief, the density and type of the vegetation and the land use and agricultural techniques. The on-site effects of soil erosion range from soil loss to a decrease in nutrients in the soil, to changes in the water balance and runoff. Off-site effects are the pollution of fresh water by delivering eroded, nutrient and heavy metal-laden sediments to rivers and lakes (Zachar, 1982, Reading et al., 1995, De Meyer et al. 2011).

Soil erosion is not only an African problem, but environmental and socioeconomic conditions provide a specific set of factors which appear to be different from those of other continents. Extensive areas of Africa encompass pericratonic and cratonic terrain-types. These areas are characterised by old landscapes, which presumably originated in the late Mesozoic or early Cenozoic periods (Wirthmann, 2000; Römer, 2007). Deeply weathered rocks, escarpments and deeply incised valleys with steep valley side slopes feature high susceptibility to soil erosion and slope failure. In some of these areas, weathering mantles and soils have survived for millions of years. This has supported the depletion of the weathering mantles and the development of sandy materials that are prone to erosion processes (Dingle et al., 1983; Areola, 1999; Partridge and Maud, 2000). In semi-arid areas of Africa such as the Sahel, the Sub-Sahelian regions or the Kalahari large tracts are covered by sand dunes and sediments that were mobilised during the Pleistocene. These materials are generally liable to soil crusting and erosion by both water and wind (Valentin, et al., 2005).

High rates of soil erosion are associated with laterised weathering layers and soils, which are characterised by clay-enriched argic horizons with weak microstructure (e.g. Acrisols) or low aggregate stability (Lixisols).

Studies from several areas in Africa imply that even under undisturbed conditions natural erosion rates may exceed the rate of soil delivery (Shakesby and Whitlow, 1991; Idike, 1992, Braun et al., 2003). However, comparative measurements are rare. In disturbed areas, soil erosion rates are estimated to range from 0.5 to 110 t ha^{-1} a^{-1}, largely depending on the eksystemic and ensystemic conditions at the site, the intensity of disturbance and the measurement methods (Reading et al., 1995; Stocking, 1995; Beckedahl, 2002). These rates appear to exceed the natural erosion rates by an order or several orders of magnitude (Thomas, 1994; Reading et al., 1995). The predicted increase in soil erosion in future is considered to have devastating consequences on the soil system, the productivity of arable land and the natural habitat.

3.2 Soil erosion and environmental conditions

Soil erosion by water is the result of a combination of several processes, which increase in intensity with the amount, duration and intensity of the rainfall events. The processes involved in soil erosion include splash, interrill, rill and gully erosion. Another group of erosion processes includes gravitative processes such as landslides and erosion by wind.

Soil erosion in Africa is not a phenomenon of a distinctive physiographic zone, though some factors support soil erosion. Soil erosion has been documented from nearly all physiographic zones of Africa (Rapp, 1976; Elwell and Stocking, 1982; Biot, 1990; Mensching, 1990; Bork, 2004). A characteristic feature of the tropical and subtropical regions of Africa is that rainfall varies in amount, intensity and structure within a year, and between years and decades (Seuffert, 1987; Hulme, 1999; Tyson and Partridge, 2000). In semi-arid tropical areas, the decrease in the annual rainfall is often associated with an increase in the intensity and variability of rainfall events and a decrease in vegetation cover of natural and cultivated areas (Seuffert et al., 1999). While rainfall intensity provides the energy for soil erosion, the vegetation cover controls the amount of rainfall that reaches the surface. Important parameters are the high spacing and the structure of the vegetation, the density of the under storey vegetation, and the ground cover. A change in the vegetation cover may result in a marked increase in the rate of soil erosion. According to Lal (1998), rainfall intensity determines the distribution of raindrop sizes while interception is a function of the total rainfall amount and of the high spacing and the structure of the vegetation cover. A decline in the under storey vegetation and the ground cover is associated with a decrease in the litter at the soil surface. This causes a decrease in surface roughness. The decline of the under storey vegetation results in a change in the size distribution and the terminal velocity of raindrops. This involves an increase in the impact energy of the raindrops as a function of the height of the canopy. The decreased roughness at the soil surface and the increase in the impact energy of raindrops promotes splash erosion and the generation of overland flow. A further effect of a decline in the vegetation cover is the depletion of organic matter at the soil surface. This enhances the physical, chemical and biological degradation processes in the soils and results in a reduced stability of the soil structure. The reduced stability of the soil structure increases the susceptibility to soil-crusting, which, in turn, affects the runoff production and susceptibility to soil erosion (Valentin et al., 2005). The decrease in the density and depth of the roots weakens the mechanical reinforcement of the soils as the root-binding effects and the apparent cohesion are reduced (Greenway, 1987).

As soil erosion is caused by individual rainfall events, the total amount and the temporal distribution of rainfall intensity are more important than the mean annual rainfall. The rate of overland flow generation depends on the infiltration rate which is a function of the physical characteristics of the soil, the vegetation cover, the relative relief, the slope gradient, the roughness of the surface and the moisture content of the soil (Bork, 2004). Susceptibility of soils to soil erosion is a results of several interacting components. At microscale level, these include the content of organic matter and physical properties such as grain-size distribution, mineralogical composition, water-holding capacity and shear strength (Zachar, 1982; Grabowski et al., 2011). The amount of water-stable aggregates appears to control the generation of overland flow and the detachability of soil aggregates (de Vleeschauwer et al., 1978). Raindrop impact during high-intensity rainstorms can have profound effects on soils with silty and clayey composition. The impact of large raindrops promotes the compaction of the soil. Small particles that have been moved by the impact block the soil pores and air is imprisoned in the pores. This impedes the infiltration of water into the soil. Even small changes in the textural composition of the surface soil can induce change in susceptibility to erosion. In the Sahel zone, the deposition of dust and the colonisation of the soil surface with blue-green algae during fallow periods promoted the development of soil crusts (Valentin et al., 2004).

Soils and deposits such as colluvium that are characterised by a high exchangeable sodium percentage are highly prone to soil erosion. Susceptibility to erosion is associated with highly expansive clays (Botha and Partridge, 2000; Grabowski et al. 2011). Clay minerals tend to adsorb more water at a high sodium-adsorption-ratio and water infiltrating between the clay units causes an expansion of the clays. Consequently, the clay minerals are pushed apart. The expansion reduces the attraction between the clay particles. Dispersive soils are prone to piping processes, when the seepage water causes the development of subsurface drainage conduits (Bryan and Jones, 1997). Piping may support the development of slope failures by undermining the slope base and the collapse of pipe roofs is frequently associated with an increase in gully erosion (Heinrich, 1998; Singh et al., 2008).

On a hillslope and drainage basin scale, the response to a change in the eksystemic components is modified by the steepness of the hillslopes, the coupling strength of hillslopes to major rivers and by the density and degree of development of the drainage net. On a regional scale lithological and structural controls, neotectonic activities and rainshadow effects caused by large escarpments or mountain chains may exert considerable influence on the rate of soil erosion. Recent studies of Fubelli et al. (2008) on the Ethiopian highlands indicate that increased rainfalls and neotectonic activity are likely to be responsible for the high rates of river incision and the frequent occurrence of landsliding, and Singh et al. (2008) emphasise the association between palaeolandslides and active seismic zones in the KwaZulu-Natal area of South Africa.

A consequence of the high number of interacting factors is that erosion rates are highly variable and tend to vary even between areas which are characterised by identical structural, lithological and geomorphological settings within the same bioclimatic zone. Accordingly, this results in a highly variable response to changes in the eksystemic components, which often obscures the distinction between the effects of human interference and naturally induced fluctuations.

The methods for predicting the rates of soil loss range from the extrapolation of test plots to the calculation by empirical and theoretical formulations. Of the latter, the "Universal Soil Loss Equation" (USLE) or related formulations of the soil erosion process are used (Stocking, 1995, Seuffert et al., 1999). However, the determination of erosion rates applying these methods remains a challenge. Beckedahl (2002) mentions a prediction error of 55% by applying the USLE on African soils. Apart from the complex interaction of factors which has to be determined on test sites, the inclusion of erosive rainfall events in soil erosion formulations remains a problem. Indices of rainfall, such as mean rainfall, wettest month or other indices, which are derived from rainfall data are rarely capable of predicting soil-erosion events and may be misleading (de Ploey et al., 1991). Methods based on the determination of magnitude-frequency relationships of individual rainfall events that are beyond the threshold of erosive rains may provide an alternative (de Ploey et al., 1991). Such methods enable the determination of the likelihood of erosive rains and provide information on the cumulative effects of individual rainfall events. This information makes it possible to deduce the Cumulative Erosion Potential (CEP) (de Ploey et al., 1991). The CEP-Index is based on the magnitude-frequency concept of Wolman and Miller (1960). According to this concept, the impact of extreme events is compensated by its lower frequency whilst the cumulative effect of more frequent events of a certain magnitude results in higher output. Figure 1 shows the magnitude-frequency relationship and the CEP-Index (table 1) for some stations in Lesotho, Kenya and Zimbabwe. The magnitude-frequency relationship and the CEP have been calculated from data provided by de Ploey et al. (1991), Calles and Kulander (1996) and Römer (2004). At stations where recurrence intervals of erosive rains are very short, the CEP-Index indicates a high potential of soil erosion. However, even if the concept of the cumulative effects of discrete rainfall events provides a reasonable approach to erosion events, the problems involved in a numerical calculation of the complex repercussions between seasonal effects, vegetation growth periods, rainfall structure and short-term clusters of intense rainfall events require further research.

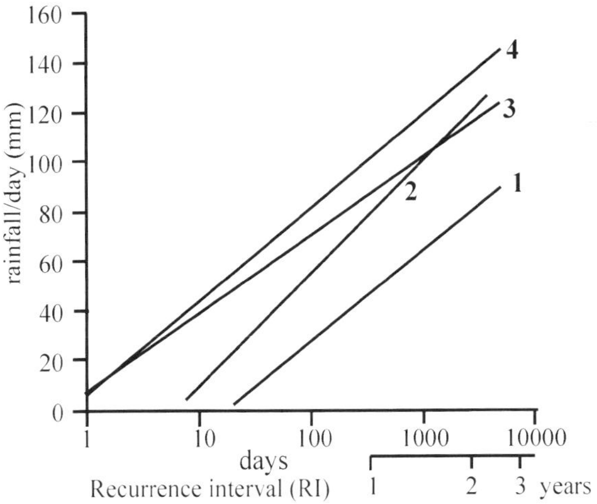

Fig. 1. Magnitude frequency relationship for stations in Zimbabwe, Lesotho and Kenya. 1 - West Nicolson (Zimbabwe) y = 36.46 log RI - 46.17; mean annual rainfall 579 mm (Römer, 2004, p. 21). 2 - Harare (Zimbabwe) y = 48.48 log RI - 32.07; mean annual rainfall 867mm (Römer, 2004, p.21). 3 - Machakos (Kenya) y = 15.85 log RI + 64.2; mean annual rainfall 1050mm (de Ploey et al., 1991, p. 404). 4 - Leribe (Lesotho) y = 26.09 log RI + 48.34: mean annual rainfall 795mm (Calles and Kulander, 1996, p. 163).

h^*	*Harare* q = 153506	*Machakos* q = 9890	*Leribe* q = 34358
0	26061	189306	73091
10	18939	100485	49822
20	14031	53235	33940
30	10189	28117	23135
40	7473	13779	15770

CEP = m! βm exp((α– h*)/β) λ; λ = exp((ln m! + m ln ß + h*/β) – h*/β)
The constants α and β are calculated from regression analysis according to the equation $y = \alpha + \beta x$. The constants correspond to the constants in the equations of the magnitude-frequency analysis in figure 1.
Constants: Harare α = -32.07; β = 48.48; Machakos α = 64.2; β = 15.85; Leribe α = 48.34; β = 26.09.
The CEP was calculated with m = 2.5 (silty to sandy soil) and different values for h*, which is a parameter for water storage. h* ranges from 0 to 10 on bare soils to 100 in areas with dense vegetation cover (de Ploey et al., 1991, p. 407, 408); q = dominant sediment transport amount. The CEP has been determined from the magnitude-frequency relationships published in de Ploey et al., (1991, p. 406); Calles and Kulander (1996; p.163), and Römer (2004; p. 21).

Table 1. Cumulative Erosion Potential (CEP) according to de Ploey et al., (1991). The CEP has been calculated for stations in Zimbabwe (Harare), Kenya (Machakos) and Lesotho (Leribe).

3.3 Soil erosion and human interference

In recent decades, intensified agriculture, livestock husbandry, clearance of forests and the increased density of settlements have contributed to the enlargement of areas affected by soil erosion. In several parts of Africa, human interference has accelerated natural erosion processes to a degree that influences the economics of extensive regions. The effects of human disturbance are generally most pronounced in hilly and mountainous terrains where steep hillslopes and high relief are conducive to high levels of erosion and a rapid response. In a study on the effects of interrill and rill erosion, Kimaro et al. (2008) demonstrated that soil loss due to deforestation and cultivation in the Uluguru Mountains of Tanzania exceeds 200 t ha^{-1} a^{-1}. The high degree of soil loss results from the steepness of the slopes, the high rainfalls but is also a consequence of continuous shallow and fine cultivation and tillage practices (Kimaro et al., 2008, p.42). In the Ethiopian highlands, changes in land use induced gully enlargement and gully incision. This resulted in a lowering of the groundwater. A concomitant effect was the decrease in soil moisture which was associated with a decline of crop yield (Nysson et al., 2004).

However, even in areas with low relief, the effects of slope gradient on soil erosion are noticeable. In an investigation conducted over a period of six years, soil erosion on maize-covered fields in Zimbabwe, Hutchinson and Jackson (1959) observed an average increase in soil loss of 3.1 t ha^{-1} a^{-1} at a slope gradient of 1.5° to 6.7t ha^{-1} a^{-1} at slope gradient of 3.5°. In the Middle Veld of Swaziland, threshold slope gradients seem to control the gully initiation on valley side slopes, whilst differences in land use or vegetation are subordinate (Morgan and Mngomezulu, 2003).

Although slope gradient is an important factor, the decline in ground cover may cause an increase in soil loss by several orders of magnitudes (Thomas, 1994, p.143,144; Reading et al., 1995). Studies of Nearing et al.,(2005) indicate that rainfall intensity and ground cover are likely to have a greater effect on soil erosion than changes in runoff and in the canopy cover alone. High soil losses of more than 200 t ha^{-1} a^{-1} are also indicated in studies of erosion in villages and on roads and in areas where heavy machines are used (Nyssen et al., 2002, de

Meyer et al., 2011). The increased runoff on roads, unpaved roads, pathways and landing sites promotes the concentration of overland flow into rills and the development of gullies by crossgrading and micropiracy. According to a study in Uganda, the soil losses range from 34 to 207t ha^{-1} a^{-1} (de Meyer et al., 2011). Despite the small percentage of total area of only 2.2 percent, de Meyer et al. (2011) emphasise that these areas are the major source areas for sediment delivery to Lake Victoria and that the total soil loss corresponds to an erosion rate of 2.1 t ha^{-1} a^{-1} (de Meyer et al., 2011, p. 97).
Patches of bare ground may induce soil erosion even on the low sloping surfaces of the basement regions of the African savannas (fig. 2). These landscapes are often characterised by a discontinuous soil cover that is interrupted by flat rock pedestals and small protrusions of bedrock. Fine-grained colluvial sediments that have been transported from the residual hills onto the gentle sloping pediments are often more prone to soil erosion than the coarser weathering products of the basement and may promote the development of large gully systems (fig. 3). However, serious and presumably irreversible effects seem to be more often the result of the interplay of several factors. This includes climatic fluctuation over a time-scale of several consecutive years or of decades, human activities and the role of inherited materials and forms in the landscape. Such conditions prevail in the Sahel, Sub-Sahelian zone, and in other transitional areas to the savannas, where extensive areas are covered with (fossil) sand dunes, sandy sediments and depleted weathering products. Highly susceptible to erosion are also savanna areas with dry seasons which last for six to eight months, where soils with a low aggregate stability or weak microstructure have been exposed by changes in the vegetation cover.

Fig. 2. Gullies formed during a heavy rainstorm at the start of the rainy season in southern Zimbabwe. Splash erosion and rill erosion affect the small slopes of the "badland" area. (Photo. Römer)

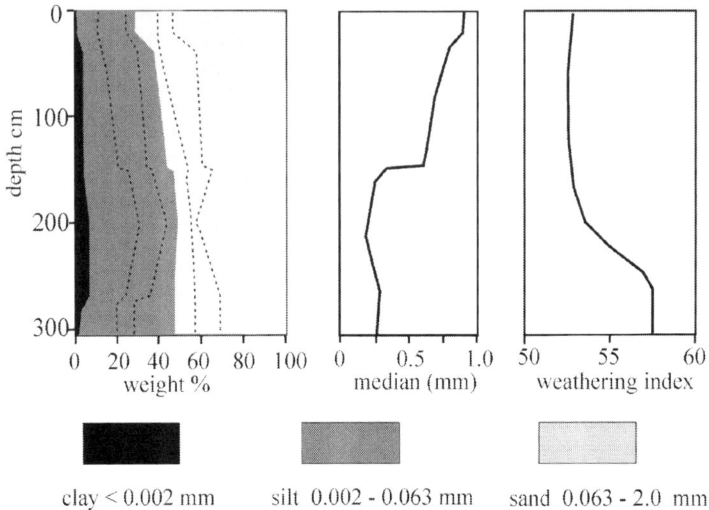

Fig. 3. Grain-size distribution, median grain-size and weathering index (CWI) of colluvium in southern Zimbabwe. The relatively fine-grained material formed the wall of a more than 3m deep gully. The decrease of the median grain-size corresponds to the increase in the weathering index and may indicate an non-conformity within the deposit (modified after Römer, 2004).

3.4 Inherited forms and materials and environmental change

Marked changes in rainfall distribution have been responsible for crisis-situations in the Sahel zone. Meteorological records of the Sahel zone show a relatively continuous decline of rainfall-levels from the 1960's to the 1970's when rainfall records attained a first minimum (Mensching, 1990; Warren, 1999). A second rainfall minimum occurred in the mid 1980's. The direct effect of these changes was an extension of areas of bare ground and a reduction in the biological diversity of the vegetation (Warren, 1999). At the same time, the variability of rainfalls increased, while there was a tendency towards a relative increase in short, high-intensity rainfall events (Gießner, 1989; Pflaumbaum et al., 1990). Studies indicate that the extension of the population and the increase in agronomic activities into a belt of formerly fixed dunes (Ooz-belt) in the west of the White Nile caused irreversible changes in the ecosystem (Mensching, 1984, 1990).

The dunes of the Ooz belt formed in the late Pleistocene and early Holocene when the climatic conditions were drier (Mensching, 1984, 1990). The development of the dune belt was interrupted by a more humid period that was characterised by greater availability of soil moisture and a denser vegetation cover. Weathering processes resulted in an encrusting of sand grains by iron-oxides. These sand layers are more resistant against wind erosion and are more impermeable than unweathered dune sands (Mensching, 1984). A further shift to dry climatic conditions provided the sands of the younger dunes of the Ooz dune belt. These dunes became inactive during the Holocene.

In the 1970's and 1980's, increased aridity, overgrazing and intense cultivation advanced the degradation of vegetation cover in the Ooz belt. This initiated the development of drifting dunes. At sites where the iron-impregnated sands of the older dunes became exposed, the lower infiltration rate increased runoff production during heavy rainfalls, which, in turn, enhanced the erosion processes (Mensching, 1984). The exposed older dune sands provide only poor soil material for further agricultural use. This caused a decline in the harvest, which was counteracted by enlarging areas for cultivation and grazing, thereby inducing a further degradation of the area. The high sensitivity of inherited sediments and soils to changes in the rainfall regime and vegetation cover is documented for several African countries.

Heinrich (1998) reports on changes in the Gongola Basin in Nigeria, where deforestation, a relatively high population density and intensified agricultural activity have caused serious erosion on Pleistocene aeolian sediments and hillwash sediments. Late Holocene (2000 BP) shifting cultivation in the Gongola Basin resulted in wash erosion on hillslopes and in the accumulation of sediments in small valleys with gallery forests, whilst the increase in population in the 20th century promoted the degradation of the savanna vegetation (Van Noten and de Ploey, 1977). Severe soil erosion resulted in a number of changes in the hydrologic regime and the nutrient balance of the soils (Heinrich, 1998). Apart from an increase in overland flow and diffuse wash erosion on low sloping surfaces, the higher runoff initiated a deeper dissection of gully-systems. The deep dissection was accompanied by a lowering of the ground water in the areas surrounding the gullies and a decline in the number of trees. High seepage gradients increased interflow and particle transport in the bedded subsoil. This resulted in the development of subsurface pipes and in increasing edaphic aridity, which strengthened the diffuse surface wash processes (Heinrich, 1998). The collapse of subsoil routes provided sites for the development of new gullies, whilst on low sloping surfaces wash erosion caused a marked decline in the clay content of the soils, which reduced the nutrient storage of the soils.

4. The role of high magnitude events

4.1 Extreme events in Africa

Climatic and hydrologic regimes on the African continent are highly variable in terms of both space and time. Rivers show the highest average extreme flood index of all continents, whilst the runoff ratios are lowest (McMahon et al., 1992). The temporal and spatial variability of rainfalls and rainstorms, as well as the repeated occurrence of periods of extreme droughts in semi-arid tropical and subtropical areas, indicate that extreme events play an important role in the African morphodynamic system. Relatively little is known about the relative work done by rare events of high magnitude when these events are compared with more frequent events with a low magnitude (Gallart, 1995). Although studies indicate that the impact of erosion increases with increasing amounts of rainfall and rainfall intensity, such relationships are not without ambiguity, as events of similar magnitude may have different effects, whilst events with a higher frequency and lower-magnitude are capable of inducing similar effects (Gallart, 1995). While the incidence of drought and rainfall events is determined by the present-day climatic system, human activities may change the magnitude of the impact by changing the vegetation cover, the hydrologic regimes and characteristics of the surface materials and forms. Therefore, an increase in the impact of smaller events with shorter recurrence intervals and lower magnitude is likely. The intensity of the response to these events is a function of the intensity of interference with the ecosystem, coupling of the subsystems and the sensitivity of the subsystems affected. However, changes in environmental conditions frequently bring with them a non-linear behavioural pattern caused by feedbacks (Thomas, 2004). These feedbacks weaken or reinforce the response to changes in different subsystems. In semi-arid areas, a decrease in vegetation cover may reinforce the decline of rainfalls as the degradation of the vegetation decreases the surface roughness and soil moisture. Consequently, the evaporation and transpiration rates decrease, which, in turn, reduces transport of vapour into the atmosphere (Warren, 1999). Further effects are likely to involve changes in the cloudiness, the spatial and temporal distribution and intensity of the rainfalls and the lower inflow of rainwater to the ground water. A decrease of the vegetation cover causes a decrease in the water-retention capacity of the soils, which, in turn, may reduce the threshold of runoff-producing storms (Gallart, 1995). The effects of such changes point towards a higher preparedness of landscape components to react to events of lower magnitude and higher frequency. This appears to bring with it an increase in the impact of events of lower magnitude. The latter is corroborated in studies of sediment yield in Kenya, which indicate that an increase in land use is associated with an increase of the relative work of events of higher magnitude (Dunne, 1979).

With respect to the impact of meteorological events on erosion-processes, the effects of continuous rainfall and short-term high-intensity rainfall events must be distinguished. Long- lasting rainfall events of exceptional magnitude determine the saturation of soils and induce saturation overland flow and liquefaction of the soil layers. High-intensity rainfalls, on the other hand, are capable of inducing Hortonian overland flow causing a rapid increase of runoff. However, the impact of such events depends strongly on the antecedent state of the ground. The role of heavy downpours increases towards the semi-arid and arid tropical areas, where daily rainfall events may exceed the mean annual rainfall by more than 40% (Starkel, 1976). Rainfall intensities ranging from 250mm to more than 400mm have been reported from Mauritania and Tunisia, and daily maximum rainfalls exceeding the annual rainfall by 50mm appear to occur several times within a decade ((Mensching et al., 1970; Starkel, 1976). Such rainstorm events are often accompanied by high discharges and floods (Starkel, 1976). Continuous rainfall events are associated with the adduction of humid air masses, which often occurs in tropical, tropical-monsoonal areas or in areas where air masses are impeded by mountains. Tropical cyclones such as the Mauritius cyclone in the Mozambique channel are also associated with high rainfall events. According to Weischet and Endlicher (2000) about 520 cyclones have been registered in 70 years, and most cyclones deposit large volumes of rainfall along the coast. An extreme event accompanied the cyclone Donoina, which occurred in the year 1984. This cyclone crossed southern Africa, and rainfall intensities achieved about 900mm in a few days. This resulted in severe flooding and intense erosion in Mozambique, Swaziland and South Africa (Goudie, 1999).

4.2 Extreme events and complex response

The response to extreme events depends not only on the magnitude of the event. Studies on flood frequencies at the Orange River in South Africa indicate that the rate of change and antecedent environmental conditions play an important role. During the last 5500 years, the lower Orange River has experienced marked changes in terms of its hydrologic regime. Zawada (2000) was able to distinguish four periods with different flood magnitudes and frequencies. Although there is a close association between high levels of discharge and warm and wet periods, the most extreme discharge events occurred during a warm interval of the "Little Ice Age", in the period 1500 to 1675 AD (Zawada, 2000). The maximum flood discharge during this brief period exceeded any historically gauged floods by a factor of three. According to Zawada (2000), the high floods cannot be attributed to the increase in rainfall, as during earlier, more humid periods the flood discharge was significantly lower, though these paleoflood discharges exceed all documented floods since the end of the 18th century. Zawada (2000) argues that the sudden onset of warming caused an intense change in the hydrologic regime. Apparently the change affected hillslopes as well as rivers within a time interval that was shorter than the time that is necessary to achieve a full adjustment of the vegetation cover to the changed conditions.

Singular events of high magnitude may result in serious damage. Rapp (1976) has documented the effects of a rainstorm in the Mgeta mountains of Tanzania,. The rainfall event achieved an intensity of 100.7mm in less than three hours and triggered more than 1000 shallow landslides in the highly weathered soils. Landsliding affected about 47% of the cultivated land, 46% of the grasslands but less than 1 % of the wooded areas (Rapp, 1976, p.92). The results highlight the link between slope stability, soil properties and changes in the vegetation cover. Trees lower the water table in the soils by transpiration and reduce the amount of rainfall reaching the slope surface as a part of the rainwater is intercepted in the canopy. Both processes counteract soil saturation and delay the development of high pore water pressure. Once deforestation takes place, these positive effects are lost. In combination with the loss of tree roots, this results in a reduced shear strength of the soils, a higher probability of high-pore water pressure and a lower threshold of stability against landsliding (Rapp, 1976).

The sensitivity to change is a further factor which appears to exert an important influence on the magnitude of events. Most landscapes in Africa have suffered progressive change through time and tend to accumulate the imprints of different environmental conditions. These imprints range from deposits and weathering layers formed during periods with different climatic conditions to hillslope forms and polyphase landscape elements. In the KwaZulu area, Singh et al. (2008) investigated extensive landslide complexes which seem to have been active in the middle and late Holocene. The volume of large individual landslides ranged from $1{\cdot}10^7$ to $2{\cdot}10^7$ m^3. Some smaller, secondary occurrences of slope failure were apparently reactivated on the larger landslide masses. However, the large landslide complexes appear to be stable. According to Singh et al. (2008), these landslides resulted from a combination of long–term rock-weathering and the location in a seismic active zone.

High-intensity rainfall events in 1987 and 1997 in Natal (Southern Africa) indicated a strong association between landsliding and colluvial deposits (Bell and Maud, 2000; Singh et al., 2008). The colluvial deposits in this area are characterised by several non-conformities resulting from differences in the intensity of weathering, the variable thickness, texture and permeability. According to Bell and Maud (2000), landslides on the hillslopes of the Natal group are closely associated with the specific behaviour of the colluvial deposits. The weathered colluvium consists of an upper sandy layer (topsoil) and an illuvial horizon, which lies above a clayey weathered layer. During heavy rainstorms, the silty and clayey layers impede the downward percolation of the water. This promotes the development of high-pore water pressure and of saturated conditions in the upper soil layers. The lateral throughflow in the more permeable layers of the colluvium and weathering layers on the upper hillslope-segments increases the flow of ground water to the middle and lower hillslope-segments. This causes the development of excess pore water pressure and artesian conditions on the lower hillslope segments, which, in turn, is accompanied by viscous flow movements and liquefaction of the soils (Bell and Maud, 2000). During the 1987 event most landslides were triggered by an extreme rainstorm episode with an intensity of 576 mm in 72h (Bell and Maud, 2000, p. 1034).

However, antecedent moisture conditions seem to play an important role, as prior to 1987 no records of larger landslide events are documented, while it is likely that rainfall events of similar magnitude have occurred several times in the past. The importance of antecedent moisture conditions and of the properties of the colluvial layers is indicated in the critical precipitation coefficients for slope failure that were calculated by Bell and Maud (2000). According to their investigations, major landslides and landslide episodes will occur when rainfall intensities exceed the mean annual precipitation by 20%. However, most landslides were triggered in the latter months of the rainy season when the colluvium was almost saturated with water. Accordingly, occurrences of slope failure in this area depend on the rainfall intensity and on the antecedent moisture conditions (Bell and Maud, 2000). On the other hand, the investigations emphasise the important role of permeability non-conformities in the colluvium and at the weathered-unweathered rock boundary. This indicates that the occurrence of landslides depends strongly on local conditions and that several factors must be kept in mind in the analysis of landslides. These factors include the association between slope parameters, mechanical parameters of the soils, rocks or sediments, and the presence of palaeolandslides.

5. The role of fires

Fires play an important role in African environments, and few areas in the African savannas appear to have ever escaped fires. In the savanna areas, fires appear to determine the volume of biomass above the ground and the turnover of herbivores and saprophytes.

Wildland fires can be induced by lightning, volcanism and rockfalls. Most fires in savanna environments are ignited in the dry season by lightning. In west Namibia, lightning ignites about 60% of the savanna fires (Held, 2006). However, in mountainous terrains, rockfalls may be also an important factor. Reports from the Cedar Hills in South Africa indicate that rockfalls contribute to the development of about 25% of the fires (Goldammer, 1993). Since the appearance of humans, the impact of fires on vegetation patterns has progressively increased. The modification of the vegetation in the savannas began in an early epoch, when hunters and gatherers used fire to make hunting easier. Evidence of the early use of fire ranges from sedimentary layers in the Swartkrans Cave in South Africa, with an age of about 1.5 Ma BP (Gowlett et al., 1981) to changes in the vegetation pattern on the Nyika Plateau in Malawi, which seems to indicate the repeated burning of the savanna vegetation at the end of the Pleistocene (Goldammer, 1993).

In more recent times, increasing demand for arable land has resulted in a regular burning of larger savanna areas and in the development of extensive grasslands. Within the moist-savanna-zone, this has caused the development of "derived savannas", which consist of grasslands and are a secondary vegetation formed by fires (Goldammer, 1993; Schultz, 2005). The effects of fires decrease from the moist savannas to the dry savannas, largely as a function of the available biomass.

The on-site effects of fires range from the immediate impact of the selective burning on the bio-diversity and vegetation structure to changes in the physical, chemical and biological components in soils (Schultz, 2005). However the impact varies as a function of the composition of the plant communities, the size and shape of the woody species, the frequency of fires, the heating temperature during burning, the length of the period of time since the last fire, the onset of the fire during the dry season, and the land-use techniques applied (Schultz, 2005). Some cultivation techniques appear to reinforce the danger of further fires by changing the composition and structure of the ground cover as in the case of "slash and burn agriculture" (Goldammer, 1988). Biomass burning affects the reserves and storage of organic matter in the ground cover and in the soils and hence induces changes in soil-nutrient levels. An immediate effect of burning is an increase in K, Ca, Mg and the pH (Singh, 1994). However, the baring of ground promotes erosion by wind and water, and the transport of ashes contributes to the distribution of nutrients over a larger area. The change in the surface colour results in a higher absorption of the solar radiation and in an increase in evaporation. A further effect involves the enrichment of condensed volatile organic substances in the topsoil. This causes the development of a thin layer, which impedes the infiltration of water (Cass et al., 1984). Accordingly, these changes tend to increase the likelihood of soil erosion by the first rainfall events.

The off-site effects of fires are changes in the sediment delivery and in the nutrient level of the rivers' draining areas which are affected by fires. Fires tend to increase the content of dust in the atmosphere as they provide aerosols. Aerosols released by smouldering fires exert control on radiation activity as they increase condensation and cloudiness. However, the surplus of condensation nuclei results in small water droplets that remain suspended in the cloud. Consequently rainfalls are less likely. A further consequence of fires is the emission of oxides of carbon and nitrogen as well as of ozone and halogenides (Helas et al., 1992; Andreae et al., 1996). Particularly methyl chloride and methyl bromide emissions appear to support ozone depletion in the upper atmosphere, though the residence-times of these compounds are shorter than 2 years (Andreae et al., 1996). The estimated amounts of methyl-chloride and methyl-bromide emissions range from 1.8 Tg a^{-1} to 7 Gg a^{-1} (Andreae et al., 1996). This indicates that these compounds are capable of contributing significantly to ozone depletion in the upper atmosphere.

With respect to the extensive areas which are affected by fires, the question arises whether fires increase the level of greenhouse gases in the atmosphere. Andreae (1991) reports that in each year about 75 % of the African savannas are affected by fires. However, during the savanna fires, parts of the biomass are converted into elementary carbon (e.g. black carbon, charcoal). The estimated amount of charcoal formed during a fire appears to account for 5 to 10 % of the total biomass (Goldammer, 1993; Kuhlbusch et al., 1996). This fraction remains in the soil or sediments or is transported by rivers to the ocean, but cannot reenter the atmospheric carbon cycle (Goldammer, 1993). Consequently, this deficit in carbon has to be compensated for by consumption of atmospheric carbon. According to this concept, savannas may become a carbon sink when the processes are balanced through vegetation regeneration. Studies of the annual gas emissions of fires indicate that in the dry savannas the emissions of carbon dioxide, ammonia and nitric oxide do not exceed the amount dictated in the biomass by processes of nitrification and photosynthesis (Schultz, 2000; 2005). However, in the moist savannas, the changes in the vegetation are more pronounced, particularly if there is no regeneration of woody plants and the vegetation structure is destroyed. Accordingly, this may counteract the compensating effects of regeneration.

However, we have a poor understanding of the turnover of carbon in quantitative terms in the savannas due to the complex interaction of weathering, soil formation, vegetation and litter production and different reaction-times. Finally, a full assessment of the climatic impact of biomass-burning depends also on the reliability of the data and on the quality of case studies.

6. Prospect and conclusions

In Africa the superimposition of climatic changes and human activities is accompanied by a serious degradation of environmental conditions generally. The impact of this change involves certain thresholds which depend on the intensity and duration of meteorological events, the condition of the vegetation cover, the physical and chemical properties of the soil system and of the geomorphic settings. High rates of change occur in regions where large areas are affected by human intervention and where factors such as a high relief, steep slopes and a strong coupling between hillslopes and rivers support a rapid response. Slope failure in colluvial deposits, and erosion of hillwash and aeolian deposits indicate the important role of forms and deposits which are inherited from the past. Long-term processes, such as deep weathering, can contribute to the humanly-induced instability of hillslopes, once intrinsic thresholds are exceeded due to a continuous lowering of the shear strength or the increase in soil thickness (Shroder, 1976).

Studies on the impact of climatic changes on erosion processes in the late Pleistocene and the early Holocene indicate that a complete adjustment to the changed conditions requires a simultaneous response of all landscape components throughout a period of time that is long enough, to overcome the inertia of the geomorphic system (Thomas, 2004). With respect to the time frame of change in the vegetation-soil systems, these adjustments are considered to have been accomplished within a period of 10^3 to 10^4 years (Thomas, 2004, 2006). The expected rates of response point to the temporal and spatial differences between natural changes and humanly-induced change. Human interference is capable of changing the vegetation cover and the hydrologic regimes of extensive areas within a relatively short time. Repeated biomass burning in the savanna and rain forest zones coupled with intensified land-utilisation activity resulted in a degradation of the vegetation-soil system in several areas and often initiated an array of self-reinforcing processes.

Predictions of IPCC (2001) on the climatic development in Africa suggest that the climate is likely to get warmer, while the total amount of rainfall will not change significantly. However, a higher number of days with heavy rainfall is likely. These changes may affect the biota, the land use pattern and the hydrologic regimes. In the alpine Usambara Mountain area of East Africa (Tanzania) the lower replacement of montane forest trees seems to have been accompanied by general global warming over the last 100 years (Binggeli, 1989, Hamilton and Macfadyen, 1989). As a result of global warming, a general decline in the extent of the Afroalpine areas is likely (Taylor, 1999). The predicted increase in heavy rains may promote the increase of runoff, whilst the decrease of soil moisture is likely to bring with it edaphic aridity and an increase in erodibility (Beckedahl, 2002). This may result in a reinforcement of soil erosion. The decline in the number of rain days, on the other hand, may promote vegetation decay and leave more areas unprotected from heavy rainfalls. However, land use changes seem to have a much greater impact on susceptibility to soil erosion (Beckedahl, 2002, Valentin et al., 2005).

The increase in population in Africa is expected to result in an extension of the area cultivated land, even in steeply sloping mountainous regions. The impact of change in the climate and the intensified land use are likely to cause a reinforcement of degradation processes in the landscapes and may result in a lowering of the carrying capacity of land. However, predictions on future rates of change also depend on socioeconomic processes and political decisions. The devastating impact of desertification in the Sahel was not only a result of drought but was also associated with one of the highest population growth rates in the world.

7. References

Andreae, M.O., 1991. Biomass burning. Its History, Use and Distribution and its impact on Environmental Quality and Global Climate. In. Levine, J.S. (Ed.). Global Biomass Burning. Cambridge, Mass., p.3-21.

Andreae, M.O.; Goldammer, J.G.;Schebeske, G.; 1996. Methyl chloride and methyl bromide emission from vegetation fires. EGS XXI General Assembly, The Hague, The Netherlands, 6-10 May 1996. Annales Geophysicae 14, Supp. II, C 595.

Areola, O., 1999. Soils. In. Adams, W.M.; Goudie, A.S.; Orme, A.R. (Eds.). The Physical Geography of Africa. Oxford Univ. Press, Oxford, pp. 134-147.

Beckedahl, H. R., 2002. Bodenerosion in Afrika: ein Überblick. Petermanns Geographische Mitteilungen (PGM) 146, 18-23.

Bell, F.G.; Maud, R.R., 2000. Landslides associated with the colluvial soils overlying the Natal Group in the greater Durban region of Natal, South Africa. Environmental Geology 39, 1029-1038.

Bingelli, P., 1989. The Ecology of Maesopsis Invasion and Dynamics of Evergreen Forest of East Usambara and their Implications for Forest Conservation and Forestry Practices. In. Hamilton A.C.; Bensted-Schmith, R. (Eds.). Forest Conservation in the East Usambara Mountains. Tanzania. Gland, 269-300.

Biot, Y., 1990. THEPROM: an erosion-productivity model. In. Boardman, J.; Foster, I.; Dearing, J. (Eds.). Soil Erosion on Agricultural Land. Wiley, Chichester, pp. 465-479

Bork, H.-R., 2004. Soil erosion during the 20th century. Examples form South Africa, the Americas, China and Europe. In: Li, Y.; Poesen, J.; Valentin, C. (Eds.). Gully Erosion under Global Change. Sichuan Science and Technology Press, Chengdu, China, pp. 3-10.

Bork, H.-R., 2006. Landschaften der Erde. WGB-Darmstadt, Darmstadt.

Botha, G.A.; Partridge, T.C., 2000. Colluvial Deposits. In. Partridge, T.C.; Maud, R.R. (Eds.). The Cenozoic of Southern Africa. Oxford Monographs on Geology and Geophysics 40, pp.88-99.

Braun, J.-J., Ngoupayou, J.R.N., Viers, J., Dupre, B., Bedimo, J.-P.B., Boeglin, J.-L., Robain, H., Nyeck, B., Freydier, R., Nkamdjou, L.S., Rouiller, J., Muller, J.-P., 2003. Present weathering rates in a humid tropical watershed: Nsimi, South Cameroon. Geochimica et Cosmochimica Acta 69 (2), 357-387.

Bryan, R.B.; Jones, J.A.A., 1997. The significance of soil piping processes: inventory and prospect. Geomorphology 20, 209-218.

Calles, B.; Kulander, L., 1996. Likelihood of erosive rains in Lesotho. Z. Geomorph. N.F., Suppl. Bd. 106, 149-168.

Cass, A.; Savage, M.J.; Wallis, F.M., 1984. The effects of fire on soil and microclimate. In. De Booysen, P.V.; Tainton, N.M. (Eds.). Ecological effects of fire in South African ecosystems. Ecol. Studies 48, Springer, Berlin, p.311-325.

Dahlke, C.; Bork, H-R., 2006. Die Höhlen und Schluchten von Inxu Drift in der Transkei (Republik Südafrika). In Bork, H.-R.. Landschaften der Erde. WBG Darmstadt, Darmstadt, pp. 50-55.

De Meyer, A.; Poesen, J.; Isabirye, M.; Deckers, J.; Raes, D., 2011. Soil erosion rates in tropical villages: A case study from Lake Victoria Basin, Uganda. Catena 84, 89-98.

De Ploey, J.; Kirkby, M.J.; Ahnert, F., 1991. Hillslope erosion by rainstorms - a magnitude-frequency analysis. Earth Surface Processes and Landforms, 16, 399-409.

De Vleeschauwer, D.; Lal, R.; De Boodt, M., 1978. Comparision of detachable indices in relation to soil erodibility for some important Nigerian soils. Pedologie 28, 5-20.

Dingle, R.V.; Siesser, W.G.; Newton, R.A., 1983. Mesozoic and Tertiary Geology of Southern Africa. Balkema, Rotterdam.

Dunne, T., 1979. Sediment yield and land use in tropical catchments. Journ. Hydrol. 42, 281-300.

Elwell, H.A.; Stocking, M.A., 1982. Developing a simple yet practical method of soil loss estimation. Trop. Agric. 59, 43-48.

Eriksson, M.G.; Olley, J.M.; Payton, R.W., 2000. Soil history in central Tanzania based on OSL dating of colluvial and alluvial hillslope deposits. Geomorphology 36, 107-128.

Fubelli, G.; Bekele Abhebe; Dramis, F.; Vinci, S., 2008. Geomorphological evolution and present-day processes in the Dessie Graben (Wolio, Ethiopia). Catena 75, 28-37.

Gallart, F., 1995. The relative geomorphic work effected by four processes in rainstorms: a conceptual approach to magnitude and frequency. Catena 25, 353-364.

Gießner, K., 1989. Die räumlich-zeitliche Niederschlagstruktur im westlichen Jebel-Marra-Vorland und deren hydrologische Auswirkungen. In. Mensching, H.G. (Ed.). Morphodynamic Processes in the Lower River Atbara Area (Nile Province, Republic of Sudan) and in the Western Foreland of the Jebel Marra (Dafur, Republic of Sudan). Akad. der Wissenschaften in Göttingen, Hamburg, p. 79-125.

Goldammer, J.G. 1988. Rural land-use and fires in the tropics. Agroforestry Systems 6, 235-252.

Goldammer, J.G. 1993. Feuer in Waldökosystemen der Tropen und Subtropen. Birkhäuser-Verlag, Basel-Boston.

Goudie, A. , 1999. The Geomorphology of the Seasonal Tropics. In. Adams, W.M.; Goudie, A.S.; Orme, A.R. (Eds.). The Physical Geography of Africa. Oxford Univ. Press, Oxford, pp. 148-160.

Gowlett, J.A.; Harris, J. W.; Walton, D.; Wood, B.A., 1981. Early archaeological sites. Hominid remains and traces of fire from Chesowanja, Kenya. Nature 284, 125-129.

Grabowski, R.C.; Droppo, I.G.; Wharton, G., 2011. Erodibility of cohesive sediments: The importance of sediment properties. Earth Science Reviews 105, 101-120.

Greenway, D.R., 1987. Vegetation and slope stability. In: Anderson, M.G.; Richards, K.S. (Eds). Slope stability. Wiley, Chichester, pp. 289-230.

Hamilton;, A.C.; Mcfadyen, A., 1989. Climatic Change on the East Usambaras: Evidence from Meteorological Stations. In. Hamilton A.C.; Bensted-Schmith, R. (Eds.). Forest Conservation in the East Usambara Mountains. Tanzania. Gland, 103-107.

Haq, B.U., 1981. Paleogene paleooceanology: Early Cenozoic oceans revisited. Oceanologica Acta 4, Suppl., pp.71-82.

Heinrich, J., 1998. Formen und Folgen der jungholozänen Bodenzerstörung in Trockensavannenlandschaften Nordnigerias. Petermanns Geographische Mitteilungen 142, 355-366.

Helas, G., J.-P.Lacaux, R.Delmas, D.Scharffe, J.Lobert, J.G.Goldammer, and M.O.Andreae. 1992. Ozone as biomass burning product over Africa. Fresenius Environ.Bull.1, 155-160.

Held, A., 2006. Global Forest Resources Assessment 2005. Report on fires in the Sub-Saharan Africa (SSA) Region. Forestry Department Working Paper FM/9/E, Rom.

Hulme, M., 1999. Climate change within the period of meteorological records. In: Adams, W.M.; Goudie, A.S.; Orme, A.R. (Eds.). The Physical Geography of Africa. Oxford Univ. Press, Oxford, pp. 88-102.

Hutchinson, N.W.; Jackson, D.C., 1959. Results achieved in the measurement of erosion and runoff in Southern Rhodesia. Proceedings of the 3rd Inter-Africa Soil Conference Dalaba, CCTA, 573-583.

Idike, F.I., 1992. On appraising the soil erosion menace and control measurements in south eastern Nigeria. Soil Technology 5, 57-65.

Intergovernmental Panel on Climate Change (IPCC), Working Group I, 2001. Climate Change 2001. The Scientific Basis. Cambridge University Press.

Kimaro, D.N.; Poesen, J.; Msanya, B.M.; Deckers, J.A., 2008. Magnitude of soil erosion on the northern slope of the Uluguru Mountains, Tanzania. Interrill and rill erosion. Catena 75, 38-44.

Kuhlbusch, T.A.J., M.O. Andreae, H. Cachier, J.G. Goldammer, J.-P. Lacaux, R. Shea, and P.J. Crutzen. 1996. Black carbon formation by savannah fires: Measurements and implications for the global carbon cycle. Special Issue TRACE-A and SAFARI, J.Geophys.Res. 101, D 19, 23,651-23,665.

Lal, R., 1998. Drop size distribution and energy load of rain storms at Obadan, western Nigeria. Soil and Tillage Research 48 (1/2), 103-114.

Landman, W. A.; Mason, S.J., 1999. Change in the association between the Indian Ocean sea-surface temperatures and summer rainfall over South Africa and Namibia. Int. Journ. Climatology 19, 1477-1492.

Lewis, C.A., 2008. Late Quaternary climatic change, and associated human responses, during the last ~ 45000 yr in the Eastern and adjoining Western Cape, South Africa. Earth-Science Reviews 88, 167-187.

Maley, J., 1991. The African rain forest vegetation and palaeoenvironment during Late Quaternary. Climate Change, 19 (S. I. Tropical Forests and Climate), 79-98.

McMahon, T.A.; Finlayson, B.I.; Haines, A.; Srikanthan, R., 1992. Global Runoff: Continental Comparison of Annual Flows and Peak Discharges. Cremlingen-Destedt, Germ.

Mensching, H.G., 1984. Veränderungen des morphodynamischen Prozessgefüges durch Desertifikation in Darfur. In.Mensching, H.G. (Ed.). Beiträge zur Morphodynamik im Relief des Jebel-Marra-Massivs und in seinem Vorland (Darfur/Republik Sudan). Akad. Wissenschaften in Götttingen. Hamburg, p. 166-177.

Mensching, H.G., 1990. Desertifikation. Wissenschaftliche Buchgesellschaft (WBG) Darmstadt.

Mensching, H.G.; Giessner, K.; Stuckmann, G., 1970. Die Hochwasserkatastrophe in Tunesien im Herbst 1969. Geogr. Zeitschrift 58, 81-94.

Morgan, R.P.C., Mngomezulu, D., 2003. Threshold conditions for initiation of valley-side gullies in the Middle Veld of Swaziland. Catena 50, 401-414.

Nearing, M.A.; Jetten, V.; Baffaut, C.; Cerdan, O.; Couturier, A.; Hernandez, M.; Le Bissonais, Y.; Nichols, M.H.; Nunes, J.P.; Renschler, C.S.; Souchère, V.; van Oost, K., 2005. Modelling response of soil erosion and runoff to changes in precipitation and cover. Catena 61, 131-154.

Nicholson, S.E., 1999. Environmental Change within the Historical Period. In. Adams, W.M.; Goudie, A.S.; Orme, A.R. (Eds.). The Physical Geography of Africa. Oxford Univ. Press, Oxford, pp.60-87.

Nyssen, J., Poesen, J.; Moeyersons, J., Luyten, E.; Veyret Picot, M.; Deckers, J.; Mitiku, H.; Govers, G., 2002. Impact of road building on gully erosion risk, a case study from the northern Ethopian highlands. Earth Surface Processes and Landforms 27, 1267-1283.

Nyssen, J.; Poesen, J.; Moeyersons, J.; Deckers, J.; Mitiku, H.; Lang, A., 2004. Human impact on the environment in the Ethiopian and Eritrean highlands - a state of the art. Earth Science Reviews 64, 273-320.

Partridge, T.C.; Maud, R.R., 2000. Macro-Scale Geomorphic Evolution of Southern Africa. In. Partridge, T.C.; Maud, R.R. (Eds.). The Cenozoic of Southern Africa. Oxford Monographs on Geology and Geophysics 40, pp. 3- 18.

Petters, S.W., 1991. Regional Geology of Africa. Lecture Notes in Earth Sciences 40, Springer Verlag, Heidelberg.

Pflaumbaum, H.; Pörtge, K.-H.; Menschung, H.G., 1990. Hydrologische Steuerungsfaktoren morphologischer Prozesse in der Nil-Provinz (Rep. Sudan). Geoökodynamik 11, 241-256.

Rapp, A., 1976. Studies of mass wasting in the Arctic and in the Tropics. In. Yatsu, E.; Ward, A.J.; Adams, F. (Eds.). Mass Wasting. 4th Guelp Symposium on Geomorphology, 1975. Geo Abstracts LTD, Norwich, England, p. 79-103.

Reading, A.J.; Russell, D. T.; Millington, A.C., 1995. Humid tropical environments. Blackwell, Oxford UK.

Richard, Y.; Fauchereau, N.; Poccard, I.; Rouault, M. ; Trzaska, S., 2001. 20th Century droughts in southern Africa: spatial and temporal variability, teleconnections with ocean and atmospheric conditions. Int. Journ. Climatology 21, 873-885.

Römer, W., 2004. Untersuchungen zur Hang- und Inselbergentwicklung in Süd-Zimbabwe. Aachener Geographische Arbeiten (AGA), 39, pp. 397.

Römer, W., 2007. Differential weathering and erosion in an inselberg landscape in southern Zimbabwe. A morphometric study and some notes on the long-term development of inselbergs. Geomorphology 86, 349-368.

Römer, W. 2012. Hillslope processes in tropical environments. In. Shroder, J.Jr.; Marston, R.; Stoffel, M., (eds.). Treatise on Geomorphology. Academic Press, San Diego, CA, vol 7. (in press).

Schmidt, P.R., 1997. Iron Technology in East Africa: Symbolism, Science and Archaeology. Indiana Univ. Press, Bloomington, James Currey, Oxford.

Schultz, J., 2000. Handbuch der Ökozonen. Ulmer, Stuttgart.

Schultz, J., 2005. The Ecozones of the World. Springer, Berlin Heidelberg.

Scoones, I.; Reij, C.; Toulmin, C., 1996. Sustaining the soil : Indigenous soil and water conservation in Africa. In. Reij, C.; Scoones, I.; Toulmin, C. (Eds.). Sustaining the Soil - Indigenous Soil and Water Conservation in Africa. London, 1-27.

Seibold, E.; Berger, W.H.; 1995. The sea floor. An introduction to Marine Geology. 3rd Ed., Springer, Heidelberg.

Seuffert, O., 1987. Desertification in the Tropics and Subtropics - past and present - . Geoökodynamik 8, 145-182.

Seuffert, O.; Busche, D.; Löwe, P., 1999. Rainfall structure - rainfall erosivity: new concepts to solve old problems. Petermanns Geographische Mitteilungen 143, 475-490.

Shakesby, R.A.; Withlow, R., 1991. Perspectives on prehistoric and recent gullying in central Zimbabwe. GeoJournal 23, 49-58.

Shroder, J.F., Jr., 1976. Mass movements on Nyika Plateau, Malawi. Z. Geomorph. N.F. 20, 56-77.

Singh, R.S., 1994. Changes in soil nutrients following burning of dry tropical savannah. Int. Journ. Wildland Fire 4, 187-194.

Singh, R.G.; Botha, G.A.; Richards, N.P.; McCarthy, T.S., 2008. Holocene landslides in KwaZulu-Natal, South Africa. South African Journ. of Geology 111, 39-52.

Skinner, B.J.; Porter, S. C., 2000. The dynamic earth. 4th Ed., John Wiley and Sons, New York.

Starkel, L., 1976. The role of extreme (catastrophic) meteorological events in contemporary evolution of slopes. In. Derbyshire, E., (Ed.). Geomorphology and Climate. John Wiley and Sons, London, p.203-246.

Stocking, M., 1995. Soil erosion in developing countries: where geomorphology fears to tread! Catena 25, 253-267.

Summerfield, M.A., 1999. Tectonics, Geology, and long-term Landscape Development. In. Adams, W.M.; Goudie, A.S.; Orme, A.R. (Eds.). The Physical Geography of Africa. Oxford Univ. Press, Oxford, pp.1-17.

Taylor, D., 1999. Mountains. In. Adams, W.M.; Goudie, A.S.; Orme, A.R. (Eds.). The Physical Geography of Africa. Oxford Univ. Press, Oxford, pp. 287-306.

Thomas, M.F., 1994. Geomorphology in the Tropics. John Wiley & Sons, Chichester.

Thomas, M.F., 2004. Landscape sensitivity to rapid environmental change - a Quaternary perspective with examples from tropical areas. Catena 55, 107-124.

Thomas, M.F., 2006. Lessons from the tropics for a global geomorphology. Singapore Journal of Tropical Geography 27, 111-127.

Tyson, P.D.; Partridge, T.C., 2000. Evolution of Cenozoic Climates. In. Partridge, T.C.; Maud, R.R. (Eds.). The Cenozoic of Southern Africa. Oxford Monographs on Geology and Geophysics 40, 371-387.

Valentin, C.; Rajot, J.-L. ; Mitja, D., 2004. Responses to soil crusting, runoff and erosion to fallwing in the sub-humid and semi-arid regions of West Africa. Agriculture, Ecosystems and Environment 104, 287-302.

Valentin, C., Poesen, J., Yong, L., 2005. Gully erosion: impacts and controls. Catena 63, 132-153.

Van Noten, F.; De Ploey, J., 1977. Quaternary research in Northeastern Nigeria. Musée Royal de L´Afrique Central, Annales IN-8, Sciences Humaines 92, 1-61.

Warren, A., 1999. Desertification. In: Adams, W.M.; Goudie, A.S.; Orme, A.R. (Eds.). The Physical Geography of Africa. Oxford Univ. Press, Oxford, 342-355.

Weischet, W.; Endlicher, W., 2000. Regionale Klimatologie. Teil 2 Die Alte Welt. Teubner, Stuttgart.

Wirthmann, A., 2000. Geomorphology of the Tropics. Springer, Berlin.

Wolman, M.G.; Miller, P., 1960. Magnitude and frequency of forces in geomorphic processes. J. Geol. 68, 54-74.

Zachar, D., 1982. Soil erosion. Developments in Soil Science 10, Elsevier Scientific Publ. Comp., Amsterdam.

Zawada, P.K., 2000. Slackwater sediments and Paleofloods. In. Partridge, T.C.; Maud, R.R. (Eds.). The Cenozoic of Southern Africa. Oxford Monographs on Geology and Geophysics 40, 198-206.

14

Assessment of the Impact of Land-Use Types on the Change of Water Quality in Wenyu River Watershed (Beijing, China)

Yuanzhi Zhang[1,2] and Yufei Wang[1]
[1]*Yuen Yuen Research Centre for Satellite Remote Sensing, Institute of Space and Earth Information Science, The Chinese University of Hong Kong (CUHK), Shatin, Hong Kong*
[2]*Laboratory of Coastal Zone Studies, Shenzhen Research Institute of CUHK, Shenzhen, China*

1. Introduction

Land use-land cover (LULC) change is one of the major environmental changes occurring around the globe. Water quality is one of such factors affected by LULC change, since it is a key component of a healthy watershed where it integrates important geomorphic, hydrologic, and some of the biological processes of a watershed (Hem, 1985). Alteration of any one of these processes will affect one or more water quality parameters (Peterjohn and Correll, 1984). Hydrologists and aquatic ecologists have long known that the surface across which water travels to a stream or a lake has a major effect on water quality. Accordingly, the relative amounts of particular types of land use-land cover (LULC) in a watershed will affect water quality as well (Griffith, 2002). Therefore, the change in land-use and management practices will give rise to the considerable impact on water quality.

The importance of the interrelationships between LULC and water quality is reflected by the increased recognition over the past two decades that non-point source (NPS) pollution has come into being the major environmental concern (Loague *et al,* 1998; Sharpley and Meyer, 1994; Griffith, 2002). Pollutants affecting water quality may come from point or nonpoint sources. Point pollution can be easily monitored by measuring discharge and chemical concentrations periodically at a single place. In the past several decades, the major efforts and funding of water pollution control programs focused on the point sources management, and the magnitude of the point source pollution problem has been reduced in many cases. However, NPS pollution presents great challenges because of their dispersed origins and the fact that they vary with the season and the weather, in addition to the fact that non-point inputs are often overlooked by human beings. Land cover influences water quality because land cover determines the type and quantity of NPS pollutants that may enter the water body.

There are a lot of studies examining non-point source pollution focused on the effects from runoff over the agricultural land and concluded that agricultural coverage strongly influenced water nitrogen (Johnson *et al.*, 1997; Fisher *et al.*, 2000; Ahearn *et al.*, 2005), phosphorus (Hill, 1981), total suspended solids (Ahearn *et al.*, 2005) and sediments (Allan *et al.*, 1997). A number of documents have illustrated the increasing urban areas were another significant contributor to the water quality deterioration, since the impervious surface coverage can alter the hydrology and geomorphology of urban streams and give the negative impacts on urban stream ecosystems (Schueler, 1995; Paul and Meyer, 2001; Morse *et al.*, 2003), and runoff from urbanized surfaces carries greater sources of pollutants, which results in the increasing loading of nutrients (Emmerth and Bayne, 1996; Rose, 2002), heavy metals (Norman, 1991; Callender and Rice, 2000), sediment loadings (Wahl *et al.*, 1997) and other contaminants to the near stream waters.

In recent years, since 1978 when China has initiated her economic reform and open-door policy, rapid urbanization and economic expansion has resulted in massive land alteration. However, people only focus on the economic growth, and always neglect this factor that economy grows at the expense of the environmental destruction. In this study, therefore, we applied Landsat TM data (2000-2008) to examine the changes of land-use and establish the relationship between land-use types and water quality variables, and give the technical support which can help propose the appropriate strategy that will permit the sustainable regional development and protection of the ecological environment, and understand how it important to assess their potential impacts of land-use types on water quality changes in the watershed scale. This study also demonstrates an example of the issue of how LULC change is linked to water quality, one of the most precious resources on earth.

2. Study area

Wenyu River watershed is a key area in Beijing (China), belongs to the water systems of the Beiyun River, which is the most intensive area of human activity in Hai River Basin (Figure 1). Wenyu River, the main stream is 47.5 km, which is originates from the south of Yan Mountain and flows from north to south though Haidian, Changping, Shunyi, Chaoyang and Tongzhou Districts, all of these districts are in the core area of Beijing City. Wenyu River is usually called "the mother river" of Beijing, because of all the main streams in Beijing City, it is the only river which originates in the border and never runs dry.

The total area of Wenyu River watershed is 2,478 km^2 and the percentage of mountain and flatland area are 40.4% and 59.6%, respectively. The ground elevation in this area is in the region of 15-1000m. And the study area has the terrain characteristics with the high terrain in the northwest and low plain in the southeast. There are many tributaries in this watershed, with the Dongsha, Beisha, and Nansha Rivers in the upper reaches of Wenyu River, meeting in the Shahe Reservoir, and the Lingou, Qing, Ba and Xiaozhong Rivers flowing into the main stream of Wenyu River. The average annual temperature in this watershed is about 11.6 degree Centigrade (for the year 1959-2000). The predominant soil type is cinnamon (53.5%) of the total area. The average annual precipitation is 624.5mm (for the year 1959-2000), more than 80% of a year's total precipitation is concentrated in the flood season from June to September, the average annual water surface evaporation is 1,175mm, and about 42% of a year's evaporation is concentrated from April to June. The average annual runoff is 450 million cubic meters.

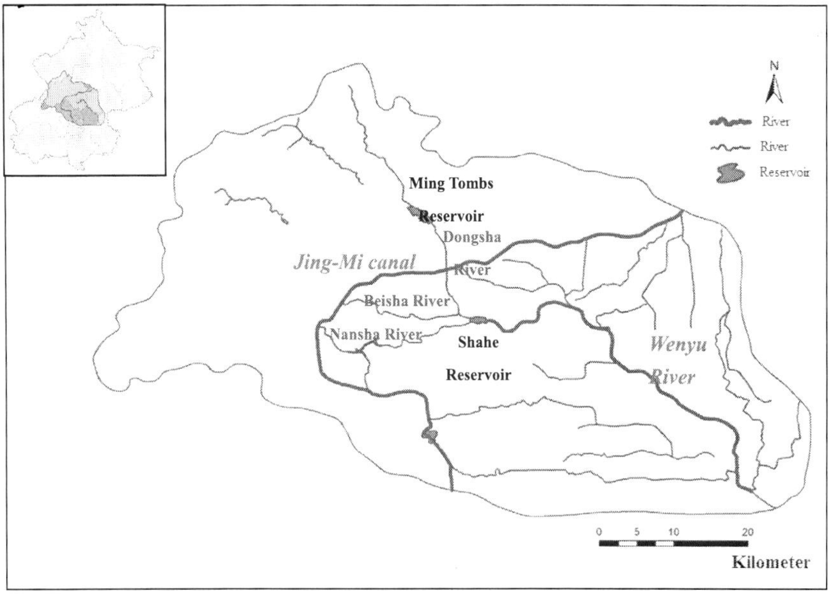

Fig. 1. Map of Wenyu River Watershed in Beijing (China).

As the main drainage canal in the Beijing City, the problems of water pollution and water ecosystems degradation in Wenyu River watershed have come along with the economic development in these years. Several documents estimated the pollution status (Wang and Song, 2008; Shi, 2008; You *et al.*, 2009; Hua *et al.*, 2010) of Wenyu River and pointed out that the water environment of this area was under sub-health; additionally, some other authors put forward the reasonable strategies to restore the ecological environment and improve the water quality in Wenyu River Watershed (Zheng *et al.*, 2007; Wang *et al.*, 2008; You *et al.*, 2009). Although there have many studies noted the water quality problems in Wenyu River Watershed, but the studies linking land use to water quality are limited.

3. Methodology

An integrated approach (involving remote sensing, geographic information systems, statistical and spatial analysis, and hydrologic modeling) is used to link the relationship of land use-land cover and water quality in a regional scale. The soft-wares used in this study include ENVI version 4.3, ArcGIS version 9.3, and SPSS version 14.0 for Windows. Figure 2 shows the flowchart of examining the relationship between land-use and water quality.

3.1 Water quality monitoring

Water samples were collected from twenty-four stations within Wenyu River watershed (see Figure 3) from May to August (on May 22, June 9, July 18 and August 18, respectively) in 2009, and each water sample collection was conducted after the rainfall. Most of these stations distribute in the mid-upper stream area of the Wenyu River watershed.

Water quality data are often collected through direct measurement in situ. To some variables cannot be measured in situ, a sample must be taken and then analyzed in a laboratory. In this research, water samples are analyzed to obtain six water quality variables, as Table 1 listed. The variable of DO is in situ measured using Portable Dissolved Oxygen Analyzer, TOC is analyzed in the laboratory using Total Organic Carbon Analyzer, and the other variables are measured according to National standardized water quality detection method (State Environmental Protection Administration of China, 2002).

Variable Name	Chemical Formula or Abbreviation	Unit
Dissolved Oxygen	DO	mg/l
Chemical Oxygen Demand	COD	mg/l
Total Nitrogen	TN	mg/l
Nitrate	NO_3- N	mg/l
Total Phosphorous	TP	mg/l
Phosphate	PO_4- P	mg/l

Table 1. Water Quality parameters selection in this study.

3.2 Sub-watershed delineation

Because the 24 water sampling points of this study locate across a range of land uses, geology types, and stream orders within the entire Wenyu River watershed. Thus, the sub-watersheds within Wenyu River watershed should be firstly delineated, and Arc Hydro Model is employed to do this job. Arc Hydro Model was developed by a consortium for geographic information systems (GIS) in water resources, integrated by the University of Texas' Center for Research in Water Resources (CRWR) and the Environmental Systems Research Institute (ESRI) during the years 1999-2002. The Arc Hydro data model is a conceptualization of surface water systems and describes features such as river networks, watersheds and channels. The data model can be the basis for a "hydrologic information system", which is a synthesis of geospatial and temporal data supporting hydrologic analysis and modeling (Maidment, 2002). The Arc Hydro tools are a set of utilities developed based on the Arc Hydro data model, and operating in the ArcGIS environment. These tools can be used to process a digital elevation model raster (DEM) to delineate sub-watersheds.

The major data used to delineate the sub-watersheds is the 30 meter DEM (Digital Elevation Model) data set for China, which is a part of ASTER (Advanced Space-borne Thermal Emission and Reflection Radiometer) Global 30m DEM topographic data set and available for download free of charge from the NASA's Land Process Distributed Active Archive Center, at URL https://wist.echo.nasa.gov/api/. Using Boundary vector of the study area, the DEM for the study area can be obtained. In this process, higher threshold will result in less dense stream network and less internal sub-watersheds; when the value of threshold decrease, a relatively dense stream network and more internal sub-watersheds will be obtained. In this research, the value of 50000 is applied as the threshold value, the resultant stream network and sub-watershed delineation rasters are displayed in Figure 4. It can also be found 42 sub-watersheds are delineated within Wenyu River watershed when 50000 is used as the threshold value.

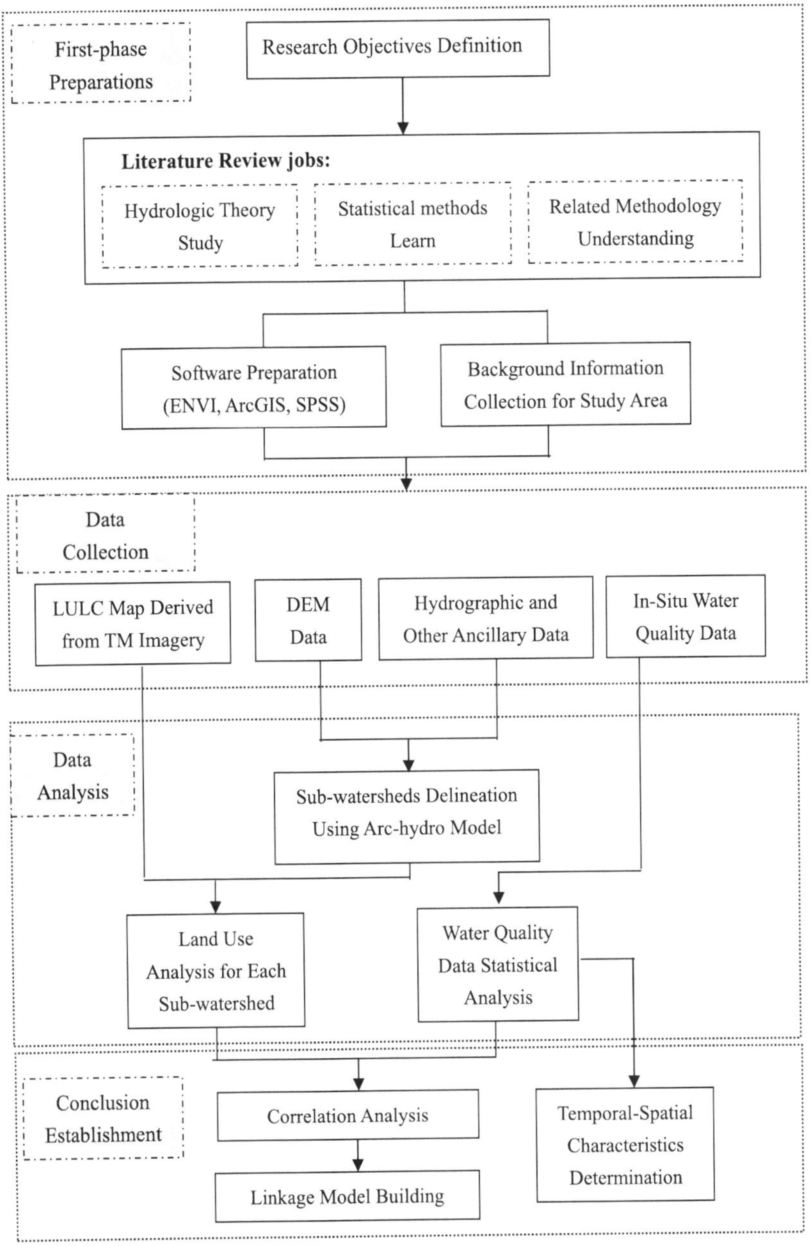

Fig. 2. The flowchart of examining the relationship between land use-land cover and water quality.

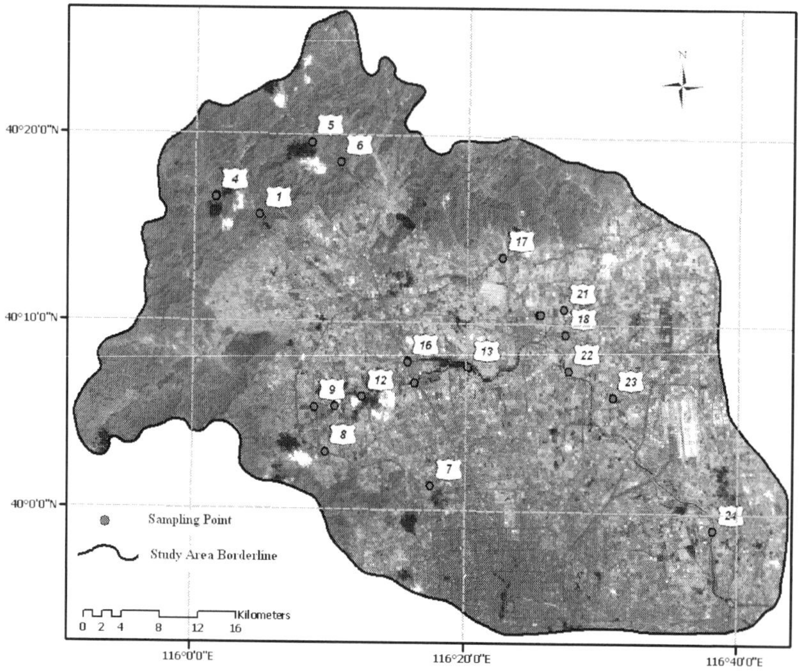

Fig. 3. Water Quality Sampling Points in Wenyu River Watershed (Landsat TM5 image).

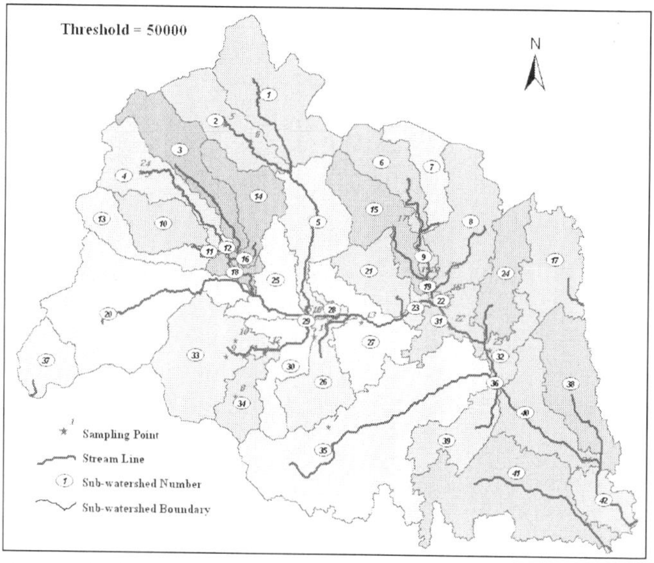

Fig. 4. The sub-watersheds delineation results generated by using the threshold value of 50000.

To those sub-watersheds containing in-situ measured water quality data, it is very clear about the water quality status there and obtain the mean values of each water quality parameters of these sub-watershed through the statistical computing process.

3.3 LULC classification in the study area

Landsat TM data are used to extract the land use-land cover information of the Wenyu River watershed. Landsat TM is appropriate for the purpose in this research because it is free online and can be downloaded easily. Its spatial resolution is 30 meter which will be appropriate to conduct land use analysis of the watershed of Wenyu River. One nearly cloud-free Landsat 5 TM image covering the study area is acquired from the USGS website, http://glovis.usgs.gov. Table 2 describes the general information of this downloaded image.

Landsat Scene Identifier	LT51230322009201IKR00
WRS Path/ROW *	123/032
Data Acquired	2009/07/20
Cloud Cover	3.58%
Corner Upper Left	41°16'19"N/ 115°53'07"E
Corner Upper Right	40°57'24"N/118°02'53"E
Corner Lower Left	39°41'38"N/115°24'26"E
Corner Lower Right	39°23'08"N/117°31'21"E

* WRS means The Worldwide Reference System, which is a global notation used in cataloging Landsat data; both Landsat 5, 7 follow the WRS-2, and Landsat 1,2,3,4 follow the WRS-1.

Table 2. The general information of downloaded Landsat 5 TM scene.

To extract land covers of Wenyu River watershed from Landsat TM 5 data, the supervised classification method is adopted in this research, which is the procedure most frequently used for quantitative analysis of remote sensing data, and the maximum likelihood algorithm is employed to detect the land cover types in ENVI software. Based on the priori knowledge of the study area and additional information from previous research in Wenyu River watershed, a classification system concerned with six land classes has been established for this study area, including forest, farmland, urban, village, bare land and the water bodies, the description of these land cover classes are presented in Table 3.

No.	Land Cover Type	Description
1	Forest land	Coniferous & deciduous forest, trees covers, shrubs with partial grassland
2	Farmland	Cropland and pasture, Orchards, other agriculture land
3	Urban area	Residential, commercial, industrial, transportation, and communications facilities; the area of intensive use with much of the land covered by structures and high population density, usually located in the center of a city
4	Village area	Located in the rural areas, surrounding the urban area and has a relatively low population density
5	Bare land	Areas with no vegetation cover, stock quarry, stony areas, uncultivated agricultural lands
6	Water body	Seas, lakes, reservoirs, rivers and wetland

Table 3. Land use-land cover classification scheme used in TM data.

During the process of supervised classification, the collection of training sites constitutes a very critical stage and it is essential that all the required classification classes are sampled. The quality of a supervised classification depends on the quality of the training sites. In order to select the accurate training sites, different band combinations are used to identify the different land categories, according to Landsat TM Band spectral characteristics. Figure 5 displays the generated land use-land cover map of Wenyu River watershed in 2009.

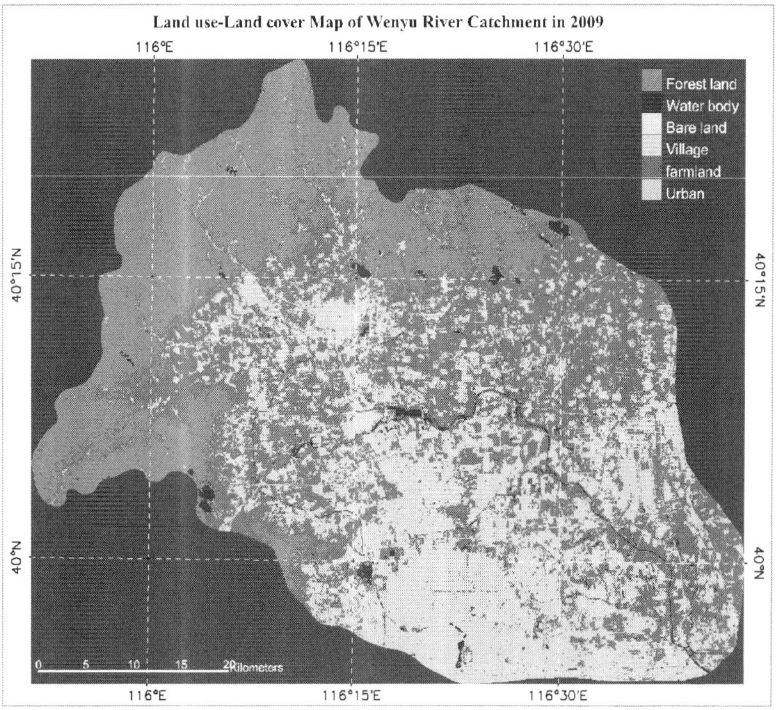

Fig. 5. Land use-land cover map of Wenyu River watershed in 2009 from Landsat TM 5 data.

The LULC map shows that, upper region of Wenyu River has significantly more forest land with higher elevation, while the middle region of the research watershed has a higher percentage of urban area and the major land types in the lower region are village and farmland. The different regions in Wenyu River watershed differ significantly in terms of percentage of forest, urban, village and farmland covers.

3.4 Spearman's rank correlation

Since most of the water quality variables do not distribute normally, the statistical analyses are confined to non-parametric statistical tests, spearman's rank correlation analyses are used to explore the relationships between land use types and water quality indicators in Wenyu River Watershed. And this statistical analyses are performed using SPSS 14.0 for Windows.

In statistical researches, Spearman's rank correlation coefficient is a non-parametric measure of statistical dependence between two variables, which allows us to easily identify the strength of correlation within a data set of two variables, and whether the correlation is positive or negative. The absolute value of the correlation coefficient, with the range from 0 to 1, indicates the strength, with larger absolute values indicating stronger relationships. The significance level (also termed as p-value) is the probability of obtaining results as extreme as the one observed. If the significance level is very small (p value is less than 0.05), the correlation is significantly raleted at 95% confidence level, and the two variables are linearly related. The data set, which are used in the Spearman's rank correlation process to determine the relationships between land use cover and water quality in this research, includes the land use-land cover variables (%) and the water quality variables (mg/L) of the delineated sub-watersheds.

3.5 An exponential model

Delivery of non-point source pollutants from discrete upstream contributing zones to a particular downstream point is a multi-step, often episodic, process (Phillips, 1989). During the rainfall event, the pollutants released from different land use types will flow through various land covers with the surface runoff, continuing to be absorbed, deposited and released, and eventually enter the nearest stream water. A first-order rate equation can be used for modeling nutrient attenuation in flow through various land uses to the nearest stream (Phillips, 1989). Thus in most cases, the concentration of nutrients or total suspended solids (NPS_i) at a sample point received from a basin i, can be described in the form of an exponential model (Fetter 1994; Basnyat *et al.*, 1999; Basnyat *et al.*, 2000) as follows:

$$NPS_i = \alpha e^{(\beta_1 Forest_i + \beta_2 Farmland_i + \beta_3 Urban_i + \beta_4 Village_i + \beta_5 Bare_i + \beta_6 Water_i)} \tag{1}$$

Where NPS_i is the dependent variable, α is the intercept β_1, β_2, β_3, β_4, β_5 and β_6 are parameters that specify the direction and strength of the relationships between each land use type and NPS_i.

Based on the linkage model, multiple regression models were applied to each of water quality variables: total nitrogen, nitrate, total phosphorous, phosphate, chemical oxygen demand and dissolved oxygen, respectively. A backwards stepping approach is employed to isolate a final model with only significant independent variables included. In Backward approach, all the predictor variables will go into the model firstly. The weakest predictor variable is then removed and the regression re-calculated. If this significantly weakens the model, the predictor variable will re-entered, otherwise it will be deleted. This procedure will repeated until only useful predictor variables remain in this model.

The purpose of multiple regression process is to predict a single variable (dependent variable) from one or more independent variables. For each model, the initial fixed independent variables are LULC variables (forest, farmland, urban, village, bare and water). The dependent data of water quality parameters and the independent data of land use variables will be natural log transformed to meet the assumptions of normality, as determined via graphical evaluation of standard diagnostic graphs. Finally, goodness-of-fit of final significant statistical models will be evaluated by scatter plot to compare the observed data against equivalent model prediction.

4. Results and discussion

4.1 Water quality temporal and spatial characteristics

The 24 water sampling points of this study were located across a range of land uses, geology types, and stream orders within the entire Wenyu River watershed (Figure 6). Thus, the Wenyu River watershed was firstly delineated into 42 sub-watersheds using DEM raster. According to the in-situ water quality measured data, water quality status of certain sub-watershed can be obtained.

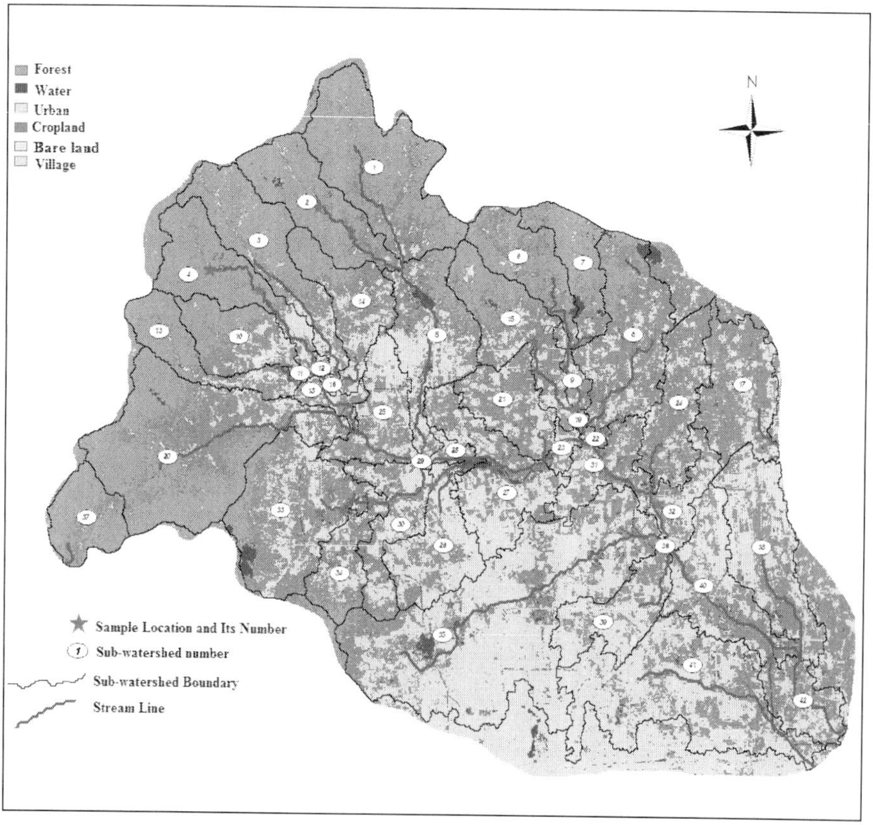

Fig. 6. There different spatial areas definition within the Wenyu River Watershed.

Considering their similarity of geographic location, topographic characteristic, land use-land cover, and human activities, the delineated sub-watersheds were generally clustered into three types in which they located (see Figure 6): Upstream Mountain Area, Midstream Urban Area, and Downstream Plain Area. Table 5 summarizes the characteristic information of these three different spatial areas within Wenyu River watershed. And only those sub-watersheds containing in-situ water quality data were considered in this research.

Different Spatial Areas	Sub-watershed Number	Water Sampling Sites	Area Characteristics
Upstream Mountain Area	w2	Sites 5, 6	• Lying in the upstream of Wenyu River Watershed and with the higher elevation; • With the only significant land use of forest; • Sparse human population; • Less influence on water quality from human activities.
	w4	Sites 1, 2, 3, 4	
Midstream Urban Area	w26	Site 14	• Lying in the midstream of Wenyu River Watershed; • With gently sloping surface; • With the notable land use of Urban and village; • High density of population; • Considerable influence on water quality from human activities.
	w27	Site 13	
	w28	Site 15, 16	
	w33	Sites 9, 10, 11, 12	
	w34	Site 8	
	w35	Site 7	
Downstream Plain Area	w8	Site 21	• Lying in the downstream of Wenyu River Watershed; • With gently sloping surface; • With the dominant land use of production agriculture; • Relatively low density of population; • Certain influence on water quality from agriculture activities.
	w9	Site 20	
	w15	Sites 17, 19	
	w22	Site 18	
	w31	Site 22	
	w32	Site 23	
	w42	Site 24	

Table 4. Three different spatial areas definition within Wenyu River Watershed.

Through the statistical computing process, water quality information in Upstream Mountain Area, Midstream Urban Area and Downstream Plain Area can be obtained based on the measured water quality data at total 24 water sampling sites. These water quality statistical information include the mean value (the sum of all observations divided by the number of observations) and the standard error of the mean (SEM, calculated by dividing the standard deviation by the square root of the sample size) of six water quality prameters's concentration, including TN, NO_3- N, TP, PO_4- P, COD and DO.

4.2 Water quality comparison between different land-use types

In order to conduct the further analysis of the relationship between land use and the water quality within Wenyu River watershed, in this section, the sub-watersheds are divided into different classes according to their different land-use structures. And the results of water quality comparison between different land-use structures tell us that land use types are significantly correlated to water quality variables in Wenyu River Watershed.

Here the total nitrogen (TN) is an example of water quality parameters to be monitored from May to August in 2008. Figure 7 illustrates that, between the four different land-use structures, the TN concentration of class III has the largest value, while the TN concentration of class I is the smallest. And the total nitrogen counts produced from class III is about three times greater than that from class I. The sub-watersheds belonging to the class III have three mixed dominant land use types, village, urban and farmland, and all of these sub-watersheds are located in the midstream urban area of Wenyu River watershed, where have the high density of population and the human activities must give rise to the considerable influence on the water quality. The sub-watersheds of w2 and w4 belonging to the class I, they locate in the upstream mountain area with the single significant land use of forest and sparse human population. The result indicates that contribution from forest is the smallest to the total nitrogen loading compared with those from farmland, urban and village.

The water quality parameters of NO_3-N concentration was also monitored in the months of May, July and August. Figure 8 shows that, between the four different land-use structures, NO_3-N concentration of class Ⅳ has the largest value, while the value of class Ⅰ is the smallest. Both class Ⅰ and class Ⅳ are the land-use structures with single dominant land use; the dominant land use of the former is forest while the latter is farmland cover. It is clear that the contribution from the farmland is larger than the forest to the nitrate loading in the surface water within Wenyu River watershed.

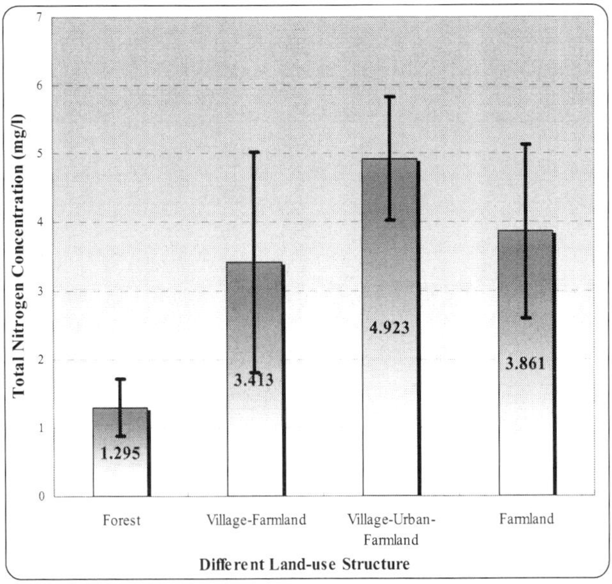

Fig. 7. TN concentration (mean ± SEM) comparison between different land-use structures.

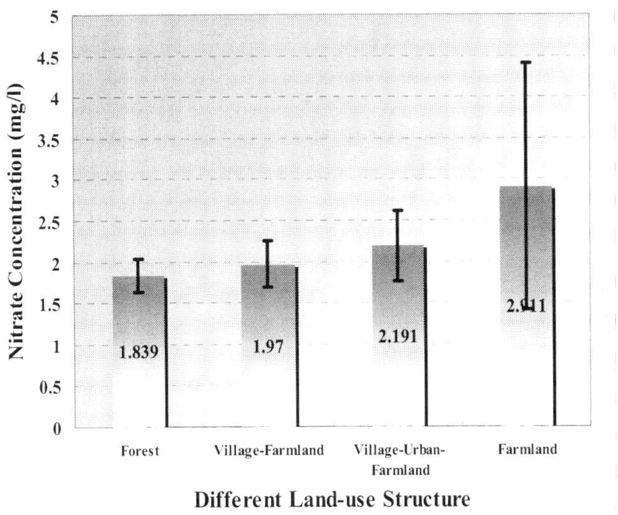

Fig. 8. Nitrate (NO_3-N) concentration (mean ± SEM) comparison between different land-use types.

4.3 Spearman's rank correlation analysis

The result from Spearman's rank correlation analysis between land use-land cover variables (%) and the water quality variables (mg/L) is shown in Table 5, which indicates that land use types are significantly correlated to many water quality variables within Wenyu River Watershed. For example, the water quality variables of total nitrogen, total phosphorous, phosphate and chemical oxygen demand have strong positive relationships with urban and village lands, while they are all present the negative correlation with the forest land use.
Except for dissolved oxygen, forest is negatively correlated with the other five variables. In comparison, farmland, urban and village have the negative relationship with dissolved oxygen, while urban and village have the strong positive relationship with five variables.

Water quality Indicators	Land use types						N
	Forest	Farmland	Urban	Village	Bare	Water	
TN	-0.401	0.181	0.456	0.462	0.187	0.632*	13
NO_3- N	-0.055	-0.209	0.181	0.033	0.275	-0.242	13
TP	-0.412	0.198	0.560*	0.681*	0.681*	0.352	13
PO_4- P	-0.725**	0.357	0.533	0.621*	0.302	0.714**	13
COD	-0.297	-0.346	0.676*	0.720**	0.187	0.22	13
DO	0.082	-0.291	-0.28	-0.396	-0.385	-0.341	13

Notes: * * indicates significance $p < 0.01$ while * indicates $p < 0.05$;
Absolute coefficient value of 1.0 is a perfect fit.

Table 5. Correlations analysis between land use types and water quality indicators based on Spearman's rank correlation coefficient.

The above results can provides insight into the linkage between land use types and stream water quality, which is just in line with the comparison results (as Table 6 listed) of water quality variables between different land-use structures.

Water Quality Variables	Order for different land-use structure
Total Nitrogen (TN)	Village- urban-Farmland > Farmland > Village-Farmland > Forest
Nitrate (NO_3- N)	Farmland > Village- urban-Farmland > Village-Farmland > Forest
Total Phosphorous (TP)	Village- urban-Farmland > Village-Farmland > Farmland > Forest
Phosphate (PO_4- P)	Village- urban-Farmland > Farmland > Village-Farmland > Forest
Chemical Oxygen Demand (COD)	Village-Farmland > Village- urban-Farmland > Forest > Farmland
Dissolved Oxygen (DO)	Village- urban-Farmland < Farmland < Village-Farmland < Forest

Table 6. The order of water quality variables for different land-use structures.

Three water quality variables including total nitrogen, total phosphorous and phosphate, have strong positive relationships with urban and village lands, while are negatively related to the forest land. This means that the observed concentration values of the three variables would increase if the persentage area of urban or village land cover increases, whereas the concentration values would decrease if the percentage area of forest land increases. Therefore, the same order exists of the three variables for different land-use structures: Village- urban-Farmland > Village-Farmland > Forest. In comparison, dissolved oxygen has the negative relationships with urban, village and farmland, so the order represents as Village- urban-Farmland < Farmland < Village-Farmland < Forest.

4.4 The linkage model

Based on the exponential model, separate multiple regression models are developed to estimate the contributions of different land types on six stream water quality variables, including TN, NO3- N, TP, PO4- P, COD and DO, in Wenyu River watershed. The resulted models are identified to well explain the water quality variables using land use types. And the goodness-of-fit of these models are reasonably satisfactory. Table 7 presents the examples of regression models developed for TN and NO_3-N in this case study, in which each model is selected with the highest R and R^2, which indicates the significant level of using land use types to explain the water quality of the watershed.

For this land regression analyses, the concentration data of total nitrogen and nitrate are respectively natural log-transformed. The use of predictive equations allows city planners to model various scenarios of landscape alterations and observe the effects on water quality. From the table, it is determined that the regression models have a reasonably high degree of "goodness of fit", i.e., the R^2 values > 0.65, but the result of total nitrogen is less than 0.65. The observed and predicted data for total nitrogen and nitrate are compared using scatter plots in Figure 9. In the figure, most data distribute around the 45 degree lines, indicating a strong linear relationship between the two concentrations. The further investigation will be performed with more water samples of *in situ* measurements in the near future.

Water Quality Parameters	Regression Equations	*R* Value	*R* Square	Std. Error of the Estimate
Total Nitrogen (TN)	*Predictors:* Forest, Urban, Village and Water *Equation:* *Ln (TN)* = -0.086Forest-0.057Village +0.301Urban+0.7Water+0.954	0.729	0.531	0.639
Nitrate (NO_3- N)	*Predictors:* Forest, Urban, Village and Bare *Equation:* *NO_3- N* = -0.083Forest-0.240Village +0.794Urban+1.209Bare+2.819	0.828	0.685	0.857

Note: The independent variables of %Forest, %Urban and %Water were natural log-transformed because their normality assumptions were not met

Table 7. Regression equations developed for TN and NO_3-N in Wenyu River Watershed.

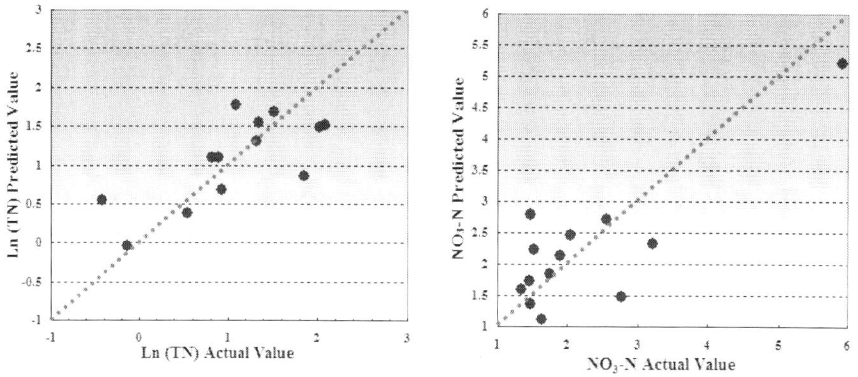

Fig. 9. Goodness-of-fit of statistical models for six water quality variables prediction.

The results can provide insight into the linkage between land-use types and stream water quality. The regression models have used in several ways by environmental planners and others interested in watershed management. The models can help examine the relative sensitivity of water quality variables to alterations in land-use types within a watershed. If the pattern of land-use changed, the levels of contaminants should be changed accordingly. Only with a better land-use planning, it is able to reduce the water quality deterioration.

4.5 Water quality changes with land-use types

In the study, we also examined the changes of water quality in relation to the changes of land-use types in the Wenyu River watershed. It is very clear that most of water quality variables were degraded from 2000 to 2008. For example, both TN and TP increased relatively high in farmland, urban and village areas, but very little change in forest areas. Since urban areas are dramatically increasing from 2000 to 2008, their impacts on TN and TP are quite obvious. These results not only provide the linkages between land-use types and stream water quality, but also show the high correlation of land-use types and water quality variables. The results indicate that water quality improvement and ecological restoration have great effects on the regional sustainable development. Thus, if the sustainable development is pursued, land management should consider the potential impacts of land-use on water quality changes in the watershed scale.

5. Conclusion

Land use-land cover (LULC) change is one of the major environmental changes occurring around the globe. Water quality is such one factor affected by LULC change. In this study, an integrated approach, involving remote sensing technology, geographic information system (GIS), statistical and spatial analysis, and hydrologic modeling, is used to conduct a comprehensive study on the relationship between land-use types and water quality in the Wenyu River watershed. Landsat TM data is used to extract the land-use information in the study area. The result suggests that this model is indeed an useful tool in hydrologic research and management. The results of water quality comparison with different land-use types show that land use types are significantly correlated to water quality variables. The Spearman's rank correlation analyses confirmed the change of water quality is impacted by land-use changes. Based on an exponential model, multiple regression models were applied to estimate the contributions of different land types on six stream water quality variables, including TN, NO_3- N, TP, PO_4- P, COD and DO, in Wenyu River watershed. The obtained results are identified well to explain the water quality variables using land-use types, with the reasonable satisfactory in the goodness-of-fit of the models.

The results can provide insight into the linkages between land-use types and stream water quality. The study offers the supporting evidence for the previous studies to serve as a reference to similar studies estimating the response of water quality to the land-use change. The models can help examine the relative sensitivity of water quality variables to alterations in land-use types within a watershed. The predicted values are close to the actual monitored values, which indicates that with little calibration and validation, the regression model can be applied in other watersheds under a different geographical scale in a different region with variable landscapes.

The results also indicate that with the integration of GIS and ecological modeling, a decision-making support system can be developed to manage land development and control non-point sources pollution at the watershed scales. This study also suggests that if a sustainable development is pursued, land management should consider the ipmacts of land-use types on water quality change in the area. The study provides a technical support for the water quality improvement and ecological restoration in the Wenyu River watershed, which has the great significance for the sustainable economic development of Beijing City.

However, in the study, only land-use related variables are considered in the models. In fact, there are other factors would be related to water quality levels in a sub-watershed, such as population characteristics, waste water treatment plants, soil types, average precipitation and other physical or biological variables. Nevertheless, the models in this study fail to reflect these variations for the sake of discussion. Hence, in the near future, other characteristics as the research background information will be helpful in the identification of the problems and developing a more rigorous linkage models between land-use types and water quality.

Several previous studies argued that the significant influences from land-use on water quality only exist within a shorter distance of the receiving water body. Hence, estimating the relationship between the buffer landscape and stream water quality will be another subject of the future study.

Estimating the links between land-use types and water quality over an extended period is crucially important task in the future works. The further study can help understand the response of water quality change to the change of land-use types, and give the environmental planners more information for the decision-making in land management. Furthermore, persistent water quality monitoring is useful to assist in identifying how land-use planning brings help in the control of water quality change in the watershed scale. This study also demonstrates an example of the issue of how LULC change is linked to water quality, one of the most precious resources on earth.

6. Acknowledgments

The authors are extremely grateful to Prof. Liding Chen and Dr. Ranhao Sun for their support of the field work and data collection at the Research Center for Eco-Environmental Sciences (RCEES) under the Chinese Academy of Sciences (CAS), and Jinrong Hu and Xiaofei Chen from the Chinese University of Hong Kong (CUHK) for their help in data processing and discussion. The authors would like to thank the two anonymous reviewers who gave helpful and critical comments to improve the original manuscript. The research is jointly supported by the CUHK Direct Grants, GRF (CUHK454909 and CUHK459210), National Outstanding Youth Science Foundation of China (40925003) and the Open Fund of State Key Laboratory of Urban and Regional Ecology (SKLURE2010-2-3).

7. References

Ahearn, D.S., Sheibley, R.W., Dahlgren, R.A., Anderson, M., Johnson, J. and Tate, K.W. (2005). Land use and land cover influence on water quality in the last free-flowing river draining the western Sierra Nevada, California. *Journal of Hydrology*, 313, 234-247.

Allan, J.D., Erickson, D.L. and Fay, J. (1997). The influence of catchment land use on stream integrity across multiple spatial scales. *Freshwater Biology*, 37, 149-161.

Anderson, B. W. and Ohmart, R. D. (1985). Riparian Revegetation as a Mitigating Process in Stream and River Restoration. In *The Restoration of Rivers and Streams: Theories and Experience* (pp. 41-80), Gore, J. (Editor). Boston: Butterworth Publication.

Basnyat, P., Teeter, L. D., Flynn, K. M. and Lockaby, B. G. (1999). Relationships between landscape characteristics and nonpoint source pollution inputs to coastal estuaries. *Environmental Management*, 23(4), 539-549.

Basnyat, P., Teeter, L. D., Lockaby, B. G. and Flynn, K. M. (2000). The use of remote sensing and GIS in watershed level analyses of non-point source pollution. *Forest Ecology and Management*, 128, 65-73.

Callender, E. and Rice, K.C. (2000). The urban environmental gradient: anthropogenic influences on the spatial and temporal distributions of lead and zinc in sediments. *Environmental Science and Technology*, 34, 232-238.

China State Environmental Protection Administration (2002). Environmental quality standards for surface water-GB 3838-2002. Beijing: China Environmental Science Press (in Chinese).

Djokic, D. and Ye, Z. (1999). DEM Preprocessing for Efficient Watershed Delineation. 1999 Esri User Conference, San Diego, CA.

Emmerth, P.P. and Bayne, D.R. (1996). Urban influence on phosphorus and sediment loading of West Point Lake, Georgia. *Water Resources Bulletin*, 32, 145-154.

Fetter, C.W. (1994). Applied Hydrogeology (3rd edition). New York: Macmillan College Publishing.

Fisher, D.S., Steiner, J.L., Endale, D.M., Stuedemann, J.A., Schomberg, H.H., Franzluebbers, A.J. and Wilkinson, S.R. (2000). The relationship of land use practices to surface water quality in the Upper Oconee Watershed of Georgia. *Forest Ecology and Management*, 128: 39-48.

Garbrecht, J., and Martz, L.W. (1999). Digital Elevation Model Issues in Water Resources Modeling. Proceedings of the 19th Esri Users Conference, San Diego, CA.

Griffith, J.A. (2002). Geographic techniques and recent applications of remote sensing to landscape-water quality studies. *Water, Air, and Soil Pollution*, 138: 181-197.

Hem, J. D. (1985). Study and interpretation of the chemical characteristics of natural water. United States Geological Survey Water-Supply Paper 2254, Washington, DC.

Hill, A.R. (1981). Stream phosphorus exports from watersheds with contrasting land uses in southern Ontario. *Water Resources Bulletin*, 17, 627-634.

Hua, Z., Du, G., Wang, D., Wu, Y. and An, Z. (2010). Analysis on Phytoplankton and Water Quality of Wenyu River in Beijing. *World Sci-Tech R & D*, 32(2), 213-215.

Johnson, L.B., Richards, C., Host, G.E. and Arthur, J.W. (1997). Landscape influences on water chemistry in midwestern stream ecosystems. *Freshwater Biology*, 37, 193-208.

Loague, K., Corwin, D.L. and Ellsworth, T.R. (1998). The challenge of Predicting Nonpoint-source Pollution. *Environmental Science and Technology*, 32, 130-133.

Luo, X., Shi, Z., Yin, W., Chen, J., Li, L. and Wu, S. (2010). Effects of Land Use Structure on Nitrogen Export in Hujiashan Watershed of Danjiangkou Reservoir Area, China. *Environmental Science*, 31(1), 56-62.

Maidment, D. R., 2002a. Arc Hydro: GIS for Water Resources. Redlands, CA: ESRI Press.

Morse, C.C., Huryn, A.D. and Cronan, C. (2003). Impervious surface area as a predictor of the effects of urbanization on stream insect communities in Maine, USA. *Environmental Monitoring and Assessment*, 89, 95-127.

Ngoye, E. and Machiwa, J. (2004). The influence of land-use patterns in the Ruvu river watershed on water quality in the river system. *Physics and Chemistry of the Earth*, 29, 1161-1166.

Norman, C.G. (1991). Urban runoff effects on Ohio River water quality. *Water Environmental Technology*, 3, 44-46.

Paul, M.J. and Meyer, J.L. (2001). Streams in the urban landscape. *Annual Review of Ecology and Systematics*, 32, 333-365.

Peterjohn, W. T. and Correll, D. L. (1984). Nutrient dynamics in an agricultural watershed: Observations on the role of a riparian forest. *Ecology*, 65, 1466-1475.

Phillips, J.D. (1989). Evaluation of North Carolina's estuarine shoreline area of environmental concern from a water quality perspective. *Coastal Manage*, 17, 103-117.

Roberts, G. and Marsh, T. (1987). The effects of agricultural practices on the nitrate concentration in the surface water domestic supply sources of Western Europe. *IAHS Publication No. 164*, 365-380.

Rose, S. (2002). Comparative major ion geochemistry of Piedmont streams in the Atlanta, Georgia region: Possible effects of urbanization. *Environmental Geology*, 42, 102-113.

Schoonover, J.E. and Lockaby, B.G. (2006). Land cover impacts on stream nutrients and fecal coliform in the lower piedmont of West Georgia. *Journal of hydrology*, 331, 371-382.

Schueler, T. (1995). The importance of imperviousness. *Watershed Protection Techniques*, 1, 100-111.

Sharpley, A. and Meyer, M. (1994). Minimizing agricultural nonpoint-source impacts: a symposium overview. *J. Environ. Qual.*, 23, 1-13.

Shi, H. (2008). Study on water pollution of Wenyu River System in Changping. *Water Resources and Hydropower Engineering*, 39(5), 10-19.

Tong, S. and Chen, W. (2002). Modeling the relationship between land use and surface water quality. *Journal of Environmental Management*, 66, 377-393.

Wahl, M.H., McKellar, H.N. and Williams, T.M. (1997). Patterns of nutrient loading in forested and urbanized coastal streams. *Journal of Experimental Marine Biology and Ecology*, 213, 111-131.

Wang, J., Cheng, W., Liu, H., Wu, Y. and Wang, F. (2008). The key points to ecological planning for the improvement of urban river: A case study of Wenyu River in Beijing city. *Sichuan Environment*, 27(1), 1-4.

Wang, S. and Song, X. (2008). Water Resource Assessment. Zhengzhou: Yellow River Irrigation Press.

You, Y., Xu, Z., Wang, P., Shen, Y., Liao, R., Liu, C. and Gu, H. (2009). Study on the Health Assessment of Ecotype River in Wenyuhe River. *Journal of Water Resource & Water Engineering*, 20(3), 19-24.

Zheng, F., Meng, Q., Wang, P. and Jin, G. (2007). Study on status and improvement strategies of water environment in Wenyu River of Beijing. *Beijing Water*, 5, 5-8.

Part 5

Indicators of Change

15

Heavy Metals Contamination of a Mediterranean Coastal Ecosystem, Eastern Nile Delta, Egypt

M. F. Kaiser[1], H. A. Aboulela[2], H. A. El-Serehy[3] and H. Ezz El-Din[3]
[1]*Geology Department, Faculty of Science, Suez Canal University*
[2]*Marine Science Department, Faculty of Science, Suez Canal University*
[3]*Department of Zoology, College of Science, King Saud University, Riyadh,*
[1,2]*Egypt*
[3]*Saudi Arabia*

1. Introduction

Coastal Zone is "The scope of marine environment which covers territorial water, and the scope of land extending in wards that can affect or be affected by the marine environment. Half of Egypt's population lives in Egypt coastal zones, where sources of food, jobs and income are available. They depend mainly on traditional fishing and to a lesser degree on automation.

Coastal areas are characterized by high organic matter and nutrients from the continent, having fragile coastal ecosystem dependent on terrestrial conditions (Yáñez-Aracibia and Sánchez-Gil, 1988). Industrialization of coastal areas is very common in countries characterized by exploitation and importation economics, causing serious damage to coastal ecosystems, e.g. contamination of metals (Cardoso et al., 2001). Moreover, anthropogenic activities are known to have a wide range of potential effects of these coastal ecosystems, particularly from point and non-point sources of pollution.

The release of pollutants into coastal environment is a major human concern worldwide. These contaminants are known to readily accumulate in bottom sediments which serve as a repository of pollutants. Sediment contaminants could be released to the overlying water, resulting in potential adverse health effects to aquatic organisms (Daskalakis and O'Connor, 1995; Long et al., 1995; Argese et al., 1997; Ross and Delorenzo, 1997; Freret-Meurer et al., 2010). Among the adverse health effects associated with these contaminants are toxicity to the kidney, nervous and reproductive systems, as well as endocrine disruption and mutations (Collier et al., 1998; Nirmala et al., 1999; Ketata et al., 2007; Liu et al., 2008; Brar et al., 2009).

In addition, trace metals are known to bioaccumulate in edible aquatic organisms (e.g., mollusks), thus, representing a health risk to top predators, including humans (Fox et al., 1991; Renzoni et al., 1998; Huang et al., 2006; Díez et al., 2009).

The Egyptian coastline extends 3000 km along the Mediterranean Sea and Red Sea beaches in addition to the Suez and Aqaba gulfs. Natural conditions on Egyptian Mediterranean coasts differ significantly from those on the Red Sea coasts in terms of salinity, sea currents and temperature. Such difference has led to different biodiversity and ecosystems in each.

The Nile Delta coastal zone located along southeastern coastal area of Mediterranean Sea at unstable shelf of the northern section of Egypt (Fig.1). It is boarded by the north western part of Suez Canal from the east and northern side of Manzala Lake from the south. It is considered as the most important vital area along the Mediterranean Sea. Many of the Egyptian coastal zones are subjected to variable significant environmental hazards including, loss of ecosystem quality, coastal erosion, seismic risk and over-exploitation. Moreover, it comprises a lot of land-use changes and modifications to development of future mitigation strategies. High profile example is the region of the eastern Nile Delta coastal ecosystem. It is one of the most beautiful recreation centers along coastal area of the Mediterranean Sea. In addition, it has several natural gas companies, recreational areas and fishing activities. A number of factors acting together on this zone has contributed to environmental and coastline changes. Natural processes and anthropogenic activities have to be considered as the most effective factors at the area of study. The pollution problems in the study sites are chiefly due to high quantity of domestic sewage and the virtually total absence of control on toxic components. Mistakes in their management can have catastrophic consequences for ecosystem integrity and human development.

Contamination by trace metals has not been extensively studied in the Egyptian coastal zones along the Mediterranean Sea which are subjected to intense discharges of pollutants. Therefore, it is important that sediment and water contamination by these pollutants be assessed for better management and protection of these valuable costal ecosystems at El-Gamil beach along the western coast of Port Said on the Mediterranean Sea. Especially, this study area represents a pronounced area for fishing, industrial development, urban extensions and more tourism activities along eastern Mediterranean Sea.

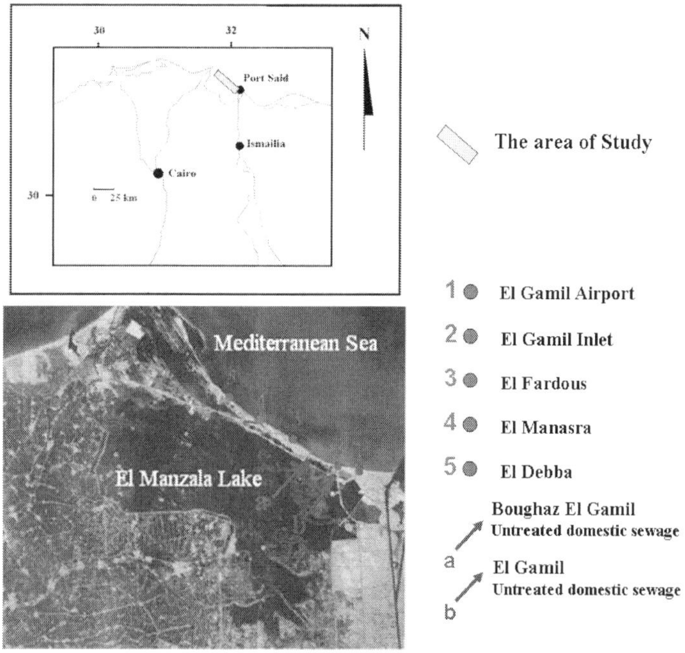

Fig. 1. Satellite image shows the location of the study area and the sampling sites

Domestic waste waters released into the eatern Nile Delta ecosystem contain fairly high concentrations of metals as Al, Cu, Fe, Pb and Zn (Stephenson, 1987). These produced from household products such as cleaning materials, toothpaste, cosmetic and human faces. In addition, pollutants resulting from industrial, sewage, agricultural discharges, many types of industrial wastes and urban runoff caused many problems for the human bodies. Furthermore, most of the present freshwater discharged into the coastal zones of Mediterranean Sea is of drainage water, which has been used for irrigation, often contaminated with sewage or industrial waste, which reached to the sea via the Delta lakes (Dowidar, 1988). The present study aims at to assess environmental status, develop and establish database required to monitor the main environmental problem along the Nile Delta ecosystem.

2. Materials and methods

2.1 Sampling sites and measurements

The study area constitutes a small part of the low lands laying west of Port Said City vise El Gamil zone, extending west wards parallel to the Deltaic Coast-Mediterranean Sea. It is boarded by the Suez Canal from the east and by Lake Manzala from the south. It is situated at nearly about 13 kilometers west of Port-Said City and extending between Latitude 31°:10' - 31°:20' N and Longitude 32°:00' - 32°:20' E with about 24 km^2 coverage area. Five sites along the El- Gamil beach, including El Gamil airport (Site 1), El Gamil inlet (site 2), El Fardous (Site 3), El Manasra (Site 4) and El Debba (Site 5) were chosen for this study (Fig.1). General features of anthropogenic activities at the study area with emphasis on descriptive features of sampling sites are listed in table (1). Heavy metals (Fe, Mn, Cd, Zn, Cu and Pb) levels were measured in bottom sediment, surface water and bivalve samples collected from each site during December 2005 and August 2006.

Sampling sites	Coordinates of sampling sites	General features of anthropogenic activities
El-Gamil Airport	31° : 16': 56" N and 32° : 14': 36" E	- El-Gamil Airport.
El-Gamil Inlet	31° :17':25" N and 32° : 12 ': 35" E	- Boughaz El-Gamil " - El-Gamil and El-Abtal tourist villages - Standing apart breakwaters
El-Fardous	31° :18': 12" N and 32° : 10 ': 49" E	- Ashtoum El-Gamil - El-Maghraby and El-Fardous tourist villages. - Company of Natural gas Petroget
El-Manasra	31° : 19': 45"N and 32° : 8 ': 14" E	- Company of Petrobel Balaeim " Natural gas". - International Factory of pipelines Industry. - Electric station power. - Small urban. - Anti-fouling paints.
El-Debba	31° :21':28"N and 32° : 0.4': 50" E	- Small urban and Traffic highway

Table 1. Locations and descriptive features of sampling sites.

2.2 Bottom sediment, surface water and bivalve analysis

A total of 30 sediment samples were collected from the five selected sites along El-Gamil coast during December 2005 and August 2006. The sediment samples were collected by pushing a plastic core (12 cm in diameter) into the bottom sediment to a depth of approximately 5 cm, then put in plastic bags and transferred to the laboratory. Sediments were dried at 100 C° for 24hrs and kept in plastic bags until analysis. A representative 50 gm of sediment sample were repeatedly treated with hydrogen peroxide (30%) for removal of organic matter and loss in weight was determined and the organic matter free samples were treated by using dilute HCl to remove calcium carbonates and loss in weight was determined according to method described by Folk, 1974. The grain size analysis of the examined samples was done according to Friedman and Johnson 1982. Dry sand was fractioned by dry sieving using sieves with openings of 2, 1, 0.5, 0.25, 0.125, 0.063 and 0.032 mm and an electric shaker. The weight percentages of the different size classes were calculated in three replicates per site sampled.

To detect the heavy metal contamination in the sediment samples, an exact weight of dry sample (0.5 gm) of each sediment sample was completely digested for about 2hrs in Teflon vessels using a mixture of HNO_3, $HCLO_4$ and HF (3:2:1 v/v 10 ml) (triplicate digestions were made for each sample). The final solution was diluted to 25 ml with distilled de-ionized water (Oregioni and Astone, 1984).

A total of 30 surface water samples were collected at a depth of 0.5 m from the five selected sites along the El-Gamil coast during December 2005 and August 2006. Water samples were collected in one liter white polyethylene bottles, which were placed in an ice-box following collection, and transferred to the laboratory for storage at 4°C until analysis.

Specimens of the most commercial bivalve *Donax trunculus* were collected from the five sampling sites during December 2005 and August 2006 by obtaining individuals with standard lengths between 1 and 3 cm. Standard length and body weights were recorded for each specimens. Live samples were left in clean water for 30 minutes to purge their guts. Only the soft tissue was kept in plastic bags and frozen until analysis. The soft parts of *Donax trunuculus* were dried at 70°C for 12 hrs before analysis. Exact dry weight of sample (0.5 g) was digested in Teflon vessels with analar nitric acid (HNO_3), tightly covered and allowed to predigest overnight at room temperature. Complete digestion and preparation of bivalve sample for trace metal analysis were done according to (UNEP/FAO/IAEA/IOC, 1984).

Total concentrations of Fe, Mn, Zn, Cu, Pb and Cd metals in bottom sediment, surface water and bivale (*D. trunculus*) samples were analyzed using an atomic absorption spectrophotometer (AAS) (Perkin Elmer, Waltham, MA, USA, model 1200 A), at the El-Fostat Center, Cairo, Egypt, according to the Standard Method 3110 (APHA 1992). All analyses were carried out in triplicate. For each run, three "blanks" were analyzed using the same procedure to check the purity of reagents and any possible contamination. A similarity test was carried out through cluster analyses from simple Euclidian distance.

3. Results and discussion

A similarity test established that sites 2 and 3 were highly similar, as well as sites 4 and 5, which were equally similar. Site 1 presented a similarity index closer to sites 2 and 3 than to the others. This test verified that El Gamil Beach can be divided into two areas. Sites 1, 2 and 3 represent the first area, and sites 4 and 5, the second.

The sediment at the five selected sites showed a highest percentage of very fine and fine sand coupled with low percentage of coarse fraction (Table 2). The maximum percentage of organic matter content was recorded at both sites of El-Gamil inlet and El Fardous with values of 4.13 ± 1.3 % and 4.23 ± 1.77 %, respectively. The high percentage of organic matter in the two mentioned sites can be attributed to the discharge of sewage wastes into the Mediterranean Sea through El Gamil outlet and Ashtoum El Gamil outlet along El Manzala Lake. Furthermore, the untreated domestic sewage discharged into the sea from four touristic villages in this coastal area *vise*: El Gamil, El Abtal, El Maghraby and El Fardous. The calcium carbonate content was very high at El Gamil airport with value of 27. 53 ± 5.10 %, while it was very low (3.03 ± 1.18%) at El Debba. The high values of calcium carbonate content at site (1) can be attributed to the accumulation of tremendous amounts of shell fragments blanketing the bottom sediment at El Gamil airport.

In the sediment samples, the concentration of heavy metals was higher during winter (Table 3). Even though there was variability among sites, the overall concentration range for a particular metal was relatively narrow, with no values that appeared to be unusual. Generally, in the sediment samples the heavy metals distribution followed the decreasing order of Fe> Mn> Zn> Pb> Cu> Cd. Moreover, El Manasra site suistained the heighest values of heavy metal contaminations recorded during the present study. This can be explained by the increasing industrial activities at this site. So, these high contaminants were associated with natural gas companies, pipeline industries and an electric power generating station operating at this area.

Sample number	Site	O.M.%	$CaCO_3$ %	Coarse trun. %	Fine trun. %	Phi (Φ)	Wentworth Mz
1	El-Gamil Airport	2	27.40	0.11	99.4	2.5	Fine sand
2		1.90	22.50	1.5	98.71	2.7	
3		2	32.70	0.06	98.06	2.8	
1	El-Gamil Inlet	3.60	7.40	0.92	97.1	3.2	Very fine sand
2		3.20	4.90	0.15	97.77	3	
3		5.60	7.30	0.13	96.25	3.06	
1	El-Fardous	2.20	6.70	0.13	98.72	2.6	Fine sand
2		5.50	2.80	0.16	98.18	2.9	
3		5	4.30	0.10	99.41	2.7	
1	El-Manasra	3.60	7.50	0.83	97.73	2.7	Fine sand
2		4.60	3.40	0.11	97.85	2.5	
3		3.40	2.50	0.10	98.50	2.8	
1	El-Debba	2.40	2.40	1.64	98.98	2.7	Fine sand
2		2	4.40	0.54	98.82	2.7	
3		3.10	2.30	0.35	99.07	2.7	

Notes: O.M: Organic matter; $CaCO_3$: Calcium carbonate; Coarse trun.: Coarse truncation; and Fine trun.: Fine truncation.

Table 2. The physical and chemical properties of sediment during summer 2006

Metal	Season	Sites									
		El-Gamil Airport Site (1)		El-Gamil Inlet Site (2)		El-Fardous Site (3)		El-Manasra Site (4)		El-Debba Site (5)	
		Mean	±S.D.	Mean	±S.D.	Mean	±S.D.	Mean	±S.D.	Mean	±S.D.
Fe (μg/g)	Winter	1936.5	45.2	1927.1	48.5	1865.7	38.3	1968.7	76.2	2030	4.8
	Summer	1830.4	68.8	1837.6	47.1	1748.2	44.1	1888.6	145.7	1918.7	24.7
Mn (μg/g)	Winter	200.8	16.5	204.2	13.5	213.5	8.3	254.3	16.6	229.4	32.7
	Summer	197.2	12.8	202.9	4	191.4	4	217.8	33	208.3	8.1
Cd (μg/g)	Winter	2.1	0.4	2	0.11	1.8	0.3	2.3	0.46	2.2	0.05
	Summer	1.9	0.41	1.4	0.23	1.8	0.3	2	0.57	1.9	0.2
Zn (μg/g)	Winter	33.9	3.4	34.6	2.1	33.1	1.3	42.2	1.5	36.9	5.5
	Summer	30.4	2.4	32.4	2.9	28	3.7	36.6	6	32.5	0.94
Cu (μg/g)	Winter	4	3.1	4.2	3.4	6.7	2.3	9.4	0.4	8.7	1
	Summer	6.6	1.6	5.7	1.1	6.6	1.6	9.4	1	7	1.3
Pb (μg/g)	Winter	23.8	4.1	18.4	2.5	20.2	4.1	22.8	3.4	24.8	1.3
	Summer	20.6	5	18.8	2.4	20.4	4.5	24.4	7.5	21	3.1

Note: ± S.D.: Standard deviation.

Table 3. Mean concentration of the total heavy metals in sediment samples during winter and summer (2005-2006)

	Fe	Mn	Cd	Zn	Cu	Pb	O.M.%	$CaCO_3$%	Sand %	Silt %
Fe	1									
Mn	0.847	1								
Cd	0.288	0.309	1							
Zn	0.782	0.981	0.219	1						
Cu	0.438	0.749	0.735	0.724	1					
Pb	0.427	0.714	0.788	0.684	0.996	1				
O.M.%	-0.317	0.11	-0.479	0.166	0.091	0.091	1			
$CaCO_3$%	-0.174	-0.373	0.138	-0.285	-0.235	-0.176	-0.694	1		
Sand %	0.238	0.157	0.955	0.024	0.562	0.62	-0.528	0.052	1	
Silt %	-0.341	-0.193	-0.904	-0.04	-0.499	-0.554	0.559	0.021	-0.983	1

Table 4. Correlation coefficient matrix between heavy metals, organic matter (O.M.), Calcium carbonate ($CaCo_3$), and grain size in sediment samples during summer 2006

Metal	Season	Sites									
		El-Gamil Airport Site (1)		El-Gamil Inlet Site (2)		El-Fardous Site (3)		El-Manasra Site (4)		El-Debba Site (5)	
		Mean	±S.D.	Mean	±S.D.	Mean	±S.D.	Mean	±S.D.	Mean	±S.D.
Fe (µg/L)	Winter	786	14.4	793.3	17	744	19	834.6	50	841.3	14
	Summer	872.6	27	832	24.9	822.6	14.7	846.6	30.2	896	15
Mn (µg/L)	Winter	170	11.1	174.6	7.5	168.6	20.8	162	15	171.3	7.5
	Summer	198	5.2	190.6	9	180	7.2	169.3	13	187.3	6.4
Cd (µg/L)	Winter	2.4	0.3	2.4	0.11	2.7	1.5	3.1	0.2	2.8	4.2
	Summer	2.6	0.2	0.8	1.3	2.7	0.11	2.6	0.2	2.4	0.2
Zn (µg/L)	Winter	281.3	21.5	287.3	16.7	242.6	6.4	277.3	16.2	269.3	18.5
	Summer	300	17.4	271.3	17.2	283.3	16.2	294.6	32.5	282.6	20
Cu (µg/L)	Winter	15.3	2.3	14	2	16	2	21.3	1.1	17.3	4.1
	Summer	13.3	3	12.6	2.3	14	2	19.3	1.1	18.6	1.1
Pb (µg/L)	Winter	34.6	7	31.3	5	35.3	6.1	44	2	50	2
	Summer	42.6	8	40	6	41.3	6.4	50.6	4.1	56	7.2

Note: ± S.D.: Standard deviation.

Table 5. Mean concentration of total heavy metals in water samples during winter and summer (2005-2006).

Metal	Season	Sites									
		El-Gamil Airport Site (1)		El-Gamil Inlet Site (2)		El-Fardous Site (3)		El-Manasra Site (4)		El-Debba Site (5)	
		Mean	±S.D.	Mean	±S.D.	Mean	±S.D.	Mean	±S.D.	Mean	±S.D.
Fe (µg/g)	Winter	52.8	0.5	53.1	1.3	57.2	3.6	55.2	2	47.2	2.5
	Summer	66.4	1.7	61.8	0.9	57.2	4.1	60.7	3.3	59.3	4.2
Mn (µg/g)	Winter	4.8	0.3	6.8	0.3	6.4	0.1	8.4	1.2	7.6	0.6
	Summer	7.6	0.4	6.8	0.4	6	0.5	7.1	1.2	6.3	1.4
Cd (µg/g)	Winter	2	0.1	2.1	0.15	1.6	0.3	2.4	0.3	2	0.3
	Summer	2	0.3	2.2	0.2	2.4	0.3	2.1	0.1	2.1	0.1
Zn (µg/g)	Winter	34.4	0.7	27.1	5.2	25.2	1.5	22	1.9	26.8	3.2
	Summer	36.4	3.8	34.8	0.8	32.8	0.7	33.4	0.5	32.9	0.1
Cu (µg/g)	Winter	3.6	0.2	3.8	0.3	4.8	0.3	3.6	0.2	3.2	0.2
	Summer	4.4	0.6	4.2	0.5	4	0.1	4.3	0.1	4.1	0.1
Pb (µg/g)	Winter	6.8	1.7	6.6	0.6	5.6	0.9	6.8	1.7	7.2	0.1
	Summer	9.2	0.2	9	0.3	8.8	0.3	9	0.2	8.9	0.1

Note: ± S.D.: Standard deviation.

Table 6. Mean concentration of total heavy metals in *Donax trunculus* samples during winter and summer (2005-2006).

Classification of pollution level of each element according to the American rules for sediment, in conformity to the lowest effect (LEL), heavily polluted category (HPC) and sever effect level (SEL). Judging by the present results, El Gamil beach can be considered a non-metal-polluted area, according to the pollution levels determined by the international rules (Ontario Ministry of the Environment-OME and United States Environmental Protection Agency - EPA).

Correlation coefficient matrix between heavy metals, organic matter, calcium carbonate and grain size of the sediment collected during August 2006 was statistically calculated (Table 4). The correlation coefficient matrix between heavy metal concentrations and physico-chemical characteristics of the sediment samples of El Gamil beach varied between negatively and significantly ones. Strong positive correlations were noticed between Cd and sand ($r = 0.995$); and Cu and Pb ($r = 0.996$). While, inverse correlations between Cd and silt ($r = -0.904$); and sand and silt ($r = -0.983$) as shown in figures 2, 3, 4, and 5 respectively.

The concentrations of heavy metals were higher in the summer than in the winter in the surface water samples collected during the present study (Table 5). Even though there was variability among sites, the overall concentration range for a particular metal was relatively narrow, with no values that appeared to be unusual. Not surprisingly, Fe concentrations were the highest, ranging from 822.6 ± 14.7 $\mu g\ g^{-1}$ at El Fardous to 896 ± 15 $\mu g\ g^{-1}$ at El Debba. Cd concentrations were the lowest and ranged from 0.8 ± 1.3 $\mu g\ g^{-1}$ at El Gamil inlet to 2.7 ± 0.11 $\mu g\ g^{-1}$ at El Fardous.

The concentrations values of heavy metal ($\mu g\ g^{-1}$ dry wt) in the soft tissue of *Donax trunculus* are shown in table (6). The concentrations of heavy metals were higher in the summer than in the winter in the bivalve samples collected during the present study, and with the decreasing order of Fe> Zn> Pb> Mn> Cu> Cd. Based on the data given in table (6), it seems that the observed variation in metal levels in *Donax trunculus* at different sites can be attributed to two mechanisms. The first is the availability of different metals in different sites, which in its turn depends on the pollution sources that usually vary among different sites. The second is the animal involves different uptake and retention mechanisms which may also vary with physiological and environmental factors (Byran, 1973) or even with the sexual state of the animal (Alexander and Young, 1976). Generally, the values of heavy metals concentration varied with insignificant range in both seasons and among sampling sites due to similar conditions affecting these sites. The highest concentration level of Zn (36.4 ± 3.8 $\mu g\ g^{-1}$) and Cu (4.8 ± 0.3 µg g-1) metals in the analyzed species are less than the maximum permissible levels (MPLs) of 100 µg g-1 and 10 µg g-1 for Zn and Cu, respectively. The (MPLs) of 2 µg g-1, and 5 µg g-1 declared by WHO 1982 and WHO 2006 for Cd and Pb, respectively are much lower than those detected in the soft tissues of D. trunculus with 4.8 ± 0.3 µg g-1 for Cu and 9.2 ± 0.2 µg g-1 for Pb. Donax trunculus inhabiting El Gamil beach along the western coast of Port Said on the Mediterranean Sea are more likely to be toxic for public health.

On the other hand the concentration factor of metals is considered as an indicator of heavy metals accumulation in the tissues of aquatic organisms in relation to their concentration in the ambient water (Sultana and Rao, 1998). The concentration factor values for the studied metals in *Donax trunculus* are delineated in table (7). For *Donax trunculus*, Cd gave the highest accumulation rate in the animal tissue with C.F. values ranging between 833.33 and 592.59 in winter and 2750 and 769.23 in summer. The order of C.F. in the soft tissues of *Donax trunculus* was Cd> Cu> Pb> Zn> Fe> Mn, respectively. This order indicates that *Donax trunculus* can be used as good indicator for the toxic metals as Cd and bio-indicator for essential metals as Cu.

Metal	Sites	Seasons					
		Winter			Summer		
		Mean concentration of water	Mean concentration of *Donax trunculus*	C.F.	Mean concentration of water	Mean concentration of *Donax trunculus*	C.F.
Fe	El-Gamil Airport	786	52.8	67.17	872.6	66.4	76.09
	El-Gamil Inlet	793.3	53.1	66.93	832	61.8	74.27
	El-Fardous	744	57.2	76.88	822.6	57.2	69.53
	El-Manasra	834.6	55.2	66.13	846.6	60.7	71.69
	El-Debba	841.3	47.2	56.10	896	59.3	66.18
Mn	El-Gamil Airport	170	4.8	28.23	198	7.6	38.38
	El-Gamil Inlet	174.6	6.8	38.94	190.6	6.8	35.67
	El-Fardous	168.6	6.4	37.95	180	6	33.33
	El-Manasra	162	8.4	51.85	169.3	7.1	41.93
	El-Debba	171.3	7.6	44.36	187.3	6.3	33.63
Cd	El-Gamil Airport	2.4	2	833.3	2.6	2	769.2
	El-Gamil Inlet	2.4	2	833.3	0.8	2.2	2750
	El-Fardous	2.7	1.6	592.5	2.7	2.4	888.8
	El-Manasra	3.1	2.4	774.2	2.6	2.1	807.6
	El-Debba	2.8	2	714.3	2.4	2	833.3
Zn	El-Gamil Airport	281.3	34.4	122.3	300	36.4	121.2
	El-Gamil Inlet	287.3	27.1	94.32	271.3	34.8	128.3
	El-Fardous	242.6	25.2	103.8	283.3	32.8	115.7
	El-Manasra	277.3	22	79.33	294.6	33.4	113.4
	El-Debba	269.3	26.8	99.51	282.6	32.9	116.4
Cu	El-Gamil Airport	15.3	3.6	235.3	13.3	4.4	330.8
	El-Gamil Inlet	14	3.8	271.4	12.6	4.2	333.3
	El-Fardous	16	4.8	300	14	4	285.7
	El-Manasra	21.3	3.6	169	19.3	4.3	222.8
	El-Debba	17.3	3.2	184.9	18.6	4.1	220.4
Pb	El-Gamil Airport	34.6	6.8	196.5	42.6	9.2	215.9
	El-Gamil Inlet	31.3	6.6	210.8	40	9	225
	El-Fardous	35.3	5.6	158.6	41.3	9	217.9
	El-Manasra	44	6.8	154.5	50.6	8.9	175.8
	El-Debba	50	7.2	144	56	8.9	158.3

Note: C.F.: Concentration factor.

Table 7. Concentration factor of metals in the soft tissue of *Donax trunculus* during winter and summer 2005-2006.

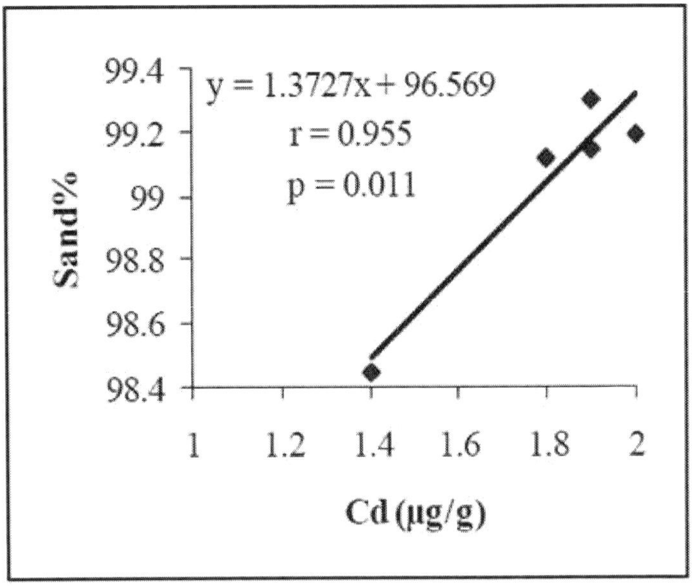

Fig. 2. Correlation between Cd and sand% in sediments

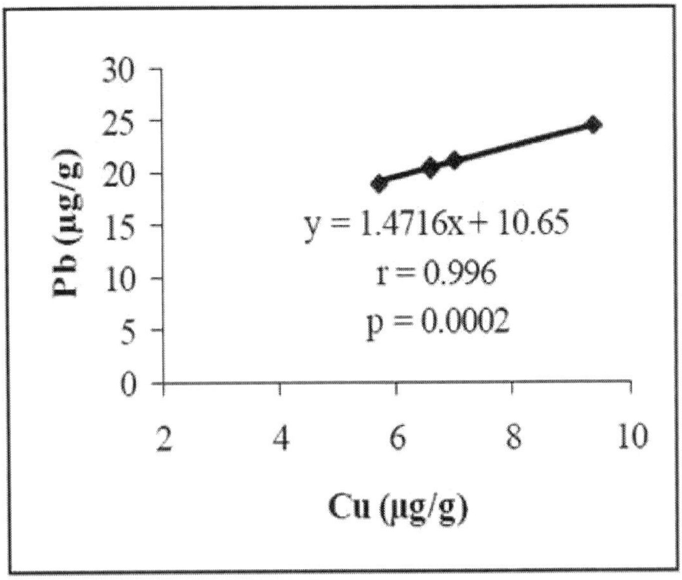

Fig. 3. Correlation between Cu and Pb in sediments

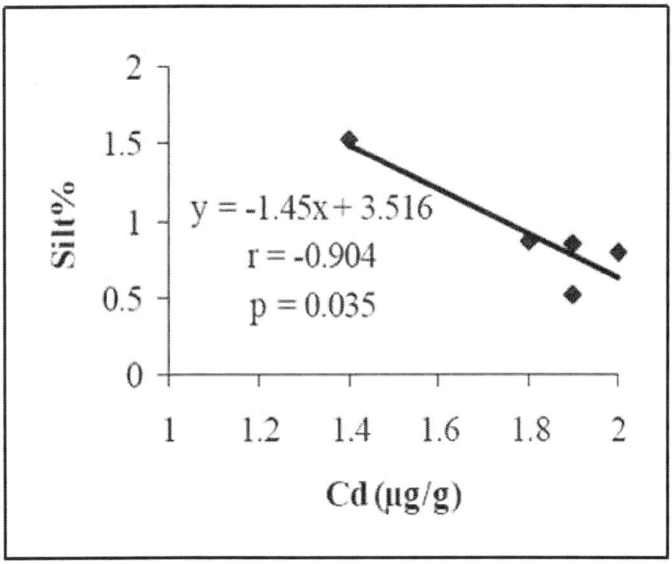

Fig. 4. Correlation between Cd and silt% in sediments

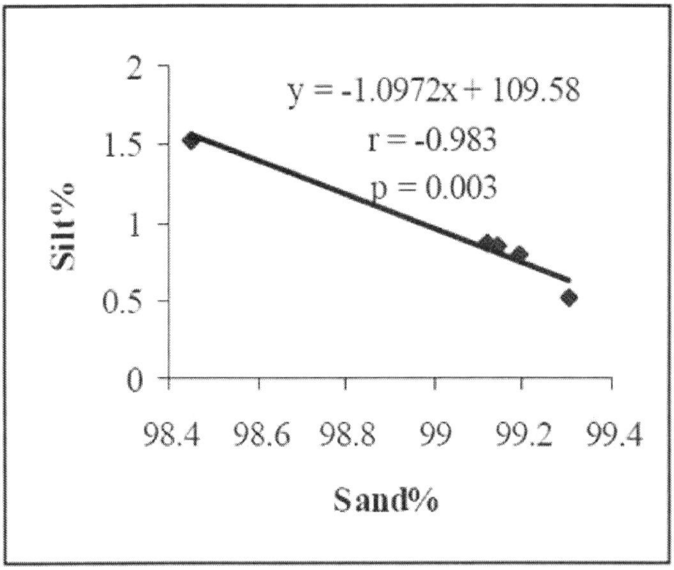

Fig. 5. Correlation between sand% and silt% in sediments

4. Conclusion

Nearly 40% of industrial development activities are practiced in Egyptian coastal zones, in addition to a number of urban and tourism development activities. Furthermore, coastal zones monopolize the seaports infrastructure, in addition to agricultural and land reclamation sectors, as well as a developed road network capable of accommodating all development aspects. The coastal zones attract increasing numbers of migrating workers from other areas and Governorates. Tourism development represents one of the main activities in Egypt's coastal zones, particularly in terms of beach development regarded as the basis of international tourist attraction.

The quality of marine and coastal environments and their environmental resources along the Nile Delta ecosystem are threatened by a number of hazards related to internal development inside the country whose impacts are carried to coastal zones via the river Nile, agricultural drainage system and air (land sources).

Results of the present study indicate El Gamil beach along the western coast of Port Said City on the Mediterranean Sea is considered a non metal polluted area, suggesting regular monitoring of metals in the sediments, water and marine organisms should be undertaken, due to the rapid growth of area. Different types of pollutants, including agricultural, industrial, organic compounds and domestic discharge were identified by analyzing bottom sediment, surface water and bivalve sampling collected from the study area of heavy metals. Contaminants originating from agricultural and domestic sources were detected along the El Fardous and El Gamil inlets. Industrial pollutants were detected at the El Manasra and El Fardous sites. These contaminants were associated with natural gas companies, pipeline industries and an electric power generating station. The proximity to various anthropogenic sources of pollutants warrants a continue monitoring program in the Egyptian coasts along Mediterranean Sea for inorganic and chemical organic compounds in sediments, water, and biota in order to have an effective coastal management program to protect the ecological integrity of this valuable ecosystem and the health of humans associated with it. The reported results could be considered as documentation of a good understanding to assess the ecosystem status of the concerned sites, and might be useful to researchers interested in the cotastal zones of Mediterranean Sea. Farther environmental studies are required to assess and improve the development planning and economic activities along of study area and its vicinity.

5. References

Alexander, G., Young, D. (1976). Trace metals in Southern California mussels. Marine Pollution Bulletin 7, 7-9.

APHA (1992). *Standard Methods for the examination of waters and wastewaters.* 16th ed., (Washington, DC; American Public Health Association).

Argese, E., Ramieri, E., Bettiol, C., Pavoni, B., Chiozzotto, E., Sfriso, A. (1997). Pollutant exchange at the water /sediment interface in the venice canals. Water, Air, and Soil pollution 99, 255-263

Brar, N.K., Waggoner, C., Reyes, J.A., Fairy, R., Kelley, K.M. (2009). Evidence for thyroid endocrine disruption in wild fish in San Francisco Bay, California, USA. Relationships to contaminant exposures. Aquatic Toxicology doi:10.101 /j.aquatox. 2009.10.023.

Bryan, G. (1973). The occurrence and seasonal variation of trace metals in scallops *Pecten maximus* and *Chlayms opercularis*. Journal of Marine Biology Association, Vol. 53, Pp. 145-166.

Cardoso, A.G.A., Boaventura, G.R., Filho, E.V.S., Brod, J.A. (2001). Metal distribution in sediments from the Ribeira Bay, Rio de janeiro, Brazil. Brazilian Chemical Society 12 (6), 767-774.

Collier, T.k.,Johnson, L.L., Stehr, C.M., Myers, M.S, Stein, j.e. (1998). A comprehensive assessment of the impacts of contaminants on fish from an urban waterway. Marine Environment Research 46, 243-247.

Daskalakis, K.D., O'Connor, T.P. (1995). Normalization and elemental sediment contamination in the coastal United States. Environmental Science and Technology 29, 470-477.

Díez, S., Delgado, S., Aguilera, I., Astray, J., Pérez-Gómez, B., Torrent, M., Sunyer, J., Bayona, J.M. (2009). Prental and early childhood exposure to mercury and methylmercury in Spain, a high-fish-consumer country. Archives of Environmental Contamination and Toxicology 56, 615-622.

Dowidar, M. (1988). Productivity of the southeastern Mediterranean. In national and man made hazard, El-Sabh, M.I. and Murty, T.S., eds., 477-498.

Folk, R. (1974). Petrography of sedimentary rocks. Herrphill, Texas. P. 182.

Fox, G.A., Collins, B., Hayakawa, E., Weseloh, D.V., Ludwig, J.P., Kubiak, T.J., Erdman, T.C. (1991). Reproductive outcomes in colonial fish-eating birds: a biomarker for occurrence and prevalence of bill defects in young double-crested cormorantsin the Great Lakes, 1979-1987. Journal of Great Lakes Research 17, 158-167.

Freret-Meurer, N.V., Andreata, J.V., Meurer, B.C., Manzano, F.V., Baptista, M.G.S., Teixeira, D.E., Longo, M.M. (2010). Spatial distribution of metals in sediments of the Riberia Bay, Angra dos Reis, Rio de Janeriro, Brazil. Marine Pollution Bulletin 60: 627-629.

Friedman, G., Johnson, K. (1982). Exercises in sedimentology. John Wiley & Sons, Inc, USA.

Huang, X., Hites, R.A., Foran, J.A., Hamilton, C., Knuth, B.A., Schwager, S.J., Carpenter, D.O., 2006. Consumption advisories for salmon based on risk of cancer and noncancer health effects. Environmental Research 101, 263-274.

Ketata, I., Smaoui-Damak, W., Guermazi, F., Rebai, T., Hamza-Chaffai, A. (2007). In situ endocrine disrupting effects of cadmium on the reproduction of Ruditapes descussatus. Comprative Biochemistry and Physiology C146, 415-430.

Liu, J., Goyer, R.A.,Waalker, M.P. (2008). Toxic effects of metals. In : Klaasen, C.D.(Ed.), Casarett and Doull's Toxicology. The Basic of Science of Poisons. McGraw-Hill, New York, pp.931-979.

Long, E.R., Macdonald, D.D., Smith, S.L., Calder, F.D. (1995). Incidence of adverse biological effects within ranges of chemical concentration in marine and estuarine sediments . Environmental Management 19, 81-97.

Nirmala, K., Oshima, Y., Lee, R., Imada, N., Honjo, T., Kobayashi, K. (1999). Transgenertaional toxicity of tributyltin and its combined effects with polychlorinated biphenyls on reproductive processes in japanese medaka (Oryzias latipes). Environmental Toxicology and Chemistry 18, 717-721.

Oregioni, B. and Astone, S. (1984). The determination of selected trace metals in marine sediments by flameless/ flame-atomic absorption sepctrophotometry. IAEA Monaco Laboratory, Internal Report.

Renzoni, A., Zino, F., Franchi, E. (1998). Mercury levels along the food chain and risk for exposed populations. Environmental Research 77, 68-72.

Ross, P., Delorenzo, M.E. (1997). Sediment contamination problems in the Caribbean islands: research and regulation. Environmental Toxicolology and Chemistry 16, 52-58.

Stephenson, M. (1987). The environmental requirements of Aquatic plants publication No. 65-Appendix A., agreement no. 8-131-499-9. A report prepared for the state Water Resources Control Biard, State of California. Sacramento, California.

Sultana, R., Rao, D. (1998). Bioaccumulation patterns of Zinc, Copper, Lead and Cadmium in Grey Mullet, *Mugil cephalus* (L.) from harbour waters of Visakhapmum, India. Bull. Environ. Contam. Toxicol., Vol. 60. Pp. 949-955.

UNEP/ FAO/ IAEA/ IOC, (1984). Sampling of selected marine organisms and sample preparation for trace metal analysis. Reference methods for marine pollution studies. Review 2:19.

WHO (1982). Toxicological evaluation of certain food additives, WHO Food Addit. Ser. No. 17, World Health Org., Geneva, Pp. 28-35.

WHO (2006). Joint FAO/WHO Expert Committee on Food Additives. Meeting (67th: 2006: Rome, Italy) Evaluation of certain food additives and contaminants: sixty-seventh report of the Joint FAO/WHO Expert Committee on Food Additives. (WHO technical report series; No. 940).

Yáñez- Arancibia, A., Sánchez-Gil, P. (1988). Ecologia de los recursos demersales marions, first edicao. AGT Editor S/A Mexico, D, P. 189.

16

HPLC Fingerprints of Porewater Organic Compounds as Markers for Environmental Conditions

Viia Lepane
Tallinn University of Technology
Estonia

1. Introduction

Lake sediments are considered invaluable natural archives that provide long-term records of past changes in climate and environment related to catchment processes as well as in-lake changes in biological communities. Moreover, lake sediments also register anthropogenic activities and man-made environmental problems. Lake sediments are known to accumulate different compounds during their formation and adsorption processes, and thus sediment investigations can be used as an important tool to assess the contamination of aquatic ecosystems. The organic matter in sediments is distributed between the particulate and dissolved phases, and usually the aquatic phase is named porewater (*pw*).

The dissolved organic matter (DOM) is an important component of aquatic ecosystems and of the global carbon cycle. It is known that changes in DOM quality and quantity have effects on the whole ecosystem. Quantitative and qualitative changes in DOM are related to precipitation, runoff, and seasons. DOM consists of a mixture of macromolecular compounds with a wide range of chemical properties and diverse origins. The DOM in lakes can serve as a molecular proxy for identification of previous inputs of organic matter. Moreover, detailed knowledge about DOM is greatly needed in order to reconstruct palaeoclimate or land-use. The biogeochemical transformation of DOM helps to elucidate past and present environmental conditions. For all those reasons, detailed DOM characterization at the molecular level is of utmost importance.

Since DOM is naturally a very complex mixture of molecules, the determination of its exact chemical composition is a complicated task. Only detailed chemical characterization using various analytical methods could be carried out. A part of DOM is optically active, enabling spectroscopic methods based on UV absorption to be used for the characterization. Another possible method of DOM analysis is chromatographic size fractionation using high-performance liquid chromatography (HPLC) with a size exclusion option (high-performance size exclusion chromatography - HPSEC). HPSEC has been widely used in studies of DOM together with spectroscopic methods. The reliability and sensitivity of this method have been reported and discussed previously (Chin et al., 1994; Hoque et al., 2003; Minor et al., 2002; Nissinen et al., 2001; Pelekani et al., 1999; Perminova et al., 1998, 2003; Specht & Frimmel, 2000; Zhou et al., 2000). Although HPSEC characterization of lake sediment *pw*DOM has demonstrated great potential for palaeolimnological research (Leeben et al., 2008a; Lepane et al., 2004, 2010a; Makarõtševa et al., 2010) it is not widely used for evaluating the long-term changes in aquatic ecosystems. At present, no comparative investigations of *pw*DOM from lake sediments are available. Coupling of HPSEC as a separation method with diode-array detection (DAD) allows DOM fingerprints and spectra of DOM molecular fractions to be obtained for qualitative and semi-quantitative analysis. The non-destructive analysis, small sample volume, and minimal sample pretreatment are great advantages of the HPSEC-DAD approach, making the method suitable for environmental studies. HPSEC-DAD has been adapted and optimized for analysis of *pw* samples under various conditions (Lepane et al., 2004; 2010a; O'Loughlin & Chin, 2004). The advantage of the usage of this chromatographic system is a better understanding of the qualitative and quantitative *pw*DOM properties by detecting aromatic fractions (chromophoric compounds). This method has recently been applied for monitoring and detection of organic matter from surface waters after oxidation treatment (Liu et al., 2010).

This study aims: (1) to investigate temporal changes in *pw*DOM components' qualitative and quantitative characteristics by exploring different sediment core records; (2) to find the similarities and differences in HPSEC-DAD fingerprints of *pw*DOM after applying the statistical data treatment methods; (3) to explore the potential impact of environmental change on *pw*DOM records in investigated sediment cores.

2. Materials and methods

2.1 Study area and sampling

The case studies were conducted at two sediment cores from Estonia: Lake Peipsi and Lake Rõuge Tõugjärv.

Lake (L.) Peipsi is the largest transboundary lake in Europe shared between Estonia and Russia. It is the fourth biggest lake in Europe. Its surface area is 3,500 km^2 with an average depth of 7 metres. The maximum depth is only 15 metres. The catchment area covers more than 47,000 km^2. The catchments area has been used for agricultural purposes for several millennia. On the northern side of the lake, extensive mining areas and several electric power plants operating on oil shale exist. As a consequence, enhanced delivery of nutrients to L. Peipsi has induced an increase in primary productivity within the lake and anthropogenic eutrophication during the last few decades. Today the lake is classified as eutrophic.

A 43-cm long sediment core was collected from L. Peipsi in March 2007 from location 58°47'13"N and 27°19'18"E. The sampling point was located in the middle of the lake. The water depth at the sampling site was 9.8 m. Sediment samples were taken by a Willner corer. The core was cut into 1-cm thick subsamples, packed into plastic bags, and transported to the laboratory. The chronology of the core was established via correlation of its loss-on-ignition (LOI) curve with that of the 2002 year core, which was previously dated by the ^{210}Pb radiometric method using gamma spectrometry (Appleby et al., 1986). For calculations of the ^{210}Pb dates the Constant Rate of Supply model (Appleby & Oldfield, 1978) was applied and the results were compared to two other independent dating approaches – the sediment distribution of artificial radionuclide ^{137}Cs and spheroidal fly-ash particles (Alliksaar et al., 1998). The methodology and results of the dating methods used and the reliability of the chronology are explained in detail in Heinsalu et al. (2007).

The second lake investigated is situated in South Estonia, where the anthropogenic pressure is not too high and is expressed mainly through the agricultural activity. L. Rõuge Tõugjärv (57°44′30″N; 26°54′20″E) is a small-size stratified hard-water mesotrophic lake with a surface area of 4.2 ha and a maximum depth of 17 m. The main source of pollutants in L. Tõugjärv sediments is the catchments area. The studied sediment core was visibly laminated, reflecting the annual changes in the lake. Annual laminations, or varves, typically consist of two visible layers (a clastic inorganic layer and a darker organic humic layer), and each varve can be considered as representing one year's deposition.

The topmost 13 cm of the sediment was loose unconsolidated dark gyttja (dated until 1986 AD), while the rest of the sediment sample was laminated gyttja with well-developed varves (dated until the year 1852 AD). The L. Rõuge Tõugjärv sediment core was taken in May 2006 with a Willner-type sampler. The core was transported in a tightly closed Plexiglas tube to the laboratory, immediately sliced into 1-cm thick sub-samples, and packed into plastic bags to maximally avoid oxygen exposure.

The age-scale for the sediment sequence was obtained by correlating marker varve horizons and LOI values with another sediment core sampled in 2001, which had been carefully dated by several parallel dating methods (varve counting, ^{210}Pb, ^{137}Cs, ^{241}Am, and spheroid fly-ash particles) (Alliksaar et al., 2005; Veski et al., 2005; Poska et al., 2008). According to this correlation the obtained sediment core covered about 150 years (1850–2005).

Pw samples for analysis were obtained by extraction of unfrozen sediments by centrifugation at 3,500 rpm for 30 minutes and filtration through 0.45 µm filters (Millex, Millipore). Samples were stored at 4 °C in the dark.

2.2 Chemical analyses

Absorbance spectra of the *pw* samples were collected using a Jasco V-530 UV/VIS Spectrophotometer (Japan), with 1-cm-pathlength fused silica cells and ultrapure water as the blank. Spectra were measured over the range of 200–500 nm with a 2.0-nm bandwidth. The dissolved organic carbon (DOC) concentration in *pw* samples was calculated from absorption spectra using the equation given by Højerslev (1988). The absorbance ratio at 250 and 360 nm (A250/A360), which reflects the aromaticity of dissolved molecules (Peuravuori & Pihlaja, 1997), was calculated from the spectra.

2.3 HPLC analyses

The molecular characteristics of DOM in sediments were determined using an HPLC system. The HPLC system comprised a Dionex P680 HPLC Pump, Agilent 1200 Series (Agilent Technologies, UK) diode array absorbance detector (DAD), and a Rheodyne injector valve with a 20-µL sample loop. A BioSep-SEC-S 2000 PEEK size exclusion analytical column (length 300 mm, diameter 7.50 mm, Phenomenex, USA) preceded by a suitable guard column (length 75 mm, diameter 7.50 mm, Phenomenex, USA) was used for separation. The applied flow rates were 0.5 mL min^{-1} (L. Peipsi samples) and 1.0 mL min^{-1} (L. Rõuge Tõugjärv samples). The column packing material was silica bonded with a hydrophilic diol coating, with a particle size of 5 µm and a pore size of 145 Å. The mobile phase consisted of 0.10 M $NH_4H_2PO_4$ - $(NH_4)_2HPO_4$ buffer at pH 6.8. The HPLC system was calibrated using five different molecular mass protein standards (Aqueous SEC 1 Std, Phenomenex, USA) (see Fig. 1). All solutions for HPLC measurements were prepared using ultrapure water passed through a MilliQ water system, filtered with 0.45-µm pore size filters

(Millipore), and degassed. Samples were analysed in triplicate. In general, the relative standard deviations for the replicated measurements did not exceed 5% (obtained by comparison of total peak areas). For quality control the aqueous protein standard was analysed each day. The chromatograms were recorded and processed by Agilent ChemStation software. Full details of the method used are described previously (Lepane et al., 2004).

Weight-average and number-average molecular masses of DOM (M_w and M_n, respectively) were determined using the formulae $M_w = \Sigma(h_iM_i)/\Sigma h_i$ and $M_n = \Sigma h_i/ \Sigma(h_i/M_i)$, where h_i is the detector output and M_i is the molecular mass, both at the i-th retention time (Mori & Barth, 1999).

As a semi-quantitative DOM characteristic, the total chromatogram peak areas, representing the total UV-absorbing fraction of the specific molecular size fraction of DOM in each sample, were used in the data analysis. The total chromatogram peak areas obtained with DAD actually represent the variations in optical intensities of DOM fractions at the chosen wavelength of 280 nm. The detector response (the height of the chromatogram at the i-th elution volume) refers to the amount of DOM in a specific molecular size fraction. The sum of all peak heights represents the total amount of DOM capable of UV adsorption in the sample (Matilainen et al., 2006; Peuravuori & Pihlaja, 1997; Vartiainen et al., 1997). Peak areas were used as a semi-quantitative characteristic to present age-related variations in the DOM fractions. To obtain qualitative DOM characteristics the chromatograms were divided into two molecular size fractions: 1) high molecular mass (HMW), and 2) humic substances (HS) (Lepane et al., 2010a, 2010b). The polydispersity M_w/M_n, describing the homogeneity or heterogeneity of organic matter, was calculated from the data obtained.

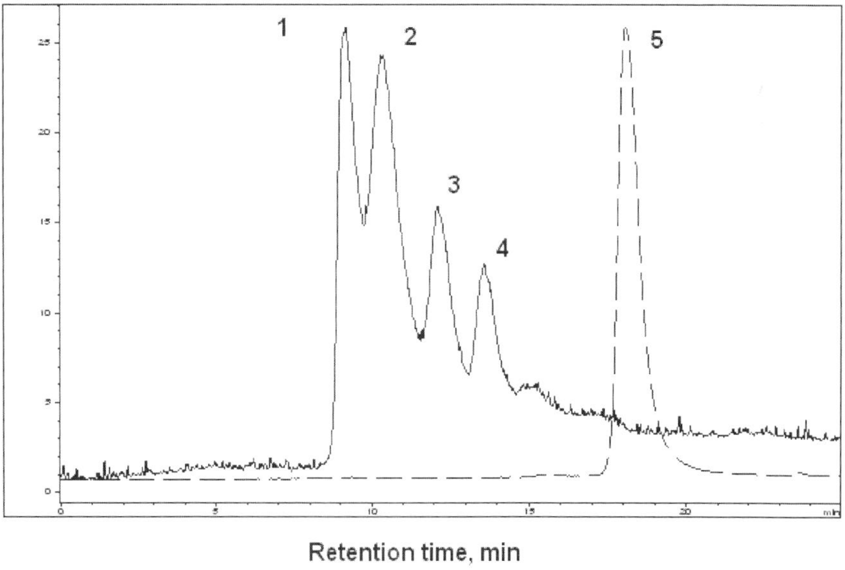

Fig. 1. Separation of calibration standards by HPLC. Protein molecular masses (1) 670 kDa, (2) 150 kDa, (3) 44 kDa, (4) 17 kDa, (5) 244 Da; detection wavelength 280 nm.

2.4 Statistical analyses

Cluster analysis using the Ward method was applied to reveal age-related periods in the analysed samples (Brereton, 2003). The analysis was performed on the chromatographic data. As descriptors of the DOM, all of the separated peak areas and total chromatogram areas, molecular masses, and their ratios, DOC and A250/A360, for all samples were included in the analysis. The Euclidean distance was used as a measure of the similarity-dissimilarity of the samples. The statistical analyses were carried out using WinSTAT for Excel software (R. Fitch Software, Germany).

3. Results and discussion

3.1 Main characteristics of the sediment profiles

The descriptions and dating values for analysed sediment cores are reported in Table 1. Both sediment cores covered roughly 150 years of sediment deposition. The characteristics of L. Peipsi sediment core solid-phase have been reported previously (Lepane, 2010b). Briefly, the organic matter content was 23 to 25% for the period 1860–1950 and increased up to 27% after the 1950s. This increase was followed by an increase in DOC values from the 1960s.

In the second case investigated, L. Rõuge Tõugjärv, the organic component of the particulate sediment matrix had low values (11-13%) between 1850 and the 1880s, followed by a distinct peak (18%) around the 1900s and a progressive rise to 27% thereafter (Alliksaar et al., 2005). The rest of the sediment consisted mainly of terrigenous mineral matter eroded into the lake from the catchment. In L. Rõuge Tõugjärv, the low organic matter values in sediment probably indicated land derived input from human-induced topsoil erosion and dilution of organic matter by addition of clastic mineral particles, while the increase in organic matter presumably reflected a lower contribution of terrestrial mineral material with a reduction in the overall rural activities. The DOC values corresponded to an increase in organic matter from 1850 to the 1890s; thereafter the decrease has been significant up to the present. The low DOC values may be due to a high amount of aliphatic organic compounds resulting from the microbial activity. Since the determination of DOC was based on the spectroscopic method, the aliphatic organic matter fraction was not determined in present study.

3.2 Multi-wavelength HPLC analyses

The sediment DOM was characterized by *pw* sample analysis using HPLC (HPSEC) with DAD. Absorbance spectroscopy with a single detection wavelength has been verified as a suitable detection method after HPLC separation (Lepane, 2010a). The UV absorbance at 250–280 nm has been widely used to provide an estimation of aromatic compounds (Filella, 2010). In the present study, multi-wavelength HPLC analysis has been carried out to detect changes in *pw*DOM composition and molecular mass profiles down the cores. The absorbance spectra were examined from 200 to 400 nm. Figures 2 and 3 display the HPLC chromatograms of *pw* from studied sediment cores and from different layers, respectively. The UV detector response range from 205 to 400 nm has been plotted against the retention times to obtain the multi-wavelength contour plots. Plot colours provide the visual representation of the relative absorbance intensity. The multi-wavelength plots allowed the most suitable detection wavelength for separated DOM components to be selected. Simultaneously, the chromatograms were registered at 280 nm. The chromatogram patterns suggested that the *pw*DOM molecules included two fractions in both lakes. The first peak, with shorter retention times of 6–7 min and of 8–9 min, corresponded to the fraction with larger molecules and was operationally named the high molecular mass (HMW) fraction. The second peak, whose maximum was at retention times of 11–12 min and of 15–16 min, was assigned to smaller DOM molecules and named the humic substances (HS) fraction.

Lake	Peipsi			Rõuge Tõugjärv			
Period	Sample no	Depth, cm	Age dating, y	Period	Sample no	Depth, cm	Age dating,y
I	1	0/-1	2006	I	1	0/-1	2006
	2	-1/-2	2005		2	-1/-2	2005
	3	-2/-3	2004		3	-2/-3	2005
	4	-3/-4	2003		4	-3/-4	2004
	5	-4/-5	2002		5	-4/-5	2002
	6	-5/-6	2000		6	-5/-6	2000
	7	-6/-7	1997		7	-6/-7	1998
	8	-7/-8	1994	II	8	-7/-8	1996
	9	-8/-9	1990		9	-8/-9	1994
	10	-9/-10	1986		10	-9/-10	1991
	11	-10/-11	1982		11	-10/-11	1989
	12	-11/-12	1978		12	-11/-12	1986
	13	-12/-13	1974		13	-12/-13	1984
II	14	-13/-14	1970		14	-13/-14	1981
	15	-14/-15	1966		15	-14/-15	1979
	16	-15/-16	1962		16	-15/-16	1976
	17	-16/-17	1958		17	-16/-17	1974
	18	-17/-18	1954	III	18	-17/-18	1971
	19	-18/-19	1950		19	-18/-19	1968
	20	-19/-20	1945		20	-19/-20	1964
	21	-20/-21	1940		21	-20/-21	1961
	22	-21/-22	1934		22	-21/-22	1958
	23	-22/-23	1928		23	-22/-23	1954
	24	-23/-24	1922		24	-23/-24	1951
	25	-24/-25	1916		25	-24/-25	1948
	26	-25/-26	1911		26	-25/-26	1944
	27	-26/-27	1907		27	-26/-27	1941
	28	-27/-28	1904		28	-27/-28	1938
	29	-28/-29	1901		29	-28/-29	1934
	30	-29/-30	1897		30	-29/-30	1931
	31	-30/-31	1892		31	-30/-31	1928
	32	-31/-32	1887		32	-31/-32	1924
	33	-32/-33	1882		33	-32/-33	1921
III	34	-33/-34	1880		34	-33/-34	1917
	35	-34/-35	1877		35	-34/-35	1914
	36	-35/-36	1875		36	-35/-36	1911
	37	-36/-37	1872	IV	37	-36/-37	1905
	38	-37/-38	1870		38	-37/-38	1899
	39	-38/-39	1867		39	-38/-39	1893
	40	-39/-40	1865		40	-39/-40	1887
	41	-40/-41	1862		41	-40/-41	1881
	42	-41/-42	1857		42	-41/-42	1875
	43	-42/-43	1852		43	-42/-43	1869
					44	-43/-44	1864
					45	-44/-45	1858

Table 1. Sediment samples numbering and depths down the profiles with dating values for Lake Peipsi and Lake Rõuge Tõugjärv.

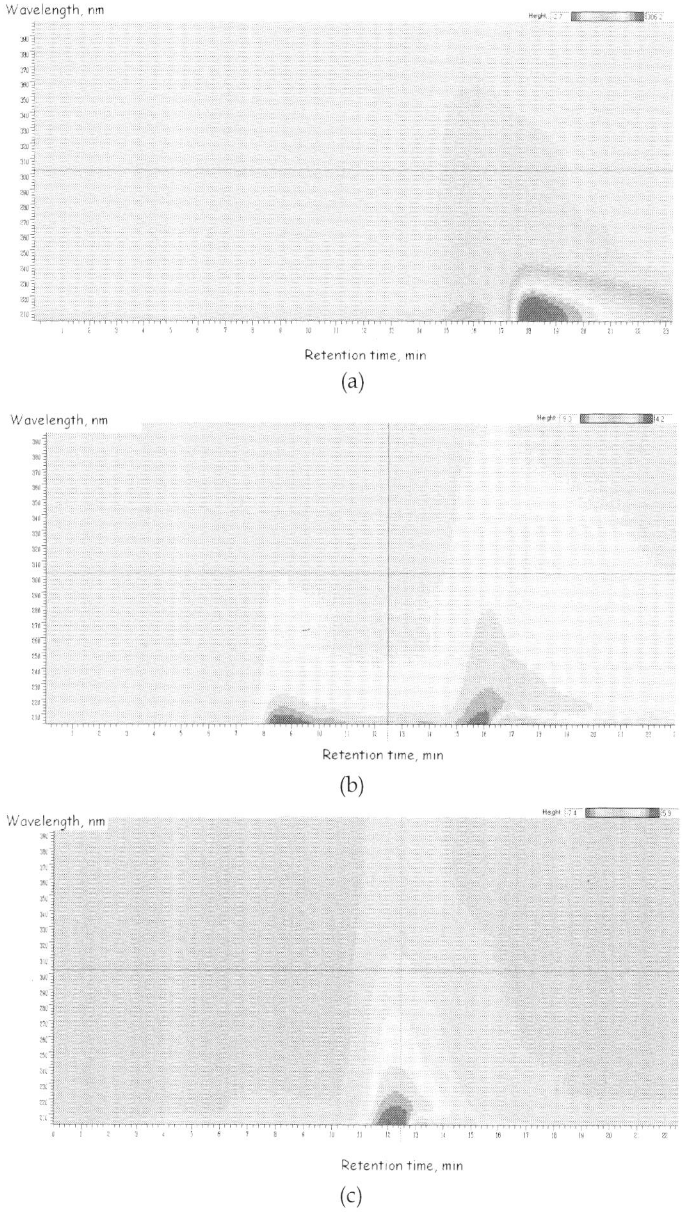

Fig. 2. Multi-wavelength HPLC chromatograms of porewater samples from Lake Peipsi sediment core at different depths: (a) 4 cm, (b) 37 cm, (c) 41 cm.

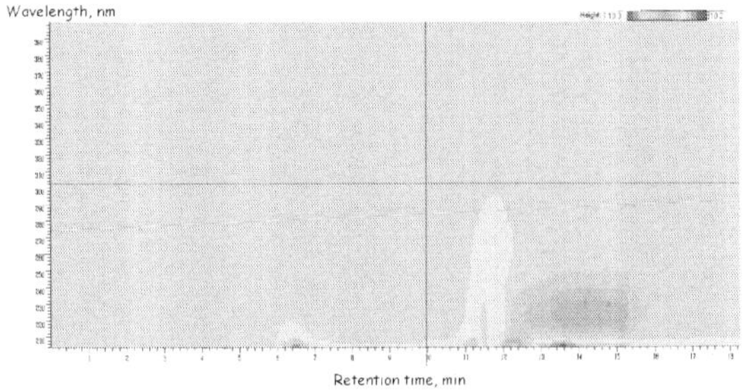

(a)

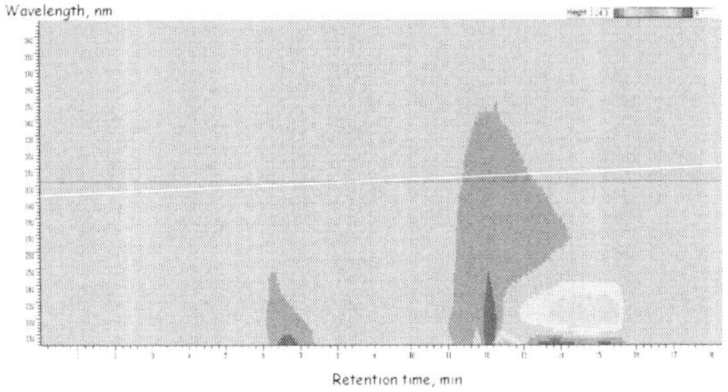

(b)

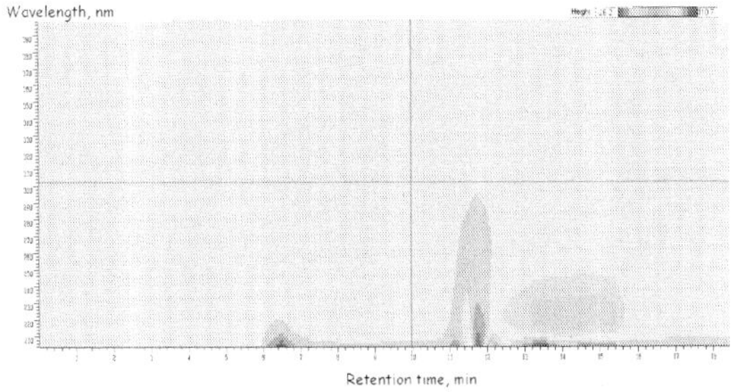

(c)

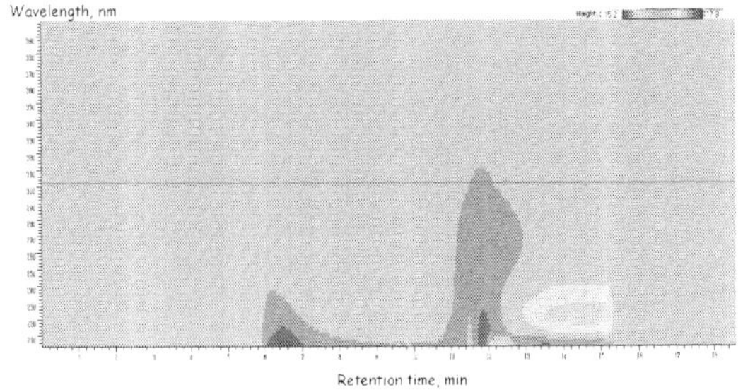

(d)

Fig. 3. Multi-wavelength HPLC chromatograms of porewater samples from Lake Rõuge Tõugjärv sediment core at different depths: (a) 6 cm, (b) 13 cm, (c) 27 cm, (d) 42 cm.

All chromatograms of L. Peipsi sediment *pws* were very similar, consisting of two main peaks representing HMW and HS fractions. The intensities and positions of the peaks (*i.e.* fractions) changed in different sediment layers reflecting age-related changes in the concentrations and transformation of organic constituents. The area of the HMW fraction was always smaller (~3% of the total area) than the second HS fraction (~97% of the total area). The calculated molecular masses for the HMW fraction varied between 200 and 270 kDa. The HMW fraction was absent from samples dating from the 1990s and from older samples from the nineteenth century. The HS fraction, with molecular masses between 700 and 3,700 Da, was dominant. The calculated average M_w was 1,500 Da, which is characteristic for aquatic humic and fulvic acids. The profiles of the determined chemical characteristics and HPLC variables are presented in Fig. 4.

The L. Rõuge Tõugjärv *pw*DOM was also separated into two peaks. The components of the second peak eluted as a broad distribution and sometimes with a partially resolved sub-shoulder. Possibly, the composition of the *pw*DOM from those sediment layers where the sub-shoulder appeared (some layers from the 1980s and 1960s) might have been somehow different from the major DOM composition. According to DAD spectra, components eluted with the first peak contained proteinaceous material, while the second peak spectra were characteristic of HS. The retention times of both peaks remained stable down the core. The calculated molecular masses for the HMW fraction varied between 800 and 1,000 kDa (M_w). The HMW fraction varied between 6 to 13% of the total peak area and was thus present in a significantly higher amount than in L. Peipsi sediment *pws*. Possibly, HMW material might have been formed from some proteins encapsulated into HS aggregates or micelles. The ability of HS to aggregate into large supramolecules has been reported previously (Havel & Fetsch, 2007; Piccolo, 2001). Generally, average M_w values of the analysed L. Rõuge Tõugjärv sediment *pw* HSs slightly exceeded 1,000 Da, and M_n was close to 400 Da. Molecular mass values of HS were in good agreement with molecular mass distributions reported for aquatic fulvic acids (Klavinš, 1997; Lepane et al., 2004).

The depth profiles for both lake cores (Fig. 4) indicated corresponding changes in M_w and M_n values. The molecular mass values for HS from L. Peipsi were slightly higher: 1,500 Da *vs.* 1,000 Da. The high fluctuations in HS molecular masses during 1870–1930s were not detected for L. Rõuge Tõugjärv. The down-core profiles of the chromatogram total peak areas and HS fraction areas were similar and exactly followed the changes in DOC.

The M_w/M_n ratio, or polydispersity, which is a measure of the homogeneity of organic matter, was mostly stable down the core, varying from 2.3 to 3.5 for L. Rõuge Tõugjärv and from 1.9 to 3.0 for L. Peipsi. This indicated the relatively homogeneous HS fraction in both lakes studied. The results showed that the molecular mass distribution and the polydispersity of DOM from L. Peipsi and L. Rõuge Tõugjärv were quite similar to those of sediment *pw*DOM from other lakes from Estonia and other regions studied by HPSEC (Fu et al., 2006; Leeben et al., 2008a; Lepane et al., 2004, 2010a; Makarõtševa et al., 2010; O'Loughlin & Chin, 2004).

The absorbance ratio of DOM at wavelengths of 250 and 360 nm (A250/A360) indicates the source of organic matter in the sediments (Peuravuori and Pihlaja, 1997). A higher ratio is related to autochthonous organic matter, which is produced within the lake, and the substances of smaller size and lower aromaticity are present in DOM molecules. Lower values of absorbance ratio reflect a higher aromaticity with an extent of allochthonous organic matter that originates outside the lake and is carried into the lake by inflows (McKnight et al., 2001). The ratio for L. Peipsi core samples was constant until the 1960s, with an average value close to 4.0. Thereafter, up to the present, it increased to 6.5, meaning that the origin of the organic matter changed to autochthonous. Constant values were also obtained for the L. Rõuge

Tõugjärv core until the 1940s, indicating higher degree of allochthonous organic matter in the lake. In the mid-twentieth century the ratio slightly increased, which coincided with the period when the lake sediments received decreased proportions of allochthonous organic compounds due to the decline in rural land-use practices and decreased sub-soil erosion. However, since 1980s there was a sharp increase in the absorbance ratio of the DOM up to 8, which also indicated the dominance of a more aliphatic autochthonous organic matter.

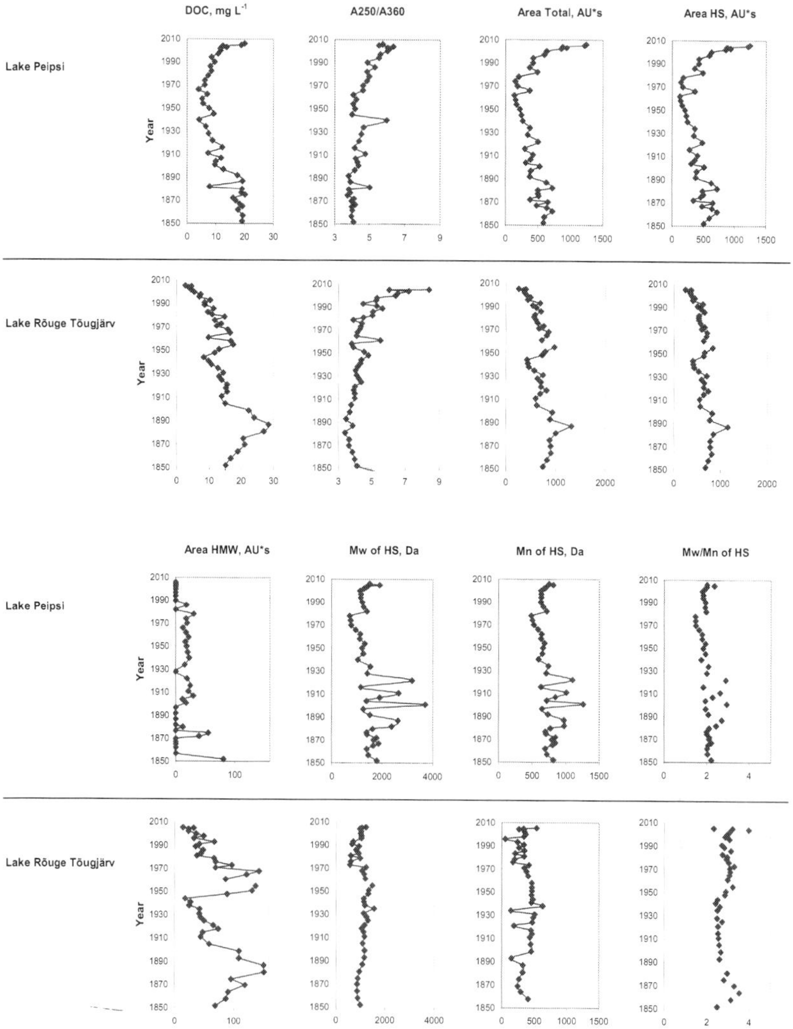

Fig. 4. Profiles of general chemical characteristics and variables by HPLC of Lake Peipsi and Lake Rõuge Tõugjärv sediment porewaters. The year denotes the year of sediment deposition.

3.3 Age-related changes in DOM characteristics of sediment cores

The statistical analysis of data was performed to reveal periods in the characteristics of separated DOM fractions. Based on DOC and absorbance ratio data, the L. Peipsi sediment core was divided into three age/depth periods: (I) 0–13 cm of sediment core depth, dated to 2006–1974; (II) 14–33 cm core depth, dated to 1970–1882; and (III) 34–43 cm core depth, dated to 1880–1852. L. Rõuge Tõugjärv sediment core was operationally separated into four age/depth periods: (I) 0–7 cm, dated to 2006–1998; (II) 8–17 cm, dated to 1996–1974; (III) 18–36 cm, dated to 1971–1911; and (IV) 37–45 cm, dated to 1905–1858 (Table 1). The mean values of the analysed variables divided into three or four periods with 95% confidence limits are shown in Figs. 5 and 6.

3.3.1 Lake Peipsi

The HMW fraction data (peak area, molecular masses, and polydispersity) were statistically similar down the core, as was the HS fraction polydispersity, and therefore did not allow the differentiation of sediment layers (Fig. 5). The DOC, total chromatogram peak area, and HS peak area changed similarly, thus proving the suitability of peak areas as semi-quantitative characteristics of DOM. The 1880–1852 dated samples had elevated DOC values. The upper 0–13 cm sediment DOM had statistically relevant differences in comparison to period III as revealed by DOC and HS molecular masses. The recent DOM accumulating into sediments has lower molecular masses and the highest absorbance ratio in comparison with preceding sediment layers. The obtained results indicate that recent *pw*DOM in L. Peipsi is more aliphatic and contains lower average molecular mass organic compounds which are likely of autochthonous origin. This might be the result of the microbial degradation of labile organic matter constituents such as carbohydrates (Zaccone et al., 2009). The absorbance ratio in L. Peipsi *pw*s shows significant differences throughout the sediment profile and can thus serve as an excellent variable for revealing the changes in sediment core.

3.3.2 Lake Rõuge Tõugjärv

As in the first lake sediment core studied, the polydispersity of HMW and HS fractions did not show any particular trend along the L. Rõuge Tõugjärv core profile (Fig. 6). The DOC, total chromatogram peak area, and HMW and HS fraction peak areas changed similarly. The obtained results indicated a general increase in all those variables with depth. However, it was not possible to differentiate between periods II and III (*i.e.* corresponding to years 1996–1911) by using DOC and semi-quantitative chromatographic data. Also, in the case of this lake the highest *pw*DOC was registered in the deepest layer 37–45 cm. The molecular masses of both HMW and HS of this undisturbed sediment core show different trends down the profile in comparison with L. Peipsi core. Similarities were found between the most recent and the oldest layers (dated to 2006–1998 and 1905–1858, respectively) and differences were found between the intermediate ones (periods II and III, dated to 1996–1974 and 1971–1911, respectively). Thus, the upper sediment layer (0–7 cm) variables indicate decreased DOM input with the characteristic high molecular mass compounds. The molecular mass data variations may reflect the influence of the watershed but also the seasonal climatic factors, like in-lake primary production. The observed distinct increase in absorbance ratio that was synchronous with a decrease in the DOC content possibly indicates the enhanced algal productivity and eutrophication of the lake, but also the lower contribution of allochthonous organic matter into the lake.

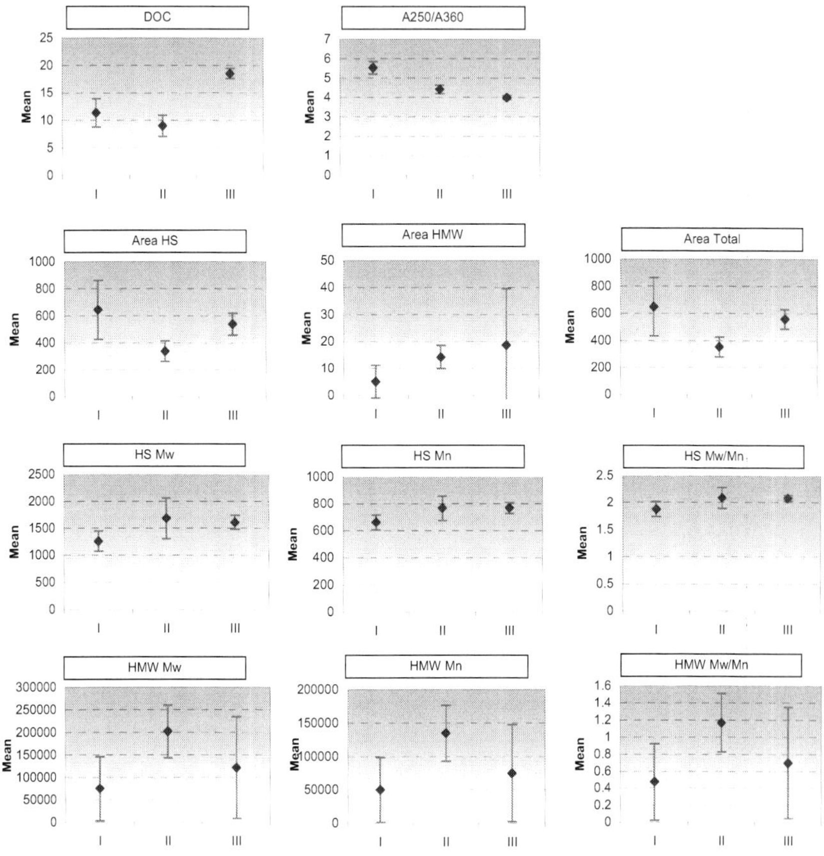

Fig. 5. Plots describing mean values of Lake Peipsi DOM semi-quantitative (areas), molecular, and spectroscopic characteristics arranged into three age/depth periods (see text). Red bars indicate confidence limits at the 95% level. DOC, mg L^{-1}; A250/A360: absorbance ratio at respective wavelengths; Mw/Mn: polydispersity; Mw and Mn: weight - and number-average molecular masses, respectively, Da; Area Total: total chromatogram peak area; Area HMW and Area HS: HMW and HS fraction peak areas, respectively, mAU*s.

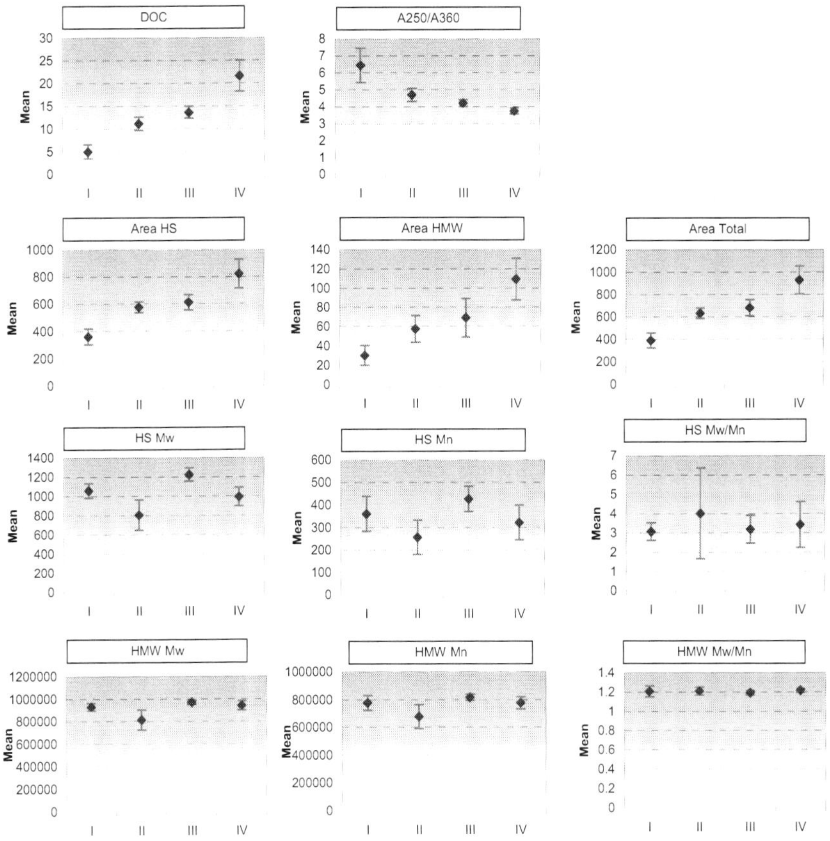

Fig. 6. Plots describing mean values of Lake Rõuge Tõugjärv DOM semi-quantitative (areas), molecular, and spectroscopic characteristics arranged into four age/depth periods (see text). For abbreviations see Fig. 5 legend.

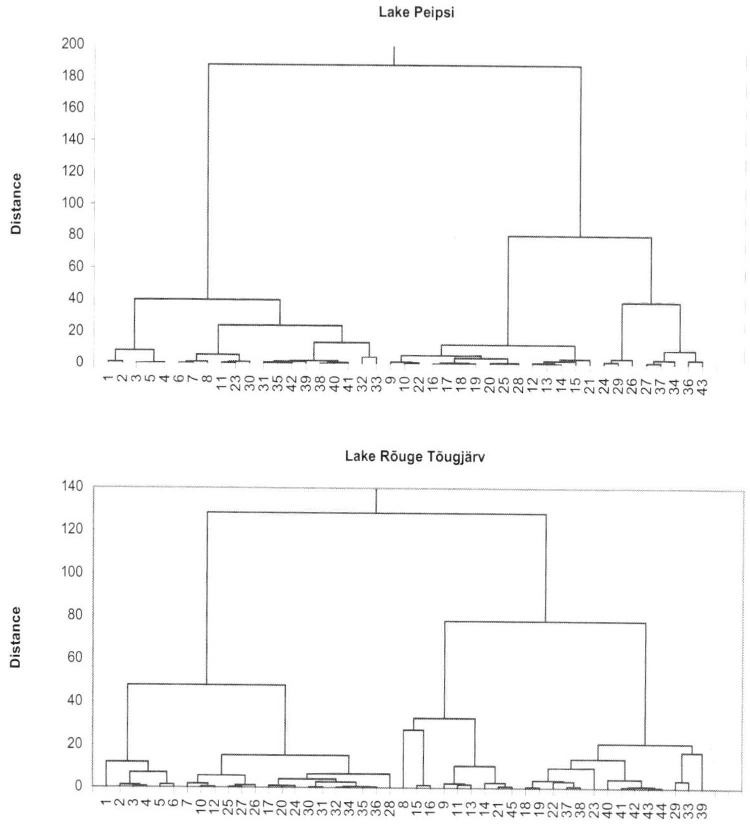

Fig. 7. Cluster analysis of Lake Peipsi and Lake Rõuge Tõugjärv sediment core samples from different depths (numbers indicate sediment depth in centimetres).

3.4 Tracking environmental change in organic compounds records

During the second half of the nineteenth and early twentieth centuries, L. Peipsi had a stable ecosystem similar to natural reference conditions as indicated by low autochthonous productivity. During the second half of the twentieth century, the ecological conditions of L. Peipsi worsened constantly. In the 1960s the lake was classified as mesotrophic. Eutrophication is the major environmental issue in the L. Peipsi basin due to the nutrient load to the lake. The main source of nutrient pollution of L. Peipsi is agriculture and municipal wastewaters. The decline in agriculture during the 1990s caused pollution to decrease and the quality of waters to improve. The lake area has been in a period of transition for more than decade.

Cluster analysis of *pw*DOM data was performed to reveal periods with similar characteristics in the studied L. Peipsi and L. Rõuge Tõugjärv sediment cores. The dendrograms are shown in Fig. 7. The aim was to identify the subgroups within the HPLC dataset and relate them to environmental changes. The L. Peipsi data allowed the layers to be grouped into two major groups. According to the results, the DOM from 2006-1994 formed a homogeneous subgroup that was in the same cluster as samples from 1982, 1928, and 1897-1857. The second major group was also divided into two subgroups. The first included sediment layers from years 1880-1872, 1911-1901, 1922 and 1940. The second subgroup covered mainly the sediment layers from 1990-1934, excluding the 1982 layer. Palaeolimnological studies state that anthropogenic impact on the lake has increased since the 1950s. Until that time, the lake was considered mesotrophic (Leeben et al., 2008b). Biomanipulation of the lake was carried out in 1993-1994 and was reported to have improved the lake ecosystem. This event can be seen in HPLC data considering the grouping results of organic compounds. Sedimentary pigment analysis indicated and thus confirmed the eutrophication of the lake since the 1980s (Leeben et al., 2008a).

L. Rõuge Tõugjärv experienced anthropogenic catchment disturbances up to the beginning of twentieth century, as indicated by extensive farming and increased drainage. During the first part of the twentieth century the development of efficient agricultural practices and reforestation improved the water quality. During the second part of the twentieth century the cultivated area declined and reforestation continued but the widespread use of mineral fertilizers caused an increase in primary production. After old agricultural practices stopped in the 1990s the lake was recovered and is reported to be mesotrophic today. The anthropogenic activities can be tracked by sediment investigations. L. Rõuge Tõugjärv sediments were annually laminated and thus possessed records with calendar year chronology. Thus, changes in this lake ecosystem and climate could be resolved seasonally. The L. Rõuge Tõugjärv HPLC organic matter data enabled the sediment layers to be classed into two major homologous groups. The recent sediment layers (2006-1998) formed a separate subgroup and were included in the same cluster as samples from 1951-1911 and some separate layers from 1991, 1986, 1974, and 1964. The second major group was also divided into two subgroups: the first one was similar to period II (1996-1976) and the second was similar to period IV (1905-1864), together with some separate layers from years 1971-1968, 1958-1954, 1934, 1921. The organic matter characteristics from period IV samples may reflect long-term agricultural impact because the lake has been mediated by human activity over hundreds of years (Heinsalu & Alliksaar, 2009). The massive utilization of fertilizers led to increased primary production in the 1960s-1980s (Alliksaar et al., 2005). Thus, one of the major clusters might reflect the eutrophication of L. Rõuge Tõugjärv. Since the 1990s the lake has been classified as mesotrophic with a decrease in diatoms and very good water quality. Historically, the same is reported for the time period 1920-1940. The above-described periods correlate well with the major cluster that included the most recent organic matter data together with data from the first part of the twentieth century.

The obtained results for both lakes show quite good agreement with some common eutrophication indicators (diatoms, fossil pigments) and thus confirm the suitability of organic compounds data for the assessment of the ecological state of the water bodies.

4. Conclusion

The results presented in the present study allowed the changes in the sediment porewater organic compounds to be assessed and related to the environmental conditions of the studied lakes. The applied HPLC method with multi-wavelength detection did not alter the nature of DOM. It was useful to reveal changes in *pw*DOM and molecular mass profiles, enabling the separation of organic high molecular mass and humic substances fractions in sediment cores of both lakes. Additionally, the qualitative analysis of DOM components based on UV-spectra can provide insights into their sources. The statistical analyses confirmed that porewater organic component variables obtained by HPLC could be used to differentiate between sediment layers and to track environmental changes.

5. References

Alliksaar, T.; Hörstedt, P. & Renberg, I. (1998). Characteristic fly-ash particles from oil-shale combustion found in lake sediments. *Water, Air and Soil Pollution*, Vol.104, No.1-2, (May 1998), pp. 149–160.

Alliksaar, T.; Heinsalu, A.; Saarse, L.; Salujõe, J. & Veski, S. (2005). A 700-year decadal scale record of lake response to catchment land use from annually laminated lake sediments in southern Estonia. *Verhandlungen der Internationalen Vereinigung für Theoretische und Angewandte Limnologie*, Vol.29, No.1, (March 2005), pp. 457–460.

Appleby, P.G. & Oldfield, F. (1978). The concentration of ^{210}Pb dates assuming a constant rate of supply of unsupported ^{210}Pb to the sediment. *Catena*, Vol.5, No.1, (April 1978), pp. 1–8.

Appleby, P.G.; Nolan, P.J.; Gifford, D.W.; Godfrey, M.J.; Oldfield, F.; Anderson, N.J. & Battarbee, R.W. (1986). ^{210}Pb dating by low background gamma counting. *Hydrobiologia*, Vol.143, No.1, (December 1986), pp. 21–27.

Brereton, R.G. (2003). *Chemometrics: Data Analysis for the Laboratory and Chemical Plant.* Wiley, Chichester, UK

Chin, Y.-P.; Aiken, G. & O'Loughlin, E. (1994). Molecular weight, polydispersity and spectroscopic properties of aquatic humic substances. *Environmental Science and Technology*, Vol.28, No.11, (October 1994), pp. 1853–1858.

Filella, M. (2010). Quantifying 'humics' in freshwaters: purpose and methods. *Chemistry and Ecology*, Vol.26, No.4, (October 2010), pp. 177–186.

Fu, P.; Wu, F.; Liu, C.-Q.; Wei, Z.; Bai, Y. & Liao, H. (2006). Spectroscopic characterization and molecular weight distribution of dissolved organic matter in sediment porewaters from Lake Erhai, Southwest China. *Biogeochemistry*, Vol.81, pp. 179–189.

Havel, J. & Fetsch, D. (2007). Humic substances. Capillary zone electrophoresis, In: *Encyclopedia of Separation Science*, M. Cooke & C. Poole, (Eds.), 3018–3025, Elsevier Science

Heinsalu, A.; Alliksaar, T.; Leeben, A. & Nõges, T. (2007). Sediment diatom assemblages and composition of pore-water dissolved organic matter reflect recent eutrophication history of Lake Peipsi (Estonia/Russia). *Hydrobiologia*, Vol.584, No.1, (June 2007), pp. 133–143.

Heinsalu, A. & Alliksaar, T. (2009). Palaeolimnological assessment of the reference conditions and ecological status of lakes in Estonia - implications for the European Union Water Framework Directive. *Estonian Journal of Earth Sciences*, Vol.58, No.4, (December 2009), pp. 334–341.

Højerslev, N.K. (1988). *Natural Occurrences and Optical Effects of Gelbstoff.* Univ. of Copenhagen, H.C.Ø. Tryk, København, Denmark

Hoque, E.; Wolf, M.; Techmann, G.; Peller, E.; Schimmack. W. & Buchau, G. (2003). Influence of ionic strength and organic modifier concentrations on characterization of aquatic fulvic and humic acids by high-performance size-exclusion chromatography. *J. Chromatogr. A*, Vol.1017, No.1-2, (October 2003), pp. 97–105.

Klavinš, M. (1997). *Aquatic Humic Substances: Characterisation, Structure and Genesis.* Maris Klavinš, Riga, Latvia

Leeben, A.; Tõnno, I.; Freiberg, R.; Lepane, V.; Bonningues, N.; Makarõtševa, N.; Heinsalu, A. & Alliksaar, T. (2008a). History of anthropogenically mediated eutrophication of Lake Peipsi as revealed by the stratigraphy of fossil pigments and molecular size fractions of pore-water dissolved organic matter. *Hydrobiologia,* Vol.599, No.1 (March 2008), pp. 49–58.

Leeben, A.; Alliksaar, T.; Heinsalu, A.; Lepane, V. & Veski, S. (2008b). Tracking changes in the organic matter in a lake paleoecosystem: a spectrophotometric approach. *Organic Geochemistry,* Vol.39, No.8, (August 2008), pp. 915–918.

Lepane, V.; Leeben, A. & Malashenko, O. (2004). Characterization of sediment pore-water dissolved organic matter of lakes by high-performance size exclusion chromatography. *Aquatic Sciences,* Vol.66, No.2, (June 2004), pp. 185–194.

Lepane, V.; Tõnno, I. & Alliksaar, T. (2010a). HPLC approach for revealing age-related changes of aquatic dissolved organic matter in sediment core. *Procedia Chemistry,* Vol.2, No.1, (January 2010), pp. 101–108.

Lepane, V.; Morriset, M.; Viitak, A.; Laane, M. & Alliksaar, T. (2010b). Partitioning of metals between operational fractions in the sediment record from Lake Peipsi. *Chemistry and Ecology,* Vol.26, No.4, (October 2010), pp. 35–48.

Liu, S.; Lim, M.; Fabris, R.; Chow, C.W.K.; Drikas, M.; Korshin, G. & Amal, R. (2010). Multi-wavelength spectroscopic and chromatography study on the photocatalytic oxidation of natural organic matter. *Water Research,* Vol.44, No.8, (April 2010), pp. 2525–2532.

Makarõtševa, N.; Lepane, V.; Alliksaar, T. & Heinsalu, A. (2010). A 10,000 year record of sediment pore-water dissolved organic matter characteristics from Lake Peipsi as revealed by HPSEC. *Chemistry and Ecology,* Vol.26, No.4, (October 2010), pp. 13–24.

Matilainen, A.; Vieno, N. & Tuhkanen, T. (2006). Efficiency of the activated carbon filtration in the natural organic matter removal. *Environment International,* Vol.32, No.3 (April 2006), pp. 324–331.

McKnight, D.M.; Boyer, P.K.; Westerhoff, P.K.; Doran, P.T.; Kulbe, T. & Andersen, D.T. (2001). Spectrofluorometric characterization of dissolved organic matter for indication of precursor organic material and aromaticity. *Limnology and Oceanography,* Vol.46, No.1, (January 2001), pp. 38–48.

Minor, E.C.; Simjouw, J.-P.; Boon, J.J.; Kerkhoff, A.E. & van der Horst, J. (2002). Estuarine/marine UDOM as characterized by size-exclusion chromatography and organic mass spectrometry. *Marine Chemistry,* Vol. 78, No.2-3 (May 2002), pp. 75–102.

Mori, S. & Barth, H.G. (1999). *Size Exclusion Chromatography.* Springer, Berlin Heidelberg, Germany

Nissinen, T.K.; Miettinen, I.T.; Martikainen, P.J. & Vartiainen, T. (2001). Molecular size distribution of natural organic matter in raw and drinking waters. *Chemosphere,* Vol.45, No.6-7 (November 2001), pp. 865–873.

O'Loughlin, E.J. & Chin, Y.-P. (2004). Quantification and characterization of dissolved organic carbon and iron in sedimentary porewater from Green Bay, WI, USA. *Biogeochemistry*, Vol. 71, No.3, (December 2004), pp. 371–386.

Pelekani, C.; Newcombe, G.; Snoeyink, V.L.; Hepplewhite, C.; Assemi, S. & Beckett, R. (1999). Characterization of natural organic matter using high performance size exclusion chromatography. *Environmental Science and Technology*, Vol.33, No.16 (July 1999), 2807–2813.

Perminova, I.V.; Frimmel, F.H.; Kovalevskii, D.V.; Abbt-Braun, G.; Kudryavtsev, A.V. & Hesse, S. (1998). Development of a predictive model of molecular weight of humic substances. *Water Research*, Vol.32, No.3, (March 1998), pp. 873–881.

Perminova, I.V.; Frimmel, F.H.; Kudryavtsev, A.V.; Kulikova, N.A.; Abbt-Braun, G.; Hesse, S. & Petrosyan, V.S. (2003). Molecular weight characteristics of humic substances from different environments as determined by size exclusion chromatography and their statistical evaluation. *Environmental Science and Technology*, Vol.37, No.11, (April 2003), pp. 2477–2485.

Peuravuori, J. & Pihlaja, K. (1997). Molecular size distribution and spectroscopic properties of aquatic humic substances. *Anal. Chim. Acta*, Vol.337, No.2 (January 1997), pp. 133–149.

Piccolo, A. (2002). The supramolecular structure of humic substances: a novel understanding of humus chemistry and implications in soil science. *Advances in Agronomy*, Vol.75, pp. 57–134.

Poska, A.; Sepp, E.; Veski, S. & Koppel, K. (2008). Using quantitative pollen-based land-cover estimations and a spatial CA_Markov model to reconstruct the development of cultural landscape at Rõuge, South Estonia. *Vegetation History and Archaeobotany*, Vol.17, No.5 (September 2008), pp. 419–443.

Specht, C.H. & Frimmel, F.H. (2000). Specific interactions of organic substances in size-exclusion chromatography. *Environ. Sci. Technol.* Vol.34, No.11, pp. 2361–2366.

Vartiainen, T.; Liimatainen, A. & Kauranen, P. (1987). The use of TSK size exclusion columns in determination of the quality and quantity of humus in raw waters and drinking waters. *Science of Total Environment*, Vol.62, pp. 75–84.

Veski, S.; Koppel, K. & Poska, A. (2005). Integrated palaeoecological and historical data in the service of fine-resolution land use and ecological change assessment during the last 1000 years in Rõuge, southern Estonia. *Journal of Biogeography*, Vol.32, No.8, (August 2005), pp. 1473–1488.

Zaccone, C.; Said-Pullicino, D.; Gigliotti, G. & Miano, T.M. (2008). Diagenetic trends in the phenolic constituents of *Sphagnum*-dominated peat and its corresponding humic acid fraction. *Organic Geochemistry*, Vol.39, No.7 (July 2008), pp. 830–838.

Zhou, Q.; Cabaniss, S.E. & Maurice, P.A. (2000). Considerations in the use of high-pressure size-exclusion chromatography (HPSEC) for determining molecular weights of aquatic humic substances. *Water Reserach* Vol.34, No.14 (October 2000), pp. 3505–3514.

17

Management Strategies for Large River Floodplain Lakes Undergoing Rapid Environmental Changes

Giri Kattel and Peter Gell
School of Science & Engineering, University of Ballarat, Victoria Australia

1. Introduction

Large river basins are the origin of ancient civilization (Barbier & Thompson, 1998; Sadoff & Grey, 2002). Floodplain lakes, situated adjacent to large river systems, are connected with river channel networks. The connectivity between river channels and wetlands makes the "boom" and "bust" ecology following the drought and flood events that continues to support diverse floral and faunal communities in the floodplains lake systems (Jenkins & Boulton, 2003). Rich biodiversity and occurrence of macro-invertebrate drifts in the Upper Paraguay River-Floodplain-System, parts of the Pantanal (Brazil) Wetland System, and dense microphyte community with regularly supplied allochthonous nutrient inputs and moderation of physical extremes in the billabongs of the Murray-Darling River Floodplain-System Australia are some examples of highly productive floodplains lake ecosystems in the world (Shiel, 1976; Wantzen et al., 2005). Being a productive ecosystem, people living across the large river basins have been greatly benefited from the resources generated by these wetlands for generations (Bright et al., 2010). For example, the indigenous people of the Orinico River Basin, South America, and Murray Darling Basin, Australia have been harvesting the specialised fish community that are adapted to the floodplains wetland systems over several centuries in the past (e.g., Lundberg et al., 1987; Humphries, 2007).
Since the productivity of the large river floodplains lake ecosystems is dependent on naturally occurring riverine flood events, any alternation of the hydrological patterns of rivers can have strong impacts on nutrient dynamics, biological diversity and assemblages of these lakes (Fisher et al., 2000). Over the past few decades the large river systems and its adjacent wetland habitats have undergone rapid environmental changes. Anthropogenic activity across the river basin has increased substantially. River regulations such as construction of dams, irrigation channels, dykes and weirs, and catchment land use activities such as deforestation, agriculture and cattle ranching and introduction of exotic flora and fauna are increased (Power et al., 1996; Kingsford, 2000, Bunn & Arthington, 2002). Rapid climate warming is further intensifying the conditions of ecosystems including the-changes in hydrology and water quality of rivers and lakes (Carpenter et al., 1992; Lewis et al., 2000; Palmer et al.; 2008). The coupled human-climate disturbances have led to an increased habitat heterogeneity and complexity of ecosystem processes of majority of floodplains lake systems worldwide (Tockner et al., 2000). Consequently, the people who

have been directly associated with these large river systems for a range of services over generations are influenced by these changes for sustainable living.

One of the critical issues today for majority of river scientists is therefore to understand the large river floodplains lake ecosystems processes that are exposed to a range of coupled human-climate disturbances. Understanding the ecosystem processes and identifying the disturbances altering ecosystem processes can help resource managers to tackle challenges of floodplains lake management and promote healthy and productive ecosystems across the large river basins worldwide. The large river floodplains lake ecosystems are longitudinally modulated by upstream processes, where the main source of organic carbon such as fine particulate matter is transported to downstream environments (Vannote et al., 1980). The use of particulate organic matter is maximized by benthic heterotrophs and micro-crustaceans because depositional structures are limited to backwater and nearshore areas (Naiman et al., 1987). However, the role of locally derived ecological processes is unknown in the longitudinal river continuum (Statzner & Higler, 1985) since locally metabolised carbon and the bottom-up control of algal communities are also important for ecosystem processes (Wehr & Descy, 1998). Metabolism and turnover rates of organic carbon in floodplain lakes can vary with the type and nature of the river system from which they are derived. Floodplain lakes associated with blackwater rivers for example have low content of suspended sediments but a high concentration of dissolved organic matter (Meyer, 1990). Metabolism of these floodplain lakes is dependent on allochthonous organic carbon with increased river size despite increases of downstream gross primary production, where riparian swamps are the source of organic carbon (Meyer & Edwards, 1990). Flood pulses in particular are the significant source of carbon for ecosystem structure and functions in large river floodplain lakes (Junk et al., 1989). Accessibility and retention of organic matter are functions of the frequency and duration of flood pulse and extent of inundations (Humphries et al., 1999; King et al., 2003). Apart from the organic matter derived from flood pulses, integration of locally derived autochthonous matter such as phytoplankton, benthic algae and aquatic vascular plants and direct inputs from the riparian zones such as abscised leaves, particulate organic matter, and dissolved organic carbon during flood pulses are also significant source for floodplains lake ecosystem structure and function (Thorpe & Delong, 1994). Autochthonous carbon and direct allochthonous inputs from the riparian zone are labile and maximally utilized by heterotrophs dominant in near-shore leaf-litters and littoral habitats (Thorpe & Delong 1994). As the large river floodplains lakes ecosystems are exposed to coupled human-climate disturbances, scientists have been facing increasingly difficulties to understand the complex ecosystem processes worldwide.

Changes in species richness, diversity and assemblages of biota such as fish, diatoms, macro-and micro-invertebrates across temporal and spatial scales have become useful for understanding rapid environmental changes of large river floodplains lake ecosystems (Ward et al., 1999). Evaluation of the large scale changes in ecosystems as a result of micro-scale changes in environments such as a small rise in surface water temperature or additional inputs of phosphorous concentrations in floodplain lakes over a temporal scale is crucial in relation to changes in biotic assemblages. Interaction between the channel and floodplain systems and between the channel and groundwater aquifers plays a significant role for rapid ecosystem changes in floodplain lakes ecosystems, since the temporal dimensions of these ecosystems are largely integrated and dynamic (Fig. 1, Ward, 1989; Lewis et al., 2000).

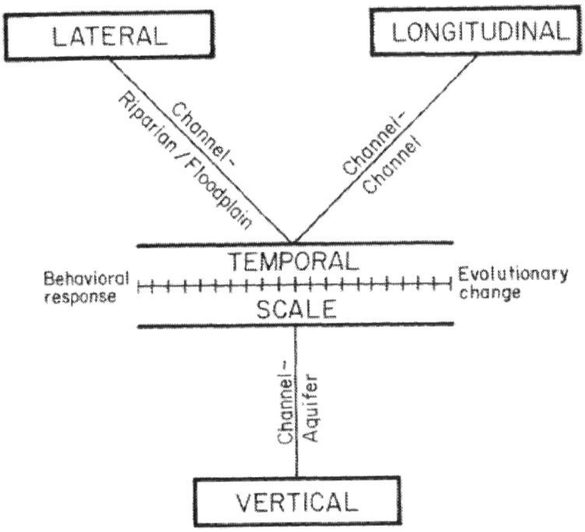

Fig. 1. Four dimensional structures of floodplain lake ecosystem (adapted after Ward, 1989). Triangular interactions (lateral-longitudinal-vertical) determine the spatial and temporal changes of floodplain lakes biota. Lateral influence occurs at riparian zone while longitudinal influence is restricted in river channels. Vertical influence occurs at ground-water aquifer.

Micro-crustaceans are one of significant indicators of rapid environmental changes of large river floodplains lakes over a range of temporal and spatial scales. Microcrustaceans prefer littoral benthic and pelagic habitats and they have wide optima and tolerance to a range of environmental variables. Micro-crustaceans such as cladocerans are one of significant components of the large river floodplains lake ecosystems. Cladocerans emerge rapidly following the inundation, feed principally on phytoplankton, bacteria and detritus and actively transfer energy across the food webs (Reid & Brooks, 2000; Jenkins & Boulton, 2003). Cladoceran exoskeletons and their ephippia are archived in floodplain lake sediment being useful indicator for a long term environmental changes (Kattel, 2011). The use of modern and sub-fossil assemblages of micro-crustaceans such as cladocerans can help floodplains lake ecologists and river scientists to understand complex ecosystem processes and develop effective management strategies for these ecosystems worldwide. In this chapter, we have identified a range of issues of rapid environmental changes of large river floodplains lake ecosystems worldwide. We have then highlighted the use of the micro-crustaceans, such as cladoceran zooplankton to improve management practices of the vulnerable ecosystems of floodplains lakes in the large river basins.

2. Materials and methods

This chapter is based on a range of case studies on large river floodplains lake ecosystems worldwide. The case studies were varying in nature either focusing on theoretical models being developed over the past decades on large river floodplains lakes ecosystem processes, or highlighting the impacts of global environmental changes on these floodplains lake ecosystems. The theoretical models were reviewed mainly on river continuum concept

(RCC), flood pulse and riverine productivity (RPM) models, where most of these models were tested in North America, Europe and Australia for understanding ecosystem processes of the large river systems (Vannote et al., 1980; Naiman et al., 1987; Junk et al., 1989; Thorpe & Delong, 1994). The case studies on critical management issues of rapid environmental changes of the large river basins were collated from various continents including the Yangtze River System, Asia (Yang et al., 2007; Chen et al., 2011), the Mississippi River System, North America (Wren et al., 2008), the Orinco, Salado and the Paraguay River Systems, South America (Lundberg et al., 1987; Claps et al., 2009), Orange-Vaal River System, South Africa (Ashton et al., 1986), Erbo River System, Europe (Gallardo et al., 2007) and the Murray Darling River System, Australia (Humphries et al., 1999; King et al., 2003). Following the identification of critical management issues of the large river systems, prevailing conditions of biotic assemblages in changes in large river floodplain lakes were reviewed from the case studies of Europe, North America and Australia (Fisher et al., 2000; Lewis et al., 2000). Then a comprehensive review was undertaken on the use of micro-crustaceans to understand the complex ecosystem processes and configure effective management strategies when they are exposed to a range of external disturbances including climate change over temporal and spatial scales.

3. Management issues for large river floodplains lake ecosystems

At least four major disturbancing factors have been identified for alteration of ecosystem processes of the large river floodplains lakes worldwide. Resource managers are increasingly concerned about management issues of river regulation, land use activity, introduction of exotic species and rapid climate warming which have been considered to make significant impacts on large river floodplains lake ecosystem processes worldwide in recent decades.

3.1 River regulation

Water abstraction, diversion and regulations are one of critical management issues of floodplains lake ecosystems of river basins. Naturally occurring physical structures strongly support biodiversity of lakes. Natural flows are important for succession of food web structure and dynamics (Fig. 2). Physical structures enhance water quality, energy budget and flow regime of rivers. Improved water quality maintains the health of floodplain lakes. Interaction between channel morphology and river discharge in up-streams is important for structure and function of the downstream river ecosystems (Bunn & Arthington, 2002). For example, braided river channels in arid and semi-arid climates are characterised by a network of constantly shifting low sinuosity water courses while meandering river channels are influenced by channel width and depth with steady discharge (Johnston et al., 1997). The alteration of these river channels influences morphology consequently the habitat of a wide range of biota in the reaches which are significant source of energy transfer across the food web. For example, fish of the Orinoco River use the main channel primarily for migration and dispersal depends largely on floodplains for growth and subsistence (Lewis et al., 2000). Diversity and length of the food chains increase in natural flow regimes of many large river systems (Maddock et al., 2004). However, construction of dams, weirs and irrigation channels can have substantial implications for physical structures and riparian ecosystems (Walker, 1985). Such consequences can be immediate and obvious or gradual and subtle depending on the nature of regulation and the occurrence, diversity and composition of biota (Power et al., 1996).

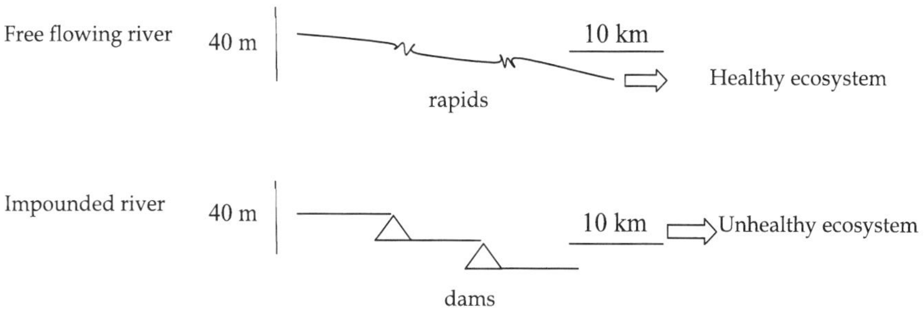

Fig. 2. Free flowing rivers consist of series of rapids and slow flowing stretches. Rapids are important for succession of food web structure and dynamics. Naturally occurring physical structures enhance water quality, energy budget and flow regime of the river. Improved water quality maintains the health of riverine floodplain lake ecosystems. In impounded rivers, rapids have been removed by erected dams. Dams can have direct implications on hydrology reducing the downstream flow variation. Dams hinder the upstream migration of biota, alter thermal environment, nutrient movement and sediment loading and predator-prey interaction in downstream food webs (adapted after Nilsson & Berggren, 2000).

Regulation of the Murray River, Australia over the past 50 years has resulted in considerable implications for ecosystem structure and functions. For example, construction of dams in the Murray River has reduced downstream flows as well as obstructed upstream migration of biota including thermal environment, nutrient movement and sediment loading and predator-prey interactions (Gehrke & Harris, 2000). Since flooding generates bio-geochemical processes, the major impact of dams is the interruption of the exchange of energy between river and riparian zone during flood events (Sam et al., 2000). Low flows events are critical for lowland fish assemblages and plant community structure (Capon, 2003). Increased water residence time increases crustacean biomass (Humphries et al., 1999). However, alternation of natural low flow patterns can influence diadromous fish populations which utilize crustaceans as their major diet. Fish species such as Murray cod (*Maccullochella peelii peelii*) and silver perch (*Bidyanus bidyanus*) which do not require special flood events in the Murray River Australia is able to utilize low-flows events for spawning (King et al., 2003. However, the growth of larvae of Murray cod (*M. peelii peelii*) is significantly influenced by construction of dams and irrigation channels across the MDB. Larvae are consistently stranding in the dam when drifting (Koehn & Harrington, 2005). In contrast, recruitment of other fish species requires floodplain inundation and increased water volume (King et al., 2003).

Alteration of riparian vegetation can influence nutrient sources of wetland biota. Composition and diversity of naturally occurring riparian forests such as river red gum trees (*Eucalyptus camaldulensis*) in MDB have declined as a result of river regulation (Robertson et al., 2001). For example, stable isotope ratios of oxygen reveal that river red gum (*E. camaldulensis*) forests are efficient for utilizing water at varying salinity gradients (Mensforth et al., 1994) through reduced transpiration rates (Costelloe et al., 2008). However, continued low flows occur as a result of a rise in ground water salinity. Absence of natural floods influences recharge of naturally occurring groundwater salinity levels and will have detrimental effects on floodplains riparian biota (Jolly et al., 2001). Die back in river red gum (*E. camaldulensis*) communities can occur at electrical conductivity as high as 40dSm^{-1} (Mensforth et al., 1994). Substantial buffering of catchment soils as a result of river regulation and subsequent release of sulphur in floodplains following the European settlements has influenced on diatom communities in the Lower Murray River (Gell et al., 2007). Irrigation of soils with low permeability is also causing saline groundwater to rise. Salt accumulates in the top soil as water continues to evaporate. Partial drying of previously inundated floodplains reduce nutrient availability such as total nitrogen (TN) and total phosphorous (TP) in the system causing negative effects on ecosystem functioning (Baldwin & Mitchell, 2000). Irrigation dams in the previously fertile Indus River floodplain (Pakistan) are also reported to have caused a massive salinity problem. Extensive abstraction of water from the Amu Dar'ya and Syr Dar'ya, the two largest tributaries of the Aral Sea has caused 80% reduction of the water volume in the Aral Sea within the last four decades resulting in a four-fold increase in salinity concentrations of the floodplain lake consequently limiting ecosystem structure and functions (Aladin & Plotnikov, 1993).

3.2 Land use

Increased land use activity across the catchment of the large river system is other significant management issue. Ecological attributes of large river floodplain lakes have been constantly modified by industrial and cultural developments. Modern farming practices have made implications for physical and hydrological features of floodplain wetlands including the changes in water quality and sediment processes. Wren et al. (2008) reported that the sediment accumulation rates of the Sky Lake in the Mississippi River system, USA has increased to 50-folds following the clearing of forests began by humans in 300 years ago. In natural flood pulse concept, river floodplains are regularly flooded and dried (Bayley, 1995). Catchment organic matter generated across spatial and temporal scales is transported to river floodplains. A high turnover rate of organic matter and nutrients are predicted to occur as a result of natural flood events. During flood events, nutrients dissolve with flood waters consequently accelerating primary production. However, under dry conditions, decomposition processes of floodplain lakes would increase relative to production. Intensification of land use including waste disposal, agriculture, grazing and forest clearance in catchments all have considerable implications for natural flood pulse events (Jansen & Robertson, 2001).

Large river floodplain wetlands are species rich habitats which connect distant ecosystem not only through the migration of river biota but also from the transport of water, sediments, nutrients and contaminants (Sparks, 1995; Fisher et al., 2000, Chen et al., 2011). The integrity of floodplain lakes, which is maintained by hydrological dynamics, biological productivity and river connectivity are significantly impeded by land use activity. Alteration in riparian vegetation in particular is detrimental for changes in species diversity and ecosystem functioning of floodplain lakes. For example, alteration of the natural riparian vegetation by humans has modified the ecosystem processes of the wetlands and its catchments across the Sacromento, USA. Modification of wetland landscape has already been noticed as a result of cultivation, soil erosion and sedimentation to down-streams and in many cases loss of productivity has also occurred (Alpert et al., 1999).

Application of nitrogen and phosphorous has increased for agriculture across the large river basins worldwide. An alteration in global nitrogen cycle has occurred in recent years by widespread use of N-fixing crops, fertilizers, habitat change and burning of fossil fuels.

Continuous use of nitrogen and phosphorous as fertilizers in agriculture and urban landscapes lead to leaking of mobile inorganic nitrate ions in the system (Turner et al. 2003). As a result of algal blooms and low dissolved oxygen at nutrient rich environment, wetland ecosystem health has reduced substantially. Widespread release of phosphorous into the Yangtze River floodplain lakes over the past decades, for example, has caused a regime shift, where a transformation has occurred in large number of lakes with macrophyte dominated states to algal dominated states (Yang et al., 2007). Although some disturbances are beneficial to habitat heterogeneity and species, the lack of disturbance events have negative impacts on these lakes. For example, the Oxbow Lake of the Middle Erbo River (Spain) and Bottle Bend Lagoon of the Murray River (Australia) are reported to have undergone increased salinisation and eutrophication followed by a loss of biodiversity as a result of land use change (Lamontagne et al. 2006; Gallardo et al., 2007).

Land use activity has also exacerbated the release of a range of toxic substances in large river floodplain lakes. Trace metals (e.g., Hg, Pb, Zn), persistent organic pollutants (POPs), and organometallic compounds are detrimental for ecosystem health. Polycyclic aromatic hydrocarbons (PAHs), polychlorinated biphenyls (PCBs), polychlorinated dibenzo-p-dioxins (PCDDs) and polychlorinated dibenzofurans (PCDFs) as well as many other organochlorine pesticides (e.g. DDTs; taxophene) and brominated flame retardants (BFRs; including polybrominated diphenyl ethers (PBDEs) can be lethal for wetland biota if their concentration is high in the system. The organometallic compound, such as methylmercury (MeHg) is the most toxic compound (Leung et al., 2007). Following the industrial revolution in Europe (1800 AD), wetland contamination by organochlorine compounds increased substantially. The DDT concentrations for example in lake sediments were the highest during the 1950s. Boating activity in floodplain lakes has influenced substantially for estuarine biota in the large river mouths as well as mollusc communities in freshwater environments due to increased organometallic toxicity (US-EPA, 2003). Methylmercury, for example although present in a small concentration (0.1-5.0 percent) of total mercury, it represents 90-100% in invertebrates and fish. Increased POPs, PCBs and PAHs toxicity can cause endocrine disruption in fish and crustaceans (Matthiessen & Johnson, 2007). The tributyl tin (TBT) can cause reproductive failure in *Daphnia* at 400 ng L^{-1} and 380 ng L^{-1} TBT levels (Brooke et al., 1986). Macrophyte density in lowland shallow river floodplain wetlands in the UK substantially reduced in the sixties as a result of recreational boating followed by TBT pollution (Sayer et al., 2006).

3.3 Introduction of exotic species

Introduction of exotic flora and fauna is another significant management issue of the large river floodplain lake ecosystems worldwide. Displacement of habitats and subsequent extinction of native populations are reported as some of the foremost impacts of introduced species in many large river floodplain lakes. River basins of the Northern Hemisphere inhabit the highest number of non-native fish species (Leprieur et al., 2008). More than 50% of the biota of the Hudson River in the USA comprises introduced species mostly from Europe where 10% of those populations have significant ecological impacts on native populations (Nilsson & Berggren, 2000). Human activities are blamed to facilitate the establishment of non-native species by disturbing natural landscapes and by increasing propagule pressures on native populations (Leprieur et al., 2008; Simões et al., 2009).

Large river floodplain lakes are increasingly sensitive to biological invasion. The extended river networks often have recurrent disturbances and enhanced invasion (Elvira, 1995; Mills et al., 1996). Dispersal of seeds and eggs are rapid at landscape level through river channel networks. Disturbance regime and floodplain productivity also enhance invasion (Chapin III et al., 2000). Non-native species once introduced in large river systems can spread rapidly (Koehn, 2004). Favouring wide ranging climates, flexible in habitat selection and increased physiological adaptation are characteristic features of non-native species for a successful colonisation in a new environment (Mooney & Cleland, 2001). Whilst the impacts of invasion on native populations has been increasing, what condition is necessary for invasion, the way the invasion progresses through space and time and the properties of invasive biota is yet to be understood fully. Under regulated environments these patterns have become pronounced, and the nature of the invading species in susceptible habitats is also becoming unpredictable (Bunn & Arthington, 2002). For example, water regulation in one of the South African large river systems (Orange-Vaal River System) has stabilized the natural flow regimes favouring the alien aquatic vegetation (e.g. *Myriophyllum* sp, *Azolla* sp) consequently reducing the water movement, light penetration and oxygenation followed by displacing the native vegetation bed (Ashton at al., 1986). Introduced fish species such as European perch (*Perca fluviatilis*) and common carp (*Cyprinus carpio*) in the Murray River Australia have successfully established populations following the European arrival causing retarded growth and development of native fish populations (Koehn, 2004). Some endemic species including Macquarie perch (*Macquaria australasica*) Murray hardyhead (*Craterocephalus fluviatilis*) and Murray cod (*M. peelii peelii*) have become critically endangered or vulnerable in recent decades (Hutchison & Armstrong, 1993).

3.4 Climate change

Rapid rate of climate warming in recent decades has become one of important management issues of large river floodplain lake ecosystems. Climate change can cause floodplain lakes ecosystems through a variety of ways such as alteration of flood events, channel morphology, nutrient dynamics and growth and reproduction of wetland and riparian biota.

Floods are essential for nutrient dynamics, primary and secondary production and growth and development of native plant and animals (Harris & Gehrke, 1993). Regular inundation provides water for riparian vegetation and continuation of ecosystem processes (Nilsson & Berggren, 2000). Runoff with organic rich nutrients create potential for the establishment of a new community. Recovery of riparian catchments after flood or drought is rapid and the diversity and abundance of flora and fauna increase substantially within a short period (Jenkins & Boulton, 2003). However, climate warming reduces annual inflows and runoff volume of the large river systems. Climate change also alters river channels, erosion, nutrient and sediment transports influencing terrestrial vegetation, soil moisture and evapo-transpiration processes in large river floodplains lakes (Palmer et al., 2008). Holocene records of floodplains in the USA show that magnitude of floods is intense in arid regions resulting in channel widening which often have sparse riparian vegetation (Carpenter et al., 1992). In Murray River, a rise of 1° C in recent decades is predicted to have caused approximately 15% reduction in the annual flows (Cai & Cowan, 2008). Since 1950s, the MDB has experienced warming of around 0.8° C with declining rainfall as low as 10 mm per decade resulting in degraded water quality across the region. Important flood-cued native fish populations such as golden perch (*Macquaria ambigua ambigua*) are significantly altered as a result of climate-induced low flow events in the MDB (Humphries et al., 1999).

Climate change influences nutrient concentrations in floodplain lakes (Spink et al., 1998). Elementary nitrogen level and biogeochemical cycles in sediments can vary with climate warming (Catalan et al., 2002). In drought phase, sulphur stored in the upper areas of the littoral zone can re-oxidise causing lakes and river floodplains in down-streams to re-acidify (e.g., Yan et al., 1996; Dillion & Lazerte, 1992). Rising temperature, longer dry spells and runoff distribution in the MDB for example have intensified vegetation patterning and concentration of dryland salinity in recent decade (Hughes, 2003). When soil with rich sulphides (or 'black ooze') characteristic of dark and soft are disturbed and oxygenated, they react rapidly resulting in environmental hazards floodplains systems (Lamontagne et al., 2003).

Climate warming can influence growth and reproduction as well as phenology of wetlands biota directly (Hughes, 2003). Increased concentration of atmospheric CO_2 intensifies photosynthetic processes of riparian trees. The leaf stomatal conductance of these trees decreases and the plant-water use efficiency increases in elevated CO_2. However, productivity of plant biomass at high level of CO_2 is short-lasted resulting in changes in energy balance in the system (Dunlop & Brown, 2008).

Whist we have identified some key management issues of the large river floodplains lake ecosystems, the next step is to find appropriate solutions for these problems. There are a range of management options available, our aim is however, how we can understand and best interpret the ecosystem processes of the floodplain lakes that are exposed to anthropogenic and climatic variability and can guide resource mangers using the best management strategy. Understanding of changes in assemblages and diversity of wetlands biota, particularly micro-crustaceans along temporal and spatial scales is one of potential tools that provides prevailing conditions of changing large river floodplains lake ecosystems.

4. Prevailing conditions of biotic assemblages in changing large river floodplain lakes

The structure of micro- and macro- invertebrate communities is dependent on factors such as water quality (e.g. nutrients salinity, pH), food resources and habitat availability. Oligotrophic (low nutrient) conditions may limit primary production, thereby limiting a key food resource (i.e. phytoplankton) for some functional groups of invertebrates (Jeppesen et al., 2000). Eutrophic (high nutrient) conditions, high temperature and stable (e.g. stratified and poorly mixed) water bodies may favour key phytoplankton groups (i.e. cyanobacteria) that are a poor quality food resources for micro-invertebrates (Jeppesen et al., 2000). Furthermore, some functional groups of invertebrates feed exclusively on either phytoplankton or macrophytes. Consequently, shifts between macrophyte and phytoplankton dominated states will lead to shifts in the composition of micro- and macro-invertebrate communities (Jeppesen et al., 2002). Changes in water column salinity may also substantially influence the composition of the micro-and macro-invertebrate community due to differences in species-specific salinity tolerances. Furthermore, it is considered highly likely that changes in soil salinity (Brock et al., 2005) or soil pH (Hall & Baldwin, 2006) may severely impact the viability of invertebrate seed banks within wetland/floodplain soils. It is also recognised that the viability of invertebrate seed banks decreases with time (Nielsen et al., 2007). Long (i.e. >10 years) dry periods may exceed the viability period, severely compromising the invertebrate community that will hatch from soil seed banks during subsequent floods (Williams, 1985).

However, the variety of conditions among floodplains lake biota at any given time is largely dependent on system processes of the river basin. Abundance and diversity of micro-crustaceans can change with the initial condition of the basin morphometry, where set-points are determined by flood inundation (Fig. 3). The ecosystem of large river floodplain lakes adjacent to the large river is primarily influenced by its position, how far the lake is situated from the river, and how long the inundation is lasted for (Lewis et al., 2000).

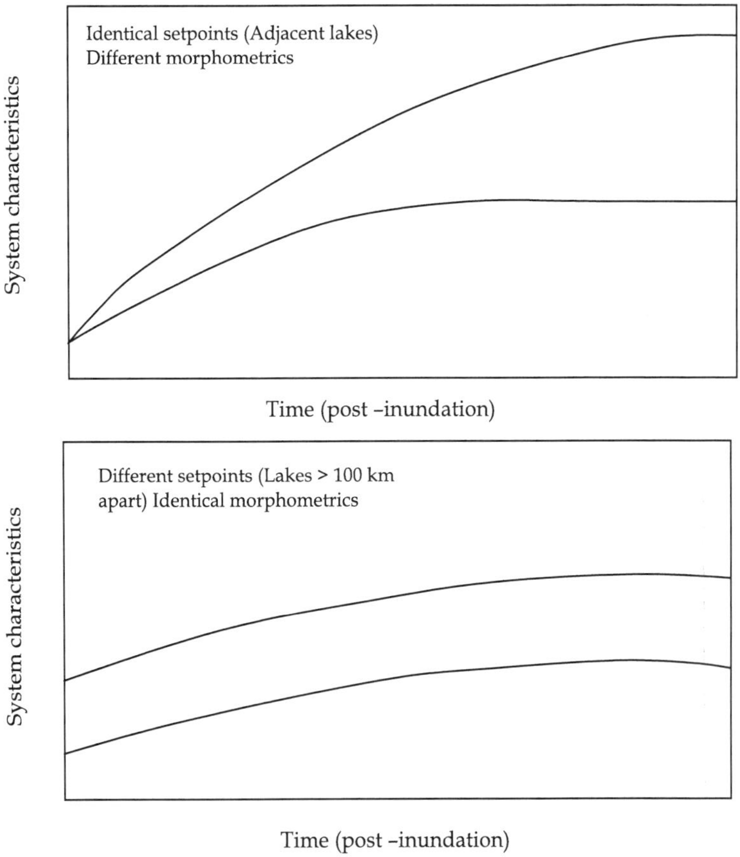

Fig. 3. Causes of the variation in the assemblages of wetland biota including micro-crustaceans, and abiotic composition amongst large river floodplain lakes. Setpoint at the time of inundation (upper diagram) is determined by position of adjacent lakes distributed within the c. 600 km of the floodplain; setpoint following the inundation (lower diagram) is determined by basin morphology (adapted after Lewis et al., 2000).

As a result, water quality of the adjacent wetlands will change following the flood inundation, consequently diversifying the biotic and abiotic assemblages across the wetland (Lewis et al., 2000). For example, in Missouri River floodplain wetlands, alteration of river corridor is reported to have reduced flood pulses significantly. As a result of the absence of flood pulses, micro-crustaceans such as copepod and *Bosmina* showed a strong sensitivity to basic habitat characteristics during and after the flood events within the naturally functioning section of the river (Fisher et al., 2000).

A comprehensive understanding of the large river floodplains lake ecosystems can only help configure effective management strategies. Information regarding diversity and assemblages of micro-crustaceans across temporal and spatial scales is useful for understanding degraded floodplains lake ecosystems and water quality. Changes in assemblages of micro-crustaceans at particular time can provide disturbances caused by external forces such as climate change, invasive species and anthropogenic release of nutrients into the systems and help resource mangers to mitigate these problems.

5. Configuring management strategies of large river floodplain lake ecosystems: Role of micro-crustaceans

Assemblage structure of micro-crustaceans such as cladoceran zooplankton across temporal and spatial scales of large river floodplains lakes can help resource managers for understanding the drivers of ecosystem changes and configuring a range of management strategies. Below how the information obtained from micro-crustaceans are useful to manage floodplains lake ecosystem is comprehensively discussed.

5.1 Management of food web

Understanding of temporal and spatial changes in diversity, composition and abundances of micro-crustacean assemblages are useful for sustainable ecosystem management in large river floodplain lakes. Seasonal production of autochthonous carbon (algae, macrophytes) and inputs derived from the riparian catchments help functioning of floodplains lake ecosystems (Thorpe & Delong, 1994; Lewis et al., 2000). The carbon derived from the riparian system is assimilated by micro-invertebrates supporting the higher trophic levels in food web. However, the degree of energy assimilation by micro-crustaceans in lacustrine food web is less understood. The physical transport of materials to biological transformation to carbon in floodplains lakes varies substantially due to alteration of river flows (Walker et al., 1995). Micro-crustaceans serve as an important role during energy transfer across the trophic levels. For example, in an arid river, Rio Grande (New Mexico, USA), recruitment of some fish occurred during high flows (spring), whereas other fish recruited during low-flows (late summer). Micro-habitats with low current velocity and high temperature were vital nursery grounds for the Rio Grande fishes. Stable isotope analyses of carbon revealed that the Rio Grande fish larvae would obtain carbon predominately from algal production in early summer, but would use organic carbon derived from emergent macrophytes when river discharge would decrease in mid-summer. The shift in carbon assimilation was facilitated by micro-invertebrates that reduced edible algae switching to macrophytes in mid-summer (Pease et al., 2006).

Some species of cladocerans have responded to flood events in the Orinoco River floodplain lakes in Venezuela by showing a varying birth, death and population rates (Twombly & Lewis, 1989). In these floodplain lakes, birth rates increase at a time of flood inundation while mortality increases when fish and invertebrate predations are high (Twombly & Lewis, 1989). In a lowland river system, fish can have size selective predation leading to small sized zooplankton dominating the system, consequently the preservation of the small-sized zooplankton such as *Bosmina* in the system. Mean size of cladoceran mandibles, remains of *Daphnia:Bosmina* ratios and the length of the carapaces and mucros of *Bosmina* can infer past fish assemblages in floodplain lakes and help understanding any changes in food web over time (Kattel, 2011). The cladocerans display morphological variability (cyclomorphosis) in food web. Vertebrate predation pressure on *Bosmina*, for example can result in variation in size of the mucro (Hann et al., 1994). In temperate arid Australia, *Daphnia carinata* show a cyclomorphic behaviour with seasonal changes in body size. Increase in *D. carinata* size indicates a low seasonal water temperature and can help infer the condition of the micro-habitat climates for growth and reproduction of these animals (Mitchell, 1978).

Prolonged drought can lead to cessation of crustacean populations and functioning of floodplain ecosystems. Intensity of floodplain drying also increases changes in algal composition and diversity intensifying the top-down predation and competition (Schneider & Frost, 1996). However, the dry-wet cycles in floodplains recharge the system contributing to the emergence of endangered species of micro-crustaceans through regeneration of egg banks (Boulton & Loyid, 1992). The ephippia of cladocerans in floodplain lakes are viable for several decades. Hatching of resting eggs through genetically advanced technology can help restoring endemic populations (Jeppesen et al., 2000). For example, *Daphnia* ephippia as old as 40 years derived from a lake in South Australia is reported to have been able to hatch in the laboratory environment (Barry et al., 2005). Recently Jeppesen et al. (2002) have successfully reconstructed the catch per unit effort (CPUE) of the planktivorous fish inferred by *Daphnia* ephippia size in a European lake. In the Murray Darling River, the patterns of micro-crustacean distribution and ecosystem processes have been altered by alteration of littoral vegetation and zooplankton egg banks (Jenkins & Boulton, 2007). Unlike in natural condition, where riparian vegetation and littoral macrophyte communities are intact, C and N production is cyclical in nature promoting the growth of littoral macrophytes and micro-crustaceans stabilizing food web structure and dynamics through improvement of the water quality (Box 1), the dieback of littoral vegetation in impounded rivers can alter entire ecological processes including the C and N balances followed by increased salinisation in the region. Zooplankton to phytoplankton ratios serves as a good indicator for grazing intensity of fish and provides the insight for food web structure and dynamics of floodplain lakes (Jeppesen et al., 2001).

5.2 Management of water quality

Micro-crustaceans are used for assessing water quality of the large river floodplain lakes extensively (Gannon & Stemberger, 1978). These organisms are classified according to their preferences for nutrient enrichments (e.g. eutrophic, mesotrophic or oligotrophic), chemistry (e.g. alkaline, acidic or saline) in water. Phophorous (P) and nitrogen (N) are two key nutrients significant for wetland ecosystems. Phosphorous is commonly the growth limiting nutrient in freshwaters exerting a strong control on species composition and primary productivity. Nitrogen can also be a limiting or colimiting nutrient with phosphorous. Anthropogenic influences especially from sewage effluences and agricultural fertilizers can enrich P and N concentrations substantially reducing the water quality (Boucherle & Züllig, 1983). For example, dramatic rise of nitrate concentrations in the Michigan River wetland system was a result of land use intensification following the European arrival. An increased grazing pressure by zooplankton on large algae resulted in increased smaller phytoplankton populations in the Michigan River wetland (Turner & Rabalais, 2003). By keeping a constant zooplankton:phytoplankton (Zp:Ph) ratio in off-shore zone has helped resource managers to maintain clear water quality of the Michigan River wetland system in recent decades (Turner & Rabalais, 2003).

Box 1	
Natural river system	**Regulated river system**
Riparian vegetation/littoral macrophytes	Dieback of littoral zone
↓	↓
Increased C and N production	No C and N balance /Increased salinization
↓	↓
Increased growth of benthic zooplankton/micro-crustaceans	Decreased growth and reproduction of littoral zooplankton/micro-crustaceans
↓	↓
Less erosion/increased water clarity	Increased erosion and turbidity
↓	↓
Increased trophic levels/ fish populations	Decreased trophic levels/less fish populations
↓	↓
Strong food web structure and dynamics	Weak food web structure and dynamics

Box 1.

5.2.1 Phosphorous (P) and nitrogen (N) management

Phosphorous compounds are measured as total phosphorous (TP) and soluble reactive phosphorous (SRP) and nitrogen is measured as total nitrogen (TN), ammonia (NH_3^+), nitrate (NO^{3-}) and nitrite (NO_2). Understanding the dynamics of P and N is essential whilst managing water quality. In shallow enriched lakes internal cycling of P can result in highly variable TP concentrations, often a strong seasonal variation occurs as well as this can usually be high in the summer when P is released from the sediment under anoxic conditions. Nitrogen concentrations however in summer are low in shallow temperate lakes due to an increased assimilation by algae. The high algal biomass leads to oxygen depletion and loss of biodiversity and fish mortality. Understanding of the environmental perturbations of P and N and toxic algal blooms has been advanced through the use of the PCR-DNA technique on cladoceran eggs. Weider et al. (1997) reported that there is a link between changes in allozyme allele and eutrophication caused by increased nitrogen concentrations and algal bloom in the European lowland wetland systems. External nutrient loading in some wetlands from the point source has been controlled by external sewage treatment. However, recovery of the shallow floodplains wetland ecosystems has been delayed due to internal phosphorous loading. Cladoceran-inferred transfer function for TP has been developed in north Europe in order to examine the relationship between zooplankton assemblages and P-induced eutrophication in lowland lake system (e.g., Brodersen et al., 1998). Some benthic chydorid cladocerans are reported to have been predominantly occurring in lowland lakes with relatively high algal productivity. Ecological effect of pollution in interconnected shallow floodplain lakes of the River Erewash system in the UK suggest that significant P and N enrichment in the catchment over the past decades have resulted in a switch from submerged macrophytes to phytoplankton dominant system which have altered macro-and-micro-invertebrates communities over a range of time scales in the past (Sayer et al., 1999; Sayer & Roberts, 2001). Evaluation of P and N concentrations in lowland large river floodplain lakes at temporal and spatial scales using the micro-crustacean assemblages provides a crucial understanding of the land use activity and ecosystem change.

5.2.2 Controlling acidification

Unlike upland lakes the effects of acidification on water quality of large river floodplain lakes is relatively less studied. The impacts of acid deposition on upland rivers and lakes of Europe are shown to have influenced negatively on ecosystem structure and function as a result of sulphur-induced acid rain in the past. In order to improve water quality, attempts were made to reconstruct acidity (lake water pH) inferred by a range of biological proxies archived in lake sediment (Battarbee, 2000). Using modern remains of zooplankton, cladoceran-based pH transfer function was developed. Cladocerans responded very well to acidification of a range of lakes (N=22) distributed across Germany and Austria over the past. The reconstruction of pH using a sediment core derived from the Lake Groβer Arbersee shows that a severe decrease in pH in this lake from about 6 to values of about 4.8 over the past decades (Krause-Dellin & Steinberg, 1986). Some acidobiontic species of cladoceran such as *Alonella exigua* preferring pH less than 5.5 are reported to have survived. Information regarding changes in micro-crustacean assemblages and diversity provides the timing of catchment modification of large river floodplain lakes by humans and help resource managers control acidification. Some endemic zooplankton species of copepods and cladocerans in Australian rivers and wetland systems are reported to have been associated with low water pH, which in turn is regarded as zooplankton preferring habitats with dominant granites and soil types (Tayler et al., 1996). Sulphidic acidification in the Murray Darling Basin is rapid (Baldwin & Mitchell, 2000). Sulpher present in floodplain sediments are exposed to reduce sulphide due to prolonged drought and river regulations enhancing acidification (Hall et al., 2006). A range of ecological effects of sulphidic acidification has been documented in the Murray River Basin. However, the timing for water quality change has not yet been tested using micro-crustaceans. Sulphidic sediment influence hatchability of micro-crustaceans and reduce diversity of acid-sensitive taxa. The use of these animals can help identifying habitat types that are exposed to sulphidic process and reconstructing acidification over various time scales.

5.2.3 Controlling salinisation

Salinity is becoming an increasingly challenging issue for managing water quality and ecosystems of many lowland riverine floodplain wetlands worldwide. Riverine floodplains of coastal zones are frequently inundated by saline water as a result of sea level rise (Schallenberg et al., 2003). Micro-crustaceans can be utilized to manage water quality in wetland since increased salinity in wetlands cause physiological stress in zooplankton resulting from limited osmoregulatory function influencing feeding rate, growth, reproduction, body size, life span and survival capacity. Cladocerans such as *Sida* and *Simocephalus* show optima very close to the mean value of salinity ranging between 0.2 and 17.4% (Aminsick et al., 2005). Amongst chydorids, *Acroperus harpae, Graptoleberis testudinaria, Alonella nana* and *E. lamellatus* prefer low salinity ranges while *Oxyurella* and *Leydigia* prefer high salinity ranges (Aminsick et al., 2005). Transfer function weighted averaging (WA) models for salinity show that cladoceran assemblages are excellent proxies for reconstructing salt concentration in wetlands and help identifying the timing of the release of salt into the system (Bos et al., 1999).

Salinity in arid and semi-arid rivers is influenced by prolonged drought, river regulation, periodic low flows and intensive land use activities in river catchments (Nielsen et al., 2003a). Unlike lowland coastal zones, arid and semi-arid rivers receive salts from groundwater and terrestrial materials via the rock weathering or from the transboundary pollutants from the atmosphere. During low flows, the combination of evaporation and groundwater intrusions assist to increase the natural salinity levels (Jolly et al., 2001). In Murray Darling River, Australia, however, the natural processes have been significantly altered by humans following the European arrival (Jolly et al., 2001). There have been noticeable differences in species richness of micro-crustaceans in low-flows and high salinity periods (Nielsen et al., 2003b, 2007). Zooplankton sampled from longitudinal gradients of the South American arid rivers such as the Salado River (Buenos Aires Province, Argentina) indicates that they have species-specific variations in salinity optima and tolerances (Claps et al. 2009). Hatching of resting eggs of zooplankton is reported to have reduced in wetland with high salinity levels (Skinner et al., 2001). Variance partitioning of benthic cladocerans response to lake water salinity in Kenya suggest that salinity explained more than 51% of the observed variations (Verschuren et al., 2000). Recently Barry et al. (2005) assessed the hatching response of *Daphnia* ephippia to the diatom inferred salinity levels of a sediment core collected from a lake in southwest Victoria, Australia, where significant differences in ephippial densities and hatching were observed with respect to varying salinity levels. Given the increased sensitivity to salinity by cladocerans these organisms can be used to quantify a threshold of salt that are appropriate for a healthy floodplain wetland ecosystems.

5.2.4 Management of toxic substances

Pollution caused by toxic substances is becoming a major threat to diversity, composition and abundances of biota in large river floodplain lakes. Most trace metals have natural mineral origins and it is essential to understand the amount of mineral inputs into wetlands. Records of anthropogenic lead pollution in European lakes are reported to have determined by ratios of $^{206}Pb/^{207}Pb$ in sediment. Natural ratios of the isotope (^{206}Pb:^{207}Pb) are generally higher than those of anthropogenically induced lead pollution and can be determined by analysis of floodplain lake sediments. Recently pyrite pollution has become one of major issues of the organic metallic toxicity across the Murray Darling Basin, Australia due to the exposure of sulpher contained sediments following the river regulations and prolong drought. The processes controlling the FeS pollution in the Murray Darling Basin floodplain lakes is unknown. Establishing a macrophyte colony tolerant to sulphur-induced acidification can be useful. Engelhardt & Ritchie (2001) examined the role of aquatic macrophytes diversity in ecosystem functioning. Greater species richness and biomass of macrophytes tend to lower the chemical activities by filtering the particulate elements from the water and assisting ecosystem functioning and enhancing the wetland management practices. Phytophylous zooplankton such as *Eurycercus* and *Graptoleberis* (Quade, 1969) are proven to be useful for reconstructing past macrophyte cover in some billabongs in Australia (Ogden, 2000). Information regarding macrophyte cover in the past can help elucidating organometallic toxicity in lakes over time. Earliest records of POPs in lake sediments are generally limited, but the PAHs are produced from the combustion of organic matter, and generally have a long term record of past events (e.g. forest fires) in sediment. Sedimentary ratio of 1,7-dimethylphenanthrene and 2,6-dimethylpheanthrene has been used as indicator of wood combustion (Fermàndez et al. 2000). Recently Kattel and Sirocko (2011) have used cladocerans subfossils to identify the range of past anthropogenic regimes including the alteration of forest catchments in a European maar lake.

5.3 Management of invasive species

The endemic floodplain lake ecosystems of the North America were invaded by exotic flora and fauna soon after their introduction (e.g., Mooney & Cleland, 2001). The invading micro-cruastaceans, *Daphnia lumholtzi* also colonised the Upper Paraná River floodplain lakes of South America soon after their introduction. Favourable temperature, water transparency and decreased nutrient concentrations supported the expansion of *D. lumholtzi* in South American wetland system (Simões et al., 2009). The actual effects of alien species on micro-crustacean assemblages are not known, but micro-crustacean assemblages are useful for understanding the impacts and timing of invasion on endemic ecosystems. Less *Daphnia* ephippia are deposited in sediments derived from introduced plants such as *Plantago* and *Pinus* in the Murray Darling River floodplain wetlands in Australia (Reid et al., 2007). Caudal remains of exotic zooplankton *Bythotrephes* sp. in sediment of a Canadian lake were useful to track the energy flow toward the higher trophic level as *Bythotrephes* sp. consistently reduced endemic crustacean populations that were important diet of fish (Hall & Yan, 1997). The timing of geographic distribution pattern of exotic *Daphnia* in North America such as *D. galeata* is unnoticed as a result of extensive hybridization with native *Daphnia.* Allozyme analysis of *Daphnia* ephippia in Europe and North America have become useful for reconstructing timing of invasion (Taylor & Hebert, 1993) and a genetic analysis of cladoceran fossil ephippia have advanced further the knowledge of global distribution patterns and impacts of exotic species on endemic ecosystems (Hairston et al., 1999).

5.4 Mitigation of climate change

Climate change exacerbates the ecological effects of large river floodplain lakes by altering the dynamics of nutrients, pH, salinity and organic toxics compounds such as PAHs and POPs. Mitigation is an action to reduce the risk and hazards of climate associated impacts on ecosystems (IPCC, 2007). Micro-crustacean assemblages are useful for understanding these impacts on large river floodplain lakes ecosystems and help configuring appropriate mitigation strategies. Cladocerans show variation in temperature optima and tolerance ranges. Subfossil cladocerans assemblages can help identifying climate change in a range of time scales in the past (Battarbee, 2000). Climate change such as amount of rainfall causes enlargement and contraction of wetland habitats leading to distinct variations in the relative abundances of littoral and planktonic cladoceran assemblages (Alhonen, 1970). The ratio of littoral:planktonic (L:P) cladocerans serves as significant indicator of climate-induced hydrological regime shifts in shallow floodplain lakes (Ogden, 2000). Cladoceran assemblages and resting eggs have responded to the termination of the last glacial maximum (LGM) and the Holocene sea level rise in coastal regions (Kattel & Augustinus, 2010). Development of a cladoceran-inferred calibration model for temperature is useful to understand the impacts of climate change on ecosystems over a range of time scales in the past and help developing effective management strategies to reduce vulnerability on time (e.g., Lotter et al., 1997; Kattel et al., 2008).

6. Conclusion

Management of large river floodplains lake ecosystems have become increasingly challenging in recent decades as a result of coupled human-climate disturbances. A range of theoretical models being developed in large river systems, have become useful to understand floodplain lake ecosystems processes and develop effective management strategies for restoration of these lakes. However, unprecedented impacts such as river regulation, land use activity, introduction of exotic species and rapid climate warming in recent decades on floodplains lake ecosystems together have intensified the effects and made the ecosystem processes complex to understand. The use of micro-crustaceans particularly the cladocerans are increasingly useful indicator to infer the changes occurring in large river floodplain lakes. Cladocerans play an invaluable role in food web structure and dynamics and they have a wide range of optima and tolerances to temperature as well as other environmental perturbations in floodplains systems. The use of cladoceran subfossils and their ephippia has further reformed our understanding of ecological processes of floodplains lakes of large river system. A long term investigation of the changes in a range of abiotic and biotic assemblages including micro-crustaceans is important to achieve conservation and management goals of large river floodplain lakes ecosystems effectively. Appropriate quantitative and qualitative assessments of these ecosystems can help understanding the past changes and developing future prediction models that provide appropriate information of risks of environmental vulnerabilities and enhances mitigation measures. However, such effort can only be achieved through wider collaborations amongst scientific communities, governments and international organisations.

7. References

Aladin, N. V. & Plotnikov, I. S. (1993). Large saline lakes of former USSR: a summary review. *Hydrobiologia,* Vol. 267, pp. 1-12.

Alhonen, P. (1970). On the significance of the planktonc/littoral ratio in the cladoceran stratigraphy of lake sediments. *Community Biology,* vol. 35, pp. 1-9.

Alpert, P., Griggs, F. T. & Peterson, D. R. (1999). Riparian forest restoration along large rivers: initial results from Sacramento River Project. *Restoration Ecology,* vol. 7, pp. 360-368.

Amsinck, S. L.; Jeppesen, E. & Landkildehus, F. (2005). Relationship between environmental variables and zooplankton subfossils in the surface sediments of 36 shallow coastal brackish lakes with special emphasis on the role of fish. *Journal of Paleolimnology*, vol. 33, pp. 39-51.

Ashton, P. J.; Appleton, C. C.; Jackson, P. B. N. (1986). Ecological impact and economic consequences of alien invasive organisms in Southern African aquatic ecosystems. In: MacDonald IAW, Kruger FJ, Ferrar AA (eds), The Ecological Management of Biological Invasions in Southern Africa, Oxford University Press, Capetown, South Africa, 247-261.

Baldwin, D. S. & Mitchell, A. M. (2000). The effects of drying and reflooding on the sediment and soil nutrient dynamics in lowland river - floodplain system: a synthesis. *Regulated Rivers: Research & Management*, vol. 16, pp. 457-467.

Barbier, E. B. & Thompson, J. R. (1998). The value of water: floodplain versus large scale irrigation benefits in Northern Nigeria. *Ambio*, vol. 27, pp. 434-440.

Barry, M. J.; Tibby, J.; Tsitsilas, A.; Mason, B.; Kershaw, P. & Heijnis, H. (2006). A long term lake-salinity record and its relationships to *Daphnia* populations. *Archiv fur Hydrobiologie*, vol. 163, pp. 1-23.

Battarbee, R. W. (2000). Palaeolimnological approaches to climate change, with special regard to the biological record. *Quaternary Science Reviews*, vol. 19, pp. 107-124.

Bayley, P. B. 1995. Understanding larger River: Floodplain ecosystems. *Bioscience*, vol. 45, pp. 153-158.

Bos, D. G.; Cumming, B. F. & Smol, J. P. (1999). Cladocera and Anostraca from the Interior Plateau of British Columbia, Canada, as paleolimnological indicators of salinity and lake level. Hydrobiologia, vol. 392, pp. 129-141.

Boulton, A. J., Lloyd, L. N. (1992). Flooding frequency and invertebrate emergence from dry floodplain sediments of the River Murray, Australia. *Regulated Rivers: Research & Management*, vol. 7, pp. 137-151.

Boucherle, M. M. & Züllig, H. (1983). Cladoceran remains as evidence of change in trophic state in three Swiss lakes. *Hydrobiologia*, vol. 103, pp. 141-146.

Bright, E. G.; Batzer, D. P. & Garnett, J. A. (2010). Variation in invertebrate and fish communities across floodplain ecotones of the Altamaha and Savannah Rivers. Wetlands, vol. 30, pp. 1117-1128, doi: 10.1007/s13157-010-0116-9

Brock, M. A.; Nielsen, D. L. & Crossle, K. (2005). Changes in biotic communities developing from freshwater wetland sediments under experimental salinity and water regimes. Freshwater Biology, vol. 50, pp. 1376-1390.

Brodersen, K. P. & Anderson, N. J. (2002). Distribution of chironomids (Diptera) in low arctic West Greenland lakes: trophic conditions, temperature and environmental reconstruction. Freshwater Biology, vol. 47, pp. 1137-1157.

Brooke, L. T.; Call, D. J.; Poirier, H. S.; Markee, T. P.; Lindberg, C. A.; McCauley, D. J. & Simonson, P. G. (1986). Acute toxicity and chronic effects of bi9tri-butyltin) oxide to several species of freshwater organism; Centre for Lake Superior Environmental Studies Report, University of Wisconsin-Superior: Superior, WI, 20pp.

Bunn, S. E. & Arthingon, A. H. (2002). Basic principles and ecological consequences of altered flow regimes for aquatic biodiversity. *Environmental Management*, vol. 30, pp. 492-507, doi: 10.1007/s00267-002-2737-0

Cai, W. & Cowan, T. (2008). Evidence of impacts from rising temperature on inflows to the Murray Darling Basin. *Geophysical Research Letters*, vol. 35, pp. 1-5, doi:10.1029/2008GL033390

Capon, S. J. 2003. Plant community responses to wetting and drying in a large arid floodplain. *River Research & Applications*, vol., 19, pp. 509-520, doi: 10.1002.rra.730

Carpenter, C. R., Fisher, S. G., Grim, N. B. & Kitchell, J. F. (1992). Global change and freshwater ecosystems. *Annu. Rev. Ecol. Syst.*, vol. 23, pp. 119-139.

Catalan, J.; Pla, S.; Rieradevall, M.; Felip, M.; Ventura, M.; Buchaca, T.; Camarero, L.; Brancelj, A.; Applby, P. G.; Lami, A.; Grytnes, J. A.; Agusti-Panareda, A. & Thompson, R. (2002). Lake Redó ecosystem response to an increasing warming in the Pyrenees during the twentieth century. *Journal of Paleolimnology*, vol. 28, pp. 129-145.

Chapin, F. S. III; Zavaleta, E. S.; Eviner, V. T.; Naylor, R. L.; Vitousek, P. M., Reynolds, H. L., Hooper, D. U.; Lavorel, S.; Sala, O. E.; Hobbie, M. M. C. & Díaz, S. (2000). Consequences of changing biodiversity. *Nature*, vol. 405, pp. 234-242.

Chen, X.; Yang, X.; Dong, X. & Liu, Q. (2011). Nutrient dynamics linked to hydrological condition and anthropogenic nutrient loading in Chaohu Lake (Southeast China). *Hydrobiologia*, vol. 661, pp. 223-234

Costelloe, J. F.; Payne, E.; Woodrow, I. E.; Irvine, E. C.; Western, A. W. & Leaney, F. W. (2008). Water sources accessed by arid zone riparian trees in highly saline environments, Australia. *Oecologia*, vol. 156, pp. 43-52, doi: 10.1007/s00442-008-0975-4

Dunlop, M. & Brown, P. R. (2008). Implications of climate change for Australia's National Reserve System: A preliminary assessment. *Report to the Department of Climate Change*. Department of Climate Change, Canberra, Australia.

Elvira, B. (1995). Native and exotic freshwater fishes in Spanish river basins. *Freshwater Biology*, vol. 33, pp. 103-108.

Engelhardt, K. A. M., Ritchie, M. E. (2001). Effects of macrophyte species richness on wetland ecosystem functioning and services. *Science*, vol. 411, pp. 687-689.

Fisher, S. J. & Wills, D. W. (2000). Seasonal dynamics of aquatic fauna and habitat parameters in a perched upper Missouri River wetland. *Wetlands*, vol. 20, pp. 470-483.

Gallardo, B.; García, M.; Cabezas, Á.; González, E.; Ciancarelli, C.; González, M. & Comín, F. A. (2007). First approach to understanding riparian wetlands in the Middle Ebro River floodplain (NE, Spain): Structural characteristics and functional dynamics. *Limnetica*, vol. 26, pp. 373-386.

Gannon, J. E. & Stemberger, R. S. (1978). Zooplankton (especially crustaceans and rotifers) as indicators of water quality. *Transactions of American Microscopical Society*, vol. 97, pp. 16-35.

Gehrke, P. C. & Harris, J. H. (2000). Large-scale patterns in species richness and composition of temperate riverine fish communities, south-eastern Australia. *Marine and Freshwater Research*, vol. 51, pp. 165-182, doi: 10.1071/MF99061

Gell, P.; Tibby, J.; Little, F.; Baldwin, D. & Hancock, G. (2007). The impact of regulation and salinasation on floodplain lakes: the lower river Murray, Australia. *Hydrobiologia*, vol. 591, pp. 135-146, doi: 10/1007/s10750-007-0806-3

Hairston, N. G. Jr.; Lampert, W.; Caceres, C. E.; Holtmeier, C. L.; Weider, L. J.; Gaedke, U. et al. (1999). Rapid evolution revealed by dormant eggs. *Nature*, vol. 401, pp. 446-446.

Hall, K. C.; Baldwin, D. S.; Rees, G. N. & Richardson, A. J. (2006). Distribution of inland wetlands with sulfidic sediments in the Murray-Darling Basin, Australia. *Science of the Total Environment*, vol. 370, pp. 235-244, doi: 10.1016/j.scitotenv.2006.07.019

Hall, R. I. & Yan, N. D. (1997). Comparing annual population growth estimates of the exotic invader Bythotrephes by using sediment and plankton records. *Limnology and Oceanography*, vol. 42, pp. 112-120.

Hann, B. J.; Leavit, P.; Chang, P. S. S. (1994). Cladocera community response to experimental eutrophication in Lake 227 as recorded in laminated sediments. *Canadian Journal of Fisheries and Aquatic Science*, vol. 51, pp. 2312-2321, doi: 10.1139/f94-234

Harris, J. H. & Gehrke, P. C. (1993). Development of predictive models linking fish populations recruitment with streamflow. *Population Dynamics of Fisheries Management, Australia Society for Fish Biology Proceedings*, 199pp.

Hughes, L. (2003). Climate change and Australia: trends, projections and impacts. Austral *Ecology*, vol. 28, pp. 423-443

Humphries, P. (2007). Historical indigenous use of aquatic resources in Australia's Murray-Darling Basin, and its implications for river management. *Ecological Management & Restoration*, vol. 8, pp. 106-113

Humphries, P.; King, A. J. & Koehn, J. D. (1999). Fish, flows and flood plains: links between freshwater fishes and their environment in the Murray-Darling River system, Australia. *Environmental Biology of Fishes*, vol. 56, pp. 129-151

Hutchison, M. J. & Armstrong, P. H. (1993). The invasion of a South-Western Australian River system by *Perca fluviatilis*: history and probable causes. *Global Ecology and Biogeography Letters*, vol. 3, pp. 77-89.

IPCC, 2007. IPCC Fourth Assessment Report - Climate Change 2007: The Physical Science Basis Summary for Policymakers.

Jansen, A. & Robertson, A. I. (2001). Relationships between livestock management and the ecological condition of riparian habitats along an Australian floodplain river. *Journal of Ecology*, vol. 38, pp. 63-75.

Jeppesen, E.; Lauridsen, T.; Mitchel, S. F.; Shristoffersen, K. & Burns, C. W. (2000). Trophic structure in the pelagial 25 shallow New Zealand lakes: changes along nutrient and fish gradient. *Journal of Plankton Research*, vol. 22, pp. 951-968.

Jeppesen, E.; Jensen, J. P.; Amsinck, S.; Landkildehus, F.; Lauridsen, T. & Mitchell, S. F. (2002). Reconstructing the historical changes in *Daphnia* mean size and planktivorous fish abundance in lakes from the size of *Daphnia* ephippia in the sediment. *Journal of Paleolimnology*, vol. 27, pp. 133-143.

Jeppesen, E.; Leavitt, P.; De Meester, L. & Jensen, J. P. (2001). Functional ecology and palaeolimnology: using cladoceran remains to reconstruct anthropogenic impact. *Trends in Ecology and Evolution*, vol. 16, pp. 191-198

Jenkins, K. M. & Boulton, A. J. (2003). Connectivity in a dryland river: short-term aquatic microinvertebrate recruitment following floodplain inundation. *Ecology*, pp. 84, vol. 2708-2723.

Jenkins, K. M. & Boulton, A. J. (2007). Detecting impacts and setting restoration targets in arid-zone rivers: aquatic micro-invertebrate responses to reduced floodplain inundation. *Journal of Applied Ecology*, vol. 44, pp. 823-832.

Jolly, I. D.; Williamson, D. R.; Gilfedder, M., Walker, G. R., Morton, R., Robinson, G.; Jones, H., Zhang, L.; Dowling, T. I.; Dyce, P.; Nathan, R. J.; Nandakumar, N.; Clarke, R. & McNeill, V. (2001). Historical stream salinity trends and catchment salt balances in the Murray Darling Basin, Australia. *Marine and Freshwater Research*, vol. 52, pp. 53-63.

Johnston, C. A.; Schubauer-Berigan, J. P. & Bridgham, S. D. (1997). The potential role of riverine wetlands as buffer zones. Buffer Zones: In: *Their Processes and Potential in Water Protection*, N. E. Haycock, T. P. Burt, K. W. T. Goulding & G. Pinay (eds), Quest International.

Junk, W. J.; Bayley, P. B. & Sparks, R. E. (1989). The flood pulse concept in river continuum systems, in: Dodge DP (ed) Proceedings of International Large Rivers Symposium, *Canadian Specieal Publication of Fisheries and Aquatic Sciences*, vol. 106, pp. 89-109

Kattel, G. R. (2011). Can we improve management practice of lakes using cladoceran zooplankton? *River Research & Applications*, doi: 10.1002/rra.1527

Kattel, G. R.; Battarbee, R. W.; Mackay, A. & Birks, H. J. B. (2008). Recent ecological change in remote mountain loch: an evaluation of cladocera-based temperature transfer function. *Palaeogeography, Palaeoecology, Palaeooceanography*, vol. 259, pp. 51-76, doi: 10.1016/j.palaeo.2007.03.052.

Kattel, G. R. & Sirocko, F. (2011). Palaeocladoceran as indicators of environmental, cultural and archaeological developments in Eifel maar lakes region (West Germany) during the Latglacial and Holocene periods. *Hydrobiologia* doi:10.1007/s10750-011-0872-4

King, A. J.; Humphries, P. & Lake, P. S. (2003). Fish recruitment on floodplains: the roles of patterns of flooding and life history characteristics. *Canadian Journal of Fisheries and Aquatic Sciences*, vol. 60, pp. 773-786, doi: 10.1139/F03-057

Kingsford, R. T. (2000). Ecological impacts of dams, water diversions and river management on floodplain wetlands in Australia. *Austral Ecology*, vol. 25, pp. 109-127

Koehn, J. D. (2004). Carp (*Cyprinus carpio*) as a powerful invader in Australian waterways. *Freshwater Biology*, vol. 49, pp. 882-894.

Koehn, J. D. & Harrington, D. J. (2005). Collection and distribution of the early life stages of the Murray cod (*Maccullochella peelii peelii*) in a regulated river. *Australian Journal of Zoology*, vol. 53, pp. 137-144.

Krause-Dellin, D. & Steinberg, C. (1986). Cladoceran remains as indicators of lake acidification. *Hydrobiologia*, vol. 143, pp. 129-134.

Lamontagne, S.; Hicks, W. S.; Fitzpatrick, R. W.; Rogers, S. (2006). Sulfidic materials in dryland river wetlands. *Marine and Freshwater Research*, vol. 57, pp. 775-788.

Leprieur, F.; Beauchard, O.; Blanchet, S.; Oberdofff, T. & Brosse, S. (2008). Fish invasions in the world's river system: when natural processes are blurred by human activities. *PLoS Biology*, vol. 6, pp. 0404-0410

Leung, K. M. Y.; Grist, E. P. M.; Morley, N. J.; Momitt, D. & Orane, M. (2007). Chronic toxicity of tributylin to development and reproduction of the European freshwater snail *Lymnaea stagnalis* (L.). *Chemosphere*, vol. 34, pp. 700-717.

Lewis, W. M. Jr.; Hamilton, S. K.; Lasi, M. A.; Rodriguez, M. & Saunders, J. F. III (2000). Ecological determinism on the Orinoco Floodplain. *Bioscience* vol. 50, pp. 681-692.

Lotter, A. F.; Birks, H. J. B.; Eicher, U.; Hofmann, W. & Marchetto, A. (1997). Modern diatom, Cladocera, chironomid, and chrysophyte cyst assemblages as quantitative indicators for the reconstruction of past environmental conditions in the Alps. I. Climate. *Journal of Paleolimnology*, vol. 18, pp. 395-420.

Lundberg, J. G.; Lewis, W. M. Jr. & Saunders, J. F. III (1987). A major food web component in the Orinoco River channel: Evidence from planktivorous electric fishes. *Science*, vol. 237, pp. 81-83.

Maddock, I.; Thoms, M.; Jonson, K.; Dyer, F. & Lintermans, M. (2004). Identifying the influence of channel morphology on physical habitat availability for native fish: application to the two-spined backfish (*Gadopsis bispinosus*) in the Cotter River, Australia. *Marine and Freshwater Research*, vol. 55, pp. 173-184.

Mensforth, L. J.; Thorburn, P. J.; Tyerman, S. D.; Walker, G. R. (1994). Sources of water used by riparian Eucalyptus camaldulensis overlying highly saline groundwater. *Oecologia,* vol. 100, pp. 21-28.

Meyer, J. L. (1990). A blackwater perspective on riverine ecosystems. *Bioscience,* vol. 40, pp. 643-651.

Meyer, J. L.; Edwards, R. T. (1990). Ecosystem metabolism and turnover of organic carbon along a blackwater river continuum. *Ecology,* vol. 71, pp. 668-677.

Mills, L. E.; Scheuerell, M. D.; Strayer, D. L. & Carlton, J. T. (1996). Exotic species in the Hudson River basin: a history of invasion and introductions. *Easturies,* vol. 19, pp. 814-823

Mitchell, B. C. (1978). Cyclomorphosis in Daphnia carinata King in (Crustacea: Cladocera) from two adjacent Sewage Lagoons in South Australia. *Australian Journal of Marine and Freshwater Research,* vol. 29, pp. 565-576.

Mooney, H. A. & Cleland, E. E. (2001). The evolutionary impact of invasive species. *PNAS,* vol. 98, pp. 5446-5451, doi: 10.1073/pnas.091093398

Naiman, R. J.; Melillo, J. M.; Lock, M. A.; Ford, T. E. & Reice, S. R. (1987). Longitudinal patterns of ecosystem processes and community structure in a subarctic river continuum. *Ecology,* vol. 68, pp. 1139-1156.

Nielsen, D. L.; Brock, M. A.; Rees, G. N. & Baldwin, D. S. (2003a). Effects of increasing salinity on freshwater ecosystems in Australia. *Australian Journal of Botany,* vol. 51, pp. 655-665.

Nielsen, D, L.; Brock, M. A.; Crosslé, K.; Harris, K.; Healey, M. & Jarosinski, I. (2003b). Effects of salinity on aquatic plant germination and zooplankton hatching from two wetland habitats. *Freshwater Biology,* vol. 48, pp. 2213-2224.

Nielsen, D. L.; Brock, M. A., Petire, R. & Crosslé, K. (2007). The impacts of salinity pulses on the emergence of plants and zooplankton from wetland seed and egg banks. *Freshwater Biology,* vol. 52, pp. 784-795.

Nilsson, C. & Berggren, K. (2000). Alterations of Riparian Ecosystems Caused by River Regulation. *Bioscience,* vol. 50, pp. 783-792, doi:10.1641/0006-3568

Ogden, R. W. (2000). Modern and historical variation in aquatic macrophyte cover of billabongs associated with catchment development. *Regulated Rivers: Research & Management,* vol. 16, pp. 497-512.

Palmer, M. A.; Liermann, C. A. R.; Nilsson, C.; Flörke, M.; Alcamo, J.; Lake, P. S.& Bond, N. (2008). Climate change and the world's river basins: anticipating management options. *Frontiers in Ecology and the Environment,* vol. 6, pp. 81–89, doi:10.1890/060148

Pease, A. A.; Davis, J. J.; Edwards, M. S. & Turner, T. F. (2006). Habitat and resource use by larval and juvenile fishes in an arid-land river (Rio Grande, New Mexico). *Freshwater Biology,* vol. 51, pp. 475-486.

Power, M. E.; Dietrich, W. E. & Finlay, J. C. (1996). Dams and downstream aquatic biodiversity: potential food web consequences of hydrologic and geomorphic change. *Environmental Management,* vol. 20, pp. 887-895

Quade, H. W. (1969). Cladoceran faunas associated with aquatic macrophytes in some lakes in northwestern Minnesota. *Ecology,* vol. 50, pp. 170-179.

Reid, M. A.; Brooks, J. J. (2000). Detecting effects of environmental water allocations in wetlands of the Murray-Darling Basin, Australia. *Regulated Rivers: Research & Management,* vol. 16, pp. 479-496.

Reid, M. A.; Sayer, C. D.; Kershaw, A. P. & Heijnis, H. (2007). Palaeolimnological evidence for submerged plant loss in a floodplain lake associated with accelerated catchment soil erosion (Murray River, Australia). *Journal of Paleolimnology*, vol. 38, pp. 191-208, 10.1007/s10933-006-9067-9

Robertson, A.I.; Bacon, P.; Heagney, G. (2001). The responses of floodplain primary production to flood frequency and timing. *Journal of Applied Ecology*, vol. 38, pp. 126-136.

Sadoff, C. W. & Grey, D. (2002). Beyond the river: the benefits of co-operation on international rivers. *Water Policy*, vol. 4, pp. 389-403.

Sam, P. S.; Palmer, M. A.; Biro, P.; Cole, J.; Covich, A. P.; Dahm, C.; Gilbert, J.; Goedkoop, W.; Martens, K. & Verhoeven, J. (2000). Global change and the biodiversity of freshwater ecosystems: impacts on linkages between above-sediment and sediment biota. *Bioscience*, vol. 50, pp. 1099-1107.

Sayer, C. D.; Hoare, D. J., Simpson, G. L.; Henderson, A. G.; Liptrot, E. R.; Jackson, M. J.; Appleby, P. G.; Boyle, J. F.; Jones, J. I. & Waldock, M. J. (2006). TBT causes regime shift in shallow lakes. *Environmental Science and Technology*, vol. 40, pp. 5269-5275.

Sayer, C. D. & Roberts, N. (2001). Establishing realistic restoration targets for nutrient-enriched shallow lakes: linking diatom ecology and palaeoecology at the Attenborough Ponds, U.K. *Hydrobiologia*, vol. 448, pp. 1-3, doi: 10.1023/A:1017597221052

Sayer, C.; Roberts, N.; Sadler, J.; David, C.; Wade, P. M. (1999). Biodiversity Changes in a Shallow Lake Ecosystem: A multi-proxy palaeolimnological analysis. *Journal of Biogeography*, vol. 26, pp. 97-114.

Schallenberg, M.; Hall, C. J.; Burns, C. W. (2003). Consequences of climate-induced salinity increases on zooplankton abundance and diversity in coastal lakes. *Marine Ecology Progress Series*, vol. 251, pp. 181-189.

Schneider, D. & Frost, T. (1996). Habitat duration and community structure in temporary ponds. *Journal of North American Benthological Society*, vol. 15, pp. 64-86.

Shiel, R. J. (1976). Associations of entomostraca with weedbed habitats in a billabongs of the Goulbourn River, Victoria. *Australian Journal of Marine and Freshwater Research*, vol. 27, pp. 533-549.

Simões, N. R.; Robertson, B. A.; Lansac-Tôha, F. A.; Takahashi, E. M.; Bonecker, C. C., Velho, L. F. M. & Joko, C. Y. (2009). Exotic species of zooplankton in the Upper Paraná River floodplain, *Daphnia lumholtzi* Sars, 1885 (Crustacea: Branchiopoda). *Brazil Journal of Biology*, vol. 69, pp. 551-558.

Skinner, R.; Sheldon, F. & Walker, K. F. (2001). Animal propagules in dry wetland sediments as indicators of ecological health: effects of salinity. *Regulated Rivers, Research and Management*, vol. 17, pp. 191-197, doi: 10.1002/rrr.616

Sparks, R. E. (1995). Need for ecosystem management of large rivers and their floodplains. *Bioscience*, vol. 45, pp. 168-182.

Spink, A., Sparks, R. E.; Oorschot, M. V. & Verhoeven, J. T. A. (1998). Nutrient dynamics of large river floodplains. *Regulated Rivers: Research & Management*, vol. 14, pp. 203-216.

Statzner, B. & Higler, B. (1985). Questions and comments on the River Continuum Concept. *Canadian Journal of Fisheries and Aquatic Sciences*, vol. 42, pp. 1038-1044.

Taylor, D. J.; Hebert, P. D. N. (1993). Habitat depdendent hybrid parentage and differential introgression between neighbouringly sympatric *Daphnia* species. *Proceedings of the National Academic Sciences*, vol. 90, pp. 7079-7083.

Thorpe, J. H. & Delong, M. D. (1994). The riverine productivity model: an heuristic view of carbon sources and organic processes in large river ecosystem. *Oikos* vol. 70, pp. 305-308.

Tockner, K.; Malard, F. & Ward, J. V. (2000). An extension of the flood pulse concept. *Hydrological Processes*, vol. 14, pp. 2861-2883.

Turner, R. E. & Rabalais, N. N. (2003). Linking landscape and water quality in the Mississippi River Basin for 200 years. *Bioscience*, vol. 53, pp. 563-572.

Turner, R. E.; Rabalais, N. N.; Justic, D. & Dortch, Q. (2003). Global patterns of dissolved N, P and Si in large rivers. *Biogeochemistry*, vol. 64, pp. 297-317.

Twombly, S. & Lewis, W. M. Jr. (1989). Factors regulating cladoceran dynamics in Venezuelan floodplain lakes. *Journal of Plankton Research*, vol. 11, pp. 317-333.

Vannote, R. L.; Minshall, G. W.; Cummins, K. W.; Sedel, J. R. & Cushing, C. D. (1980). The river continuum concept. *Canadian Journal of Fisheries and Aquatic Sciences*, vol. 37, pp. 130-137

Verschuren, D.; Tibby, J.; Sabbe, K.; Roberts, N. (2000). Effects of depth, salinity and substrate on the invertebrate community of fluctuating tropical lake. *Ecology*, vol. 81, pp. 164-181.

Walker, K. F. (1985). A review of the ecological effects of river regulation in Australia. *Hydrobiologia*, vol. 125, pp. 111-129.

Walker, K. F.; Scheldon, F. & Puckridge, J. T. (1995). A perspective on dryland river ecosystems. *Regulated Rivers & Management*, vol. 11, pp. 85-104.

Wantzen, K. M.; Drago, E. & da Silva, C. J. (2005). Aquatic habitats of the Upper Paraguay River-Floodplain-System and parts of the Pantanal (Brazil). Ecohydrology *& Hydrobiology*, vol. 5, pp. 107-126.

Ward, J. V. (1989). The four-dimensional nature of lotic ecosystem. *Journal of North American Benthological Society*, vol. 8, pp. 2-8.

Ward, J. V.; Tockner, K. & Schiemer, F. (1999). Biodiversity of floodplain river ecosystems: ecotones and connectivity. *Regulated Rivers: Research & Management*, vol. 15, pp. 125-139.

Wehr, J. D. & Descy, J-P. (1998). Use of phytoplankton in large river management. *Journal of Phycology*, vol. 34, pp. 741-749

Weider, L. J.; Lampert, W.; Wessels, M.; Coulbourne, J. K.; Limburg, P. (1997). Long-term genetic shift in a microcrustacean egg bank associated with anthropogenic changes in the Lake Constance ecosystem. *Proceedings of the Royal Society of London B*, vol. 264, pp. 1613-1618.

Williams, W. D. (1985). Biotic adaptations in temporary lentic waters, with special reference to those in semi-arid and arid regions. Hydrobiologia, vol. 125, pp. 85-110.

Wren, D. G.; Davidson, G. R.; Walker, W. G. & Galicki, S. J. (2008). The evolution of an oxbow lake in the Mississippi alluvial floodplain. *Journal of Soil and Water Conservation*, vol. 63, pp. 129-135.

Yan, N. D.; Keller, W.; Somers, K. M.; Pawson, T. W. & Girard, R. E. (1996). Recovery of crustacean zooplankton communities from acid and metal contamination: comparing manipulated and reference lakes. *Canadian Journal of Fisheries and Aquatic Sciences*, vol. 53, pp. 1301-1327.

Yang, X.; Anderson, N. J.; Dong, X. & Shen, J. (2007). Surface sediment diatom assemblages and epliminetic total phosphorous in large, shallow lakes of the Yangtze floodplain: their relationships and implications for assessing long-term eutrophication. *Freshwater Biology*, vol. 53, pp. 1273-1290

Part 6

Management and Policy for Environmental Change

18

Satellite-Based Monitoring of Ecosystem Functioning in Protected Areas: Recent Trends in the Oak Forests (*Quercus pyrenaica* Willd.) of Sierra Nevada (Spain)

M.A. Dionisio, D. Alcaraz-Segura and J. Cabello
Andalusian Center for the Assessment and Monitoring Global Change (CAESCG)
Dept. Plant Biology and Ecology, University of Almería
Spain

1. Introduction

The implementation of monitoring and early warning programs on the ecological status of natural areas is increasingly recognized as an environmental priority (Lovett et al., 2007). However, the development of such programs faces important challenges derived from the many requirements that ecological indicators should fulfill to achieve effective monitoring and alert systems (Oyonarte et al., 2010). Nowadays, ecosystem functioning characterization has become crucial for the monitoring and management of ecosystems due to several reasons (Cabello et al., 2008). First, the evaluation of functional features of ecosystems, such as the carbon gains dynamics, complements the traditional description of ecosystems based solely on vegetation structural features (like physiognomy, dominant species, or floristic composition) derived from few plot observations (Mueller-Dombois & Ellenberg, 1974; Stephenson, 1990; Alcaraz-Segura et al., 2009a). Second, ecosystem functional attributes show a much quicker response to environmental changes than structural ones (Milchunas & Lauenroth, 1995; Wiegand et al., 2004; Alcaraz-Segura et al., 2008a). Third, functional traits are related to key ecological processes that provide a direct measurement of key ecosystem services (Oyonarte et al., 2010; Paruelo et al., 2011; Volante et al., In press). Finally, remote sensing tools can be used to monitor ecosystem functional attributes over extensive areas, in different regions, and with a fast-revisiting frequency (Paruelo et al., 2005; Pettorelli et al., 2005; Baldi et al., 2008; Cabello et al., 2008; Alcaraz-Segura et al., 2009a). The use of satellite-derived information allows for tracking the integrity of key ecological processes and their spatial and temporal variability with the advantage of using common protocols throughout the Earth (Dale & Beyeler, 2001). In this sense, several works have shown the ability of time-series of satellite images to assess the existence of long-term ecosystem functional changes both at the regional (Baldi et al., 2008; Alcaraz-Segura et al., 2010b) and local (Alcaraz-Segura et al., 2008a; Alcaraz-Segura et al., 2008b; Alcaraz-Segura et al., 2009b; Cabello et al., Accepted) scales.

Remote sensing tools can be used to detect both evident functional changes produced by land-use transformations (Volante et al., In press), and other subtle and less noticeable changes including insect outbreaks (Kharuk et al., 2003), wind (Yuan et al., 2002), droughts (Tucker & Choudhury, 1987) or floods (Sanyal & Lu, 2004), fires (Riano et al., 2002), pollution (Chu et al., 2003), etc. These impacts may derive in significant changes in key ecological processes, for instance, carbon balance, microclimate, and biodiversity patterns (Turner, 2005; Lovett et al., 2006; Perry & Millington, 2008). Remote sensing has been proved to be useful for monitoring this kind of "within-state" changes (Vogelmann et al., 2009). In particular, satellite-derived spectral vegetation indices, such as the Enhanced Vegetation Index (EVI) and the Normalized Difference Vegetation Index (NDVI), are considered the most useful approach to monitor ecosystem responses to environmental changes (Pettorelli et al., 2005). Vegetation indices constitute the most feasible approach to estimate primary production at the regional scale (Paruelo et al., 1997) since they show a linear response to the intercepted fraction of photosynthetically active radiation (FPAR) (Hanan et al., 1995), which represents the conceptual basis to relate vegetation indices with net primary production (NPP) through Monteith's model (Monteith, 1972) (equation 1).

$$NPP = PAR * FPAR * RUE \tag{1}$$

Where NPP is the Net Primary Production, PAR is the amount of incident Photosynthetically Active Radiation, FPAR is the fraction of that PAR that is intercepted by vegetation green tissues, and RUE is the Radiation Use Efficiency that plants have to transform that radiation into organic carbon compounds. Given this direct relationship with NPP, the most integrative descriptor of ecosystem functioning (McNaughton et al., 1989; Virginia & Wall, 2001), vegetation indices are frequently used to derive indicators of ecosystem functioning such as the annual amount of carbon absorbed by vegetation, or the seasonality and phenology of the carbon gain dynamics (Pettorelli et al., 2005; Alcaraz-Segura et al., 2006).

To evaluate the usefulness of satellite-derived vegetation indices for monitoring functional changes within protected areas, we focused on the Sub-Mediterranean Pyrenean oak forests (*Quercus pyrenaica* Willd.) of the Sierra Nevada National Park (Spain). These forests are considered as a Natural Habitat of Community Interest (*Quercus pyrenaica* oak woods and *Quercus robur* and *Quercus pyrenaica* oak woods from Iberian northwestern, Directive 92/43/CEE) (García & Mejías, 2009). The Pyrenean oak forests are a *quasi*-endemic habitat of the Iberian Peninsula. The only non-Iberian representations are in the Central West of France and in the Rif Mountains of northern Morocco. In the South of Spain, the Pyrenean oak is considered as a vulnerable species (Blanca & Mendoza, 2000). Sierra Nevada oak populations are considered of great biogeographical importance since they constitute the southernmost Iberian representation of these forests (Molero et al., 1992) and they are considered relict deciduous forests in the Southern Mediterranean region (Blanca & Mendoza, 2000; Blanca, 2001). Several stands of these forests in the Sierra Nevada National Park have an unfavorable conservation status (Molero et al., 1992; Bonet et al., 2010). Multiple global change drivers have an impact on these southernmost woodlands of *Quercus pyrenaica* in the Iberian Peninsula. Historically, these populations have been subjected to intense human disturbances (logging, fires, grazing, agriculture, etc). As a result, these forests are highly fragmented and display low ecological maturity (García & Mejías, 2009) that threatens their long-term conservation. Currently, trends towards temperature rises and precipitation decreases have been hypothesized as the main constraining factor reducing peripheral populations in Sierra Nevada National Park (Molero et al., 1992; Bonet et al., 2010). *Quercus pyrenaica* is a winter semi-deciduous tree with high water demand during the summer. Hence, the predicted lengthening of the summer dry period associated to a reduction in the annual precipitation and the increase in the mean annual temperatures (Bonet et al., 2010) could impose a serious challenge for the regeneration of these forests (Molero et al., 1992; Blanca & Mendoza, 2000). Unfortunately, compared to the wide availability of studies of forest ecology in Europe, there is an enormous lack of knowledge of the conservation status and ecology of Pyrenean oak woodlands in the Iberian Peninsula (García & Mejías, 2009).

Our objective in this study was to use a satellite-based approach to monitor changes in ecosystem functional attributes of the oak forests of the Sierra Nevada National Park (Figure 1). This approach is based on the characterization of the seasonal dynamics and the inter-annual variability and trends of the Enhanced Vegetation Index (EVI). From the mean annual curve of EVI of each forest patch, we derived functional attributes related to primary production, seasonality, and phenology of the forests. Finally, by contrasting the baseline conditions of each forest patch with the long-term observed trends for the period 2001-2009, we identified processes of functional changes happening in these forests that could guide management actions. We propose this satellite approach as a near-real-time tool to provide managers with ecologically meaningful assessments of the ecosystem status based on low-cost but effective information.

2. Methodology

2.1 The Pyrenean oak forests of Sierra Nevada National Park

Sierra Nevada National Park is located in the southeast of the Iberian Peninsula (Figure 1). This National Park protects the best samples of high and medium Mediterranean mountainous ecosystems (MMARM, 2004). This park is a hot spot for plant species richness (Blanca et al., 1998; Blanca, 2001) and invertebrate biodiversity. Its altitude (several summits over 3000 m.a.s.l.), its proximity to Africa, and steep altitudinal gradient constitute the main ecological and evolutionary factors determining its high biodiversity.

The Pyrenean oak forests (Figure 1) of Sierra Nevada represent a conservation priority for the Park managers. There are nine locations distributed on siliceous soils both in the northwestern and southern slopes of the mountain range. In general, they are associated to major river valleys and within an altitudinal range of 1200 to 1900 m.a.s.l. (Table 1).

2.2 Monitoring forest ecological status with EVI

Our monitoring approach was based on the characterization of ecosystem functional attributes derived from the seasonal dynamics of the Enhanced Vegetation Index (EVI). The EVI calculates the normalized difference in reflectance between the red light that is absorbed in photosynthesis and the strong reflection of near infra-red light caused by the cell structure of the leaves. It also includes a third wavelength (blue) that is used to correct the influence of the atmosphere and the soil. EVI is defined according to equation 2 (Huete et al., 1997).

$$EVI = G\frac{NIR - R}{NIR + C_1R - C_2B + L} \tag{2}$$

Where NIR, R and B represent the reflectance in the near infrared, red, and blue wavelengths, C_1 (6) and C_2 (7.5) are coefficients of atmospheric resistance, G (2.5) is the gain factor, and L (1) is a soil correction factor.

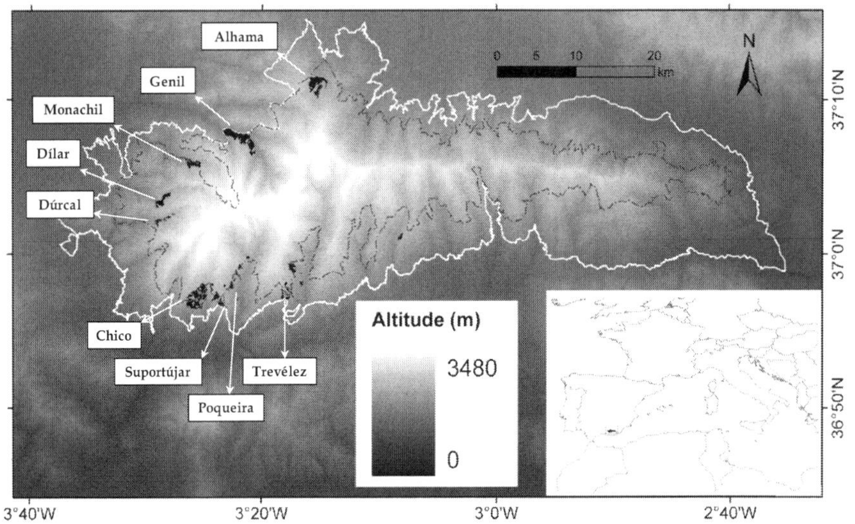

Fig. 1. Distribution of the Pyrenean oak forest patches (*Quercus pyrenaica*) in the Sierra Nevada National Park (southeastern Spain). Forest patches are named according the river basin where they are located: Alhama, Genil, Monachil, Dílar, and Dúrcal, in the northern slope; and Chico, Soportújar, Poqueira, and Trevélez in the southern slope.

Our approach uses satellite images of the Enhanced Vegetation Index captured by the Moderate Resolution Imaging Spectroradiometer (MODIS) sensor onboard the Terra satellite from 2001 to 2009 (Product MOD13Q1). These images have a temporal resolution of 16 days (23 images per year) and a spatial resolution of 231x231 m. We used the Quality Assessment information to filter out low quality data, submitting images to a purification process which removes those pixels affected by high aerosol content, clouds, snow, shadows, and water. From this dataset, we first calculated the 9-year mean EVI seasonal curve for each oak forest site (Figure 1). For this, we only used pixels with more than 75% of their surface occupied by oak woods. Then, the following descriptive attributes of the ecosystem functioning were derived (Figure 2): The EVI annual mean (EVI_mean), an estimator of primary production; the EVI seasonal (or intra-annual) coefficient of variation (EVI_sCV), an indicator of seasonality of carbon gains; the EVI maximum (MAX) and minimum (MIN) values, indicators of the maximum and minimum photosynthetic capacities respectively; and the dates when the maximum (DMAX) and minimum (DMIN) EVI values are reached, two descriptors of the phenology of vegetation greenness. These attributes are widely used and have clear biological meanings (Pettorelli et al., 2005; Alcaraz-Segura et al., 2009a).

To explore the existence of inter-annual trends of ecosystem functioning during the 2001-2009 period in the oak forests of Sierra Nevada, we followed the methodology suggested by Alcaraz-Segura et al. (2009b). In addition to the evaluation of long-term trends of the EVI_mean, we also evaluated the existence of significant trends within each of the 23 images (16-day periods) of the year by means of the Mann-Kendall trend test, a non-parametric trend test robust against non-normality, heterocedasticity, outliers, and serial dependence. For each pixel, we obtained the slope of the trends through the Sen's Method (Hirsch et al., 1982). Significant trends were considered with p-values < 0.05.

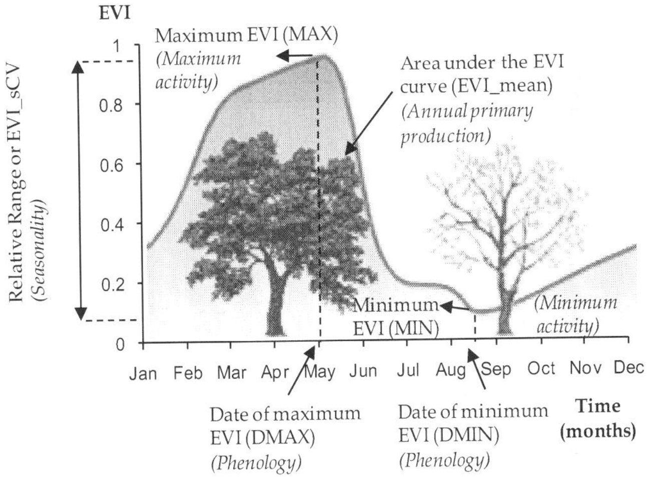

Fig. 2. Functional attributes of the EVI seasonal curve related to ecosystem primary production, seasonality, and phenology. EVI_mean: EVI annual mean, EVI_sCV: EVI seasonal Coefficient of Variation (SD/EVI_mean), MAX: Maximum EVI annual value, MIN: Minimum EVI annual value, DMAX: Date in which is reached the maximum EVI value, DMIN: Date in which is reached the minimum EVI value. These attributes have a clear biological meaning, the EVI_mean is an indicator of the fraction of the radiation used by plants and net primary productivity, MAX and MIN are indicators of the maximum and minimum photosynthetic activity, EVI_sCV is one indicator of seasonality of carbon gains, and DMAX and DMIN are indicators of phenology. Image modified from (Cabello et al., 2010) and G. Baldi from http://lechusa.unsl.edu.ar.

2.3 Statistical analyses

To evaluate whether there exist differences in the EVI attributes among the nine oak woods studied in Sierra Nevada, we performed analysis of variance (ANOVA) only when either raw or transformed attributes fulfilled the necessary parametric requirements of normality and homoscedasticity. To reach normality, for EVI_mean we applied a natural Logarithm (Ln) transformation (Shapiro-Wilk, W=0.990, p=0.266, n=177; Levene's Test F=0.474, p=0.873, n=177) and for EVI_sCV a Box-Cox transformation (Shapiro-Wilk, W=0.983, p=0.031, n=177; Levene's Test F=1.951, p=0.055, n=177). The slight but not significant deviation from normality for the EVI_sCV data did not affect results. For those attributes that even transformed did not fulfill normality (MAX, MIN, DMAX, and DMIN), the analysis was conducted using the non-parametric Kruskal-Wallis test. To determine which groups significantly differed from each other, we used multiple *post hoc* comparisons, using the Tukey test for EVI_mean and EVI_sCV, and the Bonferroni test for MAX, MIN, DMAX, and DMIN. See Figure 5.

3. Results

3.1 Functional characterization of Sierra Nevada oak woods

The Pyrenean oak forests of Sierra Nevada showed a heterogeneous spatial behavior in terms of their EVI seasonal dynamics. In general, woods of the southern slope of Sierra Nevada displayed greater annual vegetation greenness and longer growing seasons than those from the northern slope (Figures 3 and 4). The seasonal EVI curve of the oak woods in the northern slope (Figure 3) showed a gradual increase in productivity that begins around March and that reaches its maximum peak in late May - early June (Figure 5e). Then, senescence takes place at a similar but slightly lower rate than growth. In contrast, the EVI curves of southern-slope woods (Figure 4) show a later but much steeper start of the growing season in late April - early May, reaching the EVI maximum value in June, as in the northern slope woods (Figure 5e). Then, EVI maintains a slowly decreasing plateau until around November, when a less pronounced end of the growing season than in the northern woods occurs.

Statistical comparisons of the EVI attributes (Figure 5) among oak woods also revealed the former differences. In general, Northern oak woods had significantly lower EVI_mean values than southern ones (ANOVA: F=33.56; p=0.0000; n=177; Figures 5a and 6a). Dílar woods (Figure 3d) showed the lowest values and Poqueira (Figure 4c) the highest. The EVI_sCV displayed greater values in the north than in the south (ANOVA: F=29.35; p=0.0000; n=177; Figures 5b and 6b) and a much greater dispersion of data in the north. Although MAX values (Figures 5c and 6c) showed significant differences between some oak woods (Kruskal Wallis: H=36.94; p=0.0000; n=177), there were no clear differences between the northern and southern woods. In general, Max values showed little inter-woods, but large intra-wood variation. We hypothesize that this larger intra-wood variation could be related to greater altitudinal range, such as in Alhama, Genil, Chico and Trevélez (Table 1). DMAX did not either significantly differ between the northern and southern woods, happening in May-June in all oak woods but coming about later with altitude. The increase of intra-wood variability with greater altitudinal variation was also observed in DMAX (Kruskal Wallis: H=64.61; p=0.0000; n=177; Figures 5e and 7a). Regarding MIN values, southern woods showed significantly higher values than northern woods (Kruskal-Wallis: H=126.05; p=0.000; n=177; Figures 5d and 6d), which is directly related to EVI_mean (Figures 5a and 6a) and EVI_sCV (Figures 5b and 6b). Contrary to DMAX, DMIN showed great variability both within and among woods (from November to April) (Figures 5f and 7b) (May-July), with earlier DMIN values in the northern woods than in the southern ones (Kruskal-Wallis: H=86.93; p=0.0000; n=177; Figures 5f and 7b).

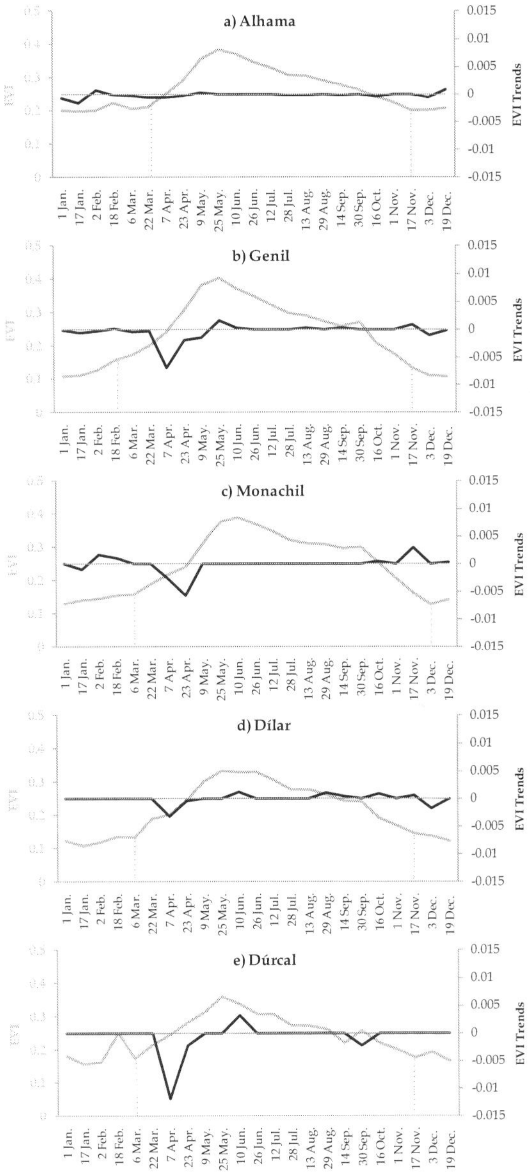

Fig. 3. EVI seasonal dynamics (left Y axis, in gray) and 2001-2009 EVI trends (right Y axis, in black) for oak forests in the northern slope of Sierra Nevada. The horizontal "zero-trend" line shows the absence of significant trends. The two vertical dotted gray lines show the beginning and the end of the growing season.

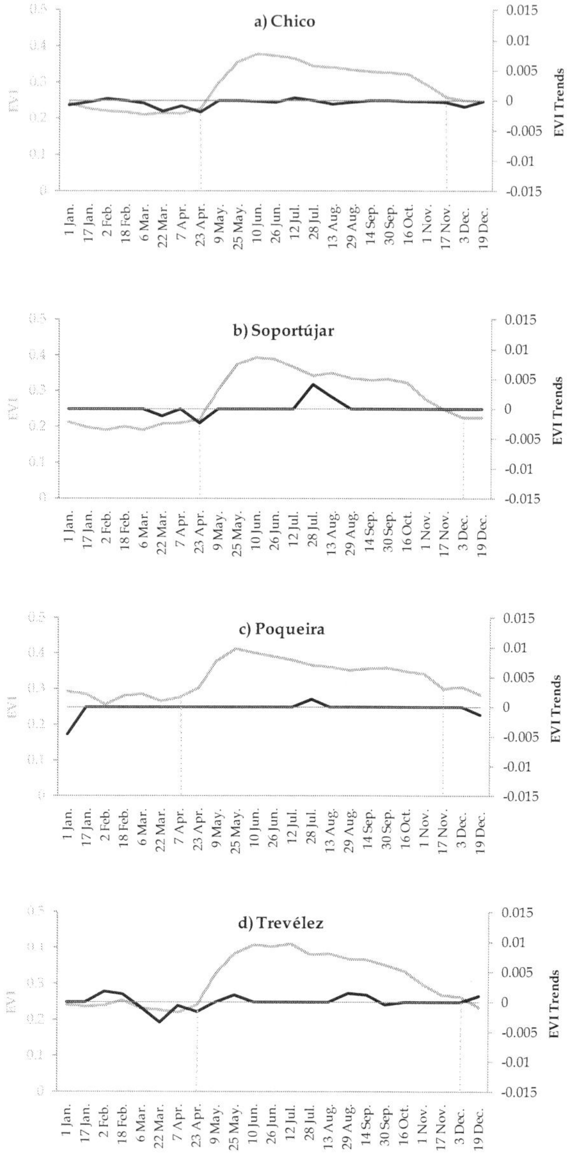

Fig. 4. EVI seasonal dynamics (left Y axis, in gray) and 2001-2009 EVI trends (right Y axis, in black) for oak forests in the southern slope of Sierra Nevada. The horizontal "zero-trend" line shows the absence of significant trends. The two vertical dotted gray lines show the beginning and the end of the growing season.

3.2 Functional changes in Sierra Nevada oak woods

We found significant functional changes happening in the Sierra Nevada oak woods during the 2001-2009 period. Though we did not observe significant long-term trends in the annual synthetic EVI attributes, particular periods of the year did show significant EVI trends. The greatest significant trends occurred at the beginning of the growing season, when strong EVI decreases were observed (March-April), particularly in the northwestern slope (Figure 3). A clearly marked downward trend in productivity was observed between 7th April - 23rd April), which took place in four out of the five northwestern oak woods (Genil, Monachil, Dílar, and Dúrcal, Figures 3b, 3c, 3d, and 3e). Alhama oak wood (Figure 3a) was the only exception, displaying no long-term trends. Some northern woods also showed small positive EVI trends in November (Genil, Monachil, and Dílar; Figures 3b, 3c, and 3d) and in the early-summer (Genil, Dílar, and Durcal).

The southern oak woods (Figure 4) also showed a decrease of vegetation greenness at the beginning of the growing season (except Poqueira, Figure 4c), but less deep than in the northern woods. In addition, EVI increases were observed in middle to late summer in three out of four southern woods (Soportújar, Poqueira, and Trevélez (Figures 4b, 4c, and 4d).

Oak woods	Environmental traits				# of pixels with EVI_mean trends		
	Area (ha)/ Pixels sampled (n)	**Altitudinal range**	**Aspect**	**Slope**	**Positive Sen's slope (+)**	**Negative Sen's slope (-)**	**M-Kendall Significant (p≤0.15) (+/-)**
Alhama	266/36	1443-1838	NE	25°	0	20	0/1
Genil	356/51	1272-1792	N	30°	0	29	0/14
Monachil	104/15	1630-1842	N	27°	0	8	0/1
Dílar	111/14	1594-1884	NW	31°	1	7	0/1
Dúrcal	58/4	1598-1833	W	28°	0	2	0/0
Chico	445/39	1459-1870	S	17°	1	21	0/9
Soportújar	46/4	1652-1755	SW	17°	1	1	0/0
Poqueira	105/5	1635-1888	SE	25°	1	2	0/0
Trevélez	167/9	1397-1880	E	24°	1	5	0/1

Table 1. Environmental traits and EVI_mean trends during the 2001-2009 period in nine *Quercus pyrenaica* oak woods of Sierra Nevada National Park. Forest patches are named according the river basin where they are located: Alhama, Genil, Monachil, Dílar, and Dúrcal, in the northern slope; and Chico, Soportújar, Poqueira, and Trevélez in the southern slope.

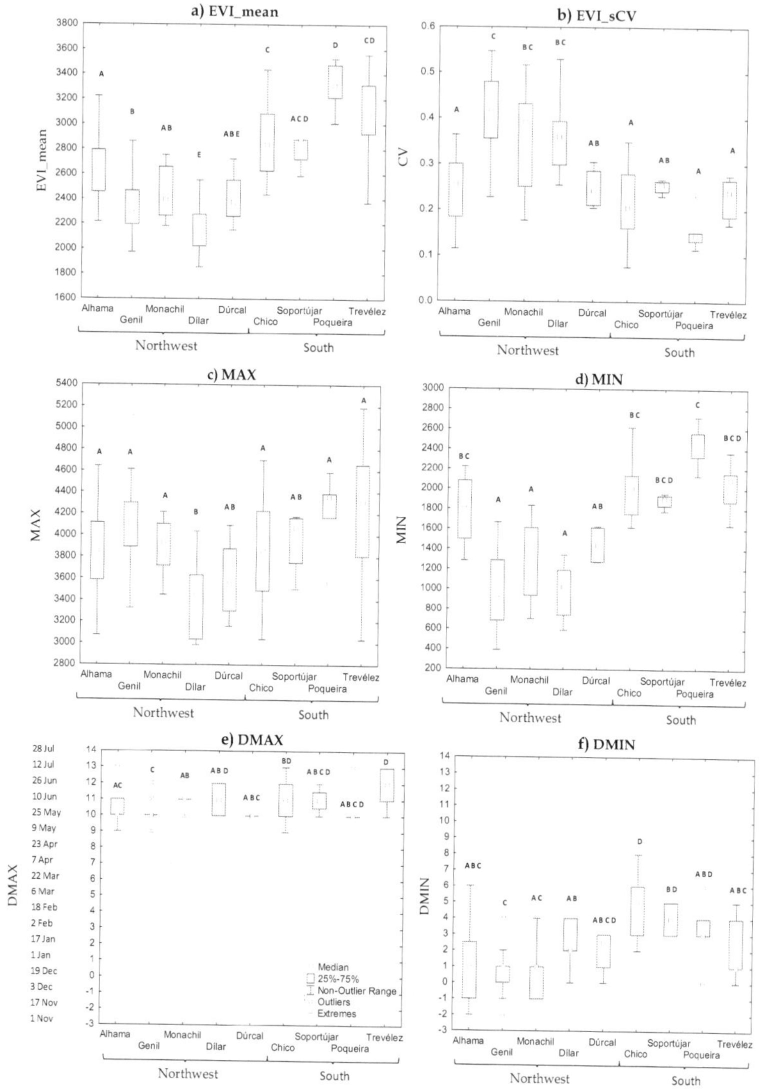

Fig. 5. Functional characterization of the oak woods of Sierra Nevada (Spain) based on the EVI attributes for the 2001-2009 period. Letters show significant differences in *post hoc* comparisons. a) EVI annual mean, an estimator of annual primary production; b) EVI seasonal Coefficient of Variation, a descriptor of seasonality; c) Maximum and d) Minimum EVI annual values, indicators of the maximum and minimum photosynthetic activity; Dates when the e) Maximum and f) Minimum EVI values are reached, indicators of phenology.

4. Discussion

4.1 Baseline conditions and trends in the ecosystem functioning of the Pyrenean oak woods of Sierra Nevada National Park

Our approach, based on a time series of satellite-derived images of the EVI, provided a description of how different attributes of ecosystem functioning change across the remaining locations of Pyrenean oak woodlands in Sierra Nevada. This reference description provides the baseline conditions of ecosystem functioning that can be used to assess the effects of environmental changes on ecosystems processes. The Pyrenean oak woodlands of Sierra Nevada showed a unimodal EVI seasonal dynamics with a unique and well-defined growing season centered in summer and winter minima, as observed in previous works (Alcaraz-Segura et al., 2009a). Differences among locations mainly occurred during the winter non-growing season and at the beginning of the growing season (spring) and were mainly related to the location in the north or south slopes of Sierra Nevada. The lower EVI_mean values in the northern oak woods (Figure 5a) are closely linked to the presence of lower winter MIN values than in the southern woods (Figure 5d) and with the more abrupt EVI decrease during the autumn. In contrast, southern woods maintained relatively high EVI values throughout their longer growing season (Figure 4). The greater annual vegetation greenness of southern woods is probably due to the greater incidence of solar radiation that favors longer growing seasons, milder temperatures during the winter, and an extra water supply from humid air masses coming from the Mediterranean sea that compensate the very high evapotranspiration rates during the summer, in comparison to the colder and more continental locations of the northern slope (Costa Tenorio et al., 2005). Contrary, summer maximum EVI values (MAX) would not cause significant differences in annual vegetation greenness between the northern and southern locations. In consequence, the northern slope shows much greater seasonality (EVI_sCV) than the southern slope since MAX values are similar in both orientations, though the northern woods showed lower MIN values than the southern ones (Figure 5d). From the analysis of the shape of the EVI seasonal curves and according to previous studies (Alcaraz-Segura et al., 2009a), the main limiting factors for vegetation greenness in the oak woodlands of Sierra Nevada are low winter temperatures and lower solar irradiation in the northern slope, which favors a longer presence of snow (Figure 5d). An important point to consider is that the greater vegetation greenness of the southern woodlands during the non-growing season is not related to the activity of the oak trees (because they are winter semi-deciduous), but to the shrubs and herbaceous vegetation occupying the undergrowth vegetation and the patches without trees (Figure 8). In the same way, since the snow melt happens faster and earlier in the southern woods, undergrowth vegetation is also responsible for the earlier and more pronounced rise in vegetation greenness during the start of the growing season than in the northern woods (Figure 3).

Our study also showed that though the oak woodlands of Sierra Nevada have not experienced significant changes of the EVI_mean during the 2001-2009 period, they have suffered seasonal functional changes that mainly affected the beginning of the growing season. In contrast to this relative stability of annual mean vegetation greenness (EVI_mean) since 2001, previous evaluations showed a significant increase in vegetation greenness throughout the eighties and nineties in Sierra Nevada (see Alcaraz-Segura et al., 2008b for the 1981-2003 period, and Alcaraz-Segura et al., 2009b for the 1982-2006 period). Such evaluations used the GIMMS-AVHRR (Global Inventory Modelling and Mapping Studies - Advanced Very High Resolution Radiometer) NDVI dataset. Though there is some debate on the existence of a long-term bias in the GIMMS dataset towards NDVI increases in some regions of the world including the Canadian Boreal forest (Alcaraz-Segura et al., 2010a) and South America (Baldi et al., 2008), the NDVI increases observed in Sierra Nevada with GIMMS during the 1980's and 1990's agreed with other independent datasets. Alcaraz-Segura et al. (2010b) showed that the positive NDVI trends that Sierra Nevada displayed in previous studies with the GIMMS dataset were observed for the 1981-1999 period using other independent datasets such as PAL (Pathfinder AVHRR Land), FASIR (Fourier-Adjustment, Solar zenith angle corrected, Interpolated Reconstructed), and LTDR (Land Long-Term Data Record) datasets. Positive NDVI trends were also observed in Sierra Nevada during the 1989-2002 period using the MEDOKADS (Mediterranean Extended Daily One-km AVHRR Data Set) archive (Martínez & Gilabert, 2009).

The EVI decrease observed at the beginning of the growing season during the 2000-2009 period in Sierra Nevada oak woodlands (Figures 3 and 4), is also in contrast with the NDVI seasonal increase in autumn, winter, and spring that was reported for the 1982-2006 period using GIMMS images of the entire Park (see Figure 2 in: Alcaraz-Segura et al., 2008a). Such contrasting trends lead to think that the increase of spring vegetation greenness that occurred throughout de eighties and nineties (Alcaraz-Segura et al., 2008a) ended around the year 2000 when the spring started to return to lower greenness values. Yet, the trends towards greater vegetation greenness in autumn and winter reached during the eighties and nineties (Alcaraz-Segura et al., 2008a) was maintained after the year 2000, since we did not find significant EVI trends in these seasons. The strong EVI decreases at the beginning of the growing season and the presence of some EVI summer increases during the senescence period lead to think that the growing season of southern oak woods (Figure 4) might be starting later but strengthening towards the summer (with the exception of Poqueira; Figure 4c).

An important outcome of our work is that significant functional changes, i.e. a significant decrease of vegetation greenness at the beginning of the growing season, took place in Sierra Nevada oak woodlands without implying significant trends in the annual averages. Despite the EVI annual mean, an estimator of annual primary production, is extensively used as an integrative descriptor of ecosystem functioning and status, our work highlights the importance of studying variables beyond the annual summaries (like seasonality and phenology) as significant trends in particular months of the year may not significantly affect the EVI annual mean but may have broad ecological consequences in critical periods such as the start of the growing season.

4.2 Application to forest monitoring and management

Since satellite images are regularly captured over large regions and under common protocols, the spectral vegetation indices represent an adequate approach to implement ecosystems monitoring programs in protected areas and to promote adaptive management actions (Alcaraz-Segura et al., 2008a; Alcaraz-Segura et al., 2008b; Cabello et al., 2008). Our work provides interesting information for the prioritization and the orientation of management actions for the Pyrenean oak forests of Sierra Nevada National Park. First, we provided a regional functional reference characterization of all oak woodlands of the Park for the 2001-2009 period. Our monitoring approach uses EVI-derived descriptors of ecosystem functioning that may allow managers to detect the spatial and temporal anomalies (Oyonarte et al., 2010), and to guide specific management actions in particular areas. The spatial and temporal deviations from the baseline conditions detected could be alerting of inconspicuous "within-state" changes in the forests as a result of cumulative impacts (Vogelmann et al., 2009). However, to improve the ecological significance of this approach for the Park management, the monitoring program should include the identification of the key ecological processes that can be related to this functional description and that are central for the maintenance of the ecological integrity. For instance, the differences in the strength of the EVI trends among different oak forest patches could be associated to the two modes of climatic variability that affect Sierra Nevada. The observed weaker start of the growing season during study period could be related to the increase of positive phases of the North Atlantic Oscillation (NAO Index), which are the main control of winter precipitation and temperature, particularly in the north-western slope (Liras, 2011). In addition, we also observed EVI increases during the summer (July-August) in the southern slope (Figure 4), which could be related to the increase of active phases of the Western Mediterranean Oscillation (WeMO), increasing late summer precipitation during the study period (Liras, 2011; Cabello et al., Accepted). In this sense, the obtained results in the EVI trends for the different woods could be used to prioritize management actions in relation to climate change adaptation in the most threatened sites. Nevertheless, this should be only one of the guiding hypotheses for adaptive management, since other processes such as insect damage and forest succession could also be taking place in the park (Sierra Nevada National Park managers, *personal communication,* Stöver et al., 1996; CMJA, 2008).

A monitoring system based on the tools and analysis shown here could embrace several monitoring objectives, as it simultaneously informs managers about the changes in productivity, phenology, and seasonality of the ecosystems. For example, changes in the EVI attributes could be directly related to changes in the amount, seasonality, and phenology of ecosystem carbon gains. In addition, linking the EVI dynamics of the Pyrenean oak woodlands to the ecology of species of conservation concern could be used to evaluate and monitor the conservation status of the habitat of such species. This could be the case of the blue tit (*Parus caeruleus*), whose reproductive success is related to the ecosystem status of *Quercus pyrenaica* forests, especially at the beginning of female reproductive period (April-May), which is associated with the start of the growing season (Arriero et al., 2006). Such association implies that delays in the start of the growing season or forest degradation would negatively affect the reproduction success of this bird. Moreover, the information derived from this monitoring approach could help guiding land-use planning to avoid overexploitation of Sierra Nevada oak woodlands. For instance, livestock pressure should be limited in those periods of the year that are experiencing strong negative EVI trends.

5. Conclusions

Our approach shows how satellite based monitoring systems can be very useful to assess the effects of environmental changes on protected areas and to orientate adaptive management actions. Overall, this study provides a reference characterization against which to assess changes in ecosystem functioning of the oak woods of Sierra Nevada, and identifies functional changes that occurred during the 2001-2009 period. Such information helps to fill the lack of knowledge about these woodlands, as demanded by the Spanish Ministry of Environment (García & Mejías, 2009). In practical terms, it allows the incorporation of ecosystem functional aspects of ecosystems to nature conservation and to the maintenance of ecosystem services, in particular those related to carbon sequestration in this protected area. Our results imply that conservation and management policies cannot be only based on static situations, since ecosystems are changing. In addition, annual summaries are not enough as monitoring indicators, since functional changes may occur at key seasonal stages without affecting the annual means.

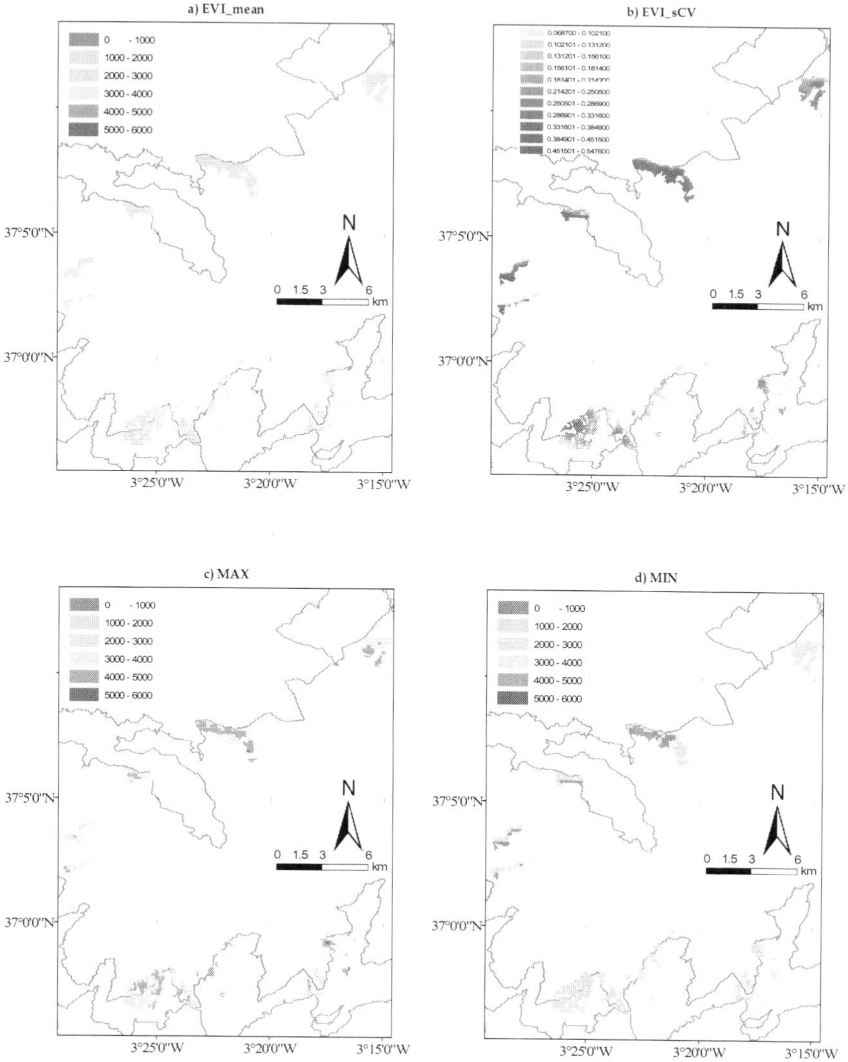

Fig. 6. Maps of the EVI attributes for Sierra Nevada Oak woods generated by the Monparq application. EVI_mean: EVI annual mean, an estimator of annual primary production; EVI_sCV: EVI seasonal Coefficient of Variation, a descriptor of seasonality; MAX and MIN: Maximum and Minimum EVI annual values, indicators of the maximum and minimum photosynthetic activity.

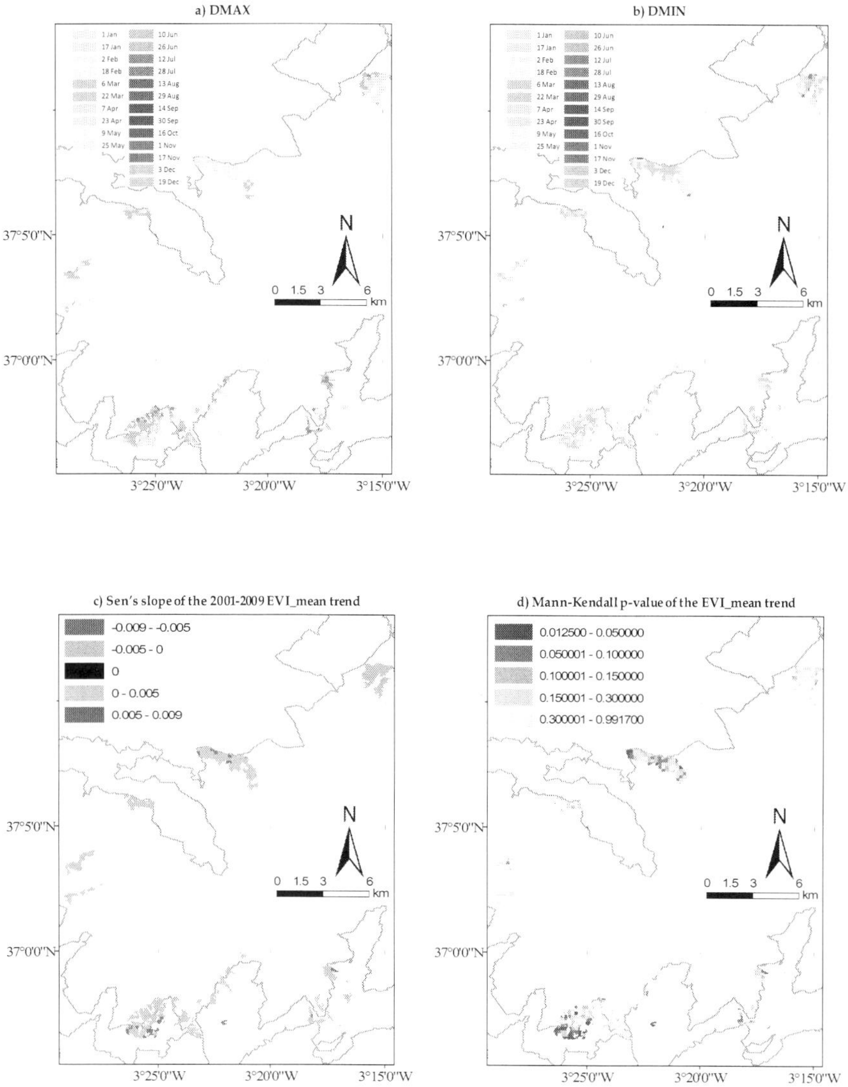

Fig. 7. Maps of the EVI attributes and trends for Sierra Nevada Oak woods generated by the Monparq application. a) DMAX and b) DMIN: Dates when the Maximum and Minimum EVI values are reached, indicators of phenology. c) Sen's slope of the 2001-2009 EVI_mean trend d) Mann-Kendall p-value of the 2001-2009 EVI_mean trend.

Fig. 8. Landscape picture showing the start of the growing season (13th April 2011) in the northernmost *Quercus pyrenaica* oak wood of Sierra Nevada National Park (Spain), the oak wood of the Alhama River at Dehesa del Camarate. The picture shows how the green sprouts of the oak trees are starting to come out while the leaves of the undergrowth shrubs are well developed.

To spread the use of our monitoring approach and to make possible for managers the exploitation of such information, we have developed a software tool named "Monparq Monitoring System for Parks" that allows a non-advance user to assess the differences between locations, to explore the different environmental controls across the northern and southern slopes, and to evaluate the inter-annual trends in ecosystem functioning. This tool provides managers with valuable information to assess management effectiveness in an adaptive management strategy. It will help managers answering questions like, what ecosystems are undergoing major changes?, or how do management actions affect ecosystem functioning stability?

6. Acknowledgments

Thanks to L. Sevilla, who helped processing the datasets, to F.J. Bonet and B. Benito from the Sierra Nevada Global Change Observatory for providing climate data, to the Park managers and J. del Río from the Andalusian Environmental Agency for their guidance and valuable information in the field, and to F. Maestre for the English revision. Financial support was given by FEDER Funds, Junta de Andalucía (GLOCHARID and SEGALERT P09-RNM-5048 projects), Organismo Autónomo de Parques Nacionales (Proyecto 066/2007), and Ministerio de Ciencia e Innovación (Proyecto CGL2010-22314, subprograma BOS, Plan Nacional I+D+I 2010). D. Alcaraz-Segura was partially covered by the Inter-American Institute for Global Change Research (IAI, CRN II 2031 and 2094) under the US National Science Foundation (Grant GEO-0452325). The trend test was run using the MATLAB code "Seasonal Kendall Test with Slope for Serial Dependent Data" provided by Jeff Burkey through the MATLAB Central file exchange (http://www.mathworks.com, accessed September 2009). Satellite images were freely provided by the MODIS Land website.

7. References

Alcaraz-Segura, D.; Baldi, G.; Durante, P. & Garbulsky, M. (2008a). Análisis de la dinámica temporal del NDVI en áreas protegidas: tres casos de estudio a distintas escalas espaciales, temporales y de gestión. *Ecosistemas*, Vol.17, No.3, pp. 108-117, ISSN 1697-2473

Alcaraz-Segura, D.; Cabello, J.; Paruelo, J.M. & Delibes, M. (2008b). Trends in the surface vegetation dynamics of the National Parks of Spain as observed by satellite sensors. *Applied Vegetation Science*, Vol.11, pp. 431-440, ISSN 1402-2001

Alcaraz-Segura, D.; Cabello, J. & Paruelo, J. (2009a). Baseline characterization of major Iberian vegetation types based on the NDVI dynamics. *Plant Ecology*, Vol.202, No.1, pp. 13-29, ISSN 1385-0237

Alcaraz-Segura, D.; Cabello, J.; Paruelo, J.M. & Delibes, M. (2009b). Use of descriptors of ecosystem functioning for monitoring a national park network: A remote sensing approach. *Environmental Management*, Vol.43, No.1, January, pp. 38-48, ISSN 1432-1009 (Electronic), 0364-152X (Linking)

Alcaraz-Segura, D.; Chuvieco, E.; Epstein, H.E.; Kasischke, E.S. & Trishchenko, A. (2010a). Debating the greening vs. browning of the North American boreal forest: differences between satellite datasets. *Global Change Biology*, Vol.16, No.2, pp. 760-770, ISSN 1365-2486

Alcaraz-Segura, D.; Liras, E.; Tabik, S.; Paruelo, J. & Cabello, J. (2010b). Evaluating the Consistency of the 1982–1999 NDVI Trends in the Iberian Peninsula across Four Time-series Derived from the AVHRR Sensor: LTDR, GIMMS, FASIR, and PAL-II. *Sensors*, Vol.10, No.2, pp. 1291-1314, ISSN 1424-8220

Alcaraz-Segura, D.; Paruelo, J. & Cabello, J. (2006). Identification of current ecosystem functional types in the Iberian Peninsula. *Global Ecology and Biogeography*, Vol.15, No.2, pp. 200-212, ISSN 1466-8238

Arriero, E.; Sanz, J.J. & Romero-Pujante, M. (2006). Habitat structure in Mediterranean deciduous oak forests in relation to reproductive success in the Blue Tit Parus caeruleus: Capsule Effects operate during laying and incubation and with less success in breeding territories characterized by a young and immature vegetation structure. *Bird Study*, Vol.53, No.1, pp. 12-19, ISSN 0006-3657

Baldi, G.; Nosetto, M.D.; Aragón, M.R.; Aversa, F.; Paruelo, J.M. & Jobbagy, E.G. (2008). Long-term satellite NDVI datasets: Evaluating their ability to detect ecosystem functional changes in South America. *Sensors*, Vol.8, pp. 5397-5425, ISSN 1424-8220

Blanca, G. (2001). Flora, In: *Parque Nacional de Sierra Nevada*, Calvo-Flores D, (Ed.), 106-148, Esfagnos, ISBN 978-84-932095-1-3,

Blanca, G.; Cueto, M.; Martínez-Lirola, M.J. & Molero-Mesa, J. (1998). Threatened vascular flora of Sierra Nevada (Southern Spain). *Biological Conservation*, Vol.85, No.3, pp. 269-285, ISSN 0006-3207

Blanca, G. & Mendoza, R.T. (2000). *Libro rojo de la flora silvestre amenazada de Andalucía: Especies vulnerables*, Junta de Andalucía, Consejería de Medio Ambiente, ISBN 978-84-89650-75-6, Sevilla, España

Bonet, F.J.; Pérez-Luque, A.J.; Moreno, R. & Zamora, R. (Septiembre 2010). Observatorio de Cambio Global de Sierra Nevada. Estructura y Contenidos Básicos. Consejería de Medio Ambiente, Junta de Andalucía, Available from http://www.scribd.com/doc/32239398/Observatorio-Sierra-Nevada-Datos-basicos

Cabello, J.; Alcaraz-Segura, D.; Altesor, A.; Delibes, M.; Baeza, S. & Liras, E. (2008). Funcionamiento ecosistémico y evaluación de prioridades geográficas en conservación. *Ecosistemas*, Vol.17, No.3, pp. 53-63, ISSN 1697-2473

Cabello, J.; Alcaraz-Segura, D.; Ferrero, R.; Castro, A.J. & Liras, E. (Accepted). The role of vegetation and lithology in the spatial and inter-annual response of EVI to climate in drylands of Southeastern Spain. *Journal of Arid Environments*, ISSN 0140-1963

Cabello, J.; Alcaraz-Segura, D.; Liras, E. & Sevilla-García, L. (2010). Monparq, una plataforma informática para el seguimiento de los Parques Nacionales. *Centro Andaluz para la Evaluación y Seguimiento del Cambio Global (CAESCG)*

CMJA. (Noviembre 2008). Consejería de Medio Ambiente. Borrador de la adecución del Plan Forestal Andaluz para el periodo 2008-2015. Dirección General de Planificación e Información Ambiental, Available from http://www.juntadeandalucia.es/medioambiente/site/web/menuitem.a5664a214f73c3df81d8899661525ea0/?vgnextoid=e45601c4bd997110VgnVCM1000000624e50aRCRD&vgnextchannel=3259b19c7acf2010VgnVCM1000001625e50aRCRD&lr=lang_es

Costa Tenorio, M.; Morla Juaristi, C. & Sainz Ollero, H. (2005). *Los bosques ibéricos. Una interpretación geobotánica*, Planeta, ISBN 978-84-08058-20-5, Barcelona

Chu, D.; Kaufman, Y.; Zibordi, G.; Chern, J.; Mao, J.; Li, C. & Holben, B. (2003). Global monitoring of air pollution over land from the Earth observing System-Terra Moderate Resolution Imaging Spectroradiometer (MODIS). *Journal of Geophysical Research*, Vol.108, No.D21, ISSN 0148-0227

Dale, V.H. & Beyeler, S.C. (2001). Challenges in the development and use of ecological indicators. *Ecological indicators*, Vol.1, No.1, pp. 3-10, ISSN 1470-160X

García, I. & Mejías, P.J. (2009). *9230 Robledales de Quercus pyrenaica y robledales de Quercus robur y Quercus pyrenaica del noroeste ibérico. Bases ecológicas preliminares para la conservación de los tipos de hábitat de interés comunitario en España*, Ministerio de Medio Ambiente y Medio Rural y Marino, ISBN 978-84-491-0911-9, Madrid, España

Hanan, N.; Prince, S. & Begue, A. (1995). Estimation of absorbed photosynthetically active radiation and vegetation net production efficiency using satellite data. *Agricultural and forest meteorology*, Vol.76, No.3-4, pp. 259-276, ISSN 0168-1923

Hirsch, R.M.; Slack, J.R. & Smith, R.A. (1982). Techniques of trend analysis for monthly water quality data. *Water resources research*, Vol.18, No.1, pp. 107-121, ISSN 0043-1397

Huete, A.R.; Liu, H.Q.; Batchily, K. & van Leeuwen, W. (1997). A comparison of vegetation indices over a global set of TM images for EOS-MODIS. *Remote Sensing of Environment*, Vol.59, No.3, pp. 440-451, ISSN 0034-4257

Kharuk, V.; Ranson, K.; Kuz'michev, V. & Im, S. (2003). Landsat-based analysis of insect outbreaks in southern Siberia. *Canadian journal of remote sensing*, Vol.29, No.2, pp. 286, ISSN 0703-8992

Liras, E. (2011). *Funcionamiento ecosistémico: Controles y patrones espacio-temporales en el SE Ibérico*, Universidad de Almería, ISBN 978-84-8240-967-2, Almería, España

Lovett, G.M.; Burns, D.A.; Driscoll, C.T.; Jenkins, J.C.; Mitchell, M.J.; Rustad, L.; Shanley, J.B.; Likens, G.E. & Haeuber, R. (2007). Who needs environmental monitoring? *Frontiers in Ecology and the Environment*, Vol.5, No.5, pp. 253-260, ISSN 1540-9295

Lovett, G.M.; Canham, C.D.; Arthur, M.A.; Weathers, K.C. & Fitzhugh, R.D. (2006). Forest ecosystem responses to exotic pests and pathogens in eastern North America. *BioScience*, Vol.56, No.5, pp. 395-405, ISSN 0006-3568

Martínez, B. & Gilabert, M.A. (2009). Vegetation dynamics from NDVI time series analysis using the wavelet transform. *Remote Sensing of Environment*, Vol.113, No.9, pp. 1823-1842, ISSN 0034-4257

McNaughton, S.J.; Oesterheld, M.; Frank, D.A. & Williams, K.J. (1989). Ecosystem-level patterns of primary productivity and herbivory in terrestrial habitats. *Nature*, Vol.341, No.6238, 14 September, pp. 142-144, ISSN 0028-0836

Milchunas, D. & Lauenroth, W. (1995). Inertia in plant community structure: state changes after cessation of nutrient-enrichment stress. *Ecological Applications*, Vol.5, No.2, pp. 452-458, ISSN 1051-0761

MMARM. (2004). Ministerio de Medio Ambiente y Medio Rural y Marino. Parque Nacional de Sierra Nevada, Available from
http://reddeparquesnacionales.mma.es/parques/org_auto/red_ppnn/parques/10_sierra.htm

Molero, J.; Pérez-Raya, F.; Tendero, F.V. & González-Tejero, M.R. (1992). *Parque Natural de Sierra Nevada: paisaje, fauna, flora e itinerarios*, Rueda, ISBN 978-84-72070-67-0, Madrid, España

Monteith, J. (1972). Solar radiation and productivity in tropical ecosystems. *Journal of Applied Ecology*, Vol.9, No.3, pp. 747-766, ISSN 0021-8901

Mueller-Dombois, D. & Ellenberg, H. (1974). *Aims and methods of vegetation ecology*, John Wiley & Sons, ISBN 978-1-930665-73-6, New York, EE.UU

Oyonarte, C.; Alcaraz-Segura, D.; Oyarzabal, M.; Paruelo, J.M. & Cabello, J. (2010). Sistema de apoyo a la gestión de reservas de la biosfera basado en el monitoreo de la productividad primaria: ensayo en Cabo de Gata-Níjar (Almería-España), In: *Reservas de la Biosfera: Su contribución a la provisión de servicios de los ecosistemas, experiencias exitosas en Iberoamérica*, Araya P, Clüsener-Godt M, (Ed.), 119-140, UNESCO, ISBN 978-956-332-417-4, Paris

Paruelo, J.; Alcaraz-Segura, D. & Volante, J.N. (2011). El seguimiento del nivel de provisión de los servicios ecosistémicos, In: *Valoración de Servicios Ecosistémicos: Conceptos, Herramientas y Aplicaciones para el Ordenamiento Territorial*, Laterra P, E. Jobbágy & J. Paruelo, (Ed.), 141-162, Ediciones INTA, ISBN 978-987-679-018-5, Buenos Aires

Paruelo, J.M.; Epstein, H.E.; Lauenroth, W.K. & Burke, I.C. (1997). ANPP estimates from NDVI for the central grassland region of the United States. *Ecology*, Vol.78, No.3, pp. 953-958, ISSN 0012-9658

Paruelo, J.M.; Piñeiro, G.; Oyonarte, C.; Alcaraz-Segura, D.; Cabello, J. & Escribano, P. (2005). Temporal and spatial patterns of ecosystem functioning in protected and areas in southeastern Spain. *Applied Vegetation Science*, Vol.8, No.1, pp. 93-102, ISSN 1402-2001

Perry, G.L.W. & Millington, J.D.A. (2008). Spatial modelling of succession-disturbance dynamics in forest ecosystems: Concepts and examples. *Perspectives in Plant Ecology, Evolution and Systematics*, Vol.9, No.3-4, pp. 191-210, ISSN 1433-8319

Pettorelli, N.; Vik, J.O.; Mysterud, A.; Gaillard, J.M.; Tucker, C.J. & Stenseth, N.C. (2005). Using the satellite-derived NDVI to assess ecological responses to environmental change. *Trends in Ecology & Evolution*, Vol.20, No.9, pp. 503-510, ISSN 0169-5347

Riano, D.; Chuvieco, E.; Ustin, S.; Zomer, R. & Dennison, P. (2002). Assessment of vegetation regeneration after fire through multitemporal analysis of AVIRIS images in the Santa Monica Mountains. *Remote Sensing of Environment*, Vol.79, No.1, pp. 60-71, ISSN 0034-4257

Sanyal, J. & Lu, X. (2004). Application of remote sensing in flood management with special reference to monsoon Asia: a review. *Natural Hazards*, Vol.33, No.2, pp. 283-301, ISSN 0921-030X

Stephenson, N.L. (1990). Climatic control of vegetation distribution: the role of the water balance. *The American Naturalist*, Vol.135, No.5, pp. 649-670, ISSN 0003-0147

Stöver, O.; Horst, D.; Engels, F.; Joachim, H. & Atzberger, C. (1996). An assessment of forest areas damaged by a Gypsy Moth infestation through satellite remote sensing and GIS, In: *Application of remote sensing in European forest monitoring*, Kennedy P, (Ed.), 359-372, Commission of the European Communities, Viena, Austria

Tucker, C.J. & Choudhury, B.J. (1987). Satellite remote sensing of drought conditions. *Remote Sensing of Environment*, Vol.23, No.2, pp. 243-251, ISSN 0034-4257

Turner, M.G. (2005). Landscape ecology: What is the state of the science? *Annual Review of Ecology, Evolution, and Systematics*, Vol.36, pp. 319-344, ISSN 1543-592X

Virginia, R. & Wall, D. (2001). Principles of ecosystem function, In: *Encyclopedia of biodiversity*, Levin SA, (Ed.), 345-352, Academic Press, ISBN 0122268652, San Diego, USA

Vogelmann, J.E.; Tolk, B. & Zhu, Z. (2009). Monitoring forest changes in the southwestern United States using multitemporal Landsat data. *Remote Sensing of Environment*, Vol.113, No.8, pp. 1739-1748, ISSN 0034-4257

Volante, J.N.; Alcaraz-Segura, D.; Mosciaro, M.J.; Viglizzo, E.F. & Paruelo, J.M. (In press). Ecosystem functional changes associated with land clearing in NW Argentina. *Agriculture, Ecosystems and Environment*, ISSN 0167-8809

Wiegand, T.; Snyman, H.A.; Kellner, K. & Paruelo, J.M. (2004). Do grasslands have a memory: modeling phytomass production of a semiarid South African grassland. *Ecosystems*, Vol.7, No.3, pp. 243-258, ISSN 1432-9840

Yuan, M.; Dickens-Micozzi, M. & Magsig, M.A. (2002). Analysis of tornado damage tracks from the 3 May tornado outbreak using multispectral satellite imagery. *Weather and forecasting*, Vol.17, No.3, pp. 382-398, ISSN 1520-0434

19

Linking Sea Level Rise Damage and Vulnerability Assessment: The Case of Greece

Areti Kontogianni[1], Christos Tourkolias[1], Michalis Skourtos[1] and Maria Papanikolaou[2]

[1]*University of Aegean, School of the Environment, Mytilini, Lesvos,*
[2]*Cambridge Quaternary Department of Geography & University of Athens, Department of Geology and Geo-Environment, Greece*

1. Introduction

Throughout the course of modern history, coasts have been a substantial means of human development and an ever-growing number of people still continue to colonize the coasts worldwide. Coasts comprise dynamic and complex socio-ecological systems, encompassing a variety of biotic and abiotic elements. Their complexity and dynamics are reflected in the multiplicity of their forms. Their dynamic nature is responsible for their high productivity, leading both to periodic changes and gradual mutation. The marine ecosystems, by storing carbon dioxide and by releasing oxygen to the atmosphere through the living processes of the phytoplankton population, play a significant role in regulating climate. The coastal areas help create and preserve microclimates. The existence of coastal forests and wetlands ensures minimization of floods, erosion and other natural disasters, and offers valuable regulating and supporting ecosystem services. The importance of coastal resources for the prosperity of coastal areas can be specified through the ecosystem services and goods, which support the human life (Daily, 1997; Turner et al., 2001; Beaumont, 2007; Kontogianni et al., 2010a). The categorization of coastal services and goods is presented in Table 1.

However, the ensuing anthropogenic activities of industrialization and economic growth have brought the coastal areas under intense pressure. Climatic change accentuates these pressures while it makes mean sea level rise (SLR) one of the most predictable and alarming impacts globally (Church et al., 2001; Nicholls, 2007). To make things worse, SLR is known to be rather inelastic against the reduction of greenhouse gas emissions (OECD, 2006), a phenomenon known as "commitment to SLR". That is, even if drastic reduction policies globally succeed in stabilizing the climate, SLR and the accompanying phenomena of coastal erosion and storm surges will continue to occur for centuries (Meehl et al., 2005; Wigley, 2005), causing possible tipping points for some systems (Tipping Points Report, 2009).

This chapter examines the impacts of SLR on the Greek coastal zone and appraises their economic dimension. Researchers engaged in studies like this face two important issues. The first is the quantification of the economic impacts (damages) caused by the losses of coastal areas due to SLR. The second is the *ex ante* estimation of welfare gains from reducing SLR risks, since this estimation constitutes an important input for decision-making regarding policy and technical measures (mitigation and adaptation measures). Cost-benefit analysis is used as a tool for prioritization among different policy goals. Therefore, methodologically, it must succeed in associating economic estimates with measurable physical indicators, so that researchers are well aware of exactly what is being appraised (Kontogianni et al., 2010a; Sonderquist et al., 2008). Changes in physical indicators mostly refer to non-tradeable environmental goods (magnitudes) (e.g. human health, biodiversity conservation, quality of ecosystems etc). Due to the difficulty in appraising their economic value, they are usually not taken into consideration in decision making, thereby they constitute an external cost. A multidisciplinary approach, in order to be integrated and successful, has to deal with the co-evolutionary aspects of both natural and socio-economic system, known together as the 'socio-ecological' system (Folke et al., 2002).

Supportive services	Regulating services
1 Biogeochemical cycling 2 Primary production 3 Food web dynamics 4 Diversity 5 Habitat 6 Resilience	1 Atmospheric regulation 2 Local climate regulation 3 Sediment retention 4 Biological regulation 5 Pollution control 6 Eutrophication mitigation
Provisioning services	**Cultural services**
1 Food 2 Inedible resources 3 Genetic resources 4 Chemical resources 5 Ornamental resources 6 Energy resources 7 Space and waterways	1 Recreation 2 Aesthetic values 3 Science and education 4 Cultural heritage 5 Inspiration 6 The legacy of nature

Table 1. Categorization of services and goods in the coastal environment (Source: Adapted from Garpe, 2008 & MEA, 2005).

As pointed out in the latest national report submitted to the UNFCCC regarding climate change (Hellenic Republic, 2006), no coordinated effort to assess the long-term impacts of SLR and to design appropriate adaptation policies has been as yet conducted in Greece. To our knowledge and to date, only two studies have calculated the monetary losses of SLR for the Greek coastal zone. Dalianis et al. (1997) calculated the total cost of impacts caused by SLR (1-m) in Greece by 2100. The total cost was estimated at €3.4 billion. The authors cite IPCC's first Assessment Report as the source of their monetary estimates. The research program PESETA estimated the future impacts on coastal areas from SLR for 22 European countries including Greece (Richards & Nicholls, 2009; Vafeidis et al., 2008). The analysis was performed with a combination of the integrated model DIVA (Dynamic and Interactive Vulnerability Assessment Tool) and the scenarios A2 and B2 of the IPCC. The calculation of damages in the Greek coastal zone was restricted to land loss due to erosion and flooding and the ensuing human migration.

Few similar attempts have been performed to date in European scale. Sanchez-Arcilla et al. (2008) examined the implications of climatic change on the Ebro delta coast (Spain). Their research focused on the effects of climatic changes in wave return periods, inundation of low-lying areas and saltwater intrusion, yet without implementing the monetary evaluation of the triggered impacts or the calculation of the necessary investment cost of adaptation policies. Pruszak and Zawadzka (2008) estimated total economic and social costs of land loss and flood risk in Polish coastal zone considering two scenarios of SLR (30 cm and 100 cm in 100 years). Kont et al. (2008) studied the impacts of SLR (1 m in 2100) on the coastal zone of Estonia without the implementation of adaptation measures. The coastal zone was studied either in the case of inundation by SLR or in the case of storm surges and the impacts were quantified in both physical and monetary terms. Sterr (2008) assessed the vulnerability (in economic terms) for five coastal states in Germany in the case of 1 m SLR and estimated the required costs for protection. Aunan and Romstad (2008) studied the potential damages from SLR to roads, bridges and port infrastructure in Norway based on possible restoration costs. Karacat and Nicholls (2008) performed a preliminary assessment of the potential costs due to SLR (1 m) in Turkey and the required investment costs for prevention. Devoy (2008) examined the physical components of coastal vulnerability to SLR in Ireland and presented available estimates for the capital value loss and the protection/adaptation costs assuming a scenario of SLR equal to 1 m until 2100.

This chapter is structured as follows: in section 2 we provide a description of the Greek coastal zone and its vulnerability. In section 3 we lay out our research hypotheses, methodology and sources of data. In section 4 we estimate the financial impacts (damages) of both long-term and short-term SLR. At last, in section 5, we summarize and conclude the chapter.

2. Ecosystem service and vulnerability assessment of the Greek coastal zone

According to the ATEAM (2004), Mediterranean is considered the most vulnerable coastal part of Europe with multiple potential impacts and low generic adaptive capacity.

Knowledge of the vulnerability and ability to adapt to climate change is valuable for adopting suitable policies for both natural and social systems.

Vulnerability holds several definitions. One of those refers to the degree to which an ecosystem service is sensitive to global change, plus the degree to which the sector that relies on this service is unable to adapt to the changes (Metzger et al., 2004).

Vulnerability is also assessed by the ATEAM (2004) as the likelihood of a specific human-environment system to experience harm due to exposure to perturbations, accounting for the process of adaptation. According to the ATEAM, high potential impact and low adaptive capacity constitutes a high degree of vulnerability for the system. Adaptive capacity according to Brooks (2003) has no direct implications to current vulnerability and can only diminish future vulnerability. IPCC (2007) defines adaptive capacity as the ability of a human-environment system to adjust to climate change (including climate variability and extremes), to moderate potential damages, to take advantage of opportunities, or to cope with the consequences.

According to IPCC, vulnerability is a function of the sensitivity of a system to changes in climate (the degree to which a system will respond to a given change in climate, including beneficial and harmful effects), adaptive capacity (the degree to which adjustments in practices, processes, or structures can moderate or offset the potential for damage or take advantage of opportunities created by a given change in climate), and the degree of exposure of the system to climatic hazards (IPCC, 2001).

Adger et al. (2004) adopt another approach by separating biophysical from social vulnerability. Vulnerability, according to Brooks et al. (2005), depends critically on context, and the factors that make a system vulnerable to a hazard will depend on the nature of the system and the type of hazard in question. Resilience is used to define two specific system attributes: The amount of disturbance a system can absorb and still remain within the same state or domain of attraction; the degree to which the system is capable of self-organization. (Klein et al., 2004). Handmer (1996) defines vulnerability generally as susceptibility to injury which may be seen as inversely related to resilience: the more resilient one system, the less vulnerable.

A typical case study of the `vulnerability` issue, described in the preceding paragraphs, is the Greek coastal zone. An assessment of coastal ecosystem goods and services in Greece and their physical geographic vulnerability are discussed below. We refer to the social vulnerability and relevant risk perceptions in section 4.3.

The Greek coastal zone has a total length of approximately 16,200 km, being one of the longest coastal zones among European countries. Almost half of the coastal zone belongs to the continental Greece while the remaining half to the 3,000 islands (or 9,800 if islets are included). The importance of the main categories of coastal goods and services (Table 1) provided by the coastal Greek area is described below (YPEXODE 2006, Zanou 2003).

About 33% of the Greek population inhabits coastal areas located at 1-2 km distance from the coast. If we consider coastal population as those inhabiting areas up to 50 km from the coast, then the percentage of Greek coastal population reaches 85% of the total. Twelve out of the thirteen Prefectures of the Greek territory are registered as coastal areas, while the largest urban centres are located in the coastal zone. About 80% of industrial activities, 90% of tourism and recreational activities, 35% of agriculture (usually of high productivity), fisheries and aquaculture, as well as an important part of infrastructures (ports, airports, roads, electricity and telecommunications network etc) are located in the coastal zone. The added value created in the coastal zone includes:

- The operation of 20 ports from which more than one million tonnes of goods are transported annually
- The total fishery production of 96,000 tonnes
- The total fishery sector fleet of 19,000 ships (constituting 20% of the total fleet of the 25 EU member-states)
- The total aquaculture production, 258,000 € worth (representing 10% of the total production of the 25 EU member-states)
- The majority of hotel beds in the tourist sector. During the tourist period, the population in some of the Greek islands increases 2 to 10-fold due to domestic and foreign tourists.

The fishery and aquaculture sectors are important due to their contribution to the Greek GDP, but mostly due to their role in fostering and preserving social and cultural cohesion of the coastal areas. The fishery sector in 1999 had 40,000 employees, with a total production of 231,000 tn, while the number of directly employed in aquacultures is 4,800 and the number of indirectly employed exceeds 7,500 employees.

The coastal zone consists of variable habitats, which contribute to the conservation of biogenetic reserves. Indicatively, over 6,000 different flora species, 670 vertebrate species and 436 avifauna species are found in coastal zones.

Over the last 20 years (1990-2010), there has been an increase in construction of summer residences at the Greek coastal areas (YPEXODE, 2006). The overall urbanized coastal zone area is estimated to be 1,315 km^2, accounting for 1.31% of the total Greek coastal zone. In

Greece, construction of summer residence occurs too close to the coast (Figure 1), increasing social vulnerability in the case of SLR. Construction near the coast happens due to the fact that tides in the Mediterranean do not exceed 40 cm. So, vulnerability rises due to the increased exposure of coastal constructions and the growing number of people colonizing the Mediterranean coasts.

Fig. 1. Storm surge in Molyvos coast, Lesvos island, Greece, December 2009 (photo T. Karabas).

All the aforementioned coastal resources contribute to the development of cultural services, such as leisure, aesthetics, and ability to perform scientific and educational activities, conservation of cultural heritage and cultural capital, also through arts, philosophy and inspirational sources. The coastal ecosystem services regulate, support and supply, in both natural and cultural terms, the Greek social capital through generations at a scale that exceeds the local and can be historically projected to a European and global level.

All the above ecosystem services provided by the Greek coastal zone lead to the conclusion that such an important natural resource should be worthy of respect and protection. The threats to the Greek coastal and marine environment stem mostly from anthropogenic driving forces (e.g. overexploitation of natural resources, urbanization, pollution, eutrophication, and invasive species). A major problem of the Greek coastal zone is the high rate of coastline erosion: over 20% of the total coastline is threatened making Greece the 4th most vulnerable country, among the 22 coastal EU member states, in terms of coastal erosion (EUROSION, 2004). Major causes for the increased erosion are the particularly strong winds and the storm surges in the Aegean Sea, the anthropogenic interventions (e.g. dams which reduce sediment input, Poulos et al., 2002) as well as the geomorphologic substrate of the coastline: the 2,400 km (15% of the total shoreline) correspond to non consolidated sediment deposits, while 960 km (6% of the total shoreline) correspond to coastal deltaic areas (Papanikolaou et al., 2010). Erosion is expected to increase in the immediate future due to (a) the foreseen rise of the mean sea level, (b) the intensification of extreme wave phenomena

and (c) the further reduction of the river sediment inflows due to changes in rainfall and construction of river management works (Emanuel, 2005; IPCC, 2007; Velegrakis, 2010).
A reliable assessment of the potential risk associated with SLR should take into account not only the trends and rates of eustatic SLR, but consider also such local factors as tectonics, sediment supply and compaction, and storm surges (Poulos & Collins, 2002; Vött, 2007). Especially the role of tectonism is important in tectonically active zones because it can counterbalance the relative SLR. Typical examples constitute the coastal zone of northern Peloponnese with an uplift rate ranging between 0.3 and 1.5 mm/year, Crete with an uplift rate between 0.7 and 4 mm/year and Rhodes between 1.2 and 1.9 mm/year. Thus, a supposed average value of 4.3 mm/year SLR would be reduced to 3.5 mm/year due to the counteraction of a mean tectonic uplift of about 0.8 mm/year (Papanikolaou et al., 2010). The expected sea level rise could also be locally offset by the increased fluvial sediment input and deposition in deltaic plains and resultant advance of the shoreline (Poulos et al., 2002). On the contrary, reduced fluvial sediment input in deltaic plains would reinforce sea inundation due to sea level rise. An important factor in the vulnerability of coastal areas to SLR is the coastal morphology (i.e. slope and lithological composition) because it is related directly to the rate of erosion. The latter can range from very high (several m/year) in the case of low-lying land to low (approximately mm/year) in the case of hard coastal limestone formations (e.g. cliffs).

Fig. 2. Coastal areas in Greece with medium (green colour) and high (red colour) vulnerability. Black colour indicates areas with altitudes below 20 m, usually of loose sedimentary deposits. (Source: Papanikolaou et al., 2010)

In Figure 2, coastal areas are subdivided into: (a) those classified as of medium vulnerability to SLR (green colour) consisting of non consolidated sediment deposits in areas with low altitude, (b) those classified as of high vulnerability to SLR including deltaic deposits in low altitude (red colour). High risk areas are deltaic areas such as Evinos in Messolonghi, Kalama in Igoumenitsa, Acheloos, Mornos at the Corinthian Gulf, Pineios, Alfeios, Aliakmonas and Axios at the Thermaic Gulf, the area of North Aegean near Platamona, Amphipolis, Strymon, Nestos (to Abdyra), the Ebros, and the deltaic areas in Malliakos, Amvrakikos, Messiniakos and Argolikos Gulfs. Black colour indicates areas with altitudes below 20 m, usually of loose sedimentary deposits. The other zones designated as coastal areas of a low vulnerability are mainly rocky and high altitude coastal regions.

Assessing the severity of the rising sea level impacts on coastal areas includes uncertainties with regard to:

a. The intensity of sea level rise, which ranges between 0.2 and 2 meters. The evolution of the sea level rise is determined by the interaction between several natural (e.g. astronomical parameters) and anthropogenic (e.g. greenhouse gas) forces. The severity of each one of these will also determine the overall development of the climate cycle we are currently in, which seems to be at the peak of today's "warm" interglacial period.
b. The relationship between the tectonic elevation and the eustatic sea level rise which, for many areas of the Greek territory is quite significant, to the extent that it may counterbalance or locally exceed the sea level rise.
c. the sedimentation of clastic materials in coastal areas, which is determined by geological and climate conditions but also by anthropogenic interventions (e.g. dams, river sand mining), which for instance in the case of river deltas, may alter their vulnerability to the sea level rise.

The estimation of the length of these three types of coastal areas shows that from a total of 16,200 km, 960 km (6%) corresponds to deltaic areas of high vulnerability (red colour), 2,400 km (15%) to non consolidated sediments of medium vulnerability (green colour) and the remaining 12,810 km (79%) to rocky coastal areas of low vulnerability. Therefore, the total coastline length characterized by medium to high vulnerability to SLR is about 3,360 km representing 21% of the Greek shoreline (Papanikolaou et al., 2010).

Typical approximate values of flooded coastal areas and shoreline retreat (excluding the tectonics and geodynamics corrections) triggered by a possible SLR equal to 0.5 m and 1 m in high risk areas are presented in Table 2. This table illustrates the impacts of SLR as estimated in 27 Greek coastal zone case studies. Available case studies were surveyed through a literature review till September 2010. The coastal land retreat for a hypothetical increase of SLR equal to 0.5 m ranges from 15 m to 2,750 m, while the range for a hypothetical increase of 1 m ranges from 400 m to 6,500 m. Figure 3 maps the geographical distribution of the examined case studies.

The selected case studies used for the economic assessment of SLR impacts on Greek coastal zone are: C1: Skala Eressos Mytilene, C2: Gulf of Nafplio, C3: Lagoon Kotichiou, C4: Hersonissos Crete, C5: Aigio Achaias, C6: Lambi Kos, C7: Kardamaina Kos, C8: Tigaki Kos, C9: Afantou Rhodes, C10: Vartholomio Ileias, C11: Acheloos River Delta, C12: Plain of Thessaloniki, C13: Abdyra Xanthi, C14: Lake Alyki Limnos, C15: Saltmarsh Kitrous Pierias, C16: Porto Heli, C17: Ermioni, C18: Evinos River Delta, C19: Mornos River Delta, C20: Kalama River Delta, C21: Penaeus River Delta, C22: Thermaic Gulf (includes Axios River Delta, Aliakmonas River Delta, Loudias-Aliakmonas Deltaic plain), C23: Kiparissiakos Gulf (includes Alfeios River Delta - northern part and Alfeios River Delta - southern part), C24: South Euboean Gulf.

Coastal area	SLR (m)	Inundated area (10^3 m²)	Length/Area of shoreline	Source
Skala Eressos Mytilene	0.3	28	2.5 km	Doukakis, 2008
Gulf of Nafplio	0.5	4,200	25 km	Doukakis, 2005a
	1	8,700		
Lagoon Kotichiou	0.5	720	27.6 km	Doukakis, 2003
	1	1,760		
Hersonissos Crete	0.5	4,700	20 km	Doukakis, 2004
	1	5,200		
Aigio Achaias	0.5	1,070	6.8 km	Doukakis, 2005b
	1	1,800		
Lambi Kos	0.5	35	0.25 km	Papadopoulou & Doukakis, 2003
	1	52		
Kardamaina Kos	0.5	19	0.615 km	
	1	33		
Tigaki Kos	0.5	161	2.7 km	
	1	322		
Afantou Rhodes	0.5	375	3 km	
	1	439		
Vartholomio Ileias	0.5	190	2.65 km	
	1	300		
Acheloos River Delta	1	72	5.8 km	Doukakis, 2007
Plain of Thessaloniki	1	37,100	41.2 km	Kanelakis & Doukakis, 2004; Doukakis, 2007
Abdyra Macedonia	1	716	7 km	Doukakis, 2007
Lake Alyki Limnos	1	2,041	4.3 km	Pliakos & Doukakis, 2004; Doukakis, 2007
Saltmarsh Kitrous Pierias	0.5	9,450	-	Stergiou & Doukakis, 2003
	1	11,800		
Porto Heli	0.5	36	38.93 km	Seni & Karibalis, 2007
	1	161		

Ermioni	0.5	19	19.903 km	
	1	278		
Evinos River Delta	0.5	12,500	92 km2	Karibalis & Gaki-Papanastasiou, 2008
	1	21,300		
Mornos River Delta	0.5	2,580	28 km2	
	1	3,710		
Kalama River Delta	0.5	7,020	78 km2	
	1	10,060		
Penaeus River Delta	0.5	6,530	69 km2	
	1	14,780		
Alfeios River Delta (northern part)	0.5	224	110 km2	Poulos et al., 2009
	1	683		
Alfeios River Delta (southern part)	0.5	35		
	1	344		
Axios River Delta	0.5	10,825	390 km2	
	1	28,482		
Aliakmonas River Delta	0.5	4,875	120 km2	
	1	8,950		
Loudias-Aliakmonas Deltaic plain	0.5	8,900		
	1	25,575		
South Euboean Gulf	0.5	7,890	18.5 km	Roussos & Karibalis, 2009
	1	12,620		

Table 2. Shoreline retreat and inundated area for potential SLR of 0.5-m and 1-m.

Apart from long-term SLR, other climate phenomena capable of causing coastal erosion , are the foreseen increase of storminess / frequency of storm surges (IPCC, 2007).

Storm surges and SLR are distinct phenomena. However, climate change may increase the risk of storm surges by changing two drivers: cyclone 's frequencies/intensities and the mean sea-level rise (McInnes et al., 2000; Emanuel, 2005). The interannual and decadal variability in time of extremes is caused by mean sea level changes (Marcos et al., 2009).Changes in mean sea level and changes in the meteorological strength of storm surges (enhanced by climate change) may cause extreme wave phenomena and, accordingly, serious damage on coastal areas. This happens because strong winds affect larger water masses which unleash more energy to storm surges, while the height of the waves increases relatively to the mean sea level rise; as a result the waves further penetrate coastal areas and have significant impacts on coastline morphology. The strong coastal waves caused by the stormy winds cause erosion, while the normal, low-mid energy waves cause sediment deposition (Komar, 1998). The impacts of storm surges include:

- Flooding of coastal areas.
- Destruction of coastal infrastructure
- Coastal erosion.
- Intrusion of salt water in coastal habitats, lagoons, rivers e.t.c.

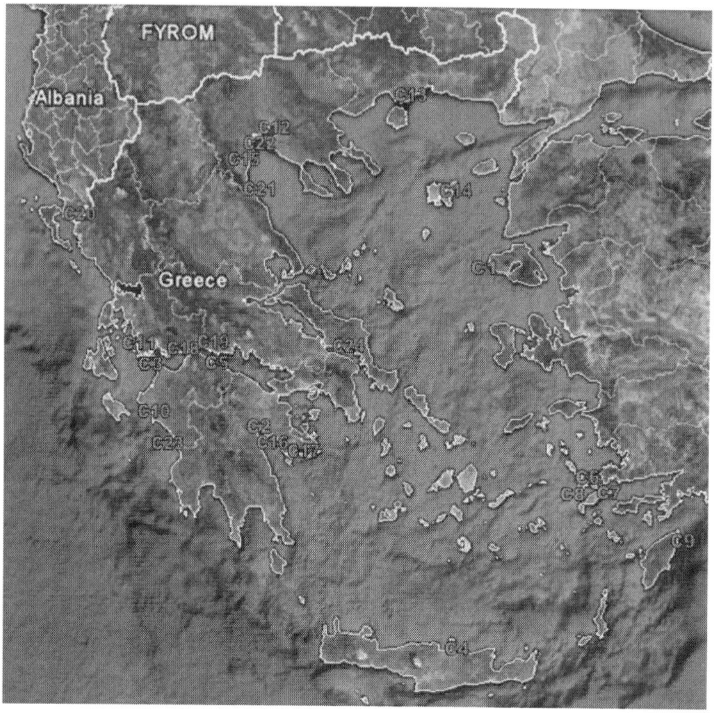

Fig. 3. Map of Greece displaying the 27 case studies (Google Earth).

3. Methodology and research hypotheses

In the present paper, we approach the assessment of economic impacts of SLR with respect to two different aspects: long-term (2100) and short-term (annual) damages. The long-term losses follow the gradual SLR as specified by the IPCC scenarios for 0.5-m and 1-m elevation. The short-term financial appraisal of losses is based on the increased frequency and intensity of storm surges, a consequence of climate change taking place in parallel to long-term SLR. The inclusion of such short-term losses in the estimation of SLR impacts follows IPCC and other experts' opinion (IIPCC CZMS, 1992; Hoozemans et al., 1993; McInnes et al., 2000; Emanuel, 2005; Velegrakis, 2010).
Referring to the long-term impacts, losses of the following land uses are quantified and evaluated:

- Housing
- Tourist
- Agriculture
- Wetlands
- Forestry

Selection was based on data availability from 27 case studies of the Greek coastal area (Table 2). Based on these studies, the total loss of land for the five uses under investigation and for 0.5-m and 1-m elevation is assessed. Then, for housing, tourist and agricultural uses, a market pricing approach is drawn on in order to estimate unit and total financial losses. For wetlands and forestry we rely on the widely used application of value transfer (Navrud & Ready, 2007). Loss of public infrastructure (airports, ports) and industrial zones were not taken into consideration. More specifically:

Housing and tourist uses

The cost assessment of these impacts - both in the 27 case studies as well as the wider coastline area - was achieved by multiplying the total area lost in each case by the mean market value of property in the specific area. Two problems were faced here: the sparse data regarding land uses in the case studies, and the wide variation of prices for land property. So the value of 1,200 €/m^2 was selected as the mean estimated market value of property, which better reflects the mean land price for housing and tourist purposes. This is equivalent to a similar figure (1300 €/year) a rough estimation by Velegrakis at al. (2008), representing the mean income from tourist activities per meter of Greek beach.

Agriculture

Assessment of the cost of loss of farmland was achieved by multiplying the lost area with the "specific basic value" (SBV) of the farmland for each location investigated. SBV represents the value of a square meter of non-irrigated farmland of yearly crop cultivations, as determined by the Ministry of Economics for property tax purposes. SBV applies only in areas facing roads or located up to 800 meters from the sea.

Wetlands

To estimate the cost of wetland losses, the total area of wetlands expected to be lost due to SLR is multiplied by their unit value. The unit value for wetlands (4.8 million €/km^2) was 'transferred' from Darwin and Tol (2001), a well-known study regarding appraisal of SLR impacts. Table 3 depicts the social values for certain Greek wetlands.

Wetland	Value
Kerkini lake (conservation of terrestrial wetland)	21.3 €/household per year
Kalloni wetland (coastal wetland)	184-300 €/household
Heimaditida lake (conservation of terrestrial wetland)	115.3-144.1 €/household
Heimaditida & Zazari lakes (conservation of terrestrial wetlands)	134.8-226.4 €/household per year
Zakynthos National Marine Park (conservation of a marine park)	0.9-4.3 €/visitor
Plomari & Vatera beaches-Lesvos island (conservation of pocket beaches)	15 €/visitor per year
Karla lake (restoration of wetland)	27.4 € per trimonth for 3 years/household

Table 3. Social values for Greek wetlands.

Forests

The cost for loss of forests was based on the unit value of Greek forests (89.25 €/ha) as estimated in the study of Kazana and Kazaklis (2005).

The estimated value of the five coastal uses indicates the (future) financial loss due to SLR. A cost index is then calculated based on the estimated cost of impacts due to loss of housing, tourist, wetlands, forestry and agriculture land uses, as well as on the total length and area of the coastline examined in each case study. This index estimates the financial cost of SLR per km or km^2 of coastline, based on data available in each case. All unit values used were adjusted across locations and time on the basis of the Purchasing Power Parity Index (PPPI) and Consumer Price Index (CPI) (Pattanayak et al., 2002).

From a socioeconomic point of view, the accompanying phenomenon of intensified storm surges (what we call here the short-term impacts of SLR) is equally interesting as the long-term impacts (over a horizon of 90 years) accelerated SLR. To our knowledge, financial impact studies regarding storm surges in Greece do not exist. Financial calculations of the impacts of past storm surges from regional authorities are limited and incomplete. To fill this data gap, a stated preference survey was designed and implemented in order to elicit social welfare losses from short term SLR (Kontogianni, 2011). Our short-term estimation of SLR impacts is based on findings of this survey.

To properly appraise the coastal system and its total economic value, the totality of ecosystem services and goods described in Table 1 has to be evaluated (Skourtos et al., 2005). Our results indicate a partial value of the coastal zone, taking into consideration the five aforementioned uses. Consequently, our appraisal constitutes a lower threshold of the future losses due to SLR. At a second level, and in order to highlight the 'true' but unknown total economic value of SLR damages, the equally important aesthetic values of these areas are also estimated on the basis of values transferred from Brenner et al. (2010).

Finally, the present value of losses was estimated by discounting total amounts with interest rates of 1% and 3%. The selection of a suitable (social) discount interest rate is a vital parameter for similar long-term estimations. Economic theory and practice are not in a position to provide a definite answer on the choice of discounting rates, since in essence the issue of discount interest rate is a moral issue related to perceptions of intergenerational justice. For example, in OECD countries, the proposed discount interest rates for long-term investments range between 3 - 12% (OECD, 2007). The European Union recommends a 4% interest rate for mid- and long-term investments but also accepts implementation of lower interest rates in the case of extended timelines, such as climate change (European Commission, 2005).

4. Costing the damages of sea level rise

This section presents our results on the base of the proposed methodological approach for the evaluation of the financial loss due to the long-term impacts of SLR as well as the monetary estimates of the short-term impacts of SLR caused by storm surges. Aesthetic values are also estimated and added up to approach the total coastal value.

4.1 Financial impacts of long-term sea level rise

The loss of coastal land according to scenarios for a sea level rise of 0.5 m and 1 m, as specified in the case studies under examination, is presented in Table 2. The financial value of land loss in the case studies is then calculated as the area to be flooded times the respective unit value for each specific land use. As a next step, cost coefficients are calculated for a SLR of 0.5 m and 1 m for housing, tourist and agricultural land uses, plus wetlands and forests. The cost coefficients are the quotient of the financial value of land loss in a specific location divided by the length/area of the coastline at this location. As a result, these coefficients comprise quantified indications of the overall financial loss expressed per km/km^2 of coastline for the five land uses examined. The values that were finally selected in terms of mean values for cost coefficients, the length and the area of the coastline per land use, are presented in Table 4.

Land use	**Average cost coefficients**		**Length/Area of Greek shoreline**
	SLR 0,5 m	**SLR 1 m**	
Housing & Touristic	144,891 10^3 €/km	262,851 10^3 €/km	2,400 km
Wetlands	138 10^3 €/km2	247 10^3 €/km2	1,000 km2
Forests	0.04 10^3 €/km2	0.13 10^3 €/km2	4,000 km2
Agriculture	222 10^3 €/km2	514 10^3 €/km2	35,511.5 km2

Table 4. Values for the average cost coefficients, the length and the area of the coastline per land use.

The estimated financial loss from the case studies is then extrapolated to the Greek territory. The total financial loss of SLR for the Greek coastal zone in 2100 is presented per land use in Table 5.

Land use	**Total financial loss 2100 (10^3 €)**	
	SLR 0.5 m	**SLR 1 m**
Housing & Touristic	347,738,400	630,842,400
Wetlands	138,000	247,000
Forests	160	520
Agriculture	7,883,553	18,252,911
Total	355,760,113	649,342,831

Table 5. Total financial loss of SLR in 2100 per land use.

The estimates of financial loss in 2100 were converted to present values using discount rates of 1% and 3%. The results are presented in Tables 6 and 7 respectively.

Land use	**Total financial loss 2010 (10^3 €)**	
	SLR 0.5 m	**SLR 1 m**
Housing & Touristic	142,013,297	257,630,475
Wetlands	56,358	100,873
Forests	65	212
Agriculture	3,219,574	7,454,328
Total	145,289,294	265,185,888

Table 6. Present value of total financial loss of SLR per land use for discount rate 1% (not including aesthetic/recreational/ storm surge damages).

Land use	Total financial loss 2010 (10^3 €)	
	SLR 0.5 m	SLR 1 m
Housing & Touristic	24,316,576	44,113,412
Wetlands	9,650	17,272
Forests	11	36
Agriculture	551,279	1,276,386
Total	24,877,517	45,407,106

Table 7. Present value of total financial loss per land use for discount rate 3% (not including aesthetic/recreational/ storm surge damages).

The aggregated results are presented in Table 8 under three discounting assumptions: 0%, 1% and 3%.

	SLR 0.5m	SLR 1m
NPV (0%)	355,760,113	649,342,831
NPV (1%)	145,289,294	265,185,888
NPV (3%)	24,877,517	45,407,106

Table 8. Total long-term financial loss of SLR in Greek coastal zone under different discount rates (10^3 €) (not including aesthetic/recreational/ storm surge damages).

At this point we need to remind the reader that the estimated loss in Tables 4, 5 and 6, are in their majority expressions of use values, with the possible exception of wetland areas, the transferred value of which might include, in part, non-use values. But non-use value components (e.g. cultural and spiritual) comprise a non-negligible part of the total economic value of coastal ecosystems in Mediterranean countries (Langford et al, 2001, Remoundou et al 2009).

To support the aforementioned position, and aiming at providing an approximate expression of the potential loss of these values, we also quantify the aesthetic/recreational and cultural/spiritual values of the Greek coastal zone. The estimation is based on transferring the corresponding values from Brenner et al. (2010) study, where the aesthetic/recreational and cultural/spiritual value of sandy and wetland areas of Katalonia, Spain, were estimated. A discussion could be raised at this point on whether adding up those values in the previous sum consists a double counting in our estimated loss due to SLR. This position could be founded on the fact that we already used market price values for housing, so one could suppose that the market had already integrated those values (at least the aesthetic/recreational) into housing prices. Ledoux et al state that 'the socio-cultural and historical contexts in which environmental assets exist provide for alternative dimensions of environmental value which may not be captured by the market paradigm'.

To minimize the possibility of overestimating our economic assessments, we abide to a strategy of using conservative estimates of financial losses while trying to avoid double counting. On the other hand, as long as we do not control for induced market adjustments, future damage estimates may be grossly overestimated. For example, the housing/tourist value of the coastal land represents a significant parameter in our damage estimates. Assuming risks regarding accelerated SLR and increasing incidents of extreme weather effects come gradually to the fore, a well functioning market for coastal land will probably internalize and discount future hazards. As a consequence, land values in coastal areas

should gradually depreciate solely as a consequence of costal risk anticipation. Therefore, future damages would also be diminished (Karageorgis et al, 2006).
Regarding the estimation of aesthetic/recreational and cultural/spiritual loss, the cost coefficients which were adopted in the current analysis are presented in Table 9.

Land use	Value	Average cost coefficients		Length/Area of Greek shoreline
		SLR 0,5 m	SLR 1 m	
Housing & Touristic	aesthetic/recreational	352 10^3 €/km	639 10^3 €/km	2,400 km
	cultural/spiritual	0.6 10^3 €/km	1.0 10^3 €/km	
Wetlands	aesthetic/recreational	0.1 10^3 €/km2	0.3 10^3 €/km2	1,000 km2
	cultural/spiritual	1 10^3 €/km2	1.8 10^3 €/km2	

Table 9. Values for the average value coefficients, the length and the area of the coastline per land use.

The estimated aesthetic/recreational and cultural/spiritual loss from the case studies is then projected to the whole Greek territory. The total loss of SLR for the Greek coastal zone in 2100 is presented per land use and value type in Table 10.

Land use	Value	Total loss 2100 (10^3 €)	
		SLR 0.5 m	SLR 1 m
Housing & Touristic	aesthetic/recreational	844,800	1,533,600
	cultural/spiritual	1,440	2,400
Wetlands	aesthetic/recreational	100	300
	cultural/spiritual	1,000	1,800
	Total	847,340	1,538,100

Table 10. Total aesthetic/recreational and cultural/spiritual value loss of SLR in 2100 per land use.

The above estimates in 2100 were converted to present values using discount rates of 1% and 3%. The results are presented in Tables 11 and 12 respectively.

Land use	Value	Total loss 2010 (10^3 €)	
		SLR 0.5 m	SLR 1 m
Housing & Touristic	aesthetic/recreational	345,009	626,309
	cultural/spiritual	588	980
Wetlands	aesthetic/recreational	41	123
	cultural/spiritual	408	735
Total		346,046	628,146

Table 11. Present value of aesthetic/recreational and cultural/spiritual value loss of SLR per land use for discount rate 1%.

Land use	Value	Total loss 2010 (10^3 €)	
		SLR 0.5 m	SLR 1 m
Housing & Touristic	aesthetic/recreational	59,075	107,241
	cultural/spiritual	101	168
Wetlands	aesthetic/recreational	7	21
	cultural/spiritual	70	126
Total		59,253	107,556

Table 12. Present value of aesthetic/recreational and cultural/spiritual value loss per land use for discount rate 3%.

The aggregated results are presented in Table 13 under three discounting assumptions: 0%, 1% and 3%.

	SLR 0.5m	SLR 1m
NPV (0%)	847,340	1,538,100
NPV (1%)	346,046	628,146
NPV (3%)	59,253	107,556

Table 13. Total aesthetic/recreational and cultural/spiritual value loss of SLR in Greek coastal zone under different discount rates (10^3 €).

4.2 Welfare losses due to storm surges: The short-term impacts

Storm surges are the short-term aspect of the SLR phenomenon, with significant annual impacts on coastal areas. Knowledge of sea level extremes is important for coastal planning purposes (Marcos et al., 2009; Krestenitis et al., 2010). We consider it necessary to include these impacts in our study, due to their economic aspects and their potential yearly repetitiveness, which may induce an increase in coastal vulnerability. But since economic data on short term damages are limited and do not allow for extrapolation of loss to the total Greek coastal zone, an additional stated preference survey was conducted in order to elicit the social cost of storm surges (Kontogianni, 2011).The social cost of storm surges is defined as the maximum Wilingness to Pay to avoid the loss. As Ledoux et al. (2001) describe it 'in environmental economics, an individual preference-based value system operates in which the damages from environmental loss are measured by social opportunity cost . The assumption is that environmental goods and services are of instrumental value and some individual is willing to pay for the satisfaction of a preference. It is taken as axiomatic that individuals almost always make choices, subject to an income budget constraint, which benefit themselves or enhance their welfare. The social value of environmental resource committed to some use is then defined as the aggregation of private values'.

In Kontogianni 2011 an open ended contingent valuation survey was designed where participants were asked their willingness to pay to fund the construction of storm surge protection works in their area. The mean willingness to pay of the respondents was statistically estimated at 200.7 € per household (standard deviation = 286 €).

According to the "Report of Greece on Coastal Zone Management" (YPEXODE, 2006), coastal populations represent 85% of the total population (10,934,097 inhabitants), that is 9,293,982 inhabitants. Assuming an average of 3 persons per household, the total number amounts to 3,674,381 Greek households, out of which 3,097,994 are located in coastal areas. Using the mean value of 200.7 € per household, and extrapolating this to the Greek coastal population, the total value for protection from short term SLR for Greek households amounts to 621,767,426 €.

4.3 Social perceptions for climate change, SLR and storm surges

Reducing vulnerability is a goal for preventative adaptation. Recent literature regarding vulnerability and adaptation, accentuates the need to take measures and plan policies on two levels: 1. Technological, 2. Institutional and behavioral. Assessment of vulnerability and risks (used as input for decision making), must also be implemented on two levels: objective and subjective (Fischhof, 1995; Slovic, 1979; Douglas, 1982; Adams, 1995; Kasperson, 1988; Kontogianni et al., 2008). By subjective assessment of risks, we mean the social perceptions of risk which are not necessarily identical to the objective assessment. For a more thorough understanding of the social perceptions of risk due to climate change in Greece, two research projects were designed and implemented, in Lesvos in 2010 and in Crete on January 2011. Their findings are comparable and demonstrate the dynamics in the respondents' perceptions, compared to a similar research conducted for the first time in Greece in South Evoia, in 2003-2004 (Kontogianni 2011). Among others, the following areas were investigated: whether respondents were aware of the climate change , whether they were aware of the causes and which they believe these causes are, their level of trust in institutions, how important they assess that the various climate change impacts are, whether they are prepared to cope with them as well as if they are willing to incur costs in order to protect themselves (from the impacts). The research in Evoia, a coastal region close to Athens, performed in 2003, shows that 87.4% out of 183 respondents regards that in a global level, climate change constitutes a very important problem, while only 2.4% believes it is of no importance. For their personal welfare, climate change represents an important risk factor (79.7%), while only 7.3% regard it irrelevant to their lives. Concerning impacts on biodiversity, 59.2% of the respondents express serious concerns, while 13.8% does not worry about it.

During July-August 2010 in Lesvos island, 312 respondents were asked about climate change. The majority (97.1%) is aware of the climate change, and 58.8% of them believe that it is directly influencing their lives. Out of the 312 respondents 27.3% believe that climate change impacts will be visible and destructive within the next 20 years, while 12.9% within the next 100 years. Only 3 out of 312 people refused the existence of climate change impacts. The survey participants (islanders) were asked to rank and assess the severity of certain impacts. Results are given in Table 14. It is quite conceivable that 91.9% rank impacts on coasts important to extremely important.

Impacts on	Of no importance	Not so important	Important	Very important	Extremely important
Water resources	0,6%	1%	13,5%	33,9%	51%
Ecosystems	1%	3,2%	20,3%	45,8%	29,7%
Food availability	0,6%	1,9%	19,5%	33,1%	44,8%
Coasts	1%	7,1%	**18,5%**	**35,7%**	**37,7%**
Health	0,6%	2,3%	9,1%	21,0%	67%

Table 14. Climate change impacts assesment by survey respondents (Lesvos island, Greece, summer 2010).

Similar results were obtained from the implementation of a similar study in Crete island (January 2011) on the perceptions of locals referring to weather extremes. A random sample of 100 people were personally interviewed. Half of the respondents (50%) is aware of the climate change, while 17.5% has no information on the subject. Regarding the time horizon of the climate change impacts 57.5% of them believe that it is directly influencing their lives. Out of 100 usable questionnaires, 17.5% believe that climate change impacts will be visible and destructive within the next 20 years, while 20% believes so within the next 100 years. Only 5% refused the existence of climate change impacts. As in the Lesvos survey, participants (islanders) were asked to rank and assess the severity of certain impacts. Results are given in Table 15. It is quite conceivable that the majority (97.5%) rank impacts on coasts important to extremely important.

Impacts on	Not so simportant	Important	Very important	Extremely important
Water resources	-	2,5%	17,5%	80%
Ecosystems	2,5%	5%	40%	52,5%
Food availability	2,5%	5%	27,5%	65%
Coasts	2,5%	**2,5%**	**37,5%**	**57,5%**
Health	7,5%	5%	35%	52,5%

Table 15. Climate change impacts assesment by survey respondents (Crete island, Greece, January 2011).

In both surveys water and food availability seem to be the highest social concerns as far specific impacts of climate change are concerned.

5. Conclusions

In this chapter we have used market and non-market values to estimate the SLR damages in Greek coastal zone. We have also combined this assessment with a preliminary socio-economic analysis of coastal vulnerability and risk perceptions. The proposed methodological approach focused on the assessment of two different categories of economic impacts: long term inundation effects of SLR and short-term extreme weather effects. In this assessment we have also incorporated a rough estimation of the losses due to cultural, recreational and other non-consumptive elements of value. In order to reaffirm the policy relevance of applying economic assessment methods in Integrated Coastal Zone Management, UNEP (1999) introduces the notion of 'resource consciousnesses' into its Regional Strategic Environmental Action Plan. It is asserted accordingly that 'raw cost information is insufficient to support investment decisions' what is needed is an investment plan where 'benefits [..] derived from the reduction or avoidance of pollution impacts on resources of social, economic and environmental value' are demonstrated. Moreover, in order for benefit estimates to be of relevance to prospective investors, their definition should include 'the conservation of resource for their existence (or non-use) value' [UNEP 1999, pp. 67-69).

Conservation (ie coastal protection measures) should be adopted if it can be demonstrated that net economic benefits are generated. So we need the total cost of SLR to compare it with the relevant conservation measures (benefit).

Tables 16 and 17 present the total SLR cost for discount rates of 1% and 3% correspondingly. Total cost here means the sum of long-term SLR, short-term SLR and non-use values (aesthetic/recreational and cultural/spiritual value loss). The total discounted SLR cost equals 2% of the Greek GDP (in 2010 prices).

Loss	**Total loss 2010 (10^3 €)**	
	SLR 0.5 m	**SLR 1 m**
Long-term SLR	145,289,294	265,185,888
Short-term SLR	621,767	621,767
Non-use values	346,046	628,146
Total	146,257,107	266,435,801

Table 16. Present value of total cost of SLR for discount rate 1%.

Loss	**Total loss 2010 (10^3 €)**	
	SLR 0.5 m	**SLR 1 m**
Long-term SLR	24,877,517	45,407,106
Short-term SLR	621,767	621,767
Non-use values	59,253	107,556
Total	25,558,537	46,136,429

Table 17. Present value of total cost of SLR for discount rate 3%.

Our study shows that there is an imperative need to study Greek coastal areas that are at high risk of flooding. This need is expanded to the detailed diagnosis/forecasting of the coastal zone's vulnerability also due to changes in the frequency/intensity of extreme weather phenomena (storm surges). From an institutional aspect, EU member states, according to Directive 2007/60/EC, must undertake a preliminary assessment of river basins flood risk (including coastal zone) by year 2011, aiming to identify areas where flooding is likely to occur. Moreover, by year 2013, member states must develop risk assessment maps for these areas, while by year 2015, member states must prepare flood risk management plans for these zones.

The principal determinant of a society's capacity to adapt to climate change is likely to be access to resources. Such access is determined by entitlements, which are often the product of external political factors. Therefore, poverty, inequality, isolation and marginalization can all undermine entitlements of individuals and groups (Adger et al., 2005). In Greece due to the imminent economic crisis the country is currently experiencing, poverty is a threatening factor; inequality is a social characteristic due to corruption; so entitlements of individuals and groups are under threat of undermining. A particularly coastal country facing SLR impacts finds itself within a vulnerable status. Is the country going to bounce back after the economic shock and prepare itself for adaptation to SLR? According to Tompkins et al. (2005) the basic preconditions for resilience are: ability to self-organize, ability to buffer disturbance and capacity for learning and adapting. Or as Handmer (1996) puts it: Stability is sought but change constantly redraws the playing field and demands redefinition of the rules.

6. Acknowledgment

This research was carried out within the IP project SESAME funded by the European Commission under Framework Programme 6 Key Action "Global change, climate and biodiversity", Contract No.2006-036949. Part of this research was carried also within the Committee for the Study of Climate Change Impacts, Bank of Greece. We thank the members of the above Committee for multiple support and ideas.

7. References

Adams, J. (1995). *Risk*, Taylor & Francis, London

Adger, W.N.; Brooks, N.; Bentham, G.; Agnew, M. & Eriksen, S. (2004). *New indicators of vulnerability and adaptive capacity*, Tyndall Center for Climate Change Research

ATEAM 2004, Final Report. PIK- Potsdam

Aunan, K. & Romstad, B. (2008). Strong Coasts and Vulnerable Communities: Potential Implications of Accelerated Sea-Level Rise for Norway, *Journal of Coastal Research*, 24, 2, pp. 403-409

Beaumont, N.J.; Austen, M.C.; Atkins, J.P.; Burdon, D.; Degraer, S.; Dentinho, T.P.; Derous, S.; Holm, P.; Horton, T.; van Ierland, E.; Marboe, A.H.; Starkey, D.J.; Townsend, M. & Zarzycki, T. (2007). Identification, definition and quantification of goods and services provided by marine biodiversity: Implications for the ecosystem Approach, *Marine Pollution Bulletin*, 54, pp. 253-265

Brooks, N. (2003). *Vulnerability, risk and adaptation: A conceptual framework*, Tyndall Centre for Climate Change Research, Working Paper 38

Brooks, N.; Adger, W.N. & Kelly, P.M. (2005). The determinants of vulnerability and adaptive capacity at the national level and the implications for adaptation, *Global Environmental Change*, 15, pp. 151-163

Church, J.A.; Gregory, J.M.; Huybrechts, P.; Kuhn, M.; Lambeck K.; Nhuan, M.T.; Qin, D. & Woodworth, P.L. (2001). Changes in Sea Level, In: *Climate Change 2001: The Scientific Basis*, J.T. Houghton and Y. Ding, (Ed), Cambridge University, Press, New York, pp. 639-693

Daily, G.C. (1997). *Nature's services: Societal dependence on natural ecosystems*, Island Press, Washington, DC, 392

Dalianis, D.; Petassis, D.; Santamouris, M.; Argiriou, A.; Cartalis, C. & Asimakopoulos, D. (1997). Social cost of electricity generation in Greece, *Renewable Energy*, 12, 3, pp. 281-289

Darwin, R. & Tol, R. (2001). Estimates of the Economic Effects of Sea Level Rise, *Environmental and Resource Economics*, 19, pp. 113-129

Devoy, R. (2008). Coastal Vulnerability and the Implications of Sea-Level Rise for Ireland, *Journal of Coastal Research* 24, 2, pp. 325-341

Douglas, M. & Wildavsky, A. (1982). *Risk and Culture*, University of California Press, Berkeley, CA

Doukakis, E. (2003). The potential consequences of climate change on Kotychi lagoon, *8th International Conference on Environmental Science and Technology*, Lemnos island

Doukakis, E. (2004). Accelerated sea level rise and coastal vulnerability in the Hersonissos coastal region (Crete, Greece), *Mediterranean Marine Science*, 5, 1, pp. 35-41

Doukakis, E. (2005a). The impacts of climate change in land area of the Gulf of Nafplion, *5o National Conference "Integrated water resources management based on river basin"*, Xanthi, (in greek)

Doukakis, E. (2005b). *Risks in coastal zone by climate change and tsunami,* Working Meeting "Risk assessment in the Corinthian Gulf" Derveni, Evrostini. (in greek)

Doukakis, E. (2007). *Natural disasters and coastal zone, Prevention - Management of natural disasters, The role of Rural and Surveying Engineering,* Technical Chamber of Greece, PSDATM, NTUA, Athens, (in greek)

Doukakis, E. (2008). The problematic definition of the foreshore and the beach at Skala Eressos Mytilene, *Conference 4MMCZ,* pp. 307-313, (in greek)

Ebersole, B.A.; Westerink, J.J.; Bunya, S.; Dietrich, J.C. & Cialone, M.A. (2010). Development of storm surge which led to flooding in St. Bernard Polder during Hurricane Katrina. *Ocean Engineering,* 37, pp. 91-103

Emanuel, K. (2005). Increasing destructiveness of tropical cyclones over the past 30 years, *Nature,* 436, pp. 686-688

European Commission, (2005). *Impact Assessment Guidelines,* SEC(2005) 791

EUROSION, (2004), *Living with coastal erosion in Europe,* Final report of the project Coastal erosion – evaluation for the need for action, DG Environment, European Commission

Fischhoff, B. (1995). Risk perception and communication unplugged: twenty years of process, *Risk Analysis,* 15

Folke, C.; Carpenter, S.; Elmqvist, T.; Gunderson, L.; Holling, C.S. & Walker, B. (2002). Resilience and Sustainable Development: Building Adaptive Capacity in a World of Transformations, *Ambio,* 31, 5

Garpe, K. (2008). *Ecosystem services provisioned by the Baltic Sea and Skagerrak,* Swedish Environmental Protection Agency, Stockholm, Report 5873

Handmer, J. (1996). Policy Design and Local Attributes for Flood Hazard Management, *Journal of Contingencies and Crisis Management,* 4, 4

Handmer, J. & Dovers, S. (1996). A Typology of Resilience: Rethinking Institutions for Sustainable Development, *Organization and Environment,* 9, 482

Hellenic Republic, (2006). *4th National communication to the United Nations framework convention on climate change,* Ministry for the Environment, Physical Planning and Public Works, Athens

Hoozemans, F.M.J.; Marchand, M. & Pennekamp, H.A. (1993). *A global vulnerability analysis, vulnerability assessments for population, coastal wetlands and rice production on a global scale,* 2nd edition, Delft Hydraulics and Rijkswaterstaat, Delft.

IPCC, (2007). *Impacts, Adaptation and Vulnerability,* Contribution of Working Group II to the Fourth Assessment Report of the Intergovernmental Panel on Climate Change, M.L. Parry, O.F. Canziani, J.P. Palutikof, P.J. van der Linden and C.E. Hanson (Ed), Cambridge University Press, Cambridge

IPCC CZMS, (1992). *Strategies for Adaptation to Sea Level Rise,* Report of the Coastal Zone Management Subgroup, Response Strategies Working Group of the Intergovernmental Panel on Climate Change, Ministry of Transport, Public Works and Water Management, The Hague

Kanelakis, A. & Doukakis E. (2004). *Impacts of climate change on the coastal areas in the plain of Thessaloniki,* Disseration, NTUA, School of Rural and Surveying Engineering, Department of Topography, (in greek)

Karabas, T.; Krestenitis, I.; Sakelariou, D.; Xasiotis, T. & Maliaros, D. (2008). Prevention and management of marine risks in the coastal zone (CORI–INTERREG III B/ARCHIMED): Risks from tsunami and flooding, *4º National Conference: Management and Improvement of Coastal Zones,* Mytilene 23-27 September, pp. 597-606, (in greek)

Karacat, M. & Nicholls, R. (2008). Potential Implications of Accelerated Sea-Level Rise for Turkey, *Journal of Coastal Research*, 24, 2, pp. 288-298

Karageorgis A., Kapsimalis V. Kontogianni A.,Skourtos M. Turner R.K, Salomons W (2006) Impact of 100-year human interventions on the Deltaik coastal zone of the inner Thermaikos Gulf (Greece): A DPSIR framework analysis. Journal of Environmental Management 38, 304-315

Karibalis, E. & Gaki-Papanastasiou, K. (2008). Geomorphological study of the deltaic areas of Penaeus, Kalama, Evinos and Mornos, *Conference 4MMCZ*, pp. 85-94. (in greek)

Kasperson, R.(1988). The Social Amplification of Risk: A conceptual framework, *Risk Analysis*, 2

Kazana, V. & Kazaklis, A. (2005), Greece. In: Merlo, M., Croitoru, L. *Valuing Mediterranean Forests: Towards Total Economic Value*, CABI International, Wallingford UK/Cambridge MA

Klein, R.J.T.; Nicholls, J. & Thomalla, F. (2004). Resilience to natural hazards: How useful is this concept?, *Global Environmental Change Part B: Environmental Hazards*, 5, 1-2, pp. 35-45

Komar, P.D. (1998). Beach processes and Sedimentation, 2nd edition, Prentice Hall, N.J., USA, 544

Kont, A.; Jaagus, J.; Aunap, R.; Ratas, U. & Rivis, R. (2008). Implications of Sea-Level Rise for Estonia, *Journal of Coastal Research*, 24, 2, pp. 423-431

Kontogianni, A. & Skourtos, M. (2008). Social Perception of Risk informing Integrated Coastal Zone Management on accidental oil – spill pollution: the reason you pollute matters, not numbers, In: *Integrated Coastal Zone Management – The Global Challenge*, R. Krishnamurthy, B. Glavovic, A. Kannen, D. Green, R. Alagappan, H. Zengcui, S. Tinti and T. Agardy (Ed), *Research Publishing*, pp. 207-225

Kontogianni, A.; Luck, G. & Skourtos, M. (2010a). Valuing ecosystem services on the basis of service-providing units: A promising solution to the 'endpoint problem' in stated preference approaches, *Ecological Economics*, 69, 7, pp. 1479-1487

Kontogianni, A. (2011). *Investigation of social vulnerability and adaptation costs of sea level rise in Greece*, Working paper, Laboratory of Applied Environmental Economics Laboratory, University of Aegean, (in greek)

Krestenitis, Y.N.; Androulidakis, Y.S.; Kontos, Y.N. & Georgakopoulos, G. (2010). Coastal inundation in the north-eastern Mediterranean coastal zone due to storm surge events, Journal of Coastal Conservation, DOI 10.1007/s11852-010-0090-7

Langford I. H., M. Skourtos, A. Kontogianni, R. Day, S. Georgiou, I. J. Bateman, (2001)Use and non-use values for conserving endangered species: the case of the Mediterranean Monk seal. Environment and Planning A, 33, pp.2219-2233

Ledoux L, Turner K. Mathieu L., Crooks S. (2001) Valuing ocean and coastal resources. Practical examples and issues for further action. Paper presented at the Rio+10: Assessing Progress,Addressing Continuing and New Challenges, UNESCO, Paris, 3-7 December 2001

Marcos, M.; Tsimplis, M.N. & Shaw, A.G.P. (2009). Sea level extremes in southern Europe. Journal of Geophysical Research Oceans, 114

Meehl, G.A.; Washington, W.M.; Collins, W.D.; Arblaster, J.M.; Hu, A.; Buja, L.E.; Strand, W.G. & Teng, H. (2005). How much more global warming and sea-level rise?, Science, 307, pp. 1769-1772

Metzger, M.J.; Leemans, R.; Schroeter, D.; Cramer, W. & the ATEAM consortium (2004). The ATEAM vulnerability mapping tool, *Quantitative Approaches in Systems Analysis*, 27

McInnes, K.L., Walsh, K.J.E & Pittock, A.B (2000), Impact of Sea-level Rise and Storm Surges on Coastal Resorts. CSIRO Tourism Research Final Report

Millennium Ecosystem Assessment, (MEA) (2005). Ecosystems and Human Well-Being: Synthesis, Island Press, Washington, 155

Ministry of the Environment, Physical Planning and Public Works (YPEXODE) (2006). Report of Greece on Coastal Zone Management. Submitted to the European Commission/DG Environment, in the context of the recommendation on integrated coastal zone management (2002/413/EC)

Navrud, S. & Ready, R. (2007). *Environmental Value Transfer: Issues and Methods*, Springer.

Nicholls, R. (2007). *Adaptation Options For Coastal Areas And Infrastructure: An Analysis For 2030*, Report to the UNFCCC, Bonn

Nicholls, R. (2009). *Adaptation costs for coasts and low-lying settlements, in Assessing the Costs of Adaptation to Climate Change: A Review of the UNFCCC and Other Recent Estimates*, M. Parry, N. Arnell, P. Berry, D. Dodman, S. Fankhauser,Ch. Hope, S. Kovats, R. Nicholls, D. Satterthwaite, R. Tiffin and T. Wheeler (Ed), International Institute for Environment and Development and Grantham Institute for Climate Change, London

OECD, (2006). *Metrics for assessing the economic benefits of climate change policies: Sea level rise*, ENV/EPOC/GSP(2006)3/FINAL, Paris

OECD, (2007). *Use of Discount Rates in the Estimation of the Costs of Inaction with Respect to Selected Environmental Concerns*, ENV/EPOC/WPNEP(2006)13/FINAL

Papadopoulou, G. & Doukakis, E. (2003). *Impact of climate change in the foreshore and the beach*, Disseration, NTUA, School of Rural and Surveying Engineering, Department of Topography, (in greek)

Papanikolaou D.; Papanikolaou, M. & Vasilakis, E. (2010), Sea level rise and impacts on the Greek coasts. Final report, Committee for the Study of Climate Change Impacts, Bank of Greece. Available in Greek from: www.bankofgreece.gr

Pattanayak, S.K.; Wing, J.M.; Depro, B.M.; van Houtven, G.; de Civita, P.; Stieb, D.M. & Hubbel, B. (2002). *International Health Benefits Transfer Application Tool: The Use of PPP and inflation indices*, prepared for Economic Analysis and Evaluation Division Health, Canada

Pliakos, G. & Doukakis, E. (2004). *Impacts of climate change on the coastal areas of the island of Lemnos*, Disseration, NTUA, School of Rural and Surveying Engineering, Department of Topography, (in greek)

Poulos, S.E. & Collins, M.B. (2002). Fluviatile sediment fluxes to the Mediterranean: Aquantitative approach and the influence of dams, *Geological Society of America Bulletin*, 191, pp. 227-245.

Poulos, S.; Ghionis, G. & Maroukian, H. (2009). The consequences of a future eustatic sea-level rise on the deltaic coasts of Inner Thermaikos Gulf (Aegean Sea) and Kyparissiakos Gulf (Ionian Sea), Greece, *Geomorphology*, 107, pp. 18-24

Pruszak, Z. & Zawadzka, E. (2008). Potential Implications of Sea-Level Rise for Poland, *Journal of Coastal Research*, 24, 2, pp. 410-422

Remoundou K, Ph.Koundouri, A. Kontogianni, P.A. Nunnes, M. Skourtos (2009) Valuation of natural marine ecosystems: an economic perspective Environmental Science and Policy 12, pp. 1040-1051

Richards, J. & Nicholls, R. (2009). *Impacts of climate change in coastal systems in Europe, PESETA-Coastal Systems study*, European Commission Joint Research Centre Institute for Prospective Technological Studies, Scientific-Technical report

Roussos, D. & Karibalis, E. (2009). *Coastal geomorphological mapping and impact assessment in the coastal zone of South Euboean Gulf between the areas Eretria and Manides by a possible future sea level rise*, Disseration, Harokopion University, Department of Geography, (in greek)

Sachez-Arcilla, A.; Jimrnez, J.; Valdemoro, H. & Gracia, V. (2008). Implications of Climatic Change on Spanish Mediterranean Low-Lying Coasts: The Ebro Delta Case, *Journal of Coastal Research*, 24, 2, pp. 306-316

Seni, A. & Karibalis, E. (2007). *Identifying the vulnerability of coastal areas by sea level rise using GIS. The case study of Porto Heli and Ermioni (Peloponnese)*, Dissertation, Harokopion University, Department of Geography, Postgraduate Program: "Applied Geography and Area management", (in greek)

Skourtos, M.; Kontogianni, A.; Georgiou, S. & Turner, R. K. (2005). Valuing Coastal Systems, In: R.K. Turner, W. Salomons, J. Vermaat (Ed), Managing European Coasts: Past, Present and Future, Springer Verlag, pp. 119-136

Slovic, P.; Fischof, B. & Lichtenstein, S. (1979). Rating the Risks, *Environment*, 21, 3

Sonderquist, T. & Hasselstroom, L. (2008). *The economic value of ecosystem services provided by the Baltic Sea and Skagerrak.* Existing information and gaps of knowledge, Swedish Environmental Protection Agency

Sterr, H. (2008). Assessment of Vulnerability and Adaptation to Sea-Level Rise for the Coastal Zone of Germany, *Journal of Coastal Research*, 24, 2, pp. 380-393

Stergiou, N. & Doukakis, E. (2003). *The impacts of sea level rise on the coastal zone of saltmarsh in Kitros Pierias*, Disseration, NTUA, School of Rural and Surveying Engineering, Department of Topography, (in greek)

Tipping Points Report (2009) WWF and Allianz SE

Tompkins, E.; Nicholson-Cole, S.; Hurlston, L.; Boyd, E.; Brooks Hodge, G.; Clarke, J.; Gray, G.; Trotz, N. & Varlack, L. (2005). *Surviving Climate Change in Small Islands - A Guidebook*, Tyndall center for Climate Change Research, UK

Turner, R.K.; Bateman, I.J. & Adger, W.N. (2001). *Economics of Coastal and Water Resources: Valuing Environmental Functions*, Dordrech: Kluwer Academic Publishers

UNEP (1999), Strategic action programme to address pollution from land-based activities. UNEP/MAP Athens

Velegrakis, A.F. (2010). *Coastal systems*, unpublished paper, Department of Marine Sciences, University of Aegean

Velegrakis, A.F.; Vousdoukas, M.; Andreadis, O.; Pasakalidou, E.; Adamakis, G. & Meligonitis, R. (2008). Impacts of dams on their downstream beaches: A case study from Eresos coastal basin, Island of Lesvos, Greece, *Marine Georesources and Geotechnology*, 26, pp. 350-371

Vafeidis, A.T.; Nicholls, R.J.; McFadden, L.; Tol, R.S.J.; Hinkel, J.; Spencer, T.; Grashoff, P.S.; Boot, G. & Klein, R.J.T. (2008). A new global coastal database for impact and vulnerability analysis to sea-level rise, *Journal of Coastal Research*, 24, 4, pp. 917–924

Velegrakis, A.F.; Vousdoukas, M.I. & Meligonitis, R. (2005). Beach erosion: Phenomenology and causes of the degradation of the greatest natural resource of the Greek Archipelago In: Tsaltas, The Greek Archipelago in the 21st Century, Vol. I, Sideris Publications, pp. 243-262 (In Greek with English Abstract)

Vött, A. (2007). Relative sea level changes and regional tectonic evolution of seven coastal areas in NW Greece since the mid-Holocene, *Quaternary Science Reviews*, 26, pp. 894-919

Wigley, T.M.L. (1995). Global-Mean Temperature and Sea Level Consequences of Greenhouse Gas Concentration Stabilization, *Geophysical Research Letters*, 22, pp. 45-48

Zanou B.,A. Kontogianni, M.Skourtos (2003) A classification approach of cost effective management measures for the improvement of watershed quality. Ocean and Coastal Management 46, pp. 957-983

20

Strengthening Regional Capacities for Providing Remote Sensing Decision Support in Drylands in the Context of Climate Variability and Change

Humberto A. Barbosa and T. V. Lakshmi Kumar
[1]*Laboratory of Analysis and Processing of Satellite Images (LAPIS), Universidade Federal de Alagoas (UFAL),*
[2]*Atmospheric Science Research Laboratory, SRM University, Kattankulathur,*
[1]*Brazil*
[2]*India*

1. Introduction

Dryland ecosystems cover one third of the earth's total land surface, comprise areas with a ratio of average annual rainfall to evapotranspiration of less than 0.65 (MEA, 2005). These regions are fragile environments characterized by unreliable rainfall patterns and support livelihoods of over 2.5 billion people (Reynolds et al., 2007). Widespread episodes of drought, heavy precipitation and heat waves have been reported as a consequence of global sea level increase (Verdin et al., 2005). However, the projections of the impacts of global warming on regional climate are largely uncertain due to the complex and site-specific interdependencies among landscape properties, environmental traits and policy decisions (Boulanger et al., 2005). The predictions of climate changes and their impacts in those dry lands are important because of their characteristics affecting economic activity based on agriculture and the role of natural ecosystems in carbon sequestration and water budget, which could lessen or mitigate the impacts of global changes in the weather system of these regions.

Climate variability and change play a significant role in dryland decision making, at various time scales. Decisions affected by climate considerations include both dryland hardware (infrastructure and technology) and software (management, policies, laws, governance arrangements). Strategic (decadal scale) and tactical (seasonal or interannual scale) decisions regarding such matters as infrastructure for storing water and dryland conservation measures must be made in the face of uncertainty about interdecadal, intraseasonal and interannual flows.

There is a need to understand changes that have occurred in the resources in dryland ecosystems that contain a variety of plant species that have developed special strategies to cope with the low and sporadic rainfall and extreme variability in temperatures A better understanding of various dynamics at work in drylands will put us in a better position to predict the future of the ecosystems. Sustainable land use under climate change requires detailed knowledge of the system dynamics. This is particularly pertinent in the management of domestic livestock in semiarid and arid grazing systems, where the risk of degradation is high and likely climate change may have a strong impact. Although these drylands are of environmental and socio-economic importance, they are faced with serious management challenges. Hence, their sustainable management requires an evaluation of the magnitude, pattern, and type of land-use/cover changes and the projection of the consequences of these changes to their conservation. It is also important to precisely describe and classify land cover changes in order to define sustainable land-use systems that are best suited for each place (FAO 1998). Monitoring the locations and distributions of land cover changes is important for establishing links between policy decisions, regulatory actions and subsequent land-use activities. In this regard, there is a need to consider both the socio-economic environment (Giannecchini et al., 2007) and other environmental factors. In this context, there is a need for agriculture administrators and policy makers to better understand the intraseasonal-to-interannual variability of climate and its effects on the landscape properties. The comprehension of interactions of weather variability and those landscape properties could lead to improved understanding of those landscape vulnerability to global changes, enhanced natural-resource management and to a better emergency planning to withstand the effects of extreme episodes on the natural and agricultural systems at regional scale (Rosenzweig et al., 1994). Nonetheless, the climatic data, at adequate spatio-temporal resolution at the regional level is scarce, representing an obstacle to researchers.

Prognostic numerical models are one of the main research tools used to predict past and future states of the Earth system (Cramer et al., 2001), yet persistent problems limit their acceptance in ecological and global change research. Aber (1997) posed the question "Why have models failed to penetrate the heart of ecological sciences?" and found that all too often model predictions are made prior to parameterization, validation, sensitivity analysis, and description of model structure. While today models are more accepted in a wide variety of fields, these issues are still prevalent and still ignored too often. With the advent of global monitoring systems based on satellites, it became possible to understand the nature and response of these ecosystems and drylands to day- to-day fluctuations in weather. In particular, the spatial-temporal analysis of vegetation dynamics (i.e., the response of vegetation to climatic conditions) in the semi-arid tropical region is important in the context of climate change where these dynamics show quicker response in short term climate indices such as the Southern Oscillation Index (SOI) and the Nino Sea Surface Temperatures (Fischer, 1996). The present study aims to cover certain regions across the tropics of arid (R/PE is 0.05 to 0.20) and semi-arid (R/PE is 0.20 to 0.50) nature (where R represents Rainfall and PE, Potential Evapotranspiration). Such regions of this type are Northeastern Brazil, West Sahel in Africa and Andhra Pradesh in India. The rainfall in India is mainly by south-west monsoon (June to September). In Sahel, it is primarily from June to August and to a lesser extent in September. During the positive phase of El Nino Southern Oscillation (ENSO) which is the sudden rise of Pacific Sea Surface Temperatures, an increase is observed the intensity of drought in Northeastern Brazil, and the Sahel (Africa) rainfall changes are also found due to global ocean circulation and patterns of SSTs. This ENSO phases are explicitly seen in inter-annual variability of south-west monsoon in India and play a major role in the agricultural sector of the country. The evaluation of El Nino and La

Nina barely showed that in many cases, La Nina had positive impact and El Nino, a negative one. Due to climate change and variability, the disasters like droughts became frequent in the above said regions. Semi arid Asia is experiencing an increase in the frequency of severity of wild fires. African rainfall changed substantially over last 60 years due to land cover changes and forest destruction. Though, India is not showing any significant trend in its annual rainfall, an increase in extreme weather events are evidenced (Lakshmi Kumar et al., 2011). So it is must to address these issues from the remote sensing perspective, that too in assessing and monitoring droughts. The present chapter aimed to study the land surface and vegetation and their response to climate in the context of climate change and climate variability.

2. Monitoring the ground vegetation – Soil wetness by satellites – Previous studies

2.1 Normalized Difference Vegetation Index (NDVI)

The study of vegetation cover over a region which can be formed either by native or by cultivation attained a great significance. Barbosa et al. (2006), reported that the NDVI is a reliable index to study the ground vegetal cover and to monitor the changes occur in the vegetation due to climatic abnormalities. Study of spatiotemporal variations of NDVI is of great importance now a days in the context of increased greenhouse gases that modulate the global climate systems in terms of short term climate signals such as El Nino and La Nina. The NDVI variations on both space and time scales not only important in view of varying crop stages but also prominent in vegetation-climate feedback mechanism thus giving a challenge to policy makers in proactive and reactive measurements of risk Cihlar et al., 1991; Davenport & Nicholson, 1993; Al-Bakri & Suleiman, 2004; Kazuo & Yasuo, 2005, Ma & Frank, 2006 & Nagai et al., 2007. Global scientific community focused on NDVI as the indicator of agricultural droughts where in crop growth is known by NDVI value and found that the NOAA Advanced Very High Resolution Radiometer (AVHRR) NDVI is one of the best among the other vegetation indices derived from the other satellites. The NDVI can be defined as the ratio of difference between Channel 1 (red) and Channel 2 (near Infrared) which is based on the more absorbance for healthy vegetation and more reflectance for the poor vegetation. In other way, NDVI measures the changes chlorophyl content (via absorption of visible red radiation) and is sponzy mesophyll (via reflected NIR radiation) with in the vegetation canopy, thus NDVI from AVHRR can be written as

$$NDVI = (\rho_{857} - \rho_{645})/\ (\rho_{857} + \rho_{645})$$

This NDVI varies from -1 to + 1 and the category in classifying the density of vegetation cover is given below.

NDVI < 0.2 ------------- Low vegetation

NDVI < 0.4 ------------- Medium vegetation

NDVI > 0.4 ------------- High vegetation

Relevant research in changes in vegetation cover in the Sahel demonstrates that the NDVI happens to correlate particularly closely with rainfall, as high as 0.84. The cause of this significant correlation is two-fold: first, it is commonly known that vegetation growth is limited by water; second, the climate in Sahel, rainfall in particular, is very sensitive to changes in vegetation Charney et al., 1977. Sarma and Lakshmi Kumar,2006 derived the NDVI from NOAA AVHRR for the state Andhra Pradesh and saw how it varies in accordance with the crop growing periods such as moist, humid, moderate dry and dry as suggested by Higgins and Kassam (1981) and found a good agreement i.e prevalence of good NDVI is subjected to moist/humid periods. Barbosa et al, 2011 studied the vegetation indices such as NDVI and Enhanced Vegetation Index to understand the underlying mechanism of vegetation dynamics in Amazon forests.

2.2 Brightness Temperature (BT)

Studies on the direct measurement of soil moisture are few. Remotely sensed data in terms of brightness temperature is useful in the study of spatiotemporal variability as well as verifying land surface processes (Rao et al., 2001). Soil moisture can be retrieved by making use of remote sensing observations. Pathak et al, 1993 reported the estimation of soil moisture using land surface temperature retrieved from the INSAT - VHRR data. Microwave sensors provide a great opportunity to measure soil moisture because these microwave radiations can penetrate the clouds and vegetation over the land surface. Microwave brightness temperature can be used to measure soil wetness under different surface roughness and vegetation cover conditions (Ahmed, 1995). Thapliyal et al. (2003) reported soil moisture over India using microwave brightness temperature of IRS-P4. Sarma and Lakshmi Kumar (2007), explained the variations in brightness temperature of different soil types in Andhra Pradesh. The data retrieved from the Multichannel Scanning Microwave Radiometer (MSMR) carried by Indian Remote Sensing (IRS) - P4 satellite, is made use in understanding the nature of relation between soil moisture and BTD. The BTD of 6.6GHz frequency channel of MSMR, taken at 1830hrs Indian Standard Time (IST) for the horizontal polarization is used for the estimation of soil wetness which in turn portrays the drought prevailing conditions over that place.

The brightness temperature depends on the angle of incidence and the plane of polarization, vertical as well as horizontal. It is reported that the horizontal polarization is more sensitive to soil moisture and hence the same is used here. The microwave polarized temperature (MPT) is defined as

$$MPT = TB\ (1,H)$$

Here 1 refers to wavelength and is related apart from other factors to moisture content of the soil horizon.

Brightness Temperature (BT) at the microwave frequencies can be written as

$$BT = eT_S$$

Where e is the emissivity of the surface and T_S is the surface temperature.

As BT is a function of emissivity and surface temperature, the lands having less emissivity (wet lands) exhibit low signatures and is a good indicator of soil wetness status. Similarly, the lands having high emissivity (dry soils) give high BT signals showing low soil wetness status from which one can assess the moisture condition over the soil to estimate the drought condition (Sarma and Lakshmi Kumar, 2006).

3. Case studies, methodologies and findings

3.1 Analysis of the NDVI temporal dynamics in semi-arid ecosystems: Brazilian Caatinga and African Western Sahel

The Caatinga and Savanna vegetation covers are likely the most sensitive to changes in climate. Satellite observations show that changes in vegetation greenness follow rainfall variability. Because water availability is a key factor in the abundance of vegetation, changes in precipitation are most critical for continued biodiversity and human livelihood opportunities in arid and semi-arid environments. In earlier studies (Barbosa 1998; Nicholson and Farar, 1994) mostly held in the atmospheric dynamics context have incorporated long time series of NDVI data taken by the National Oceanic and Atmospheric Administration (NOAA) AVHRR to monitor the dynamics of the temporal structures of vegetation responses to climatic fluctuations across the Northeastern Brazil and the West African Sahel's landscapes. These investigations have found clear and positive linear relationships between NDVI and rainfall thanks to different analyses across the semi-arid tropical ecosystems where rainfall is below an absolute amount of rainfall of 50-100 mm/month. In this study we have the objective to investigate the NDVI responses to rainfall oscillations at seasonal scale over the last two decades of the 20th century.

Temporal analyses performed in this research were based on the monthly NDVI imagery from the Goddard Distributed Active Archive Center (GDAAC) for the 1982 to 2000 period. The NDVI images were originally in the Goode's Interrupted Homolosine projection, and they were geo-referenced to a geographical coordinate system (latitude and longitude). The 20-year series of monthly NDVI data for Brazilian semi-arid and West African Sahel regions were extracted from NDVI images with a resolution spatial of 7.6 km.

Aiming to characterize the seasonal variability of land cover types in Caatinga and Savanna Biomes to the understanding of their responses to the seasonal rainfall variability, we verified how available GDAAC NDVI are able to capture the climatic variability, and how it could be used in ecological studies, at the local level. Based on the vegetation map published by the Brazilian Institute for Geography and Statistics (IBGE, 1993) and by author's local knowledge, as a basis, four homogeneous vegetation sites covering semi-arid Caatinga in Northeastern Brazil were selected from vegetation classes, and located by ground meteorological stations (sites): site#1-caatinga arbórea aberta (open arboreous shrubbery) (4^{0}31'S; 40^{0}12'W), site#2-caatinga arbustiva densa (dense shrubbery) (4^{0}37'S; 42^{0}7'W), site #3-caatinga arbórea densa (dense arboreous shrubbery) (8^{0}37'S; 42^{0}7'W), and site #4-caatinga arbustiva aberta (open shrubbery) (9^{0}25'S; 41^{0}7'W). For the semi-arid Sahelian region, four vegetation classes were conducted over the UNESCO map produced by White (1983). Representatives from the following land cover types dominated in this classification: site#1-woodland (10^{0}55'N; 14^{0}19'W), site#2-woodland (110^{0}26'S; 7^{0}25'W), site#3-woodland (11^{0}04'N; 7^{0}42'E), and site#4-wooded grassland (11^{0}04'N; 39^{0}47'E) (Figure.1).

The 20-year integrated series of monthly NDVI data were extracted by averaging the NDVI values for a window of 3 by 3 pixel arrays at selected locations within each land cover type in order to characterize the seasonal variability in land cover type for each series. The database consisted of land cover classes from the vegetation maps (1:5,000,000) that were used to guide site locations by using the geo-referenced meteorological stations on the ground in conjunction with NDVI data.

Fig. 1. The vegetation dynamic of the Brazilian Caatinga and the African Savanna is directly connected to the climatic conditions (photos). Site#1: (Nordeste Lat=-4°53′S; Long= 40°20′ W) e (Sahel Lat=11°76′N, Long=34°35′E).

Seasonal variations in the NDVI of Cattinga and Savanna vegetation types are illustrated in Figure 2. For each location, the average monthly values of NDVI were extracted from September 1981 to September 2001 (20-years) at a 519.84 km^2 (averaged area of nine pixels) for each site. The average monthly values of rainfall for the specific location within each land cover type are representative over the 30-year climatology period of 1961 to 1990. Both NDVI and rainfall series for the four caatinga types show a clear unimodal seasonal cycle. While the differences in magnitude of the NDVI are different over each site, the NDVI series show that phenological behavior varies slightly from north (site#1) to south (site#4). The annual total rainfall gradually varies from site#4 (549.10 mm) to site#3 (657.2 mm), to site#1 (839.13 mm) to site#2 (1046.12 mm). At these four sites, the beginning of the vegetation season is mostly driven by rainfall, while mean maximum temperature is quite constant during year and around 31^{0}C. Moreover, the NDVI related to dry and rain seasons are very pronounced, with a minimum of 0.21 ± 0.02 over site#4 in August to October and maximum of 0.68 ± 0.02 over site#2 in April to June. On average, NDVI and rainfall have a similar pattern (both increase and both decrease together) with a lagged response due to the differences between the onset of the rainy season occuring in October or November, and the vegetation growth in November or December. Contemporaneously, from May to October there is a downward trend in NDVI values for all four Caatinga types, with the lagged response of the vegetation to rainfall being two months. While the upward trend in NDVI values is certainly due in part to stored soil moisture during the rainy season, the downward trend is likely to be more closely related to the different soil types and their physical properties.

For the West African Sahel, the seasonal pattern of NDVI for all four vegetation types closely responds to the seasonal cycle of rainfall as illustrated in Figure 2, with a peak in rainfall followed by a peak in NDVI. The majority of NDVI and rainfall series in these figures are dominated by an indistinct unimodal pattern, except for site#4 that shows two peaks. These two NDVI peaks capture the effects of two seasonal monsoons that are related to the north-south movement of the ITCZ. The first peak occurs in early May, when the ITCZ is at its southernmost extent. The second peak occurs when the ITCZ is at its northernmost extent in early October, and is also higher than the first peak. These seasonal variations in rainfall time series represent both meteorological and geographical factors resulting a bimodal greenness pattern in the NDVI time series. The unimodal cycle of the other three selected sites show a similar phonological behavior (peaking in October), but they exhibit differences in amplitude of the NDVI time series among them.

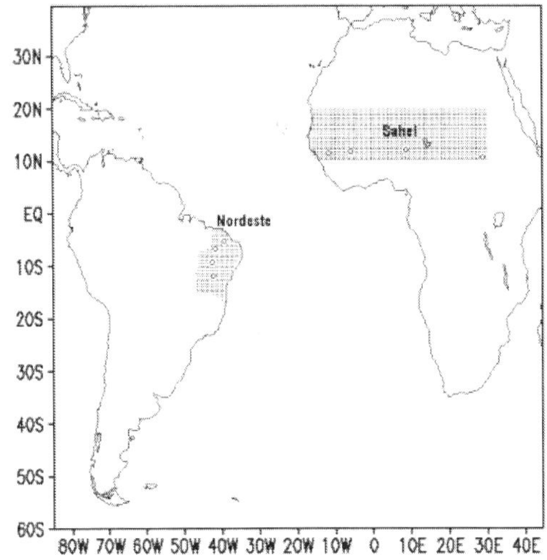

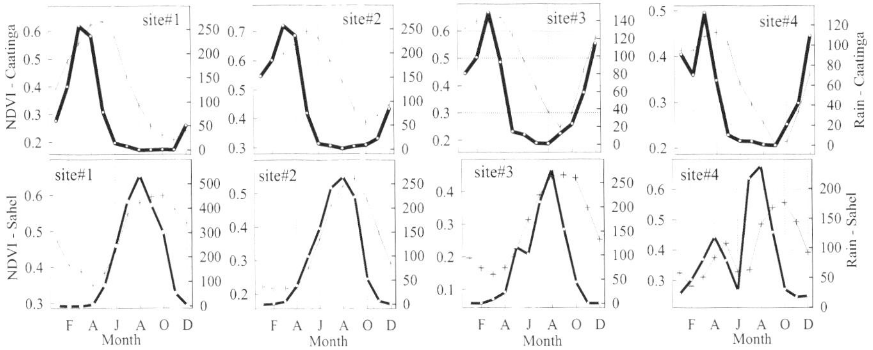

Fig. 2. Time series of monthly composites of NDVI (thin solid line) and rainfall (thick solid line). The monthly composites of NDVI relative to the 20-year Pathfinder data period from 1981 to 2001. The mean monthly rainfall values relative to the 30-year climatological period of 1961-1990.

The figures for the Caatinga and Savanna Biomes discussed above underline the relationships between surface greenness and rainfall (varying from $r^2=+0.1$ to $r^2=+0.6$, n=360 (rain), n=240 (NDVI), $P<0.05$) and lend support to the time lag between the rainfall and NDVI. The time response of the rate green up due to the rainy season (rainfall) in NDVI profile is longer for semi-arid Sertão caatinga types than Sahelian vegetation types. This is likely because the semi-arid Brazilian is dominated by deeper-rooted arboreal formations. The lagged response of vegetative activity of these deeper-rooted caatinga types to absorb the stored soil moisture is longer than the Sahelian vegetation types, which are dominated by wooded and bushed grassland (herbaceous). In contrast, the time duration of the rate of senescence in the NDVI profile, which is due to the dry season (rainfall deficit), is shorter for Sahelian vegetation type than caatinga types. This might indicate that the caatinga types, which are very drought resistant, have an additional water supply besides rainfall. It is important to note that while the same twenty-year NDVI and thirty-year rainfall periods were used for both the Sertão and Sahel regions, there is a ten-year time shift between the referenced period of the NDVI and rainfall, however, this shift was taken into consideration in our analysis.

A "see-saw" pattern can be also observed from figure 3: while the values of NDVI over the Nordeste present the peak in May, which is associated with peak of humid month, most of values of NDVI over the Sahel present the peak in August, which is associated with peak of humid month. The comparison of the Nordeste curves with the Sahel curves suggest that during the last two decades of twentieth century their Nordeste amplitudes are about half order of magnitude larger than those of Sahel. Considering the whole 1982-2001 period, the Nordeste NDVI increase during the 1980s must be primarily driven by the increase in precipitation during this period. In contrast, the most 1980s and 1990s the NDVI Sahel must be dominated by precipitation decrease. Due to its intrinsically different dynamics, Nordeste and Sahel atmospheric circulations are often regarded separately. For example, in Northeastern Brazil, the inter-annual variability of the atmospheric circulation is predominantly influenced by Sea Surface Temperature (SST) in both the Atlantic and Pacific Oceans, and has exhibited negative anomalies in rainfall during the warm phase of El Niño-Southern Oscillation (ENSO), and positive anomalies during the cold phase (La Niña). The West African Sahel, on the other hand, is mainly influenced by SST in the Indian Ocean with a portion, known as the rain belt being affected by SST in the Atlantic Ocean. Although warm events in the eastern equatorial Pacific and Indian Oceans are known to induce climatic extremes over much of Africa; the tropical Atlantic Ocean often exhibits an opposing response to ENSO which may further enhance these impacts.

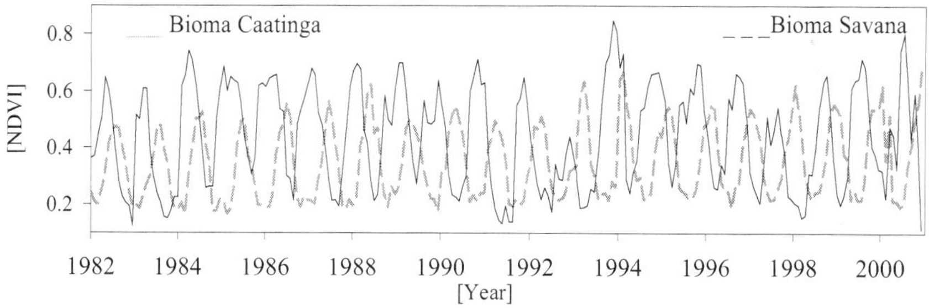

Fig. 3. Time series plot of monthly composites of NDVI over the NEB (solid line) and the Sahel (dashed line) period January 1982 through September 2001.

3.2 Linking sustainable indices and climate variability in the state of Ceará, Northeast Brazil

The rain-fed agricultural production in Northeast Brazil has experienced persistent drought episodes during the last three decades of the 20th century. However, it is necessary to assess the vulnerability of its agricultural production to precipitation variability due to interrelated global-scale fluctuations of sea-surface temperature (SST) in the tropical Pacific and Atlantic Oceans. Most of evidence presented by the recent studies on the influence of a strong ENSO event on agricultural climate of NEB is qualitatively similar to those of the Southern Africa. It is found that, there is a joint effect between the influence of a positive El Niño SST anomaly and positive gradient (SST North Atlantic warmer than South Atlantic), which tends to dramatically decrease rainfall during austral summer and autumn in the region (Alves and Repelli, 1992). While several drought spells were recorded over this region in recent history, the 20th century drought is unprecedented for its severity. Throughout the recorded history of occurrence of drought (meteorological) over this region, there were three in the 17th century, eleven in the 18th century, and twelve in the 19th and 20th centuries. It seems that in the past, the environment was more resilient to climatic variations.

In the Northeast Brazil, (Barbosa 2006, Barbosa 2004) present recent findings from satellite images which reveal a consistent upward trend in vegetation density for the period 1984–1990 and a downward trend for the period 1991–1998, but which demonstrate that such short-term vegetation changes, with a period of 7-8 years, were associated with episodes of unusually wet (years of La Niña activity) and dry (years of El Ninõ activity) climate oscillations. Trends in global SST patterns explain the recent period of desiccation in the Sahel, but do not present an exact explanation for rainfall in particular years. More strikingly, trends in tropical SST patterns on multi-year to decadal timescales explain the Sahelian desiccation during the last three decades of the 20th century, but do not provide an exact explanation for rainfall in individual years after 1970 (Nobre and Shukla, 1996). Although the challenge remains in confronting of climate variability for a range of locally-specific climate impacts, the understanding of the causes of this variability is still unfolding. Understanding the impact of the occurrence of extreme weather and climate events on the rain-fed agricultural production in Northeast Brazil is crucial to establishing an effective and comprehensive monitoring and early warning system as one component of an effective drought preparedness plan. Indeed, an ideal area of study is the state of Ceará in Northeast Brazil, which is already experiencing significant climate variability. This study, building on findings of previous studies about the impact of rainfall variability crop production and yields in the state of Ceará, carries the analysis of this impact one step further to evaluate connections between SST variability and the attendant impacts on its crop agriculture.

The geographic focus of this study is the state of Ceará located in the Northeast region of Brazil (Nordeste) known as an anomalous area within the equatorial zone because in contrast to other areas such as located in the Amazônia e central equatorial Africa, this state has a semi-arid tropical climate. The Brazilian semi-arid comprises approximately two thirds of the Northeast region (4° and 16° S and 33° and 46° W), which is subject to extreme climate variability and recurrent drought. The main rainy season in the region has an annual average of less than 600 mm, which is typically concentrated between February and April. The interannual precipitation variability is very high, which is usually around +/- 40% from the long term annual average. These fluctuations have motivated numerous studies that collectively documented the high spatial and temporal variability in the region's precipitation to both large-scale oceanic forcing (i.e., region-global SST anomalies and its distribution) and atmospheric circulation patterns (Hastenrath and Heller, 1977). During strong El Niño conditions, precipitation tends to decrease (i.e., causing drought) on the state. On the other hand, the strong La Ninã conditions are the opposite (i.e., causing flood). Because the high frequency and intensity of El Niño years have increased rapidly since the end-1990s, a repeat of such drought episodes may have severe consequences not only for the region's fragile ecosystems but also for the region's seasonal grain production. The great drought of 1958, for instance, forced 10 million people to emigrate from the region (Namias 1972). After 1958, droughts continued to cause severe impacts, but consequences like death casualties have been avoided by policy responses.

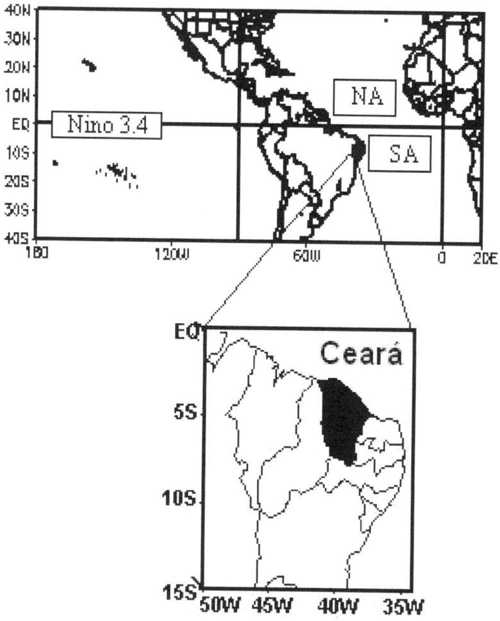

Fig. 4. Location of the state of Ceará in Northeast Brazil, Niño 3.4 and Atlantic Dipole regions (NA - SA).

Nonetheless, as the El Niño drought of 1998 demonstrated, the vulnerability of the rural population remains critically high. On the state of Ceará, around 95% of the state territory (147 thousand square km) is classified as semi-arid (Figure. 4). The state's rural population, most rain-fed farmers, is mostly located beneath the poverty line and suffers extremely vulnerability to drought (Chimelli et al., 2002).

The overall agricultural land use in the Brazilian semi-arid ecosystem is characterized by smallholder crop production and extensive livestock farming. Smallholders in the region produce about 70% of the grain production supplying the market-based agriculture, which include maize, beans and manioc. Approximately 90% of all agricultural areas are smaller than 100 ha, but they cover together only about 30% of the total agricultural production area. Since the 1950s, agricultural community in the state of Ceará has undergone profound changes, including the growth of small farms but also the development of precarious occupation of land, which cause the impoverishment of small farmers, who have faced increasing difficulty in access to land. These changes have also aggravated the conditions for the social reproduction of this community. In contrast, in the northeast of Ceará, the agricultural community produces cash crops such as cashew, cotton, fruits and vegetables, involving various irrigation projects along the Jaguaribe River.

Total monhtly maize and bean production (ton) and yields (kg/ha) from 180 municipalities in the entire state of Ceará were the raw data employed as agricultural observations. Total monthly SSTAs utilized as a climate indicator were delimited by the latitudes 170°W-120°W and longitudes 5°S-5°N (Niño-3.4 region), and also by the gradient between tropical Atlantic North (5°N-25°N) and South Atlantic (5°S-25°S) (known as the Atlantic Dipole index). Agriculture and climate time series were compiled from digital records. The former are available on the databases of the FUNCEME and the latter were obtained from a file of the Comprehensive Atmospheric Ocean Data Set (COADS) for the entire period 1971-2000. The COADS file has data of monthly averages in grade points of 1° x 1° latitude-longitude for a period of 1971 to 2000. Total monthly data are averaged from February to April.

To assess the behavior of the maize and bean production and yields in response to SSTA fluctuations in the study area, the Vulnerability Index (VI) was designated. To measure how much variation at the same rate and scale, the variables in question have deviated from the maximum and minimum values from the long-term record. This index allows a direct comparison among the different variables in question for a given period. It is calculated using the following equation, $VI_j = [(DEV_j - DEVmin)/(DEVmax - DEVmin)]*100$, which DEV represents the deviation that is employed as a measure of variability relative to mean value. It is calculated as the difference between the variable in question for the current time step and the long-term mean for a given period (DEV_j = variable $value_j$ – variable $value_{mean}$). And the DEVmax and DEVmin are measured from the long-term record for a given period and j represents the index of the current time step. DEV represents the deviation that is employed as a measure of variability relative to mean value. The VI is measured in percentage (%) and it varies between 0 to 100%. It reflects, effectively, how close the VI of the current period is in relation to the long-term minimum and maximum. In addition to that, the linear correlation was applied among the variables utilized.

The averaged trimester deviations of maize and bean production from monthly values (February-April) on Ceará for the period 1971-2000 are displayed in Figure 5. The anomaly cycles of maize and bean production vary substantially during the last three decades of the 20th (Figure. 5a). The amplitude of these cycles has increased rapidly by 37% since 1983. This result is particularly striking in relation to the increasing level of maize and bean production in the study area. Nevertheless, the most dramatic decline in maize and bean production occurs before 1981, with concomitant increase in maize and bean yields (Figure. 5b). Over the entire period, the frequency distribution of anomaly for the bean production (bean yields) is in phase with the distribution of maize production (maize yields), but in less magnitude. As is indicated in Figure 5c, there is significant connection between SSTA climate patterns and crop production on the state during the rainy season. Despite the year-to-year changes coherency between extreme SSTAs and crop production, the strength of the correlation is relatively weak (average of $r= -0.42$, $n=30$). Of particular interest is the decline in maize and bean production beginning in 1971 with concomitant increased maize and bean yields that has continued through 1981 with only slight relief in 1975 and 1977. These results provide the basis for linking seasonally changing SSTAs in Ninõ 3.4 and Atlantic Dipole regions directly to bean and maize production in the rainy season, when over half of the interannual change in crop production on Ceará is explained by changes in SSTAs. More substantively, changes in maize and bean production for the averaged February-April crop year are closely contemporaneous with well-known drought, specifically 1972, 1979,1980, 1981, 1982, 1983, 1990, 1992 and 1998. Particularly, the drought of 1982-1983, a +1.85°C and +0.47°C deviations in the Niño 3.4 and Atlantic Dipole regions decreases the maize and bean production by steadily more than 135 ton, while the maize and bean yields increased steadily more than 1.5 kg/ha. The drought story for bean production on Ceará has high similarities to that for maize, but differs in that beans crop is planted and harvested earlier than maize in the rainy season, and it can also substitute partially for bean plantings in drought years. Although increasing in its overall economic importance, maize is still the second most important food staple for most of Cearense's people.

The results in Figure 6 illustrate the vulnerability of the bean production and yield in response to SSTA variability as expressed by the vulnerability index (VI). This index was able to capture the agricultural drought in response to changing in SSTAs. The larger VI for climate (Niño 3.4 plus Atlantic Dipole SSTAs), the stronger is the drought agricultural severity, which indicated by the smaller VI for crop production. In this study, when the VI for SSTAs is generally close to the long-term maximum of 160% during 1971-2000 indicate severe drought agricultural conditions (a VI of 0%). Particularly, the period 1971-1973, the value of VI for bean production decreased sharply from +89 to +49%, while the bean yield increased from +0.1 to 21%. And the value for VI climate varies from moderate humid conditions (+42%) to normal conditions (73%). It is interesting to note that the time series of VI associated with the bean production and yield show distinct differences among the early 1970s, the late 1970s, the late 1980s, and the late 1990s. Separating the fluctuations of VI from varying lengths and intensities, the period 1977-1983 is clearly the worst agricultural drought of the last three decades of 20th. The period 1984-1989 was the optimal agricultural production, broken only by the intense agricultural drought of 1987. Total seasonal maize and bean production are inversely correlated with Niño 3.4 ($r= -0.49$ and $r=-0.36$, $n=30$, $p<0.05$) and Atlantic Dipole indices ($r= -0.42$ and $r=-0.41$, $n=30$, $p<0.05$), while maize and bean yields are directly correlated with Niño 3.4 ($r= +0.32$ and $r=+0.34$, $n=+30$, $p<0.05$) and Atlantic Dipole indices ($r= +0.32$ and $r=+0.46$, $n=30$, $p<0.05$), but inversely related with maize and bean yields ($r=-0.67$ and $r=-0.74$, $n=30$, $p<0.05$) averaged from February to April.

As suggested by Figure 7, seasonal changes associated with VI for Niño 3.4 and Atlantic Dipole SSTA regions show distinct differences from the early 1970s to the late 1970s and from the late 1980s to mid-1990s, with a transition during 1982-1983. Prior to 1982-1983, year-to-year changes in bean production and yields are primarily associated with North Atlantic SSTA variability (a positive gradient - North Atlantic warmer than South Atlantic- which leads to a decrease in precipitation on the state) that is often, but not always, reduced by La Ninã variability (a cold ENSO variability - La Niña conditions - which leads to an increase in precipitation on the state). After 1982-1983, year-to-year changes in bean production and yield reflect the influence of seasonally changing SST associated with ENSO variability (an El Niño and a La Nina variability) that is irregularity amplified (reduced) by North Atlantic variability (South Atlantic variability- a negative gradient-North Atlantic cooler than South Atlantic which leas to an increase in precipitation on the state).

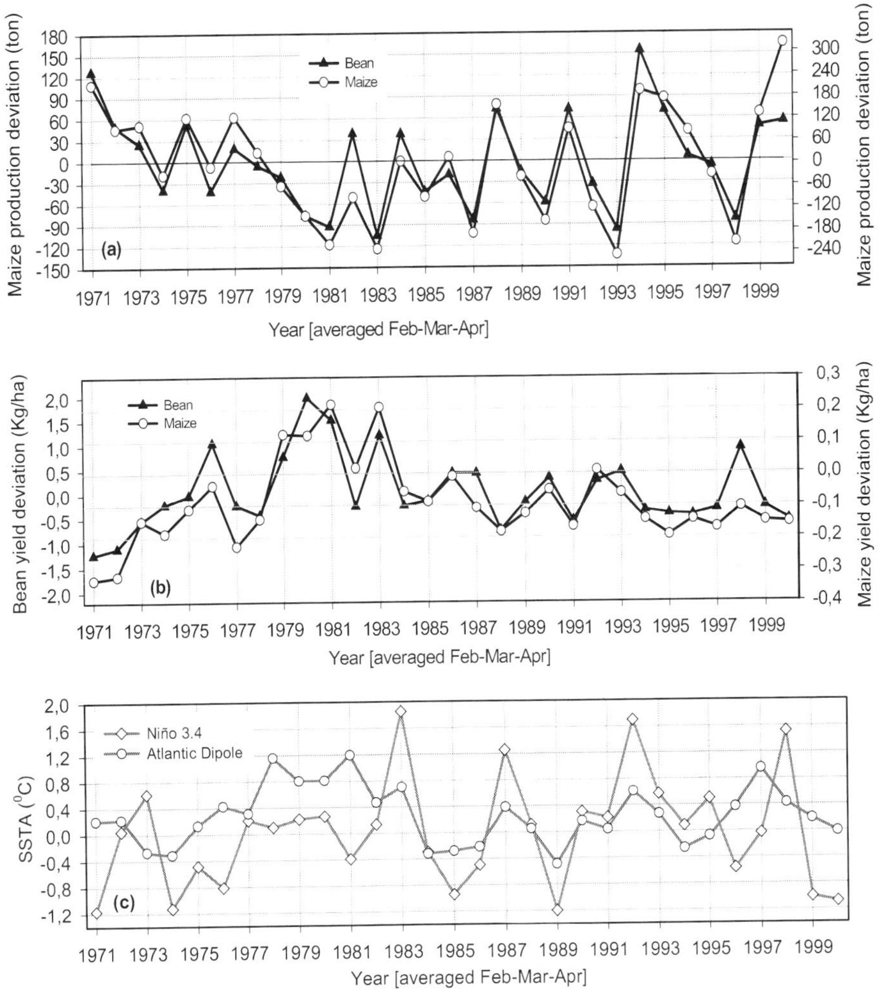

Fig. 5. Normalized deviations (anomaly). (a) Maize and crop production (ton), (b) Maize and crop yields (Kg/ha), and (c) SST (°C) in Niño 3.4 and Atlantic Dipole regions for the period 1971-2000, state of Ceará.

These large fluctuations associated with VI for Niño 3.4 [VI values close to 0% (or 100%) lead to a La Niña (or an El Niño)] and Atlantic Dipole [VI values close to 0% (or 100%) lead to a negative gradient (or a positive gradient)] vary almost in-phase for the period from 1985 to 1997, apart from the period 1998 to 2000, when the two VIs are out-phase. The warm phase of the North Atlantic beginning around 1977 is readily associated with the agricultural drought beginning in 1977. The worst drought is the1983 (a VI of 160%) from 1971 to 2000, which was strongly affected by persistent anomalous warming SST in the

North Atlantic Ocean beginning in 1977. Despite the droughts, the increase in bean production between 1984 and 2000 is striking. Severe droughts (1987, 1992, and 1998) alternated with relatively humid years (1985, 1989, and 1995). This indicates that recovery of rain-fed agricultural production from persistent droughts has been enhanced since about 1983. Moreover, these long-term fluctuations were in phase with global climate variability, specifically, ENSO and the North Atlantic variability. Therefore, the vulnerability of rain-fed agricultural production on Ceará to changing SSTAs and its recoverability after persistent droughts suggest that the long-term rain-fed agricultural production of the Ceará state may be predictable.

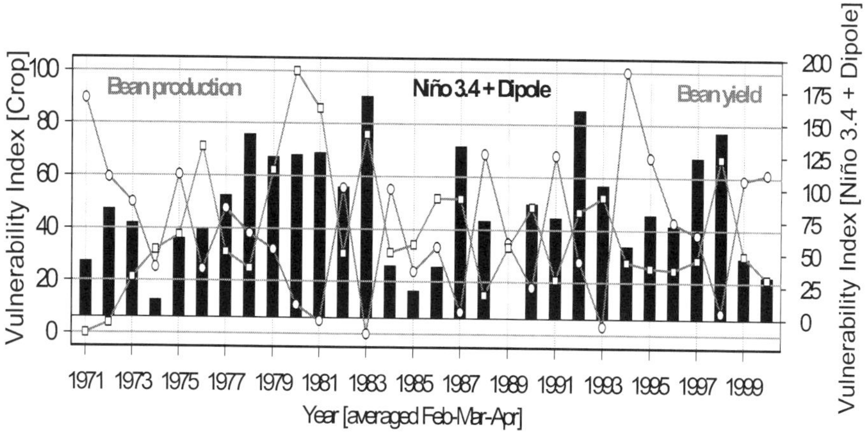

Fig. 6. Vulnerability Index (VI). (a) Bean production (ton), (b) Bean yield (Kg/ha), and (c) Niño 3.4 plus Atlantic Dipole SSTA (°C).

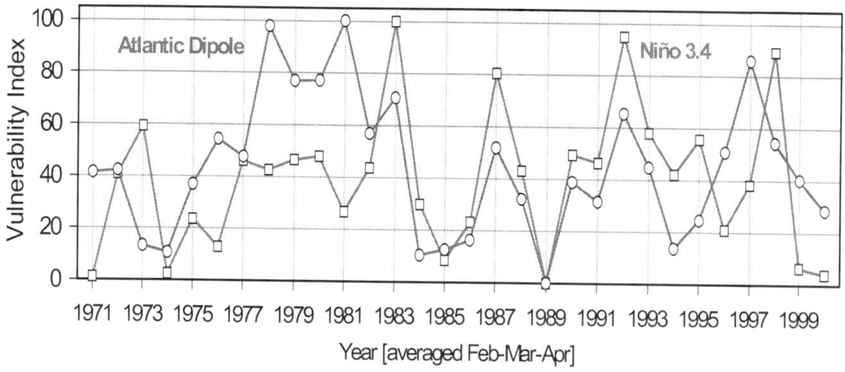

Fig. 7. Vulnerability Index (VI). Niño 3.4 SSTA (°C) and Atlantic Dipole SSTA (°C).

3.3 NDVI as an indicator of crop performance

10- day composite NDVI with 8km resolution retrieved from AVHRR which are supplied by Space Application Centre, Ahmedabad, Indian Space Research Organization (ISRO), Govt of India are used here from 1999 to 2001 during south west monsoon period (June, July, August and Septmber) over Andhra Pradesh, India. Since nearly 80% of annual rainfall is from south west (S-W) monsoon in Andra Pradesh (Sarma & Lakshmi Kumar, 2010), the present study is focussed during this period.

The study has been carried out at different test sites of Andhra Pradesh state (13S-20S; 77E-84N), covers the parts of Eastern Ghats and Deccan Plateau of India, namely Anantapur (ANT), Ramagundam (RGD), Hyderabad (HYD), Nizamabad (NZB), Kurnool (KRN), Nellore (NLR), Ongole (ONG), Kakinada (KKN), Machilipatnam (MPT) and Visakhapatnam (VSK) that are located in Figure.8.

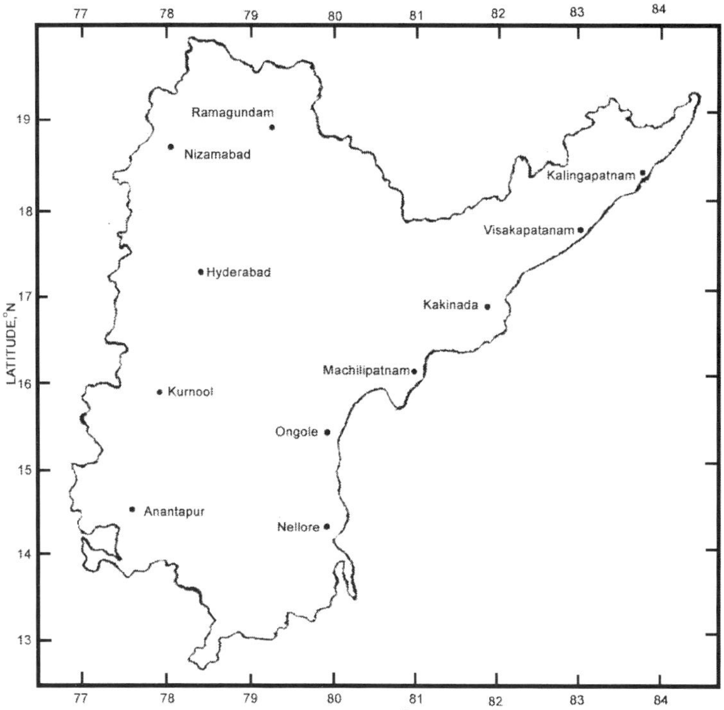

Fig. 8. Study region and locations of test sites - Andhra Pradesh State, India.

The rainfall and potential evapotranspiration data on daily basis were collected for the period 1999 to 2001 and are subjected to determine the crop growing periods. The model identifies the growing period when rainfall exceeds 0.5 times the water need (potential evapotranspiration) and ends with the utilization of assumed quantum of stored soil moisture and the categorization within the growing period is given below.

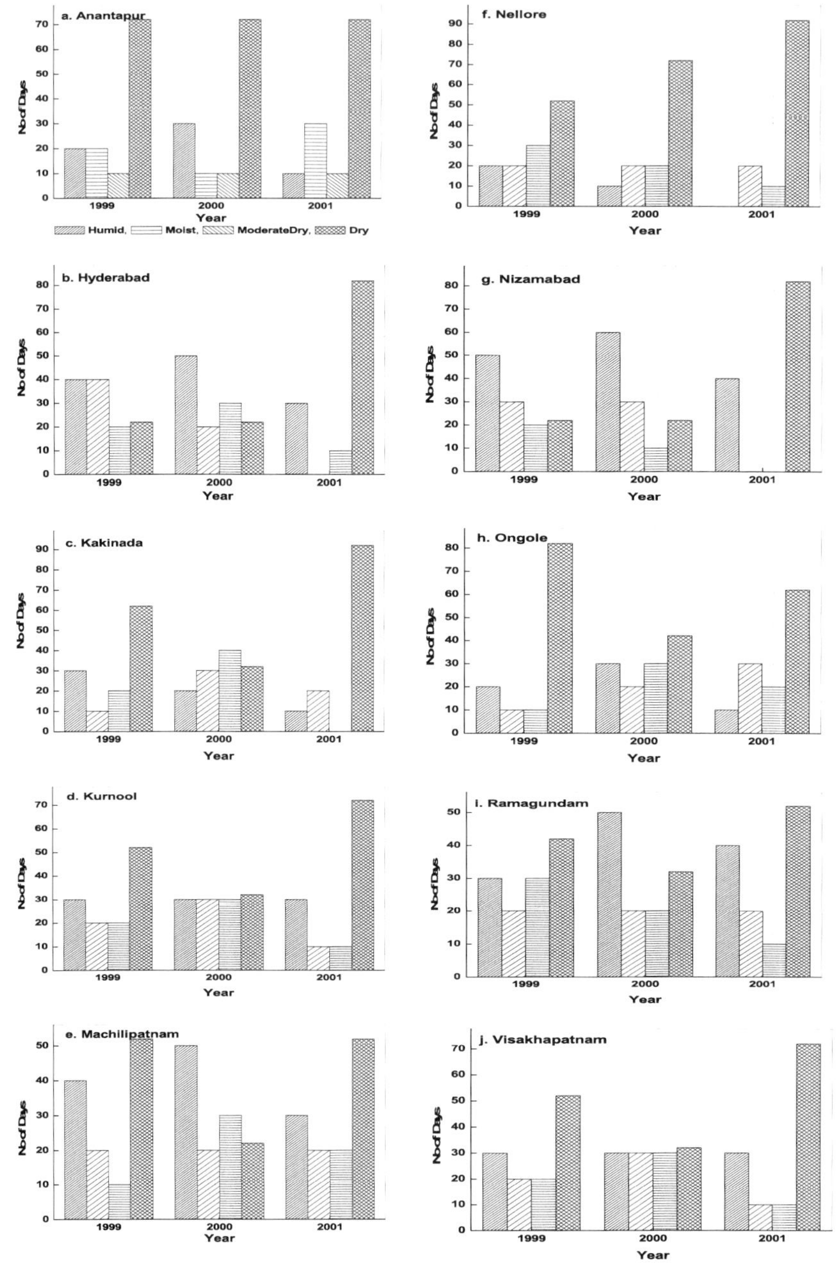

Fig. 9. Crop growing periods of different test sites during southwest monsoon.

If rainfall (P) is

i.	> PE	-----Humid period
ii.	½ PE to PE	-----Moist period
iii.	¼ PE to ½ PE	-----Moderate Dry period
iv.	< ¼ PE	-----Dry period

The agroclimatic potentiality, soil moisture adequacy is derived using the revised water balance model (Thoronthwaite and Mather, 1955) with the inputs of rainfall and potential evapotranspiration and is calculated by the following formula

$$S_{AD} \text{ in } \% = AE/PE \times 100,$$

Where AE is Actual Evapotranspiration and PE is Potential Evapotranspiration.

The following part dwells first in obtaining the crop growing periods by using rainfall and potential evapotranspiration and secondly the comparison of NDVI with agroclimatic potentialities such as rainfall and soil moisture adequacy to understand how sensitive is NDVI to the crop performance.

3.3.1 Crop growing periods from Higgins & Kassam model

Figures 10 (a-f) show the number of humid, moist, moderate dry and dry days during the south west monsoon season where the crops are subjected in for the years 1999 to 2001. In almost all the years, the dry days dominated the entire season which is expected and the relevance here is about the humid and moist days which play major role in the growth of the crop. The climatology of Andhra Pradesh infers that Anantapur, Kurnool, Nellore, Ongole are in dry subhumid zone where as the other test sites such as Mahaboobnagar, Hyderabad, Kakinada, Nizamabad, Machilipatnam and Visakhapatnam are in moist subhumid zones (Sarma & Lakshmi Kumar, ISRO RESPOND Report, 2005).

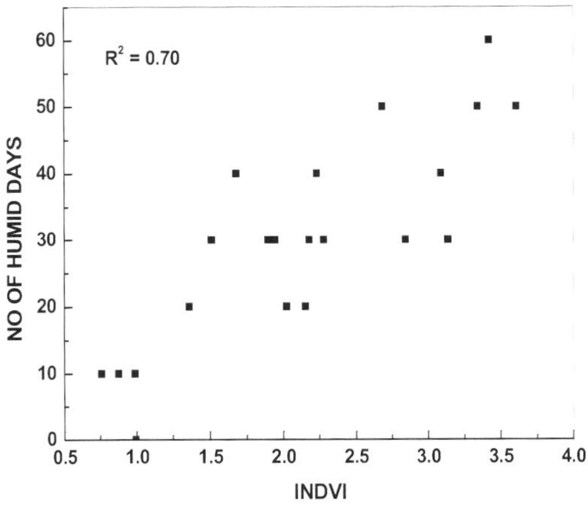

Fig. 10. Scatter plot of INDVI and the number of humid days for all the test sites - Andhra Pradesh.

In accordance with the climatology, the maximum number of humid days where the rainfall dominates the pontential evapotranspiration is high in Hyderabad (40days), Nizamabad (50days) and Machilipatnam (40days) while compared with Anantapur (20days), Ongole

(20days) and Nellore (20days). There is no much difference observed in the moist and moderate dry days during the study period.

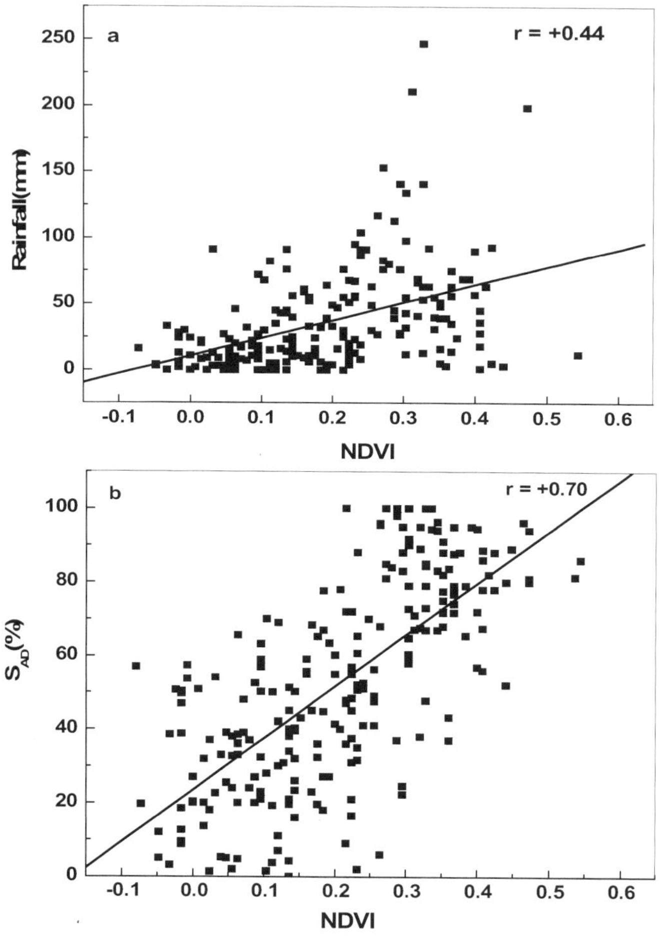

Fig. 11. Correlation of NDVI with a) rainfall and b) Soil moisture adequacy.

Among three years of study, it is unraveld that most of the stations recorded more number of humid days such as Nizamabad (60 days), Ramagundam (50 days) and Machilipatnam (50 days) in the year 2000. This is the reason in the year 2000, Andhra Pradesh is found as the wet year during 1999 to 2001 by means of daily water balance analysis. The number of humid, moist and dry (moderate dry+dry) days are subjected to scatter plot for all the stations with the Integrated NDVI (INDVI) values to understnd the mode of response of NDVI during these days. The INDVI is nothing but the sum of the all 10 -day composites of NDVI for that particular test site, and the scatter plots infered the alignment of INDVI with the number of humid days, for the other categories, such as moist and dry days, the plots are much scattered which tells the poor relation with the vegetation indices. The scaatter plot for INDVI and number of humid is given in Figure.10 along with the variance which is found to be high ($R^2 = 0.70$).

Test site	Rainfall and soil moisture adequacy during the year						Average during	
	1999		2000		2001		1999 -2001	
	P	S_{AD}	P	S_{AD}	P	S_{AD}	P	S_{AD}
Anantapur	0.67	0.83	-0.35	-0.02	0.78	0.76	0.50	0.59
Hyderabad	-0.09	0.32	-0.67	0.63	-0.27	0.84	-0.28	0.80
Machilipatnam	0.08	0.57	0.50	0.63	0.61	0.70	0.28	0.58
Kurnool	0.67	0.63	-0.30	0.87	0.50	0.43	0.46	0.79
Nellore	0.27	0.34	0.69	0.47	0.21	0.49	0.54	0.63
Nizamabad	-0.3	0.58	-0.34	0.36	-0.15	0.69	-0.15	0.65
Ongole	0.66	0.66	0.13	0.40	0.56	0.83	0.50	0.63
Ramagundam	-0.09	-0.19	0.02	0.43	0.53	0.71	0.34	0.67

Table 1. Correlations of rainfall and soil moisture adequacy (%) with NDVI.

3.3.2 Relation of NDVI with agroclimatic potentialities

An attempfst has been made to relate the Normalizaed Difference Vegetaion Index (NDVI) with the rainfall to undesrand the vegetation growth with the changes in rainfall. The Pearson correlation is obtained between 10-day NDVI and rainfall values and are tabulated in Table. 1. The positive correlation from the table tells us the vegetation is dependant on rainfall. Test sites such as Anantapur, Nellore and Ongole showed considerable correlation compared to other sites. The negative correlation in a few case may be due to heavy rains that manifest the higher rates of run-off, particularly when soil is moist and also because of irrigated areas. The overall correlation for entire Andhra Pradesh is +0.44 (Figure 11(a)), which is also not substantial. So, to reinstate the NDVI and climate relationship, an effort was put up with the soil moisture adequacy, which is the more appropriate parameter since it actually gives the amount of water that crop takes for its growth and is a measure of crop growing season. The correlation is very strong in the cases of both All Andhra Pradesh($r = +0.70$) (Figure 11(b) and individual test sites. Also, the plots of NDVI with the soil moisture adequacy for each test site is given in Figure (12). From the Figure (12), it can be known that the soil moisture adequacy maintained a very good agreement with NDVI. A maximum of 0.54 of NDVI with the moisture adequacy of 80% in moist subhumid region and a maximum of 0.45 (NDVI) with the corresponding S_{AD} of 88% in dry subhumid region are observed respectively. Among all stations, Nellore and Ongole registered low NDVI values during the three years of study period. It can be seen that the moisture adequacy of Nizamabad, Ramagundam and Nellore has not responded to NDVI very well during the year 1999. The reason for this poor relation might be not only the lack of inter coupling between vegetation and land surface but also the moisture recycling phenomena that paves way in keeping the wet condition over the land. This shows that the soil moisture adequacy is the robust parameter, since it signifies the extent of meeting the water requirement of the place and on which the crop/vegetation performance has bearing.

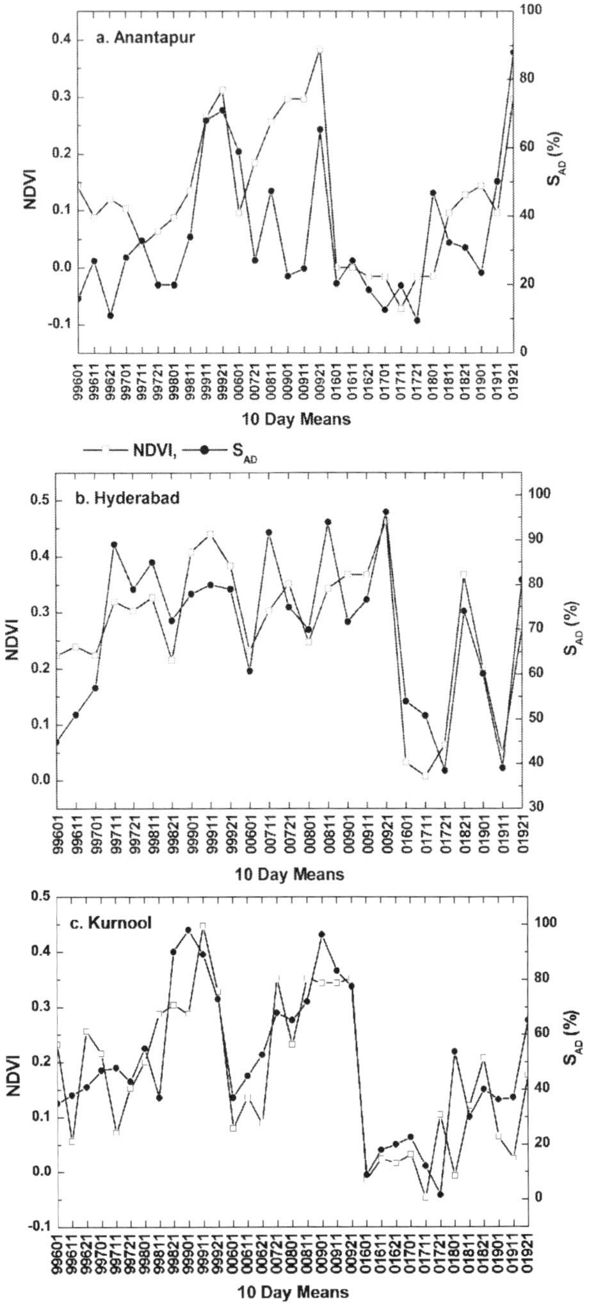
a. Anantapur
NDVI
S_AD (%)
10 Day Means
NDVI, S_AD
b. Hyderabad
NDVI
S_AD (%)
10 Day Means
c. Kurnool
NDVI
S_AD (%)
10 Day Means

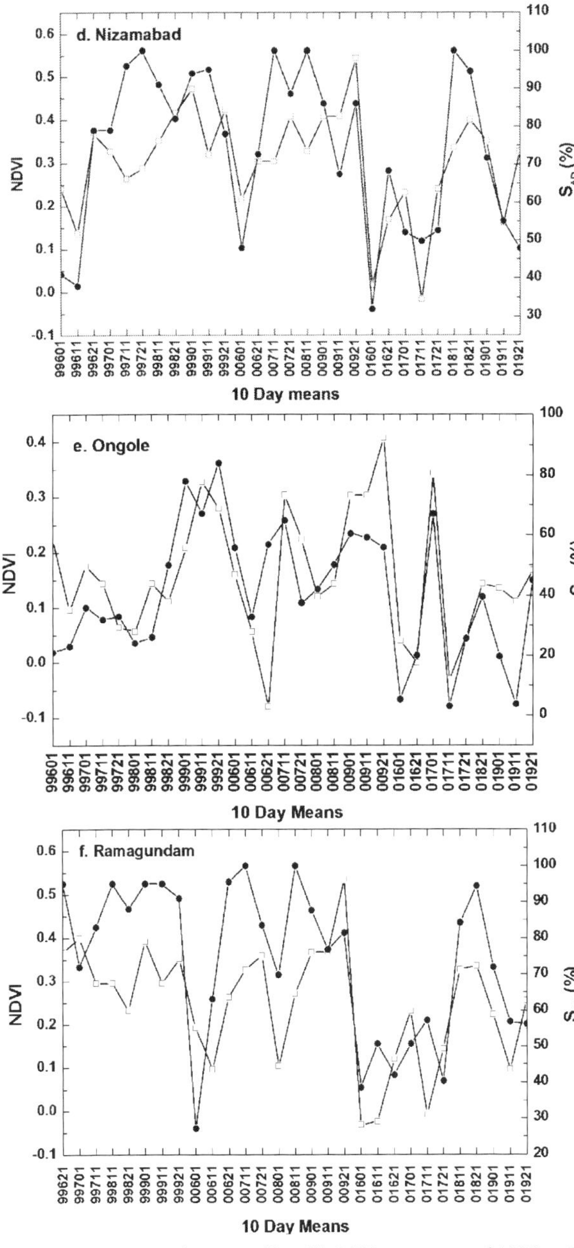

Fig. 12. NDVI and soil moisture adequacy (S_{AD} %) S-W monsoon of 1999 to 2001 (x-axis represents year/month/day).

3.4 Drought monitoring from brightness temperature data

The aim of this study is to estimate soil wetness using remote sensing data. Since the microwave penetrates the earth's surface, this can be used to measure the moisture contnent available in the soil. The BT data has been collected from the Space Application Centre for the June, July and August months and accordingly the study is made. The established relation which was mentioend in earlier scetion that the Brightness Temperature is inversely proportional to the soil wetness present in the soil, and understanding the variations in Brightness Temperature leads to assess the soil moisture. Here also the study has been carried out in Andhra Pradesh at different test sites and a linear regression model is developed to estimate the soil moisture from brightness temperature so as to understand the drought conditions.

3.4.1 Deriving soil wetness – Water balance model

Water balance analysis is accounting of water received in the form of rainfall and expending for evaporation, recharging the soils, surface and subsurface run offs. The modified water balance model as suggested by Sarma et al. (1999) is followed for the land phase of the hydrological cycle and is used in obtaining the surface hydrological fluxes. The budget considers the amount of available water stored in the soil root zone as well as any change in the amount of this storage. It also calculates any surplus of water that is not evaporated, stored or transpired. Water deficit is calculated as part of the budget, because it represents the additional amount of water that plants could have used if it had been available. While these are important parameters associated with the water budget, the primary concept behind the budget is to determine potential evapotranspiration (PE) from the revised concept of Thoronthowaite and Mather (1955) and to estimate the soil wetness. Once the storage of water is determined, the percentage of soil wetness can be calculated by following expression.,

$$S_{WT} = S_T/F_C \text{ X } 100,$$

where S_T is storage and F_C is field capacity of that particular region.

3.4.2 Comparison of soil wetness and brightness temperature data

The temporal variations of Brightness Temperature and soil wetness for the test stations are presented in Figure (13). Ramagundam (Figure 13(e)) and Kurnool (Figure 13(b)) showed the maximum BT of 257K for the 1% soil wetness in the year 2001 and the minimum of 192K and 212K for the soil wetness of 100% in 1999 and 2000 respectively. Ongole (Figure 13(d)) experienced maximum BT of 265K for a soil wetness of 19% in 2001 and a minimum BT of 195K for 100% soil wetness in the year 2000. The BT of Anantapur (Figure 13(a)) showed the highest at 278K for a soil wetness of 5% in the year 2001 and lowest at 216K for soil wetness of 89% in 1999. Nellore (Figure 13(c)) registered a maximum BT of 255K for a soil wetness of 1% in the year 1999 and a minimum at 197K for the corresponding soil wetness of 80% in the year 2000. From the present study, it is known that the BT is mostly varying from 190K to 225K for more than 75% soil wetness and from 250K to 278K for the poor soil wetness (1 to 20%). Both undisputedly, have inverse relation. So, it is evident that the BT of the wet lands with high soil wetness can vary from 192K to 225K and for dry lands, from 250K to 280K.

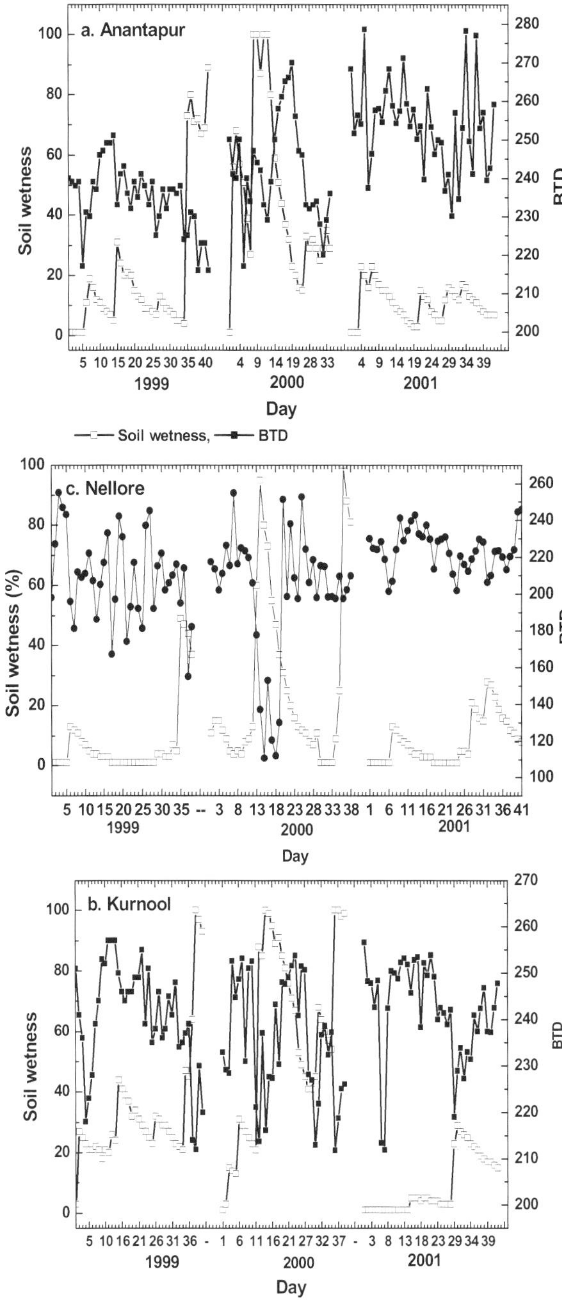
a. Anantapur
Soil wetness
BTD
1999
2000
2001
Day
Soil wetness, BTD
c. Nellore
Soil wetness (%)
BTD
1999
2000
2001
Day
b. Kurnool
Soil wetness
BTD
1999
2000
2001
Day

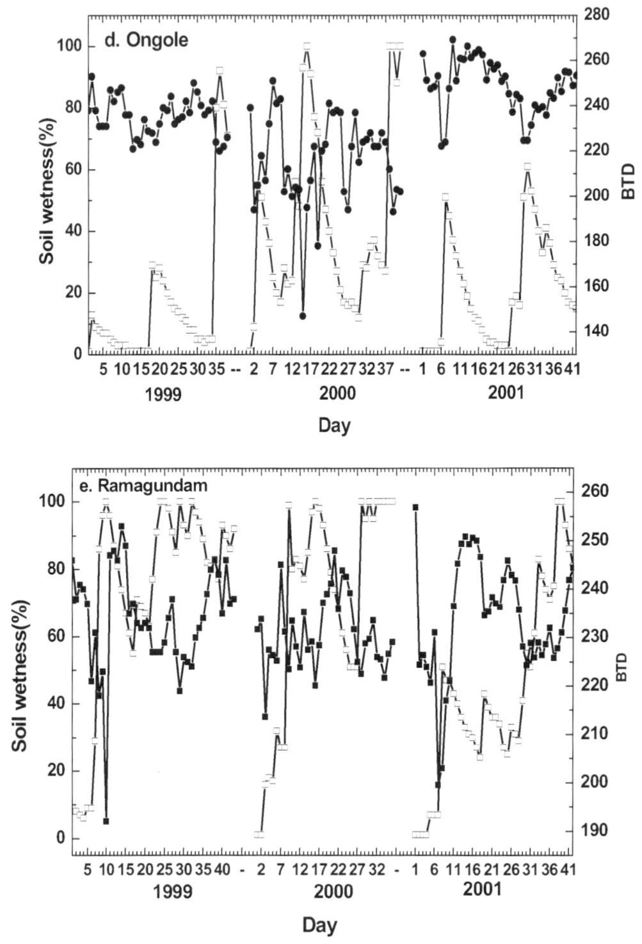

Fig. 13. NDVI and soil wetness – S-W monsoon season of 1999 to 2001

The correlation co-efficients between soil wetness and brightness temperature along with the regression expressions for the selected stations for south-west monsoon period (June to September) of 1999 to 2001 are given in Table.2. The degree of relation between soil wetness and BT is an inverse one i.e as the soil wetness increases, BT decreases and vice versa. Anantapur showed minimum correlation of -0.42 while Ongole recorded the maximum correlation of -0.63 compared to the remaining.

A regression model is developed by taking all the data points of BT and soil wetness from June 1999 to August 2001 to determine the soil wetness using BT over Andhra Pradesh(Figure 14). the correlation in this case is -0.61 which is at 0.01 level of significance and the regression expression is given in the table.

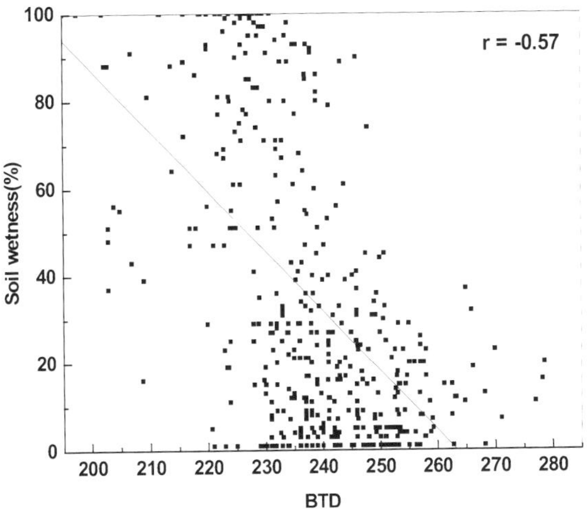

Fig. 14. Linear fit of Soil wetness and BTD - Andhra Pradesh

Test site	Correlation (r)	Regression for 1999 to 2001	
		SWT as independent Variable	BT as independent variable
Anantapur	-0.42	$BT=249.5 - 0.28(S_{WT})$	$S_{WT} = 174.7 - 0.63(BT)$
Kurnool	-0.55	$BT=247.9 - 0.2(S_{WT})$	$S_{WT} = 398.8 - 1.5(BT)$
Nellore	-0.52	$BT=220.6 - 0.68 (S_{WT})$	$S_{WT} = 96.0 - 0.39(BT)$
Ongole	-0.63	$BT=245.7 - 0.46 (S_{WT})$	$S_{WT} = 231.1 - 0.87(BT)$
Ramagundam	-0.43	$BT=242.7 - 0.14 (S_{WT})$	$S_{WT} = 376.0 - 1.3(BT)$
Andhra Pradesh	-0.56	$BT=246.4 - 0.24(S_{WT})$	$S_{WT} = 362.4 - 1.4(BT)$

Table 2. Correlation and regression statistics for BT with soil wetness (S_{WT}) for the test sites in Andhra Pradesh.

4. Conclusions

It is reported by the scientific community that there is a significant climate change and variability from which one has to learn lessons on how to tackle it. The importance of the rate of climate change can be understood by comparing the affected systems. Satellite-based observations provide a key source of data at global scales of the earth's environment, climate change, and the provision of climate services. However, observational data collected from satellite should be integrated with in-situ data. In many developing countries, a key constraint is the lack of professional and institutional capacity to make the best use of available information and knowledge for decision making. A particular difficulty is providing incentives to attract qualified staff to remote areas, far away from capital cities, where good decision making often is most critical. Local/national networks are useful for taking cognizance of local hotspots and making operational decisions on issues that relate to climate variability and change, while global change studies with satellite-based measurements are useful for international comparative assessments. In this context, it ought to strengthen the regional capacities towards decision making about the forthcoming frequent disasters due to climate change. As a result, decision making in the land surface resources could be improved through: i) developing information systems on areas that are prone to drought, and vulnerable to disasters, ii) long term understanding on land degradation because of deforestation and increased urbanization, iii) developing disaster preparedness in view of risk management, iv) development of early warning systems by utilizing the real time satellite data, to mitigate disasters like floods etc, v) assessment of crop failures during early, mid and late seasons, so as to prefix the mitigation measures and vi) educating communities about climate change and variability for better linkage of satellite data with the ground level ones for effective monitoring of drylands.

The use of satellite data into land resources decisions must be driven by the needs of the decision makers. Incorporation of satellite data by the land surface resources community requires an understanding of the particular decisions that are faced and the relevant timescales and skill needed to provide decision support. This can only be accomplished through close collaboration between operational land managers and decision makers. Researchers will not have a sense for whether this is true without understanding the needs of the user community which is achieved through close collaboration. Case studies of the incorporation of satellite land surface techniques, in combination with in-situ data, at international level are needed. In view of the above, the present chapter deals with the utilization of satellite data in i) understanding the vegetation dynamics, ii) vegetation response to climate, iii) connection with the agroclimatic indices and iv) underlying land surface processes. The established relations drawn from this chapter are of immense use in studying arid lands from the remote sensing point of view. Since the satellite indices (NDVI & BTD) are proven as the best variables, in accordance with the agrometeorological indices such as rainfall, soil moisture adequacy and soil wetness, they serve as inputs for policy makers. The relation of Brightness Temperature with soil wetness can be applicable in deciding the water supplement of a region. The crop phenological stages that can be studied by NDVI are of great use in assessing the crop health (fair / optical / poor). The response of NDVI to weather can guide in the designing of the agrometeorological advisories. Thus, the provision of remote sensing decisions over drylands will be strengthened by analysing the satellite data carefully can help in the improvising of systems where such satellite derived data can be used for multiple operations. The "lessons learned" from such studies provides critical guidance for enhancing the monitoring of the effects of climate change on land resources, through exploitation of satellite data. Another valuable impact is the enhancement and broadening of international research partnerships in order to encourage scientific exchange.

5. References

Aber, J.D (1997) Why don't we believe the models, *Bulletin of the Ecological Society of America*, *78*, 232-233.

Ahmed N.U (1995) Estimating soil moisture from 6.6GHz dual polarization and / or satellite derived vegetation index, *International Journal of Remote Sensing*, 16, 4, 687 – 708.

Al-Bakri, J.T & Suleiman, A.S (2004) NDVI response to rainfall in different ecological zones in Jordan, *International Journal of Remote Sensing*, 25(19), 3897-3912.

Alves, J.M.B., Repelli, C.A (1992) A variabilidade pluviométrica no setor norte do nordeste e os eventos El Niño-Oscilação Sul (ENOS), *Revista Brasileira de Meteorologia*, São Paulo, 7(2), 83-92.

Barbosa, H. A., Huete, A. R., & Baethgen, W. E., 2006: A 20-year study of NDVI variability over the Northeast region of Brazil, Journal of Arid Environ., 67, 288-307.

Barbosa, H.A. Interannual variability of vegetation dynamics in the semi-arid Northeast region of Brazil and its relationship to ENSO events. *8th International Conference on Southern Hemisphere Meteorology and Oceanography*, 24-28th April, 2006, Foz de Iguaçu, Brazil, 2006.

Barbosa, H.A. Vegetation dynamics over the Northeast region of Brazil and their connections with climate variability during the last two decades of the twentieth century. *Doctoral degree dissertation in Soil, Water, and Environmental Science at the University of Arizona.* Tucson, Arizona, 2004.

Barbosa, H.A. Spatial and temporal analysis of vegetation index derived from AVHRR-NOAA and rainfall over Northeastern Brazil during 1982-1985. *Master degree dissertation in Remote Sensing* [in Portuguese]. Divisão de Sensoriamento Remoto, Instituto Nacional de Pesquisas Espacias, São José dos Campos-SP, Brazil, 1998.

Boulanger, J.-P., J. Leloup, O. Penalba, M. Rusticucci, F. Lafon and W. Vargas (2005) Observed precipitation in the Paraná-Plata hydrological basin: long-term trends, extreme conditions and ENSO teleconnections, *Climate Dynamics* 24: 393–413.

Charney J. G., Quirk W. J., Chow S. H., Kornfield, J. J (1977) A comparative study of the effects of albedo change on drought in semi-arid regions, *Journal of Atmospheric Sciences*, 34:1366-1385.

Cihlar, J., St. Laurent, L & Dyer, J.A (1991) *The relation between normalized difference vegetation index and ecological variables, Remote sensing of Environment*, 35, 279-298.

Chimelli, A.B., Mutter, C.Z., Ropelewsk, C (2002) Climate fluctuations, demography and development: insights and opportunities for Northeast Brazil, *Journal of Inernational Affairs*, 56, 213-234.

Cramer, W., A. Bondeau, F.I. Woodward, I.C. Prentice, R.A. Betts, V. Brovkin, P.M. Cox, V. Fisher, J.A. Foley, A.D. Friend, C. Kucharik, M.R. Lomas, N. Ramankutty, S. Sitch, B. Smith, A. White, and C. Young-Molling (2001) Global response of terrestrial ecosystem structure and function to CO2 and climate change: results from six dynamic global vegetation models, *Global Change Biology*, *7*, 357-373.

Davenport M.L & Nicholson, S.E (1993) On the relation between rainfall and the Normalized Difference Vegetation Index for diverse vegetation types in East Africa, *International Journal of Remote Sensing*, 14, 2369.

FAO. World Reference Base for Soil Resources. *FAO, ISRIC and ISSS*, Rome, 1998.

Giannecchini M., Twine W., Vogel C (2007) Land-cover change and human–environment interactions in a rural cultural landscape in South Africa, *The Geographical Journal*, 173 (1), 26–42.

Hastenrath, S & Heller, L (1977) Dynamics of climate hazards in northeast Brazil. *Quart. J. Roy. Meteor. Soc.*, 103, 77-92.

Higgins G M & Kassam A H (1981), *The FAO* agro-ecological zone approach to Determination of land potential, Pedelogie (Germany) XXXI 147.

Barbosa, H. A., Mesquita, M. d. S. & Lakshmi Kumar, T.V., 2011: *What do vegetation indices tell us about the dynamics of Amazon evergreen forests?*,Geophysical Research Abstracts, 13, EGU 2011 - 12894.

IBGE - Instituto Brasileiro de Geografia e Estatística. Recursos naturais e meio ambiente: uma visão do Brasil. *Departamento de Recursos Naturais e Estudos Ambientais*, 154, 1993.

Lakshmi Kumar, T.V., Humberto Barbosa, K. Koteswara Rao & Emily Prabha Jothi, 2011, *Studies on the freuency of extreme weather events over India*, Journal of Agriculture Science & Technology, Accepted for publication.

MEA (Millennium Ecosystem Assessment). Ecosystem and Human Well-Being: Desertification Synthesis: Island. Press World Resources Institute, Washington, DC, 2005b.

Namias, J (1972) Influence of northern hemisphere general circulation on drought in northeast Brazil, *Tellus*, 4, 336-342.

Nobre, P., Shukla, J (1996) Variations of sea surface temperature, wind stress and rainfall over the tropical Atlantic and South America, *Journal of Climate*, 9, 2464-2479.

Nicholson S (2005) On the question of the "recovery" of the rains in the West African Sahel, *Journal of Arid Environments*,63, 615–641.

Nicholson, S. E & Farrar, T. J (1994) The influence of soil type and the relationships between NDVI, rainfall and soil moisture in semiarid Botswana, *Remote Sensing. Environment.* 50, 107-120.

Pathak P.N., Rao B.M and Potdar M.B (1993) Large area soil moisture estimation using INSAT - VHRR infrared data, *Indian Journal of Radio & Space Physics*, 22, 156-164.

Rao B M., Thapliyal PK., Manikiam B., Avinash Dwivedi (2001) Large scale soil moisture estimation using microwave radiometer data, *Journal of Agrometeorology*, 3(1&2), 179 - 187.

Reynolds, J. F. et al. Natural and human dimensions of land degradation in drylands: causes and consequences. In: Canadell, D. P. L. F. P. (ed.), Terrestrial ecosystems in a changing world. *Springer*, 247-258, 2007.

Rosenzweig, C., Parry, M.L (1994) Potential impacts of climate change on world food supply. *Nature*, 367,133-138.

S. Nagai, K.Ichii., H.Morimoto (2007) Interannual variations in vegetation activities and climate variability caused by ENSO in tropical rainforests, 28(6)1285-1297.

Sarma, A.A.L.N & Lakshmi Kumar, T.V (2006) Studies on crop growing period and DVI in relation to water balance components, *Indian Journal of Radio & Space Physics*, 35, 424-434.

Sarma A.A.L.N & T.V. Lakshmi Kumar, 10/4/375, ISRO RESPOND REPORT, 2005.

Sarma A.A.L.N & T.V. Lakshmi Kumar (2007), Rainfall and soil wetness response to ENSO and soil wetness estimation from remote sensing data, *Journal of Indian Geophysical Union*, 11(2), 91-100.

Sarma A.A.L.N & T.V. Lakshmi Kumar (2006) Studies on agroclimatic elements and soil wetness estimation from MSMR data, *Journal of Agrometeorology*, 8 (1), 19-27.

Sarma, A.A.L.N., Padma Kumari, B., Srinivas S (1999) Studies on hydrological extremes - ENSO signal, IAHS publication, No.255, 73-80.

Sarma A.A.L.N., Lakshmi Kumar, T.V., Koteswara Rao, K (2010). Extreme value analysis of summer monsoon rainfall over Andhra Pradesh, *Indian Journal of Radio & Space Physics*, 39, 32-38.

Thapliyal P.K., Rao B.M., Pal P.K and Das H P, 2003 Potential of IRS-P4 microwave radiometer data for soil moisture estimation over India, *Mausam*, 54, 277 - 286.

Thoronthwaite C W & Mather J, The water balance , Publ Clim USA, 8, 1955, 1.

Verdin, J. P., C. Funk, G. Senay, and R. Choularton (2005) Climate science and famine early warning, *Philos. Trans. R. Soc. London*, 360(1463), 2155–2168.

White, M. A., Thornton, O. E., and Running, S. W (1983) A continental phenology model for monitoring vegetation responses to interannual climate variability. *Global Biogeochem. Cycles*, 11, 217-234.

21

Using Fuzzy Cognitive Mapping in Environmental Decision Making and Management: A Methodological Primer and an Application

Elpiniki Papageorgiou[1] and Areti Kontogianni[2]
[1]*Technological Educational Institution of Lamia, Department of Informatics and Computer Technology, Lamia,*
[2]*University of Aegean, Department of Marine Sciences, Mytilini, Lesvos, Greece*

1. Introduction

Widespread concerns over the integrity of natural ecosystems worldwide have initiated numerous attempts at developing new tools of monitoring present conditions, assessing future risks and visualizing alternative futures. Reports on the 'state of the world' abound and so do policy proposals and sustainability strategies. Amidst this plenty of ideas, our ability to reverse the trend and secure a safe, minimum stock of valuable natural capital seems counterproductive. A better understanding of ecosystem dynamics at both the quantitative (biochemical cycling) as well as the qualitative (ecological structure of food webs) levels, without artificial divisions between them, is needed. We also need to understand better the institutional failures leading to a growing number of 'tragedies of the commons'.

To tackle these challenges appropriately, current environmental management strategies need to 'navigate' through an apparent tension: On the one hand they must meet the demand for scientific knowledge-based policy, expressed under the motto 'science speaks to policy'. On the other hand, the very same strategies urge for stakeholder involvement and sponsor initiatives to elicit lay-people attitudes, beliefs and visions for the future. This tension seems to reflect the ever lasting stand-off of bottom up and top down approaches.

The motivation for this chapter comes from the authors' uneasiness with the present methodological arsenal in the domain of environmental stakeholder analysis. Previous research on non-market valuation of environmental assets has shown the importance of complementing the neoclassical microeconomic framework of choice in stated preference surveys with qualitative - both ex ante and ex post – analysis of individual mental processes, perceptions and beliefs (Kontogianni et al, 2001, 2005, 2008). Especially applications of contingent valuation have benefited from in-depth interviews and focus groups conducted ex ante in order for the researcher to understand the cultural, social and psychological background of choices elicited through structured interviews. (Desvousges & Smith 1988, Brouwer, 1999). In spite though of the importance of stakeholder qualitative analysis in stated preferences valuation techniques, we still lack a coherent, standardized approach to analyze environmental perceptions and beliefs. The need to fill the gap becomes apparent when we recognize the fact that the way non-experts articulate complex relationships, such as those governing marine ecosystem functions, have their own special weight in influencing policy design and implementation: they transcend the fact/value divide (senso Putnam 1985) and offer valuable insights on the ways cause and effect relationships in nature are perceived (Karageorgis et al., 2006).

Fuzzy Cognitive Mapping (FCM) was thus selected as a suitable method for semi-qualitative analysis to achieve our research goal. In this chapter we introduce the reader to the concept of Fuzzy Cognitive Maps (FCMs) and their theoretical background. In section 2 we summarize the state-of-the-art in qualitative, stakeholder analysis for environmental management. We then present the structure of FCMs (section 3) and the analytical use of graph theory in defining relevant indices (section 4). We proceed with the development of FCMs (section 5) and the FMC inference and simulation processes (section 6). After presenting the theoretical structure, the practical steps involved in the design and implementation of a FCM exercise are codified (section 7). We then illustrate the concepts discussed so far with a practical application implemented by the authors in The Black Sea (section 8) before we summarize and conclude in section 9.

2. The many facets of stakeholder analysis in environmental management

Integrated approaches to environmental planning with proper stakeholder involvement offer a possible way forward. Such an approach needs to facilitate communication within multidisciplinary research teams; it needs to recognize the functional continuity from watersheds to the coasts to the open sea, thereby helping to locate the scale of intervention less on the base of traditional jurisdictions and more towards appropriate ecosystem scales. Last but not least, it must encompass participatory management schemes which promise a substantive change in the exploitation of local knowledge. By enhancing stakeholder involvement, participatory management strengthens policy relevance, diminishes uncertainties, improves monitoring and raises enforcement rates (NRC 1996, OECD 2005).

Participatory (or deliberative) approaches to environmental management are usually grouped under the general term of stakeholder analysis (Grimble and Wellard 1997, Bryson 2004, Reed et al., 2009). Stakeholder analysis in turn can be divided into what we opt to call macro-stakeholder and micro-stakeholder analysis. The former category includes all those qualitative approaches that refer to the interaction of social groups and their dynamics: social networks analysis (Scott, 2000, Carrington et al 2005, Turnpenny et al., 2005), analysis of conflicts (Howard, 1989, Hjortso et al. 2005, 2010; Stoney & Winstanley, 2001), and actor analysis (Hermans, 2008). The latter category refers to qualitative or semi-quantitative approaches, which explore individual perceptions, values and attitudes. These include: fuzzy cognitive mapping of social perceptions and values (Bots et al., 2000, Stone 2002), perceptions mapping (Bots, 2007), mind mapping (Buzan, 1993), concept mapping (Novak, 1993), focus groups and in-depth interviews.

Approaches in stakeholder analysis as described above share some common characteristics: they are 'eclectic but pragmatic' approaches with varying degree of sophistication, requiring in average a low in-depth academic investigation, but able to manipulate a vast quantity of soft information. Their strength lies primarily with thinking about problems than solving them. The present paper aims at contributing to a refinement of participatory management tools by applying fuzzy cognitive mapping (FCM) to the exploitation of local knowledge. FCM fits the requirements stated above better than any of the other conceptual modelling techniques analyzed here. Most other methods are either too difficult for the type of stakeholders we are aiming for, or take too much time. Yet, FCMs have their own set of specific disadvantages as Kok (2009) overview them.

In the case of stakeholders' analysis for ecological modeling and environmental management, the FCMs have found a good number of applications. At first, Hobbs et al., 2002 applied FCM as a tool to define management objectives for complex ecosystems (Hobbs et al., 2002). Next, Ozesmi and Ozesmi (2003, 2004) proposed a multi-step FCM and participatory approach of Stakeholder Group Analysis in Uluabat Lake, Turkey, for ecosystem observation. The multi-step fuzzy cognitive mapping approach analyzes how people perceive a system, and compare and contrasts the perceptions of different people or groups of stakeholders (Ozesmi & Ozesmi, 2004).

After the pioneering work of Ozesmi & Ozesmi (2003, 2004), in environmental and ecological management topics, other researchers followed with more implementations of FCMs in this area. FCMs have been employed in a number of studies including a FCM for rapid stakeholder and conflict assessment for natural resource management (Hjortsø et al. 2005; Robson & Kant, 2007), a FCM for modelling a generic shallow lake ecosystem by augmenting the individual cognitive maps (Tan & Ozesmi, 2006), FCM for predicting the effects of perturbations on ecological communities, thus to control on the fledging rate of an endangered New Zealand bird (Ramsey & Vetman, 2005), FCM for assessing local knowledge use in agroforestry management (Isaac et al., 2009), FCM for modelling of interactions among sustainability components of an agro-ecosystem using local knowledge (Rajaran & Das, 2009), FCM for predicting modelling a New Zealand dryland ecosystem to anticipate pest management outcomes (Ramsey & Norbury, 2009), FCM for cotton yield management in precision agriculture (Papageorgiou et al., 2009, 2010).

3. The structure of Fuzzy Cognitive Maps

Fuzzy Cognitive Mapping methodology is a symbolic representation for the description and modeling of complex systems. Fuzzy Cognitive Maps (FCMs) describe different aspects of the behavior of a complex system in terms of concepts. Each concept represents a state or a characteristic of the system and interacts with each other showing the dynamics of the system. FCMs have been introduced by Kosko, (1986) as signed directed graphs for representing causal reasoning and computational inference processing, exploiting a symbolic representation for the description and modeling of a system.

In fact, FCM could be regarded as a combination of Fuzzy Logic and Neural Networks (Kosko, 1992). Graphically, FCM seems to be an oriented graph with feedback, consisting of nodes and weighted arcs. Nodes of the graph stand for the concepts that are used to describe the behavior of the system, connected by signed and weighted arcs representing the causal relationships that exist between the concepts (see Figure 1). It must be mentioned that all the values in the graph are fuzzy, so concepts take values in the range between [0,1] and the weights of the arcs are in the interval [-1,1]. Observing this graphical representation it becomes clear which concept influences other concepts by showing the interconnections between them. Moreover, FCM allows updating the construction of the graph, such as the adding or deleting of an interconnection or a concept. FCMs are used to represent both qualitative and quantitative data. The construction of a FCM requires the input of human experience and knowledge on the system under consideration. Thus, FCMs integrate the accumulated experience and knowledge concerning the underlying causal relationships amongst factors, characteristics, and components that constitute the system.

A FCM consists of nodes or concepts, C_i, i = 1...N, where N is the total number of concepts. Each interconnection between two concepts C_i and C_j has a weight, a directed edge W_{ij}, which is similar to the strength of the causal links between C_i and C_j. W_{ij} from concept C_i to concept C_j measures how much C_i causes C_j. In simple FCMs, directional influences take on trivalent values {-1; 0; +1}, where -1 indicates a negative relationship, 0 no causal relation, and +1 a positive relationship. In general, Wij indicates whether the relationship between the concepts is directed or inverse. The direction of causality indicates whether the concept C_i causes the concept C_j or vice versa. Thus, there are three types of weights:

- $W_{ij} > 0$ indicates a positive causality between concepts Ci and Cj. That is, the increase (decrease) in the value of Ci leads to the increase(decrease) on the value of Cj,
- $W_{ij} < 0$ indicates a inverse (negative) causality between concepts Ci and Cj. That is, the increase (decrease) in the value of Ci leads to the decrease (increase) on the value of Cj
- $W_{ij} = 0$ indicates no causality between Ci and Cj.

It is important to note that $W_{ij} \neq W_{ji}$ in that causal relationship are not necessarily reversible. In Figure 1, an example FCM representation of the public health system is illustrated which has seven generic vertices (C_1 to C_7) and the weights (weighted edges) showing the relationships between concepts.

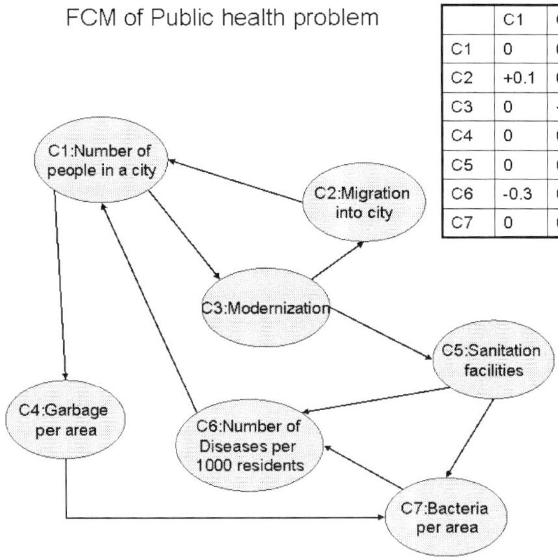

	C1	C2	C3	C4	C5	C6	C7
C1	0	0	+0.6	+0.9	0	0	0
C2	+0.1	0	0	0	0	0	0
C3	0	+0.7	0	0	+0.9	0	0
C4	0	0	0	0	0	0	+0.9
C5	0	0	0	0	0	-0.9	-0.9
C6	-0.3	0	0	0	0	0	0
C7	0	0	0	0	0	+0.8	0

Fig. 1. Example of FCM model of the public health system: (a) FCM graph, and (b) connection matrix (adapted from Montazemi & Conrath, 1986).

4. Data analysis using graph theory indices

Graph theory methods help analyzing the structural properties of cognitive maps (Ozesmi & Ozesmi, 2003). During the interviews, participants develop a FCM of the critical variables by drawing and circling the considerations they believe are important in relation to the topic under consideration. Then the main factors are defined and coded as concepts. The fuzzy directional arrows to one or more preceding factors are represented by fuzzy linguistic weights (see section 6), which after defuzzification produce a representative numerical weight. Using the defuzzification method of Centre of Gravity (COG) (Zadeh, 1976) a numerical weight is produced for each connection between concepts. CoG is computed from the following equation:

$$COG(B) = \frac{\sum_{q=1}^{Nq} \mu_B(y_q) y_q}{\sum_{q=1}^{Nq} \mu_B(y_q)} \quad (1)$$

where Nq is the number of quantization used to discretize the membership function $\mu_B(y_q)$ of the fuzzy output B.
Thus they make-up a continuity map whereby concept C_i is preceded by concept C_j indicating a cause-and-effect relationship. Each individual map is analyzed in relation to the number of concepts, connections, connection-to-concept ratio, and density (calculated by dividing the number of connections in the map by the square value of concepts). To allow for identification of key criteria within the process of cognitive mapping, an analysis of domain and centrality is also conducted. The complexity level of each individual concept is revealed through a number of structural measures of cognitive maps, e.g. the centrality index borrowed from social networks analysis.
According to graph theory an effective way to better understand the structure of complex cognitive maps is condensing them. *Condensation* is achieved by replacing subgraphs (consisting of a group of variables connected with lines) with a single unit (Harary et al. 1965). Once the individual cognitive maps are drawn, they are qualitatively aggregated using clustering concepts to produce a condensed map named the *collective* FCM. Due to the complexity in FCM graphs (as the number of nodes and connections is often very large) the most central variables with their weighted connections are usually illustrated.
Analyzing the structure of cognitive maps is to look how connected or sparse the maps are. This is expressed by an index of connectivity, called *density* of a cognitive map (D). The density is equal to the number of connections divided by the maximum number of connections possible between N variables, thus N^2. If the density of a map is high then the interviewee sees a large number of causal relationships among the variables.
The structure of a cognitive map apart from number of variables and connections can best be analyzed by finding *transmitter variables* (forcing functions, givens, tails, independent variables), *receiver variables* (utility variables, ends, heads, dependent variables) and *ordinary variables* (Bougon et al. 1977; Eden et al. 1992; Harary et al. 1965). These variables are defined by their *outdegree* and *indegree*. Outdegree is the row sum of absolute values of a variable in the adjacency matrix and shows the cumulative strengths of connections (E_{ij}). It is a measure of

how much a given variable influences other variables. Indegree is the column sum of absolute values of a variable and shows the cumulative strength of variables entering the unit.
Transmitter variables are units whose outdegree is positive and their indegree is 0. *Receiver variables* are units whose outdegree is 0 and their indegree is positive. Other variables, which have both non-zero outdegree and indegree, are *ordinary variables* (means) (Eden et al., 1992, Ozesmi & Ozesmi 2004). This type of variables reveals how people think about the causal relationships. For instance, if someone views a variable as a transmitter, this means that he perceives of the relative causal relationship as forcing function, which cannot be controlled by any other variables. In contrast, a receiver variable is seen as not affecting any of the other variables in the system. The total number of receiver variables in a cognitive map can be considered an index of its complexity. Larger number of receiver variables indicates that the cognitive map considers many outcomes and implications that are a result of the system (Eden, 1992). Many transmitter units show the "flatness" of a cognitive map where causal arguments are not well elaborated (Eden et al. 1992).
Centrality is the most important measure for map complexity, borrowed from social networks analysis, and is the summation of variable's indegree and outdegree (Bougon et al., 1977; Eden et al., 1992). Actually the centrality shows how connected the variable is to other variables and what the cumulative strength of these connections is. Another structural measure of cognitive maps is the *hierarchy index* (h), which is a function of the out-degrees and number of variables in a given map and represents the type of system as fully hierarchical, or democratic (see Ozesmi & Ozesmi, 2004, pp. 50–51 for formulas).

5. Development of Fuzzy Cognitive Maps

The design of a fuzzy cognitive map is a process that heavily relies on the input from experts and/or stakeholders (Hobbs et al., 2002). This methodology extracts the knowledge from the stakeholders and exploits their experience of the system's model and behaviour. FCM is fairly simple and easy to understand for the participants, which opens up the possibility for involving lay people as well as planners, managers and experts (Isaac et al. 2009). Even though the cognitive nature of a FCM makes it inevitably a subjective representation of the system, Tan & Özesmi (2006) emphasize that the model is not arbitrary as it is built carefully and reflexively with stakeholders (in groups or individually).
According to the FCM development process, at the first step of the construction process, the number and kind of concepts are determined by a group of experts and/or system stakeholders that comprise the FCM model. Then, a domain expert and/or stakeholder describe each interconnection either with an if-then rule that infers a fuzzy linguistic variable from a determined set or with a direct fuzzy linguistic weight, which associates the relationship between the two concepts and determines the grade of causality between the two concepts.
For example, someone can assign the strength of influence of concept C_j on concept C_i using the following form: "The strength of influence of concept C_j on concept C_i is T{*influence*}" where the variable *T*{*influence*} declares the causal inter-relationships among concepts (i.e. the degree of influence from concept C_j to C_i). Its term set *T*{*influence*} is suggested to comprise thirteen variables and takes values in the universe U=[-1, 1]. Using thirteen linguistic variables, an expert can describe in detail the influence of one concept on another and can discern between different degrees of influence. The thirteen variables used here are: *T(influence)* =

{negatively very very strong, negatively very strong, negatively strong, negatively medium, negatively weak, negatively very weak, zero, positively very weak, positively weak, positively medium, positively strong, positively very strong, positively very very strong}.
The corresponding membership functions for these terms are shown in Fig. 2 and they are μ_{nvvs}, μ_{nvs}, μ_{ns}, μ_{nm}, μ_{nw}, μ_{nvw}, μ_z, μ_{pvw}, μ_{pw}, μ_{pm}, μ_{ps} μ_{pvs} and μ_{pvvs}.

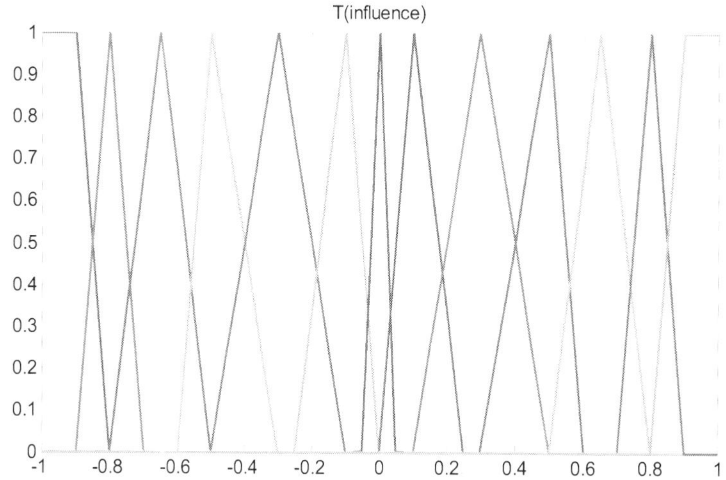

Fig. 2. The thirteen membership functions describing *T(influence)*

The inference of the rule *T{influence}* means that the linguistic weight y (w_{ij}) *is* μ_B, where μ_B is a linguistic variable from the set *T*. Also, experts and/or system stakeholders can directly assign the fuzzy linguistic weight y (that describes the strength of connection between concepts C_i and C_j) with no use of fuzzy rules. For example, someone can assign the strength of the connection between concepts C_i and C_j as follows: The influence from concept C_i to concept C_j is *positively very high*.
Finally, the linguistic variables ***D*** from the set *T(influence)* - proposed by the experts for each interconnection - are aggregated using the SUM method and so an overall linguistic weight is produced (Papageorgiou & Stylios, 2008). Finally, the Center of Gravity (CoG) defuzzification method (Zadeh, 1986) is used for the transformation of the linguistic weight to a numerical value within the range [-1, 1]. This methodology has the advantage that experts are not required to assign directly numerical values to causality relationships, but rather to describe qualitatively the degree of causality among the concepts. Thus, an initial matrix $W^{initial} = [W_{ij}]$, i, j = 1,...,N, with $W_{ii} = 0$, i = 1,...,N, is obtained. Using the initial concept values, A_i, which are also provided by the experts, the matrix $W^{initial}$ is used for the determination of the steady state of the FCM, through the application of the rule of Eq. (2) or (5).

6. The FCM inference and simulation processes

Using artificial intelligent techniques, the dynamics of a fuzzy cognitive map can be traced analytically through a specific inference and simulation process. Each one of the C_j concepts can take values in the unit interval [0,1], also called the 'activation level'. The activation level can be interpreted as relative abundance (Hobbs et al. 2002). More rigorously, the activation level can represent membership in fuzzy set describing linguistic measures of relative abundance (e.g. low, average, high) (Kosko, 1986).

Values of the concept C_i in time t are represented by the state vector $A_i(t)$ while the state of the whole fuzzy cognitive map can be described by the state vector $A(t) = [A_1(t),...,A_n(t)]$ representing a point within a fuzzy hypercube that the system achieves at a certain point. The whole system with an input vector $A(0)$ describes a time trace within a multidimensional space I^n that can gradually converge to an equilibrium point, or a chaotic point or a periodic attractor within a fuzzy hypercube. To which attractor the system will converge depends on the value of the input vector $A(0)$.

The value A_i of each concept C_i in a moment $t+1$ is calculated by the sum of the previous value of A_i in a precedent moment t with the product of the value A_j of the cause node C_j in precedent moment t and the value of the cause-effect link w_{ij}. The mathematical representation of the inference process of a fuzzy cognitive map has the following matrix form (Papageorgiou & Stylios, 2008):

$$A^{(k)} = f(A^{(k-1)} + \sum A^{(k-1)} \cdot W) \tag{2}$$

Thus, the value A_j for each concept C_j is calculated by:

$$A_i^{(k+1)} = f\left(A_i^{(k)} + \sum_{\substack{j \neq i \\ j=1}}^{N} A_j^{(k)} \cdot w_{ji} \right) \tag{3}$$

where $A_i^{(k+1)}$ is the value of concept C_i at simulation step $k+1$, $A_j^{(k)}$ is the value of concept C_j at step k, w_{ji} is the weight of the interconnection between concept C_j and concept C_i and f is a threshold (activation) function (Bueno & Salmeron, 2008). Sigmoid threshold function gives values of concepts in the range [0,1] and its mathematical type is:

$$f(x) = \frac{1}{1+e^{-mx}} \tag{4}$$

where m is a real positive number and x is the value $A_i^{(k)}$ on the equilibrium point. A concept is turned on or activated by making its vector element 1 or 0 in (0,1). The sigmoid threshold function is used to reduce unbounded weighted sum to a certain range, which hinders quantitative analysis, but allows for qualitative comparisons between concepts (Bueno & Salmeron, 2008).

A modified FCM inference algorithm, which updates the common FCM simulation process as initially suggested by Kosko (1986) can be used to avoid the conflicts that emerge in cases where the initial values of concepts are 0 or 0.5, thus overcoming the limitation present by the sigmoid threshold function. This rescaled algorithm is implemented especially for the cases where there is no information about a certain concept/state or the expert/stakeholder cannot describe efficiently the initial state of a variable (Papageorgiou et al., 2010, Papageorgiou, 2011). Thus, the eq. (2) is transformed to the eq. (5).

$$A_i(k+1) = f((2A_i(k)-1) + \sum_{\substack{j\neq i \\ j=1}}^{N} (2A_j(k)-1)\cdot E_{ji}) \qquad (5)$$

The FCM simulation process is initialized through assigning a value between 0 and +1 to the activation level of each of the nodes of the map, based on experts/stakeholder opinion for the current state; then the concepts are free to interact. The value of zero suggests that a given concept is not present in the system at a particular iteration, whereas the value of one indicates that a given concept is present to its maximum degree. Other values correspond to intermediate levels of activation. The activation level of each concept depends on its value at the preceding iteration as well as on the preceding values of all concepts that exert influence on it through non-zero relationships. The simulation, which with regard to its content is mainly qualitative, is not intended to produce exact quantitative values. It aims at identifying the pattern of system's behaviour via the achieved values of the concepts of the FCM, which are progressively formed according to given considerations.

After defining all variables and necessary values, as well as the relationships between them, the simulation is carried out by use of the simulator consisting of the following five steps:

Step 1. Definition of the initial vector $\boldsymbol{A}$ that corresponds to the concepts identified by suggestions and available knowledge.

Step 2. Multiply the initial vector $\boldsymbol{A}$ and the matrix $\boldsymbol{W}$ defined through equation (2) or (5)

Step 3. The resultant vector $\boldsymbol{A}$ at time step k is updating using eqs. (2) or (5) and (4).

Step 4. This new vector $\boldsymbol{A}^k$ is considered as an initial vector in the next iteration.

Step 5. Steps 2–4 are repeated until $\boldsymbol{A}^k - \boldsymbol{A}^{k-1} \leq e = 0.001$ (where e is a residual describing the minimum error difference among the subsequent concepts) or $\boldsymbol{A}^k = \boldsymbol{A}^{k-1}$. Thus A_f $= \boldsymbol{A}^k$.

In each step of the cycling the values of concepts change according to the equation (2) or (5). This interaction between concepts continues until: i) a fixed equilibrium is reached, ii) a limited cycle is reached or iii) a chaotic behavior is exhibited. Actually, in most cases, the iteration stops when a limit vector is reached, i.e., when $\boldsymbol{A}^k = \boldsymbol{A}^{k-1}$ or when $\boldsymbol{A}^k - \boldsymbol{A}^{k-1} \leq e$; where e is a residual, whose value depends on the application type (and in most applications is equal to 0.001). Thus, a final vector A_f is obtained.

In the previous analysis, all type of information has numerical values. FCM allows us to perform qualitative simulations and experiment with a dynamic model. Simulations allow for analysis of several aspects of FCMs, such as concepts activation levels at the final state (if there are any) and changes/trends in the activation levels throughout the simulation concerning either all concepts or a subset of concepts that is of interest to the user, and discovery of cycles (intervals, concepts activation levels within the cycle). This type of analysis allows investigating "what-if" scenarios by performing simulations of a given model from different initial state vectors. Once an FCM has been subjected to an initial stimulus, it is possible to gain insight into a system's behaviour by studying the resulting stable state or cycle of states. Simulations offer description of dynamic behaviour of the system that can be used to support decision-making or predictions about its future states (Stach et al., 2010).

7. Steps in designing and implementing fuzzy cognitive mapping

As with many other interview techniques, it is helpful to produce systematic guidelines describing the single steps of FCM before starting with the interviewing. These interview guidelines should function as a guidance/inspiration for how to conduct the interviews, and how to create FCMs over the case study areas. In this section we summarize the practical steps needed to design and conduct a FCM exercise. At first, how to draw a FCM must be explained to the interviewee(s)/stakeholder(s) using a cognitive map and its related FCM as an example. Once the interviewees understood the process of constructing a fuzzy cognitive map, then they are able to draw their own map of the issue under investigation following the steps below:

Step 1: Identification of factors.

Based on the guidelines, each one of the interviewee is asked to identify the main factors which come to his/her mind when he/she is asked about the topic been investigated, e.g. the future environmental risks in the Black Sea, seen as a system where humans, marine animals and plants are all living together. After the identification of the main factors affecting the environmental topic under investigation, each stakeholder is asked to describe the existence and type of the causal relationships among these factors and then assesses the strength of these causal relationships using a predetermined scale, capable to describe any kind of relationship between two factors, positive and negative. Thus, a FCM from each interviewee is established presenting the main factors/variables and the relationships among them and illustrating the individual's perceptions about the topic under investigation.

Step 2: Clustering of individual issues in more general concepts.

After the individual perceptions are elicited by interviews, a number of individual maps are produced. It is essential that the original concepts-variables, as described by lay people from interview, be clustered in more generic or more specific concepts, because most of them present the same meaning with a different word. Using experts' judgement, the importance of the original factors is discussed and they are then clustered. This can also be done through the construction of an ontological tree. This process of condensation enables aggregation of variables into high-level concepts, which then feed into the construction of the collective FCM.

Step 3: Estimation of causal link strengths in collective FCMs

The individual maps are then turned into a representative, collective map. To achieve this, all the suggested strength relations by lay people are transferred into linguistic variables using the aggregation method of SUM to obtain an overall linguistic weight. Following this, defuzzification turns linguistic weights into numerical weights in the range of [-1,1]. This condensed map is analyzed using the established indicators of out-degree, in-degree, centrality, density, hierarchy as well as the transmitter, receiver, and ordinary variables. Through this analysis the collective map is explained demonstrating its usefulness for identifying policies vis-à-vis individual FCMs.

Step 4: FCM simulations

Next a number of simulations are performed using the inference process given by equations (2) and (5). The calculated output of the FCM model shows how the system reacts under the assumptions given by the stakeholders or related users. Usually, the calculated output is different from the expected one, thus presenting a potential added value of Fuzzy Cognitive Map as a decision support tool.

8. An application of FCM in the Black Sea

In order to illustrate in brief the methods described so far, we present a FCM exercise designed and implemented in the Northern Black Sea with Ukrainian stakeholders (for details see Kontogianni et al., forthcoming). In this application we were interested in investigating how the citizens perceive the future prospects and risks of the Black Sea marine environment; creating and analyzing their FCMs this can be achieved. We employed 29 in-depth lay people interviews (see Appendix A). Based on specific guidelines, each interviewee was asked to identify at first the main factors which come to his/her mind when he/she is asked about the Black Sea as a system where humans, marine animals and plants are all living together. During the interviews participants developed thus a FCM of the critical variables (important considerations) by drawing and circling the considerations they believe are important for environmental health of marine state ecosystem in the Black Sea area.

After the identification of the main factors affecting the environmental health of the marine ecosystem in the Black Sea, each stakeholder was asked to describe the existence and type of the causal relationships among these factors and then, the strength of the causal relationships-influences that may exist between these factors. This phase was implemented 13 grades scale, numbering from -6 to + 6, capable to describe any kind of relationship between two factors, positive and negative (see Table 1).

Thus, a FCM from each interviewee was established presenting the main factors/variables and the relationships among them illustrating the individual's perceptions about the future prospects and the risks about the ecological health of the marine environment in the Black Sea. Figure 3 illustrates the produced FCM defined by an individual/stakeholder from Ukraine for further assessment.

The initial number of important factors identified by stakeholders was 52. Since it was decided to produce a collective FCM providing detail for future risks-related issues we limited the number of factors having the same meaning through clustering. Using marine experts' judgement, the importance of the original 52 factors was discussed and then clustered in a total of 26 concepts (Table A in Appendix). The mean number of variables in the individual cognitive maps of the Black Sea ecosystem drawn by the 29 respondents was 7.86 ± 1.7, with 11± 6.513 connections on average between the variables that they defined. There were a total of 26 variables with 145 connections in the collective cognitive map obtained by clustering and augmented the 29 individual FCMs.

The process of condensation enabled aggregation of variables into high-level concepts, which then feed into the construction of a collective FCM for Ukrainian stakeholders. The collective FCM (consisting of 26 concepts and 145 relationships among concepts) is thus obtained (see Figure 4 developed in pajek software [http://vlado.fmf.uni-lj.si/pub/networks/pajek/]).

The collective FCM was then coded as adjacency matrix $E=[e_{ij}]$ and its structure was analyzed using the indices derived from graph theory (see section 4). Due to the complexity of the collective FCM graph (as the number of nodes and connections is very large) Figure 5 illustrates the most central variables with their weighted connections.

It is observed from graph indices calculations that the density of collective FCM is high and a mentioned complexity is present. A relatively high complexity is considered in the cases where the receiver variables are more than the transmitter variables, and in our case, the complexity is equal to 1.5 (complexity>1 means relatively high complexity). The most frequently mentioned (> 3 times) variables that were recurrent in the 29 fuzzy cognitive maps are: 'Coastal Development', 'Biodiversity', 'Tourism', and 'Municipal Solid Waste'.

Strength connection by lay people	Sign and Strength of relationship (Linguistic weight)	Interpreted crisp weight
-6	Negatively very very strong	-1
-5	Negatively very strong	-0.9
-4	Negatively strong	-0.75
-3	Negatively medium	-0.5
-2	Negatively weak	-0.3
-1	Negatively very weak	-0.1
0	Zero	0
1	Positively very weak	0.1
2	Positively weak	0.3
3	Positively medium	0.5
4	Positively strong	0.75
5	Positively very strong	0.9
6	Positively very very strong	1

Table 1. Interpretation of lay people's strength connections among concepts to crisp weights in the range [-1,1].

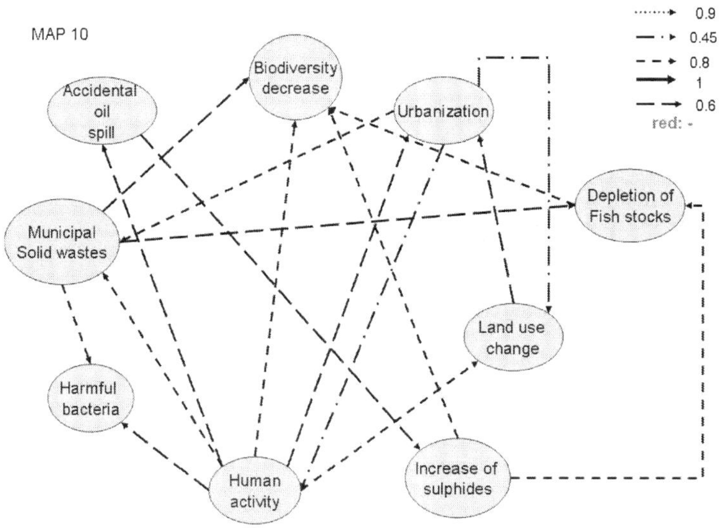

Fig. 3. The individual FCM defined by an individual/stakeholder from Ukraine.

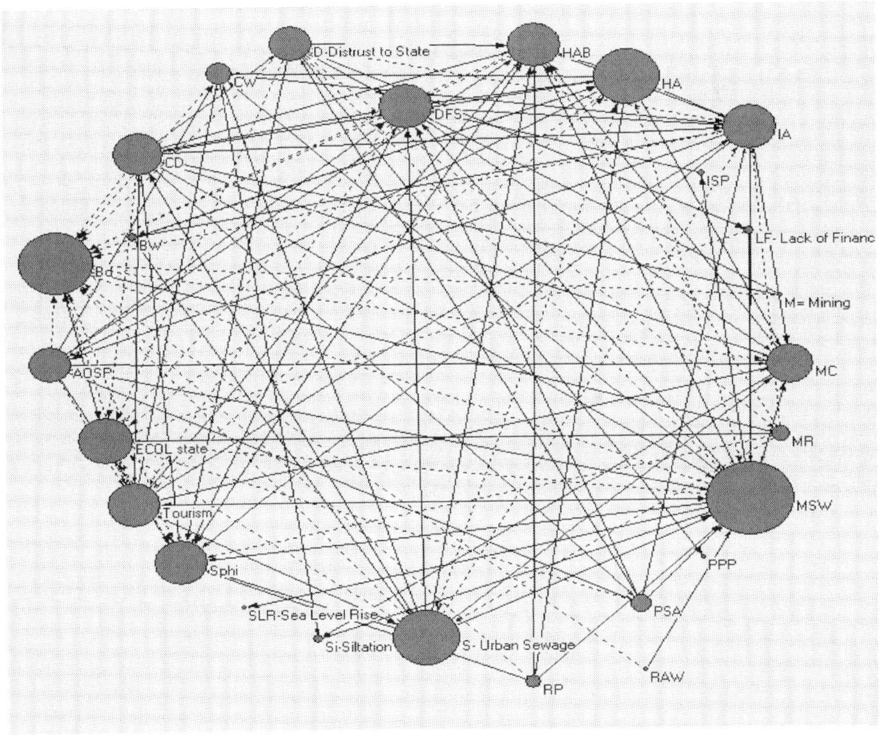

Fig. 4. The collective fuzzy cognitive map of Ukrainian stakeholders.

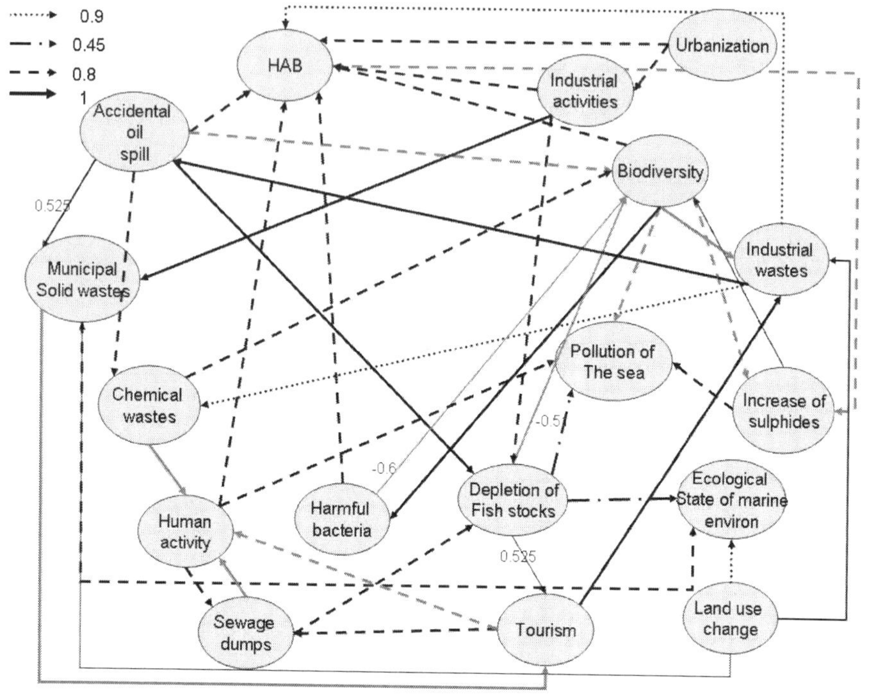

Fig. 5. The collective FCM with most central variables and strong connections.

Table 2 gathers the calculated graph theory indices for the collective map and the sum of the 29 individual maps.

Indices	Individual Maps	Collective FCM
Maps	29	1
Variables (N)	7.86 ± 1.574	26
Number of connections (W)	11± 6.513	145
No. of transmitter variables (T)	2.21 ± 1.544	2
No. of receiver variables (R)	1.96 ±1.267	3
No. of ordinary variables (O)	3.21 ±1.445	21
Connection/Variable (W/N)	1.33 ±0.617	4.577778
Complexity (Receiver/Transmitter)	1.02 ±1.198	1.5
Density (D=W/N^2)	0.167 ± 0.065	0.101728
Hierarchy index (h)	0.017 ± 0.005	0.012944

Table 2. Average (± SD) graph theoretical indices of the individual FCMs and the indices of the collective FCM.

The collective FCM was then used for analyzing system behaviour and to run management simulations. Simulations were generated using both inference equations, eq. (2) and rescaled eq. (5), by taking the product of the vector of initial states of variables ($\mathbf{A_0}$) times the square matrix E of the collective map. Simulations were generated using the five steps described previously implemented in Matlab R2008a environment for Windows. At first, we determine the steady state condition of the system's convergence before we consider any management options. We are doing this in order to be able to see the perceived tendency of the Black Sea ecosystem based on the collective cognitive map and test for its internal coherence. For this purpose, we run the generic FCM model with 50 different random initial states for all variables between 0 and 1 drawn from a uniform distribution. In all of these no-management simulations, the system reached a steady state after 25-30 iterations, where some of the produced final steady states for each initial condition were different. By excluding a chaotic behavior, we accordingly confirm a dynamically stable and coherent mental structure of our sample.

Then, we consider a number of assumptions for the concepts and risk factors, which affect the future state of the environmental health of marine ecosystem in the Black Sea to analyze the system performance and its decision-making capabilities. Initially, we considered a case where all concepts are set to zero. This means that all concepts are not activated for this specific consideration. After 23 iterations the FCM system reaches a final state where the concepts "Biodiversity" and "Ecological State" exhibit very high values and therefore increase the environmental health of marine ecosystem. This is an upper bound for the Black Sea ecosystem health, conditioned by the basic structure of respondents' cognitive reality.

In a next step, we consider a case where all concepts are set to one, meaning that the 24 factors are fully activated under this assumption. After 20 iterations the system reaches a final state where "Biodiversity" and "Ecological State" respectively clamp to zero. This is a lower bound for the Black Sea ecosystem health, conditioned by the basic structure of respondents' cognitive reality.

In-between these two extreme cases, we develop a number of policy scenarios, based on a number of ad hoc interventions for the Black Sea ecosystem conservation. In these interventions, individual concepts and groups of concepts consequently were considered to be activated and the final state of FCM system under these scenarios determined. Each simulation runs under two different versions where the activated concepts are set either one or 0.5. The rationale for this approach is to test the influence of uncertainty in the functioning of the other concepts. The value of 0.5 means there is uncertainty concerning the true state of the impact of the other concepts. The calculated output of the FCM model shows how the system reacts under the assumptions given by the stakeholders or related users. Usually, the calculated output is different from the expected one, thus presenting a potential added-value of Fuzzy Cognitive Map as a policy making tool.

One sample policy making scenario is presented. The five most central concepts- MSW (Municipal Solid Wastes), HA (Human Activities), Urban Sewage, IA (Industrial Activities), HAB (Harmful Algal Blooms)- are considered as the only de-activated concepts whereas all the other 21 concepts are considered as activated concepts that take values: (a) equal to 1 and (b) equal to 0.5, depicting strong activation or uncertainty state. Table 3 depicts the calculated values of all concepts in the final state for the considered scenarios (a) and (b), main observations being the high values of Biodiversity (Bd) and Ecological State of marine environment (ECOL). A significant increase to Biodiversity and ECOL state is a potential outcome of conservation policies regulating the input of those five most central concepts acting as risk factors. Thus the future state of the marine ecosystem could be improved if the five most central concepts might decrease at a significant amount.

Concepts	Initial values-Scenario (a)	Final state -eq. (5) Scenario (a)	Final state -eq. (2) Scenario (a)	Initial values-Scenario (b)	Final state - eq. (2) Scenario (b)	Final state- eq. (5) Scenario (b)
AOSP	1.00	0.7103	0.8808	0.5	0.8800	0.2835
Bd	**1.00**	**0.0004**	**0.0005**	**0.5**	**0.0007**	**0.9974**
BW	1.00	0.7290	0.8069	0.5	0.8063	0.3024
CD	1.00	0.8113	0.9097	0.5	0.8868	0.3855
CW	1.00	0.8558	0.9721	0.5	0.9718	0.1635
D-Distrust to State	1.00	0.8284	0.9303	0.5	0.9295	0.3751
DFS	1.00	0.9903	0.9974	0.5	0.9972	0.0174
HAB	0	0.9938	0.9986	0	0.9986	0.0162
HA	0	0.7297	0.3376	0	0.3346	0.2689
IA	0	0.5125	0.7095	0	0.7049	0.4638
ISP	1.00	0.7903	0.8367	0.5	0.8365	0.2359
LF	1.00	0.7103	0.8162	0.5	0.8161	0.4114
M- Mining	1.00	1.0000	1.0000	0.5	0.5000	0.5000
MC	1.00	0.9842	0.9924	0.5	0.9922	0.0235
MR	1.00	0.3199	0.4806	0.5	0.4808	0.5738
MSW	0	0.9664	0.9964	0	0.9950	0.0955
PPP	1.00	0.7270	0.8281	0.5	0.8280	0.4026
PSA	1.00	0.7581	0.8754	0.5	0.8737	0.3187
RAW	1.00	1.0000	1.0000	0.5	0.5000	0.5000
RP	1.00	0.6315	0.7144	0.5	0.7140	0.3678
S- Urban Sewage	0	0.7332	0.8591	0	0.8572	0.3698
Si-Siltation	1.00	0.9169	0.9379	0.5	0.9378	0.0933
SLR	1.00	0.8314	0.8655	0.5	0.8653	0.1948
Sphi	1.00	0.9970	0.9953	0.5	0.9952	0.0061
Tourism	1.00	0.2055	0.1535	0.5	0.1520	0.7777
ECOL	**1.00**	**0.0026**	**0.0055**	**0.5**	**0.0056**	**0.9913**

Table 3. Initial and final concepts' state after 25 iterations for Scenario (a) and (b).

9. Summary and conclusions

In this chapter, the FCM methodology was presented and analyzed for the elicitation and understanding of individual and collective knowledge, preferences and beliefs. The aim was to present to the reader both a theoretical underpinning of FCMs as well as a grasp of their empirical modalities.

A cognition model, like FCM, represents a system in a form that corresponds closely to the way humans perceive it. Therefore, the model is easily understandable, even by a non-professional audience and each parameter has a perceivable meaning. The model can be easily altered to incorporate new phenomena, and if its behavior is different than expected, it is usually easy to find which factor should be modified and how. In this sense, a FCM is a dynamic modeling tool in which the resolution of the system representation can be increased by applying a further mapping. The FCM methodology developed makes it possible, if the initial mapping of the risk factors and future prospects of marine ecosystem is incomplete or incorrect, to make further additions to the map, and to predict the effects of the new parameters considered.

FCMs have some specific advantageous characteristics over traditional mapping methods: they capture more information in the relationships between concepts, are dynamic, combinable, and tunable, and express hidden relationships (Kosko, 1986, 1992). The resulting fuzzy model can be used to analyze, simulate, and test the influence of parameters and predict the behavior of the system. Summarizing, FCM helps describe the schematic structure, represent the causal relationships among the elements of a given decision environment, and the inference can be computed by a numeric matrix operation. With FCM it is usually easy to find which factor should be modified and in which way.

To illustrate the FCM methodology, an empirical application for modelling lay people perceptions is presented. We describe the main features of a FCM exercise designed to elicit the Black Sea stakeholder views/ perceptions about the risks that the Black Sea may face in the future 20 years. A generic model for environmental management is constructed by augmenting the individual FCMs drawn by lay people-stakeholders from Ukraine. The graph theoretical indices were calculated out of the individual cognitive maps and the collective cognitive map produced by augmentation. A number of scenarios were run using the FCM inference process to enable us to understand the complex structure of the Black Sea problems and the risks mainly affecting its marine ecosystem. This knowledge is further used to design policies that contribute in environmental management. The results show its functionality and demonstrate that the use of FCMs is reliable and efficient for this task.

10. Acknowledgment

This research was carried out within the IP project SESAME funded by the EU under Framework Programme 6, Key Action "Global change, climate and biodiversity", Contract No. 2006-036949. We thank the Coordinator Dr. E. Papathanassiou for his encouragement to apply early in 2008 Fuzzy Cognitive Mapping in marine governance.

11. Appendix A

Interview protocol

Stage (1): A Formal Introduction

Which is the Aim of the study?
What is cognitive mapping methodology? Example cognitive map and fuzzy cognitive map from the public health study is depicted. Explanation on main factors and how these factors are interrelated is presented.
Stage (2) Creation of individual FCM
Which factors, which things, come into your mind spontaneously if I mention to you the Mediterranean (Black) Sea as a system where humans, marine animals and plants are all living together?
Is there any positive or negative relationship between these factors?
How strongly a factor A influences another factor B? A scale having 12 grades capable to describe any kind of relationship between two things is given.
Stage (3): Conclusion
Strong words/phrases they used, general comments of the interview.

12. Appendix B

Concepts	Abbreviation	Description of concepts	Concepts	Abbreviation	Description of concepts
C1	**AOSP**	Accidental oil spill pollution	C14	**Sphi**	Sulphide Increase
C2	**Tourism**	Tourism	C15	**IA**	Industrial Activities or Industrial Pollution
C3	**PST**	Pollution of Sea Trade	C16	**S**	Urban Sewage
C4	**CD**	Coastal Development	C17	**ISP**	Invasive species
C5	**BD**	Biodiversity	C18	**MSW**	Municipal solid waste pollution to the sea
C6	**ChemW**	Chemical wastes	C19	**MR**	MR= Marine Research
C7	**D**	Distrust to State & Institutions	C20	**RP**	Riverine Pollutants
C8	**DFS**	Depletion of Fish Stocks	C21	**SLR**	Sea level rise
C9	**RAW**	Radio-active	C22	**BW**	Ballast

		Waste			Waters (ship dumps)
C10	**PPP**	Polluter Pays Principle	C23	**Si**	Siltation
C11	**LF**	Luck of Financing	C24	**M**	Mining
C12	**MP**	Microbio-logical Pollution	C25	**HAB**	Harmful Algae Blooms
C13	**HA**	Human activity	C26	**ECOL**	Ecological State of the marine environment

Table A. 26 clustered concepts describing stakeholders' perceptions.

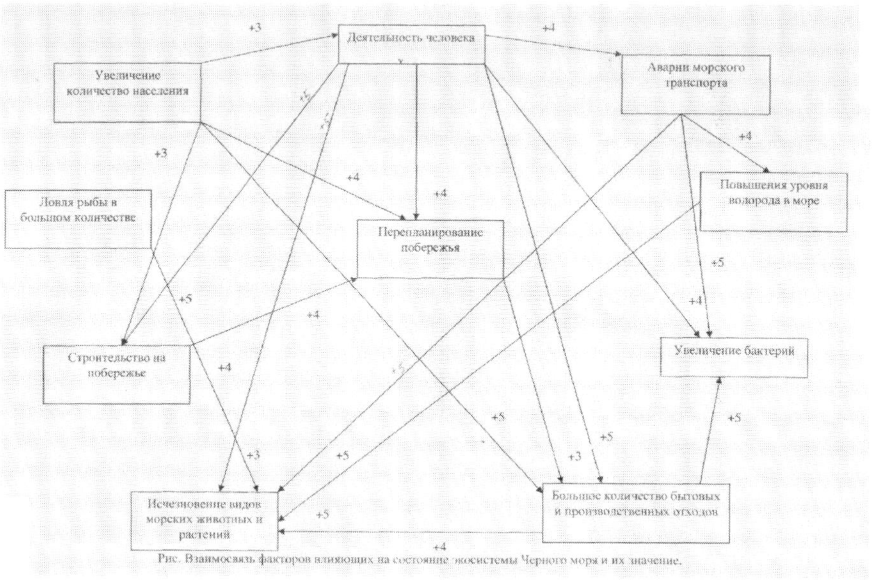

Fig. B1. A cognitive map defined by an individual/stakeholder from Ukraine.

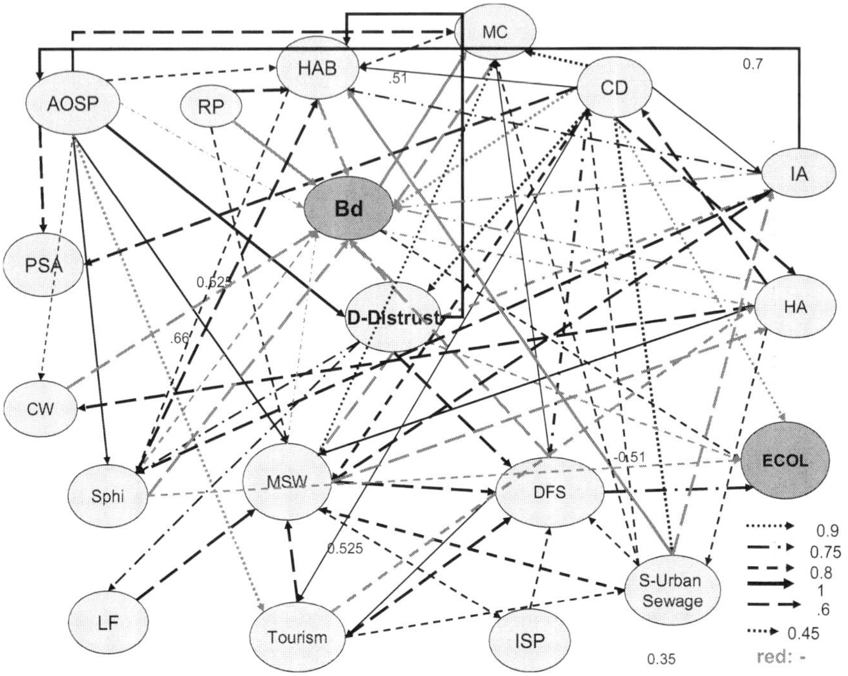

Fig. B2. The collective FCM with the eighteen most mentioned concepts and their related interconnections.

13. References

Biloslavo, R.; & Dolinšek, S. (2010). Scenario planning for climate strategies development by integrating group Delphi, AHP and dynamic fuzzy cognitive maps. *Foresight,* vol. 12, no.2, pp. 38-48.

Brouwer, R.; Powe, N.; Turner, R.K.; Bateman, I.J. & Langford, I.H. (1999). Public Attitudes to Contingent Valuation and Public Consultation. *Environmental Values* vol. 8 no. 3, pp. 325-347

Bougon, M., Weick, K., and Binkhorst, D. (1977). Cognition in organizations: An analysis of the Utrecht jazz orchestra. *Administrative Science Quarterly*, vol. 22, pp.606-639.

Bueno S., & J.L. Salmeron (2008). Fuzzy modeling Enterprise Resource Planning tool selection. *Computer Standards & Interfaces,* vol. 30 137–147.

Buzan T. (1993). *The Mind Map Book.* London: BBC Books

Carley, K. & M. Palmquist, 1992. Extracting, representing, and analyzing mental models. *Social Forces* vol. 70, pp.601–636.

Contreras J., Juan P. Paz, David Amaya & Pineda, A. (2007). Realistic Ecosystem Modelling with Fuzzy Cognitive Maps. *International Journal of Computational Intelligence Research.* ISSN 0973-1873 Vol.3, No.2, pp. 139-144

Craiger, J. P. & Coovert, M. D. (1994). Modeling dynamic social and psychological processes with fuzzy cognitive maps. *Proceedings of the Third IEEE World Conference on Fuzzy Systems*, vol. 3, pp. 1873-1877.

Desvousges W. H. and V. K. Smith (1988). Focus Groups and Risk Communication: The "Science" of Listening to Data. *Risk Analysis* vol. 8, no. 4, pp.479-484

Eden C. (1992). On the nature of cognitive maps. *Journal of Management Studies* vol. 29, pp. 261–265.

Eden C; Ackermann F, & Cropper S (1992). The analysis of cause maps. *Journal of Management Studies* vol. 29, pp. 309–324.

Eisenack K.; M.K.B. Lüdeke; G. Petschel-Held, J. Scheffran & Kropp, J.P. (2006). Qualitative modeling techniques to assess patterns of global change. In: J.P. Kropp and J. Scheffran, Editors, *Advanced Methods for Decision Making and Risk Management in Sustainability Science*, Nova Science Publishers, Hauppage, NY, USA.

Grimble R. & Wellard, K. (1997). Stakeholder Methodologies in Natural Resource Management: a Review of Principles, Contexts, Experiences and Opportunities. *Agricultural Systems*, vol. 55, No. 2, pp. 173-193.

Grosskurth J. & Rotmans, J. (2005). The scene model: getting a grip on sustainable development in policy making, *Environment development and sustainability* vol. 7, no. 1, pp. 135–151.

Grosskurth J. (2008). *Regional Sustainability: tools for integrative governance*, Maastricht University, Maastricht.

Harary, F.; Norman R. Z. & D. Cartwright, (1965). Structural Models: An Introduction to the Theory of Directed Graphs. John Wiley & Sons, New York.

Hermans, L. M. (2008). Exploring the promise of actor analysis for environmental policy analysis: lessons from four cases in water resources management. *Ecology and Society*, vol.13, no. 1, pp. 21.
[online URL: http://www. ecologyandsociety.org/vol13/iss1/art21/]

Hjortsø CN; Christensen SM; & Tarp P (2005). Rapid stakeholder and conflict assessment for natural resource management using cognitive mapping: the case of Damdoi Forest Enterprise, Vietnam. *Agriculture and Human Values* vol. 22, pp. 149–167.

Hobbs, B. F., S. A. Ludsin, R. L. Knight, P. A. Ryan, J. Biberhofer & Ciborowski, J. J. H. (2002). Fuzzy cognitive mapping as a tool to define management objectives for complex ecosystems. *Ecological Applications* vol. 12, pp.1548–1565.

Isaac M. E., Dawoe, E. & Sieciechowicz, K. (2009). Assessing Local Knowledge Use in Agroforestry Management with Cognitive Maps. *Environmental Management* vol. 43, pp. 1321–1329.

Karageorgis A., Kapsimalis V. Kontogianni A., Skourtos M. Turner R.K, Salomons W (2006) Impact of 100-year human interventions on the Deltaik coastal zone of the inner Thermaikos Gulf (Greece): A DPSIR framework analysis. Journal of Environmental Management 38, 304-315

Kardaras, D., & Karakostas, B. (1999). The use of fuzzy cognitive maps to simulate the information systems strategic planning process. *Information and Software Technology*, vol. 41, pp.197-210.

Khan, M.S., Quaddus, M.A., & Intrapairot, A. (2001). Application of a Fuzzy Cognitive Map for Analysing Data Warehouse Diffusion. Applied Informatics: Artificial Intelligence and Applications; Advances in Computer Applications. *Proceedings of IASTED International Symposia*. 19-22 February, Innsbruck, Austria, pp.32-37.

Kontogianni A., Skourtos M., Langford I., Bateman I., Georgiou S. (2001), Integrating Stakeholder Analysis in non- market valuation of environmental assets, Ecological Economics 37: 123-138, Elsevier Publishers

Kontogianni A., Tziritis I., Skourtos M. (2005), Bottom-up environmental decision making taken seriously: Integrating stakeholder perceptions into scenarios of environmental change. Human Ecology Review 12.2, 87-95

Kontogianni A., M. Skourtos (2008), Social Perception of Risk informing Integrated Coastal Zone Management on accidental oil - spill pollution: 'the reason you pollute matters, not numbers', in book: "Integrated Coastal Zone Management - The Global Challenge" (eds) R. Krishnamurthy, B. Glavovic, A. Kannen, D. Green, R. Alagappan, H. Zengcui, S. Tinti, T. Agardy. Research Publishing pp. 207-225.

Kontogianni A., Papageorgiou E., Salomatina L., Skourtos M., Zanou B., Assessing perceptions of marine environmental futures through FCM in Ukrain (forthcoming)

Kok K., (2009). The potential of Fuzzy Cognitive Maps for semi-quantitative scenario development, with an example from Brazil, *Global Environmental Change* vol.19, pp. 122–133.

Kosko, B. (1986). Fuzzy cognitive maps. *International Journal on Man-Machine Studies*, vol. 24, pp.65-75.

Kosko, B. (1992). *Neural Networks and Fuzzy Systems: A Dynamical Systems Approach to Machine Intelligence*. Prentice-Hall. New York.

Kouwen F. V.; P.P. Schot & M.J. Wassen, (2008). A framework for linking advanced simulation models with interactive cognitive maps. *Environmental Modeling & Software* vol. 23 no. 9, pp. 1133–1144.

Montazemi, A.R.& Conrath D.W. (1986) The use of cognitive mapping for information requirements analysis *MIS Quarterly*, March, 45-46

National Research Council (NRC). (1996). Understanding Risk: Informing Decisions in a Democratic Society. Washington, DC: National Academy Press.

Novak J. D. (1993). How do we learn our lesson? Taking students through the process. The Science Teacher 60: 50-55

OECD (2005). Evaluating Public Participation in Policy Making. Paris: OECD.

Ozesmi, U. & S. L. Ozesmi, (2003). A participatory approach to ecosystem conservation: Fuzzy cognitive maps and stakeholder analysis in Uluabat Lake, Turkey. *Environmental Management* vol. 31,pp. 518–531.

Ozesmi, U. & S. L. Ozesmi, (2004). Ecological Models based on People's Knowledge: A Multi-Step Fuzzy Cognitive Mapping Approach. *Ecological Modeling* vol. 176,pp. 43–64.

Ozesmi, U., (1999). Modeling ecosystems from local perspectives: Fuzzy cognitive maps of the Kizilirmak Delta Wetlands in Turkey. *Proceedings of 1999 World Conference on Natural Resource Modelling*, Halifax, Nova Scotia, Canada

Papageorgiou E.I. & Stylios C.D. (2008). Fuzzy Cognitive Maps, in book: *Handbook of Granular Computing*, editors: Witold Pedrycz, Andrzej Skowron and Vladik Kreinovich, Chapter 34, John Wiley & Sons, Ltd, pp. 755-775.

Papageorgiou E.I.; Markinos, Ath. & Gemtos, Th. (2009). Application of fuzzy cognitive maps for cotton yield management in precision farming, *Expert Systems with Applications*, vol. 36, no. 10, pp.12399-12413.

Papageorgiou E.I.; Markinos Ath. & Th. Gemtos, (2010). Soft Computing Technique of Fuzzy Cognitive Maps to connect yield defining parameters with yield in Cotton Crop Production in Central Greece as a basis for a decision support system for precision agriculture application, in book: *Fuzzy Cognitive Maps: Advances in Theory, Methodologies, Tools, Applications*, edited by M. Glykas, Springer Verlag, pp. 325-362.

Papageorgiou E.I. (2011). A new methodology for Decisions in Medical Informatics using Fuzzy Cognitive Maps based on Fuzzy Rule-Extraction techniques, *Applied Soft Computing*, vol. 11, pp. 500–513.

Petschel-Held G.; Block, A.; Cassel-Gintz, M.; Kropp, J.P.; Lüdeke, M.K.B.; Moldenhauer O.; Reusswig F. & H.J. Schellnhuber, (1999). Syndromes of global change: a qualitative modeling approach to support environmental management, Environmental Modeling and Assessment vol. 4, no. 4, pp. 295–314.

Putnam, R. A. (1985). Creating facts and values. *Philosophy* vol. 60, pp. 187-204

Rajaram T., & A. Das, (2009). Modeling of interactions among sustainability components of an agro-ecosystem using local knowledge through cognitive mapping and fuzzy inference system, *Expert Systems with Applications*, vol. 37, no. 2, pp. 1734-1744.

Ramsey D. & Veltman, C. (2005). Predicting the effects of perturbations on ecological communities: what can qualitative models offer?*Journal of Animal Ecology* vol. 74, pp. 905–916.

Ramsey, D. & Norbury, G. L. (2009). Predicting the unexpected: using a qualitative model of a New Zealand dryland ecosystem to anticipate pest management outcomes. *Austral Ecology* vol. 34, pp. 409–421.

Reed M. S. (2008). Stakeholder participation for environmental management: A literature review. Biological Conservation 141: 2417–2431

Reed M. S.; Graves, A. Dandy, N. Posthumus, H. Hubacek, K. Morris, J. Prell, C. Quinn, C. H. & Stringer L. C. (2009). Who's in and why? A typology of stakeholder analysis methods for natural resource management. *Journal of Environmental Management* vol.90, pp.1933–1949

Robson, M. & Kant, S., (2007). Structure of causation and its influence on cooperation: A comparative study of forest management in Ontario, Canada. *Forest Policy and Economics*, vol. 10(1-2), pp.70-81.

Rodriguez-Repiso, L., Setchi, R., and Salmeron, J.L. (2007). Modelling IT projects success with Fuzzy Cognitive Maps. *Expert Systems with Applications*, vol. 32, pp. 543-559.

Sharif, A.M. and Irani, Z. (2006). Exploring Fuzzy Cognitive Mapping for IS Evaluation. *European Journal of Operational Research*, vol. 173, pp.1175-1187.

Skov, F. & J. -C. Svenning, (2003). Predicting plant species richness in a managed forest. *Forest Ecology and Management* vol. 6200, pp. 1–11.

W. Stach, L. A. Kurgan, and W. Pedrycz, Expert-based and Computational Methods for Developing Fuzzy Cognitive Maps, In: Glykas, M., *Fuzzy Cognitive Maps: Advances in Theory, Methodologies and Applications*, Springer (ISBN-10: 36-42032-19-2), 2010.

Stoney, C., & Winstanley, D., 2001. Stakeholding: confusion or utopia? Mapping the conceptual terrain. *Journal of Management Studies* vol.38, pp.603–626.

Strickert, G.E., S. Samarasinghe, & T. Davies, (2009). Resilience models for New Zealand's Alpine Skiers based on people's knowledge and experience: a mixed method and multi-step fuzzy cognitive mapping approach, Proceedings of *18th World IMACS / MODSIM Congress, Cairns*, Australia 13-17 July 2009, http://mssanz.org.au/modsim09.

Taber, W. R., (1991). Knowledge processing with fuzzy cognitive maps. *Expert Systems with Applications* vol. 2, pp.83–87.

Tan C. O. & U. Ozesmi, (2006). A generic shallow lake ecosystem model based on collective expert knowledge, Hydrobiologia, vol. 563, pp.125–142.

Vliet van M.; Kok, K., Veldkamp, T. (2010). Linking stakeholders and modellers in scenario studies: The use of Fuzzy Cognitive Maps as a communication and learning tool. *Futures* vol. 42, no.1, pp. 1-14

Zadeh (1986). http://www-bisc.cs.berkeley.edu/zadeh/papers/1986-CWW.pdf

Zhang, J., S. E. Jrgensen, C. O. Tan & M. Beklioglu, (2003). A structurally dynamic modeling - Lake Mogan, Turkey as a case study. *Ecological Modelling* vol. 164, pp. 103–120.

Zorrilla P., G. Carmona, A. de la Hera, C. Varela Ortega, P. Martínez Santos (2011). Bayesian Networks as Tools for Participatory Water Resources Management: an Application to the Upper Guadiana Basin, Spain. *Ecology and Society*, vol. 15, no.3, [online URL: http://www.ecologyandsociety.org/vol15/iss3/art12/].

22

Wind Farming and the Not-in-My-Backyard Syndrome: A Literature Review Regarding Australia's Challenge in Relation to Climate Change and CO2 Emissions

Ingrid Muenstermann
Charles Sturt University, Wagga Wagga, Australia

1. Introduction

Let me begin with a personal note. When several Australian newspapers reported about people objecting to the establishment of wind farms in rural and regional Australia, the Not-in-my-Backyard syndrome (NIMBYism) entered my thoughts. Knowing that Australians were emitting more than reasonable amounts[1] of greenhouse gasses into the atmosphere, this literature review was started. It is the result of trying to understand the objections to wind farming. I must admit that I like the turbines, their imposing height, the way they enhance the landscape, and their capacity to produce electricity. In 2002, I was standing under a turbine on the Isle of Fanø in Denmark, it was noisy but not overwhelming, I was in awe, admired a manmade product fitting perfectly into the landscape. Reading years later the objections to wind farming in Australia triggered concern and an interest into researching NIMBYism.

According to the Commonwealth Scientific and Industrial Research Organisation (CSIRO) (2009, pp. 1-5) climate change is the greatest ecological, economic and social challenge of our time. Globally, CO2 emissions, temperature and sea levels are rising faster than expected and average temperatures are increasing (CSIRO, 2009, p. 3). These trends are recorded on all continents and in the ocean. "Since the Industrial Revolution global CO2 concentration has risen by 37%" which "is mainly due to fossil-fuel use and land-use change" (CSIRO, 2009: p. 5). CO2 is a contributing factor in the enhanced greenhouse effect which is resulting in climate change.

Approximately 25% of the CO2 emitted in the atmosphere is absorbed by the ocean and another 25% is absorbed by the natural environment on land. In water, CO2 makes the oceans more acidic. Ocean acidification interferes with the formation of shells and corals, and has far reaching implications for the health and productivity of the world's oceans (CSIRO, 2009, p. 5).

The CSIRO also finds that "the likelihood of observed warming being due to natural causes alone is less than 5%" (p. 5). And they continue:

---

[1] What constitutes a reasonable amount?

> Evidence of human influence also has been detected in ocean warming, sea-level rise, continental-average temperatures, temperature extremes and wind patterns. This conclusion is consistent with the observed melting of glaciers and ice sheets (CSIRO, 2009, p. 5).

One of the solutions to increased greenhouse gas emissions is the renewable energy of wind power. Wind power is currently the most rapidly growing energy source in the world (Acciona Energy, 2006; BTM Consult ApS, 2009; Firestone, Kempton, Krueger & Loper, 2005, p. 75; Liebmann, 2003, p. 4; Mercer, 2003, p. 10). Wind farms have been established in many countries to reduce dependence on fossil fuels and, at the same time, to increase the production of renewable energy. Wind turbines can be installed onshore and offshore. In spite of positive effects on the environment, i.e. reducing CO2 emissions (see Section 8.), wind farms have been opposed for environmental reasons: Paradoxically, according to Lothian (2008, p. 196), "while addressing one environmental concern, the system is being opposed because of another environmental concern, namely the perceived negative impact on the environment, particularly on landscape quality". There is, however, a strong commitment of many countries to increase the share of renewable energy and this has resulted in a significant increase in the number of wind turbines. The World Wind Energy Association (WWEA) provides a table showing the added capacity of approximately six months (end of 2009 to June 2010):

Position	Country	Total capacity end 2009 (MW)	Added capacity June 2010 (MW)	Total capacity June 2010 (MW)
1	USA	35.159	1.200	36.300
2	China	26.010	7.800	33.800
3	Germany	25.777	660	26.400
4	Spain	19.149	400	19.500
5	India	10.925	1.200	12.100
6	Italy	4.850	450	5.300
7	France	4.521	500	5.000
8	UK	4.092	500	4.600
9	Portugal	3.535	230	3.800
10	Denmark	3.497	190	3.700
Rest of the world		21.698	2.870	24.500
Total		159.213	16.000	175.000

www.wwindea.org/home/index2.php?option+com_jce&task=popup&img=imag

Table 1. Wind Power Worldwide June 2010

The table demonstrates that the use of wind power is on the increase overall. The changes in the table refer to a time of approximately six months. Looking at the near past, in the 1990s, Germany had the fastest growth in the industry, followed by Spain, Denmark and India (European Commission, 1997, table 2.2). The German wind farming industry in 2000 was the second largest in the world (Johnson & Jacobsson, 2000, p. 2), generating the highest amount of wind powered energy in the world (Nelson, 2005, p. 8; Reeves & Beck, 2003, p. 9; Rodriguez et al., 2002, p. 1089). The growing trend for Germany has continued (Macintosh & Downie, 2006, p. 1), however, Table 1 demonstrates that China, the United States of America and India have overtaken Germany during the last (at least) six months.

This literature review relates to wind farms on land and is written from an Australian perspective, including some international comparison. In 2010, four projects (worth 477 MW) were abandoned or suspended in South Australia, Victoria, New South Wales, and in Tasmania[2] (Wikipedia; Clean Energy Australia 2010; Clarke, 2008/2010, p. 16). The planning and siting of wind farms often create conflict in the local population. For that reason a look at newspaper articles should set the scene regarding the public's attitudes towards wind turbines. Problems relating to biodiversity (birds and bats are being killed); reliability (wind power can only be used as intermittent source of power); economy (wind turbines are costly but contribute very little to the reduction of CO2); economy (wind power is not cost-effective); health (the noise of turbines has adverse affects on residents' health); and the aesthetics of the environment will be discussed. Characteristics of NIMBYism will be studied, and issues that may help to find a solution to the polarised debate will be explored. In order to make sense of the disputes, the connection to, and identification with place will be looked at, and improved collaboration between the public, developers and governments will be considered; so will be trust and social distrust of the community towards governments. Scientific arguments regarding climate change and CO2 emissions will be used to strengthen the claim that changes are necessary. It will be argued that several non-rational theories (relating here to biodiversity, reliability of wind energy, economic costs, health, aesthetics of the environment) are the basis of NIMBYism. These non-rational concepts have very rational consequences, i.e. abandonment or suspension of the project.
The discussion around NIMBYism and wind farming is not new but this chapter should add weight to arguments in support of the development of wind farming.

2. Findings

2.1 The Australian wind power industry – Some facts

In 2008/09 Australia produced 438 million tonnes (Mt) of raw black coal, 334 Mt of black coal was available for the domestic use and for export (2008 Australian Coal). The CSIRO 2003-2011 finds that

> ...coal is Australia's largest export and a major contributor to the national economy. It is the primary fuel for power generation worldwide and provides more than 80% of Australia's electricity supply (CSIRO 2003-2011).

Coal is, however, a major contributor to the world's greenhouse gas emissions. According to the CSIRO 2203-2011, current power technologies account for more than one third of Australia's emissions alone. This means that, on the one hand, the profitable coal export industry has to be considered and, on the other hand, the environment. Taking a long term view, the risk of further CO2 pollution is high and our options are limited. Clarke, who has researched climate change and Australia's wind energy projects for many years, contemplates the belief of many that 'Australia can't make any difference' and summarises their negative attitude as follows:

> Australia only produces about 1.5% of the world's greenhouse gasses. If we were to cut our emissions to nothing tomorrow, it would make very little difference to the world. Therefore we would be foolish to risk crippling our economy (Clarke, 2008, p. 8).

Bond (2009) is more specific than the CSIRO regarding the production of electricity and finds that in 2006/2007, "83% of Australia's electricity is produced using coal" (p. 2).

[2]Please refer to Section 3 for some examples.

Australia "is the rich world's worst per capita greenhouse gas emitter due to a heavy reliance on ageing coal-fired power stations for electricity generation" (Taylor & Grubel, 9.7.2011, p.1). This is a worrying aspect. Data published by the United Nations Statistics Division (2008a) on carbon dioxide emission shows that Australia is in sixteenth position when the percentage of global total CO2 emission is considered: as a nation, 399,219.00 metric tonnes of carbon dioxide are emitted annually. Per capita Australians emit 18.94 tonnes of CO2 per capita (United Nations Statistics Division, 2008b).

Despite challenges, the present Australian Federal Government is committed to generate 20% of Australia's electricity from renewable energy by the year 2020 for a transition to a low-carbon economy (Combet, 2010, pp. 1-4). Interestingly, Australia was listed twenty third in the world in 2004-2005 for capacity of wind generation (Nelson, 2005, p. 8), but it was the first country to implement a mandatory requirement of energy retainers (MRET), i.e. that there be a certain amount of renewable energy target (Kann, 2009, p. 1; Macintosh & Downie, 2006, p. 1; Mercer, 2003, p. 17; Nelson, 2005, p. 3). Since the inception of MRET, wind power has been the fastest growing source of energy in Australia (Macintosh & Downie, 2006, p. 1; Nelson, 2005, p. 3). Wind power accounts for 16% of the renewable energy market in Australia (Clean Energy Council, 2009), but only 1.3% of its electricity comes from wind power (IEA, 2009, p. 79). Australia has various state based national incentive programs to promote wind energy, there is, for instance, the Green Power national accreditation program, and there are initiatives of the Department of Climate Change (IEA, 2008, pp. 79-84). According to the International Energy Agency, in 2008:

- the capacity of Australian wind turbines was 1,306 MW,
- the average number of Australian households powered by wind energy was 487,537,
- the number of wind energy projects with two or more turbines was 37,
- the annual CO2 emissions displaced by wind energy were 3,530,744 tonnes per year; equivalent to 784,610 cars been taken off the road per year,
- total capital investment was A$2.207 billion Agency (IEA, 2009, p. 81).

Since 1997, Australian installed wind power has grown by 75% (Clarke, 2008, p. 4) which implies that things are moving into the direction of renewable, clean energy. Garnaut (Update 2011a) recommends that "Australia should be ready to calibrate its emissions reductions proportionately to the global mitigation effort" (p. 13) and that "low emissions technology" should be a vital part of the mitigation effort (p. 39). And Diesendorf (2003/2004), promoting Australian sustainable energy for about twenty-five years, argues that

> ... in replacing a coal-fired power station with a mix of energy efficiency, wind power, bioenergy and gas, the economic savings from energy efficiency will be so large that they could pay for the additional costs of renewable energy (p. 2).

Australia is a land of wide open spaces, its size is 7.7 million square kilometres, its population is 22,650,000, and in 2001, almost 85% of the population lived within 50 km of the coast line (Australian Bureau of Statistics, 2006). The inside of the country is vast, but this vastness has disadvantages and can, according to Clarke (2010/2011, pp. 4-5), present limits to the development of wind farms:

- The size and cost of cranes that are used for turbine erection,
- Size and weight of components that must be transported by road,
- Potential for aviation and radar interference,
- Material fatigue.

Clarke (2009) also considers the lack of electricity transmission lines, and anticipates a "shortage of working capital due to the financial break-down" (p. 3). Despite these impediments, it is argued here that 16% wind power representing renewable energy could be improved and that the 1.3% of electricity produced from wind power should be increased. The international comparison regarding electricity produced by wind power is as follows: Denmark produces 19%, Spain and Portugal 9%, Germany 6%, Ireland 6%, USA 1.9% (IEA, 2009, p. 10). What are the barriers in Australia to develop and to use to a greater extend facilities that can provide clean, renewable energy? In 2010 four projects, worth 477 MW of wind power, were abandoned or suspended. Here are some examples of challenges developers and governments face when planning a wind farm.

3. Case studies – Some challenges when wind farming is considered in Australia

In 2008, wind energy provided 487,537 Australian households with electricity, which presented 1.3% of national electric demand (IEA, 2009, p. 79), however, as mentioned earlier, more than 80% of electricity (CSIRO) or, more specifically, 83% (Bond (2009, p. 2) was coal-fired. Let's look at some of the challenges associated with the development of wind energy projects.

In February 2009 Acciona Energy began generating green power in Waubra (approximately 150 km west-north-west of Melbourne, the capital city of the State of Victoria, and 35 km north-west of Ballarat, Victoria). The wind farm consists of 128 turbines of 1.5MW, the installed capacity is 192MW, offsetting 635,000 tonnes of CO_2-e annually (Acciona Project Snapshot). Acciona then wanted to establish another wind farm, Waubra North, close to the township of Evansford. But some challenges emerged: According to Radical Green Watch (February 13, 2010), the Pyreness Landscape Guardians organised a meeting in early 2010. This meeting "brought together sixty wind farm opponents from across the state". At the meeting a petition was signed requesting a moratorium on wind farms until health studies were undertaken. A representative of Acciona, present at that meeting, stated that "the company was still conducting investigations of the site at Evansford". Acciona (Newsletter, June 15, 2011) announced that "its early feasibility work on the site at Waubra North …has indicated a wind farm in that location would not be viable"(p. 1). The company had undertaken investigations in relation to the environment, flora, fauna and, most importantly, wind resource.

> The further north you travel from Waubra we found the wind resource to be less viable, primarily because of different topography (Acciona Newsletter, June 15, 2011, p. 1).

Interesting here are some political connotations. According to Courtice (July 30, 2011), a former member of the Liberal Party sits on the board of the anti-wind power Waubra Foundation, and a businessman, who was involved in oil, gas and mineral exploration companies, set up the foundation. Did Acciona Energy know about these powerful opponents?

Some other issues are worth noting because they relate to issues of NIMBYism. Firstly, there is the deliberate creation of "fear, uncertainty and doubt" (FUD) (Courtice, July 30, 2011) by the opponents of wind energy. FUD undermine an objective debate. Secondly, here are some facts about Acciona and its involvement in the local community of Waubra: Waubra has, according to information on the net, a population of 494, it is not far away from Melbourne and Ballarat, and its football team, the Waubra Kangaroos, has won several

Australian rule football finals. Expanding on these limited facts, it can be assumed that the Waubra wind farm provides employment for local residents. Acciona is also contributing annually $64,000 to a Community Benefit Fund ($500 per turbine per annum) and, interestingly, arrangements have been made for post 2035, when the Waubra wind farm ceases to operate: 5% of the Community Benefit Funds are allocated "to provide the community with ongoing project support" (Acciona Newsletter, June 15, 2011). Further, Acciona is involved in social events, such as the Waubra Corporate Footy Day, and is producing a quarterly newsletter, delivered to local residents and available at local shops and offices of the Shire and the City of Ballarat. All of this indicates that Acciona Energy is contributing in a positive way to this regional society (employment, community funds, social events), however, local residents "did not want more Acciona in the district. I think it is a huge win for the local community", and "it wasn't a good idea in the first place" (Australian Broadcasting Corporation, Ballarat, May 25, 2011).

I was not able to find out whether the "sixty opponents to wind farms from across the state" (Radical Green Watch, February 13, 2010) or the two politically influential executives played a part in Acciona's decision to abandon the Waubra North project, however, the remarks by residents are underpinned by NIMBYism.

Another case, also relating to Acciona, further demonstrates the force of NIMBYism. In an article entitled "Man vs Wind farm", Impey (ABC South East SA, June 22, 2011, pp. 1-3) reports about a decision that "has rocked the renewable energy industry": courts have ruled in favour of a man who lodged an appeal against the development of a proposed wind farm in Allendale East on the basis of visual amenity.

In a landmark ruling Eight Mile Creek dairy farmer Richard Paltridge lodged an appeal with the State's Environment Resources and Development Court against the decision by the Grant District Council's independent planning assessors to go ahead with a 46-turbine project (Impey, ABC South East SA, June 22, 2011, p. 1)

This project was meant to deliver substantial economic and environmental benefits to Mount Gambier and the South East region of South Australia. The A$175 million investment by Acciona was expected to generate fifty construction jobs and eleven highly-specialist full-time jobs. It was expected that the forty-six turbine project would meet the needs of about 43,000 households and reduce carbon dioxide emissions by 181,568 tonnes annually (The Border Watch, June 23, 2011, p. 3). The Grant District executive officer "was surprised ... given the company's extensive public consultation, that was the most comprehensive of any developer in the past 10 years". This officer also mentioned that he was not aware "of any other wind farm development that has been refused through the courts due to visual amenity". In contrast to these remarks, a member of the Concerned Residents Group, which had been formed to oppose the wind farm, stated that "many residents and adjoining landholders were celebrating the ruling". While claiming that the group was not against wind farms, "... multinational companies should not be allowed to place these farms 'willy-nilly'" (The Border Watch, June 23, 2011, p. 3).

The two news items, broadcast by the ABC (Australian Broadcasting Corporation), and the article in The Border Watch demonstrate the rational consequences of NIMBYism. The Clean Energy Council policy manager finds that "the ruling was subjective", that "visual amenity is subjective [and that] they rejected it purely on that basis" (Impey, ABC South East SA, June 22, 2011, p. 1). The argument of subjectivity is substantiated by Finlay-Jones and Kouzmin (2004) who find that "visual amenity is a subjective matter", that it is entirely dependent on the experience of individuals and that "acceptance ... increases with time" (p. 3). To take the analysis a little further, visual amenity, the destruction of the view, the aesthetics of the landscape can be related to sense of place (place attachment and place identity to be discussed in Section 7).

Another very interesting issue emerged in relation to the facilitation of wind turbines:

The proposed 26-turbine wind farm at Glen Innes that was the first to be approved in New South Wales's six newly created 'wind precincts' has sparked legal action against the government. Neighbouring landowners are not happy with the turbine set back distances from their properties and that the income of the participating landowners is not shared with the affected neighbouring properties. Three local families are challenging Ms Kenneally's decision in the NSW Land and Environment Court (Infigen, April 8, 2011).

The issue of "income of the participating landowners" not being "shared with the affected neighbouring properties" must cause a great deal of angst amongst those landowners who are leasing their property to developers of clean, renewable energy.

NIMBYism in relation to wind farming[3] is encouraged by proponents of the coal industry. While the present Australian government (Labor, The Greens, three independents) are committed to create a more sustainable environment, a survey of wind companies ,commissioned by the Clean Energy Council in 2010, found that between 50 to 70% of proposed wind farms would be abandoned if the Coalition (Liberal Party, National Party) would pursue its policies. According to the National Times (February 22, 2011), the Coalition has promised to support the reduction of CO2 emissions by 20% by 2020, however, "their anti-wind farm policy threatens to do the opposite". This finding cannot be when underestimated NIMBYism is considered.

4. The pervasiveness of the Not-in-my-Backyard syndrome (NIMBYism)

A short overview of national and international newspapers[4] will demonstrate the all-pervasiveness of NIMBYism: Kagkelidon (April 13, 2007) reports from Greece: "Serifos island opposes gigantic wind park plans". Siegel (November 27, 2007) also reports about the island of Serifos: "NIMBYism - Global obstacle to a renewable energy future". Howden (June 12, 2007) writes about "Conservationists fight to keep wind farms off Skyros". Russel (January 1, 2008) looks at the situation in Great Britain: "Local planning logjams are preventing renewable power projects being given the green light, research reveals two out of three applications for onshore wind farms are being rejected". He further comments that "Britain could fulfil its full potential for land-based wind power if the proposals currently going through the planning system were built". Walker (January 19, 2011) considers Great Britain's *Localism agenda* which seems to oppose low-carbon strategy: "Wind industry warns Localism bill could spark yet more planning delays for new renewable energy projects". And Woods discusses conflicting environmental visions of the rural:

> The proposal to construct the 39 turbine power station provoked considerable controversy from its announcement in early 2000. By the time that the application was approved by the local planning authority in July 2001, it had generated a public debate that split communities, political parties and pressure groups, and prompted public meetings, demonstrations, petitions and letters to the press (Woods, 2003, p. 271).

[3] For further discussion see Sections 5 and 6.

[4] An internet search regarding Australia and NIMBYism established 659,000 entries on the web but only two books in our library; these two books deal with the NIMBYism and waste disposal.

Attitudes towards wind turbines are similar in the United States: Lynley (May 24, 2011) reports "Maine wind farm could be the latest 'not in my backyard' casualty". The establishment of wind farms are being delayed (Wisconsin) or rejected (Connecticut). "Wind farm? Not off my back porch" is an interesting article by Schoetz (March 30, 2007) dealing with a "major battle in the politics of alternative energy [which] has moved to a final phase in Washington: Senator Edward Kennedy with a waterfront view and a bone to pick". But Seelye wrote (April 28, 2010) "after nine years of regulatory review, the federal government gave the green light to the nation's first offshore wind farm, a fiercely contested project off the coast of Cape Cod" (vicinity of the Kennedy property).

NIMBYism could also be detected in New Zealand: Cross (July 5, 2010) reports about "wind farms and the NIMBY Phenomenon", that New Zealand's "climate and geography lend themselves to the efficient use of wind power", however, "public objections to wind farm development and Environment Court rulings highlight some controversial issues associated with the modern energy source".

Objections to the establishment of wind turbines are very similar across the world and in Australia: "NIMBYs 'a threat to proper planning'" and "'Not in my backyard' sentiments pose one of the biggest obstacles to good planning and infrastructure development" writes Hurst (March 17, 2011). Interesting are the comments by Ryder (April 30, 2011): "I may have to buy a new filing cabinet to record all the events people claim will decimate their property values". And he continues:

> One family attracted media attention by claiming a wind farm on distant hills would destroy every thing they had worked all their young lives for. The views of wind turbines waving in the breeze would kill their property's value. What about our property rights, they said (Ryder, April 30, 2011).

Jopston (April 2, 2010) argues "Wind farm approval blows town apart" and "Tilting at windmills: why families are at war"; and Strong (May 22, 2010) reflects on "Towns split on which way the wind blows". These stories are about residents in rural Australian communities who object to the establishment of wind farms in their area. Clarke (2010/2011) finds that the anti-power-movement could present a threat to wind power. He argues:

> While it is not strong and the NIMBY principle ... is involved, it does have some justification and some potential to harm the industry. There is at least anecdotal evidence that a few people's health may be adversely affected by sound and infrasound from turbines. As there seems to be no known mechanism for the health effects, it seems likely that there is a large psychogenic factor involved (p. 2).

How can these strong objections towards wind farming, which produces clean, renewable energy, be explained? Involved in the debate are non-rational concerns, which are produced by anti-wind farm proponents by deliberately creating fears, uncertainty and doubt (FUD) (Courtice, July 30, 2011). The practical consequences of FUD are suspension or abandonment of wind energy projects. There are also the non-rational concerns regarding place attachment and place identity. The arguments presented in different newspapers are sensationalised[5], nevertheless, they capture the attitudes of local residents. In order to try and better understand NIMBYism, here is a look at its characteristics.

[5] Qualitative research (discussions, focus groups) may have produced a different outcome.

5. Characteristics of the Not-in-my-Backyard syndrome (NIMBYism)

NIMBYism is defined by Wolsink (2006) as "an attitude ascribed to persons who object to the siting of something they regard as detrimental or hazardous in their own neighbourhood, while by implication raising no such objections to similar developments elsewhere" (p. 86). NIMBYism can refer to the establishment of nuclear power plants, nuclear waste or general waste disposal, housing the underprivileged, or to racism. Here NIMBYism is related to wind farming in Australia, referring to public opposition to unwanted local developments. NIMBYism "is not new, but it has never been as pervasive and so quickly ignited as today", it is "a triumph of self-interest over principle" (Offor, 2002, p. 2). Offor contemplates on "how quickly one disgruntled landowner appears to become a major campaign with a ground swell of support that could derail an entire project" (p. 2). An insight into the development of NIMBYism may provide an understanding why it is such a powerful tool in opposing something that has potentially very positive effects.

Futrell (cited in Glickel, 2011) argues that true NIMBYism reactions include a "shift in awareness based on a sense of injustice and obligation to act". Freudenberg and Pastor (also cited in Glickel, 2011) find that three theories can explain NIMBYism in response to perceived or actual risks:

- The community is either ignorant or irrational, unwilling to accept any risk for the benefit of society: Proponents of this theory conclude that the public is uninformed and thus unreasonable. This argument, however, does not recognise that issues of uncertainty will always be a part of any assessment dealing with impacts on the natural world; experts and residents have to consider the same principles of assessment, including uncertainty.
- The community is selfish: The concept of public selfishness is embedded in the free-enterprise economic model of a market society; it is based on the assumption that it is rational for individuals to look out for their own interests. Scholars now have the difficult task to determine which self-interested attitudes can be justified and which can not.
- The community is prudent: Recent research has recognised some public opposition as valuable to an impact assessment: Organised protests challenging scientific assessment can reveal elements of the bigger picture scientists may otherwise not consider. Going beyond either blaming or understanding specific opposing views can lead to an understanding of the broader system that creates opposition in the first place.

The characteristics of NIMBYism often overlap (Glickel, 2011). Some further explanation and support of the syndrome is provided by Esaiasson (2010). He finds that in the 1980s the concept of NIMBYism gained importance among frustrated politicians and developers who argued that "narrow-minded citizens were a hindrance for societal development" (p. 27). At that time, the "simplistic NIMBYism did not take into account the needs and values of local communities" (p. 27). Esaiasson finds that "self-interest and local concerns are important in relation to individuals' responses to planned facility sitings" and that "the basic idea of NIMBYism should be part of our understanding of the complexities of public facility sitings" (p. 1). The author concludes that "without informed consent of affected individuals, the decision to site a public facility is an expression of contested legitimate power" (Mansbridge cited in Esaiasson, 2010, p. 28). This may be so, but somehow the civic good has to be fitted into the equation. Gibson (2005) raises the issue of the civic good and is critical of NIMBYism. The traditional view of NIMBYism is an "opposition between the rational/civic interest, embodied by public authorities, and the irrational/self-interest, embodied by local opponents", it "reduces land-use disputes to a moral struggle between the two groups" (Gibson, 2005, p. 385). As a consequence:

> ... more and more local opponents to much needed but controversial facilities... are successful in killing ... important projects. What you face as a state or industrial planner, then, is not just an isolated case of NIMBY opposition, but rather an accelerating NIMBY syndrome - that is, an emotional, irrational, and systematic distrust of public and corporate expertise that threatens to undermine the state's ability to solve important environmental and social problems (Gibson, 2005, p. 381).

Trust or social distrust will be discussed later in this chapter. Here are, at first, the most cited arguments against wind farming in relation to NIMBYism.

6. Most common disputes against wind farming associated with NIMBYism

6.1 Biodiversity

The threat to biodiversity may not be a significant factor for land owners to object to the establishment of a wind farm on or near their property, however, strong movements exist protecting the environment and endangered species (Clarke, 2008; Macintosh & Downie, 2006). Danger to birds and bats are often the cause of protest against wind power. An article by Maris and Fairless (2007) states that each wind turbine kills an average of 4.27 birds per year. But Macintosh and Downie (2006) find that "all available evidence indicates that, provided wind farms are located in appropriate areas, the risks to biodiversity are likely to be small" (p.22). Taking the overseas figures as benchmark they determine that approximately 2,550 birds and 2,550 bats are being killed each year in Australia as a consequence of wind turbine collisions[6], "however, these risks should be put into perspective as there are numerous other issues that pose a far greater threat to birds and bats than wind farms" (p. 22).

6.2 Reliability of wind power

Reliability of wind energy is another strong argument against the development of wind turbines. This issue, like the previous one, may not be of direct concern to property owners, but in the bigger scheme of things, i.e. promoting fossil-fuel electricity, as well as manipulating public opinion, it is an important issue. The Citizens Electoral Council of Australia (7.6.2010) claims that wind 'power' is a fraud: The article finds that "while the average high-income, inner-city Green voter voluntarily pays a premium for their 'green' electricity", current data from the Australian Energy Market Operator shows that "the fabulous windmills that are conveniently well out of their urban eyesight and earshot are usually producing only a fraction of their installed capacity". Here is their argument:

> Wind farms across New South Wales, Victoria, Tasmania and South Australia, have a theoretical 1,609MW capacity but electricity generation data shows that the wind power generated from 13th to 20th May [2010] for much of the time was next to zero. Why? Simply, the power won't flow if the wind doesn't blow. Wind power proponents claim that this doesn't matter because if the wind is not blowing in one location, it will be blowing elsewhere. However, the actual power generation data shows this to be one big

---

[6] Macintosh and Downie (2006) provide a comparison: an estimated 8.5 million birds died each year in Queensland alone in the late 1990s as a result of land clearing.

lie. Weather systems often extend over 1,000 km and no wind means no power (Citizens Electoral Council of Australia, 7.6.2010).

This critique captures the sentiments of many residents in rural areas. An article by Leaske (March 25, 2010) points out that "the production figures of individual wind farms have to be treated with caution as output can vary sharply because of breakdowns. The revelation that so many wind farms are performing well below par ... will reinforce the view of objectors who believe that many turbines generate too little power to justify their visual impact." Macintosh and Downie (2006) look at the issue from a more pragmatic point of view, suggesting that wind energy is efficient because "the variability associated with wind energy is managed by the measures that are ... in place to address fluctuations in the supply of, and demand for, electricity" (p. 12). Diesendorf (2003/2004) confirms this: he was able to refute the arguments that wind power can only be used as an intermittent source of power more than thirty years ago: "... wind power, like coal power, is a partially reliable source of power,. ...wind power has 'capacity credit'[7]" (p. 2):

> ... large blocks of wind power, with rapid-response back-up either from hydro or gas turbines and slow-response from intermediate load stations, can provide reliable base-load power and substitute for some coal power. This is not just theory, but is actually happening in countries that have made a major commitment to wind generation. Last year Denmark generated 18%[8] of its electricity from wind power and still plans to increase this substantially (Diesendorf, 2003/2004, p. 3).

6.3 The economic costs of wind farming

Arguments that wind power is not cost effective often strengthen NIMBYism, "wind farms are expensive and require government subsidies" (Government Victoria, 2007, p. 8). Government Victoria (2007) finds that wind farms are becoming increasingly cost effective, however, since the renewable energy industry is only just developing, "government assistance is necessary to allow the industry to improve new technologies and become commercially competitive" (p. 8). Clarke (2010/2011, pp. 18-19) calculates the cost of wind power, based on official data available from Australian wind farms: to produce electricity by wind power costs between A$53 and A$76 per MWh. Clarke then compares this to earlier data from a US congressional report: "costs of electricity in the USA (per MWh) generated by wind is A$67, and generated by pulverised coal A$64" (p. 19). Macintosh and Downie (2006) find that comparison between wind and other sources of energy is difficult because of the cost profiles associated with wind developments.

The most important costs are upfront capital costs, the operating costs are relatively low. ... Most of the data indicate that wind energy is one of the most cost efficient sources of renewable energy and that, when the costs associated with pollution are factored in, it is competitive with coal- and gas-fired power stations (Macintosh & Downie, 2006, p. 3)

Bond (2009) argues that wind energy is "an economically viable form of renewable energy that effectively displaces fossil-fuel electricity generation" (p. 28). This position and an agreement that wind energy can help lower CO2 emissions is supported by various scientists, economists, governments and concerned individuals (i.e. Diesendorf, 2003/2004, 2006; Macintosh and Downie, 2006; Government of Victoria, 2007; Krohn et al., 2009; Bond,

[7] Fuel saving with a need for back-up.

[8] Australia's contribution was 1.3% in 2007 (International Energy Agency (2009).

2009; Clarke, 2010/2011; Combet, 2010; Gillard, 2011). Interestingly, a study by Bond (2009) of residents in two townships in Western Australia determined that both groups agreed that wind farms would be a "boost to tourism / local economy" (p. 23).

6.4 Health issues

> There is rapidly growing world-wide professional realisation that reported health issues in people living close to established wind farms have commonality and substance. Wind energy proponents dismiss this evidence as being symptomatic of 'NIMBYism' and claim that peer reviewed studies have not conclusively linked these health issues to wind turbines, while medical professionals call for urgent scientific research to establish the facts. The concern is that development approvals are being fast tracked ahead of the outcomes of research (Birrell, n.d. p. 1).

Residents living close or relatively close to wind turbines have raised several health issues, including "headaches, tinnitus, dizziness, nausea and sleep disturbance", and "elevated blood pressure" (Clarke, 2010/2011, p. 5). The most prevalent issue is sleep deprivation caused by the noise of turbines (Government Victoria, 2007; Kamperman & James, 2008 a, b; NHMRC, 2010; Clarke, 2010/2011). Studies undertaken by Kamperman and James (2008a) show "significant health effects associated with living in the vicinity of industrial grade wind turbines" (p. 1). Their findings are based on data from industrial wind turbine developments and their findings show "that some residents living as far as 3 km from a wind farm complain of sleep disturbance from the noise", and "that many residents living only 300 m from the wind farm are experiencing major sleep disruption and some serious medical problems from night-time wind turbine noise" (p. 4). Kamperman and James (2008b) propose sound limits that are "standardised and available on all sound level meters", i.e. the "C-weighting or dBC" (pp. 8-10).

Macintosh and Downie (2006, pp. 19-20) find that modern wind turbines create very little noise. At around 40 m the noise created by a single turbine is the equivalent of conversational speech, which is around 50 to 60 decibels (adjusted using an A filter or the A scale) (dBA). "A wind farm comprising of 10 turbines would create a sound pressure of 35 to 45 dBA at 350 m if the wind was flowing from the turbine to the observer" (p. 20). The authors conclude that wind turbines are not a significant cause of noise pollution.

The NHMRC (2010) made the following statement: "Based on current evidence, it can be concluded that wind turbines do not pose a threat to health if planning guidelines are followed" (p. 6). But the Senate (Commonwealth of Australia, 2011) conducted an inquiry into *The Social and Economic Impact of Rural Wind Farms*, and it was recommended that "any adverse health effects for people living in close proximity to wind farms" as well as "concerns over the excessive noise and vibrations emitted by wind farms" (p. 1) should be determined because "the Commonwealth has responsibility for certain aspects of the development of wind farms" (p. 3).

Overall the literature shows that, at this point in time, there is no scientific evidence "indicating a direct link between wind turbines and ill health", and that "there is no known mechanism by which turbines could make people ill" (Clarke, 2010/2011, p. 2).

6.5 Aesthetics of the environment

The destruction of the aesthetic view of the environment is the most cited complaint against wind farming, it interlinks with people's concern that their property value will decrease if the view is destroyed (Sims and Dent, 2007). But the visual amenity, the aesthetics of the environment, "is a subjective matter, dependent upon the experience and beliefs of the individual" (Gipe cited in Finlay-Jones & Kouzmin, 2004, p. 3). Can this bias be counter-argued by a rational point of view? Bond (2009) claims that overseas studies show no statistical evidence that wind turbines within an 8 to 12 km radius of a home have a negative impact on price. Macintosh and Downie (2006) find:

> ...the available evidence indicates that wind farm developments are unlikely to have a significant negative impact on property price. Initial concerns about visual ... impacts could temporarily reduce prices, but these affects are likely to be small and dissipate quickly (p. 27).

Arguments regarding the visual impact are closely linked to NIMBYism. Wind farms face public opposition during the planning process (Johansson & Laike, 2007; Danish Wind Industry, 2007; Macintosh & Downie, 2006; Bond, 2009; Clarke, 2010/11; Government Victoria, 2007) and here are some examples of NIMBYism, starting with a case in point from South Australia. Lothian (2008) conducted a survey of three hundred and eleven participants and found that

> .... wind farms generally have a negative effect on landscapes of higher scenic quality but a positive effect on landscapes of lower scenic quality. The negative visual effects of a wind farm did not decrease with distance (p. 196).

Apart from having conducted his own research, Lothian (2008, pp. 197-207) also cites surveys regarding proposed wind farms in different countries: (1) Residents in North Carolina and Scotland found that spoiling the view or scenery was of greatest concern. (2) In Denmark, Germany and Sweden, local residents are generally supportive of wind farms, but there are some publications which describe the destruction of scenic beauty as a 'catastrophe'. (3) On the Isle of Wight opponents argued that the cause of the problem is based on conflicting national policies: renewable energy targets advocating wind farms contradict policies to preserve, enhance and protect the landscapes from major developments.

Johansson and Laike (2007, pp. 435-451) surveyed eighty people in Sweden regarding the "intention to respond to [to oppose] local wind turbines" and "the role of attitudes and visual perception" (p. 435). These researchers found that the intention to oppose was based on people's attitude regarding *perceived* (emphasis added) effects of wind turbines on landscape aesthetics. Interestingly, opinions regarding "the effects of wind turbines on people's daily quality of life were of minor importance" (p. 435). The authors conclude that it is important for "developers to convince the public that turbines can be integrated into the landscape without threatening the beauty and the recreational value of natural and cultural landscapes" (p. 449).

Coleby, Miller and Aspinall (2009) undertook research in the UK to establish the relationship between public opinion on wind power and public participation in turbine site planning. Most of the critiques related to the proximity of wind turbines to respondents' homes, and it was suggested that "the turbines should remain out of sight". Two interesting issues emerged in the study: (1) younger participants are more accepting of wind turbines than older respondents, and (2) city dwellers are more accepting of wind power than people living in rural areas. But all participants wanted "more public input and participation in local land use for wind power" (p. 1).

Bond's (2009) comparative Western Australian study regarding the visual impact of planned wind farms determined that people would "not want to live near a wind farm" and that

they anticipate "to pay 1% to 9% less for their property due to the presence of a wind farm nearby" (p. 1).
The next section will explain reasons for the objections.

7. Two ways to understand objections to wind farms

In order to make sense of the concept of NIMBYism and its practical implications, consideration is given here to attachment to, and identification with place, and to the concept of trust or distrust.

7.1 Place protective action

Devine-Wright (2009, p. 426-441) explains objections to wind farming by taking place protective action into account, by linking the concept of NIMBYism to place attachment and place identity. Place attachment is "the process of attaching oneself to a place and a produce of this process" (Guiliani in Devine-Wright, 2009, p. 427). As product, "place attachment is a positive emotional connection with familiar locations, which can, if disrupted, lead to action, both at individual and collective levels" (Manzo in Devine-Wright, 2009, p. 427). "Place identity refers to ways in which physical and symbolic attributes of certain locations contribute to an individual's sense of self or identity" (Proshansky et al. in Devine-Wright, 2009, p. 428). Any disruption of place can show "the emotional bond between a person and location" and create feelings of "anxiety and loss"; and it not only affects "the physical aspects of place but also the social networks which are sources of support to individuals" (Fried in Devine-Wright, 2009, p. 428).

7.2 Place-protective action further developed

Hindmarsh (2010, pp. 1-23) picks up the concept of place-protective action, suggesting that place-protection is important to communities targeted by developers and governments to site wind turbines. The major problem regarding wind farm location in Australia is inadequate community engagement. Wind energy presents "the most viable form of renewable energy" (p. 1), but "not enough recognition has been given to conflicts surrounding wind farm sitings" (p. 2). Hindmarsh looks critically at the Australian government's policies regarding wind farming and planning these sites. The current policy responses encourage an inform - consult - participatory engagement which is not sufficient, the community should have the final decision regarding the establishment of a wind farm, and the method of obtaining consent should be inform - consult - involve - collaborate - empower, with emphasis on collaboration and empowering, giving residents choices. Hindmarsh also proposes facilitating social mapping of local community qualifications and boundaries about wind farm location.

7.3 Social trust and distrust

There is indication in the literature that trust in politicians, policy makers and governments is diminishing (Kasperson, Golding & Tuler, 1992; Offor, 2002; Marquart-Pyatt & Petrzelka, 2008). Kasperson et al. (1992) explain the meaning of trusts, namely social trust is "a person's expectation that other persons and institutions in a social relationship can be relied upon to act in ways that are competent, predictable, and caring", and "social distrust is a person's expectation that other persons and institutions in a social relationship are likely to act in ways that are incompetent, unpredictable, uncaring, and thus probably inimical" (p. 169). Kasperson et al.'s (1992) research demonstrates that siting of hazardous facilities "have often led to an impasse" because of the public perceptions of risk and overall uncertainty (p. 163). They recommend "risk communication" (p. 162) which should include a needs assessment, risk debate, monitoring and evaluation.

> Initiatives based upon the explicit recognition of high social distrust may, through empowerment, risk clarification, and negotiation, ultimately prove to be more effective in the long-term recovery of social trust than approaches that assert that such trust is merited a priori ... (Kasperson et al., 1992, p. 184).

How can we make sense of these arguments and relate them to the rational issue of greenhouse gas emissions? The next section will try and explain.

8. Australia's greenhouse gas emissions – Isn't it time to act?

> The three countries which have been the largest drags on the global carbon reduction effort are the three highest per capita emitters amongst the developed countries - Australia, Canada and the United States (Garnaut, 2011b, p. 2).

"Scientific evidence points to increasing risks of serious, irreversible impacts from climate change associated with the business-as-usual attitudes regarding greenhouse gas emissions" (Stern Review, 2006, p. 3). Looking at Australia's CO2 emissions, earlier in this chapter it was established that Australia is in sixteenth position when the percentage of global total CO2 emission is considered (United Nations Statistics Division, 2008 a), emitting 399,219.00 tonnes of carbon dioxide in 2008 (United Nation Statistics Division, 2008a). Also in 2008, Australians emitted 18.94 tonnes of CO2 per capita annually (United Nations Statistics Division, 2008b, p. 1). Australians are "the worst per capita greenhouse gas emitter" (Taylor & Grubel, 9.7.2011, p. 1) of the developed world. And it is projected that the country's CO2 emissions will further rise (CSIRO, 2009; Garnaut, 2011b; Nolan, 2011; Knott, 2011; Chubby, 2011), which is "due mainly to growth in the resources sector" (Garnaut, 2011b, p. 2). Australia is one of the countries most at risk from climate change, partially because of the size of its agricultural sector and long coastline. Garnaut (2011a) urges Australia to follow the recommendations of global communities addressing climate change. Because of the risk factors, the country should "offer to reduce 2020 emissions by 25% in the context of strong international agreement" (Garnaut, 2011a, p. 63).

Bond (2009) looks at the present 83% of electricity which is produced by coal and contemplates "the growing domestic demand which forces investments into clean renewable energy" (p. 2). Looking at the numerous predictions that Australia's greenhouse gas emissions will further increase, what role could wind energy play?

> One typical 2MW wind turbine can be expected to produce over 6,000MW hours of electricity each year. If this replaces coal-fired power, the CO2 released into the atmosphere will be reduced by 6,000 tonnes each year; if it replaces oil or gas-fired power, CO2 released each year is reduced by about 3,000 tonnes (Clarke, 2009, p. 4).

Changing progressively from coal- or gas-fired power to wind power would be a positive step when we look at CO2 emissions. And, as mentioned earlier, the present Australian government promotes the reduction of CO2 emissions and has set its sight on 20% of renewable energy by 2020. In practical terms, there should be no difficulty to achieve an even higher goal. Clarke (2010/2011) calculates that "if the best wind resources of Australia were developed at least 90GW of power is possible" (p. 2). Looking at different coastal regions of Australia, he estimates, that "91,760MW installed capacity, using a capacity factor of 34%, would give an annual electricity generation of 273TWhr which would be greater than the total Australian electricity consumption for the 2006-07 year which was 262TWhr" (Clarke, 2008/2011, p. 3).

Considering the data presented in this chapter, Australia's CO2 emissions, Garnaut's advice, the estimates and evidence provided by Clarke and by Diesendorf that wind power could replace coal-fired electricity and therefore would minimize CO2 emissions, it is argued that NIMBYism has to be critically assessed.

9. Discussion

Global warming is increasing and there is scientific evidence (CSIRO, 2009) that the present problems are, at least in part, anthropogenic and that CO2 emissions contribute to global warming. While wind farming has been accepted increasingly in European countries (Damborg, 1997/2003.; Damborg & Krohn, 1998; Johnson & Jacobsson, 2000; Krohn et al, 2009), in Canada and in the United States (Brown, 2000; Firestone et al., 2005), many residents in rural and regional Australia are suspicious (see Diesendorf, 2003/2004; Macintosh & Downie, 2006; Clarke, 2010/2011), questioning the advantages of wind energy and legitimacy of government and developers' proposals (Davis, 2008; Barr, 2009; Jopson, 2010; Hindmarsh, 2010; Strong, 2010; Ryder, 2011). However, several scientists, environmentalists and (some) governments find that Australia needs wind power (Diesendorf, 2003/2004; Clarke, 2010/2011; Combet, 2010; Gillard, 2011), and that it could produce all of its electricity from wind energy (Clarke, 2008/2011). Nevertheless, NIMBYism is powerful and persisting.

Lothian's study (2007) finds that negative visual effects, which are one of the main reasons to oppose wind turbines, could be reduced if scenic locations were excluded. This would interlink with claims that NIMBYism should be part of understanding the complexities of public facility sitings (Johansson & Laike, 2007; Esaiasson, 2010; Glickel, 2011). Wanting to understand NIMBYism, some scholars mention issues of social trust and distrust in governments as a major issue when public facilities are planned (Kasperson et al., 1992; Marquart-Pyatt & Petrzelka, 2008). Improving or alleviating social distrust is important. Australians are unconvinced of politicians' promises: elections take place every three years, politicians' time in office may extend to six years (some times more). Compared to this, citizens tend to plan their lives and places of residence for longer periods of time. Politicians, governments as well as developers have to convince residents in areas of public facility sitings about the advantages of change, i.e. creation of new jobs and reduction of CO2-e; they have to be honest and visionary and not self-interested and short-sighted, i.e. take into account the residents' sense of place.

In order to counteract NIMBYism, scholars point out that environmental aims must be balanced (Haggett & Toke, 2006), and that residents must be empowered and be given a decisive voice (Hindmarsh, 2009). Both these concerns can be related to re-thinking NIMBYism (Devine-Wright, 2009), where place attachment and place identity play the most important part. But there is also the notion of the civic good that needs to be taken into consideration. According to Gibson (2005), the civic good can be achieved by (1) publicly funding local political campaigns to promote an issue; (2) persuading local activists to form associations to counteract those organisations which pursue self-interest; (3) promoting critical scholars and community activists to allocate more of their resources to the development and support of non-profit, alternative media institutions, including community newspapers and radio stations (p. 399).

All of the above issues interlink and are important consideration in relation to CO2 emissions and the advancement of wind energy in Australia, but in what way and how much do they help in understanding NIMBYism? Having considered the main characteristics of NIMBYism, it is argued that the syndrome does not have a rational basis, that it is non-rational, and that we need to assess its validity because the rational consequences include the abandonment or suspension of facilities that could present a cleaner, renewable energy future.

In Australia four projects, worth 477MW of wind power, were abandoned or are suspended in 2010. While only some motives for the abandonment or suspension of these projects could be studied[9], this literature review was triggered by objections of people to wind farming in rural and regional Australia. People's attachment to, and identification with place are powerful arguments when opposing the development of a wind farm. The literature shows that environmental disruption causes concerns which can lead to anxiety. Therefore this chapter looked at the most important concerns (biodiversity, reliability, cost, health, aesthetic of the environment) and challenged them. But it must be mentioned that a literature review has its limitations because people and their personal stories are not given a voice. There is always the possibility that personal stories let a researcher be less critical and more understanding. Despite these shortcomings, three important points remain: (1) Australia needs to reduce its CO2 emissions; (2) wind power is one of the most efficient sources of renewable energy at this point in time and could help achieve this task; (3) NIMBYism in Australia requires further research.

We are confronted here with some rational facts (consequences) and with a sense of non-rationality, which makes it challenging to find a way out of the dilemma. All articles on NIMBYism state that improved communication between governments, developers and the public is vital. Hindmarsh (2010) provides some directives how this could be achieved, namely greater empowerment of residents in areas where wind turbines are to be installed. If instigated, will it work? Its allowance for the non-rational factor is not apparent. Consideration also has to be given to the fact that most communities are guided by people with leadership qualities. Will they not try to convince other residents, not so powerful individuals, to follow their line or reasoning? Section 3 demonstrated the involvement of a (previous) Liberal Frontbencher as well as a businessman, (previously) involved in oil, gas and mineral exploration, as being instrumental in setting up the anti-wind-lobby of the Landscape Guardians. This is an important point because every community exists of leaders and followers. Hindmarsh's (2010) plan regarding facilitating social mapping of local community qualifications and limits, about wind farm location in conjunction with technical mapping of wind resources, seems appropriate. However, the concern here is will such a rational approach work? Where conflict arises, rationality is often not the line of action, as all research on NIMBYism shows. If, for instance, the anticipated or real decrease of the value of the property (rational) against the sense of, and identification with place (non-rational) is weighed up: what is more important to the residents, which issue takes priority, will the rational or non-rational argument prevail?

It is believed that similar arguments can be made when we look at the notion of civic good (Gibson, 2005). For instance, who will publicly fund local political campaigns to establish a wind farm? Developers? The government? The civic good is clearly important, but then the

[9] See Section 3.

question arises, whose civic good, and who determines it, which would lead to non-rational arguments. And those who fund the debate, will they not influence the outcome?

The issue of social trust and distrust and the suggested risk communication (Kasperson et al., 1992) also presents some challenges: Who is to be empowered to lead the risk communication? Clearly not every individual can be empowered. Will the community as a whole make a decision? Then we are presented again with leaders and followers and the difficulties described above.

Overall there does not seem to be an easy answer – NIMBYism is and remains a contentious issue; whichever way one looks at it, it is like opening Pandora's Box. An idea to improve or lessen NIMBYism in relation to wind power would be to consider employment opportunities and the escape from poverty. The estimated number of jobs in the world wide wind farm industry is as follows: Germany 90,000; United States 85,000; Spain 40,000; Denmark 25,000; Italy 18,300; United Kingdom 16,000; and Australia 1,600 (IEA, 2009, p. 18). Diesendorf (2003/2004) gives an account of community development and employment, arguing that wind power and other sustainable energy sources can provide more local employment than coal:

> Currently wind farms in Australia have about 40% Australian content and create 203 times as many local jobs per kWh generated as coal power. However, as wind power expands, Australian content is expected to rise to 80% and so the number of local jobs per kWh will rise to 4-6 times those of coal. Already job creation is under way: the world's largest wind turbine manufacturer, Vestas, is building a component manufacturing plant in Tasmania (p. 47).

Taking the world-wide estimated employment figures as well as the Australian predictions by Diesendorf into consideration, one should be able to assume that Australia will progress to a clean, renewable future.

Believing in the premise of a politician? This discussion is concluded with the words of an independent member of parliament (Tony Windsor, 2011) who argues that "34,000 new jobs in the renewable energy sector in regional Australia" (p. 2) will be created over the next two decades. Will residents in rural and regional Australia take note?

10. Conclusion

In conclusion it is argued that the Not-in-my-Backyard syndrome in relation to wind farming is alive and thriving, and that the notion needs to be critically assessed because non-rational arguments have rational consequences, i.e. threaten to undermine plans to facilitate wind turbines in regional and rural Australia. As this literature review shows, NIMBYism comes from many sources and there is no easy answer to the non-rational arguments embedded in the concept, there is no easy way to convince local residents in rural and regional Australia that wind farming presents a viable option to reduce greenhouse gas emissions. The influence of the powerful coal industry cannot be underestimated. Evidence exists that people in that industry create fears, uncertainty and doubt. Looking at climate change and at Australia's CO2 emissions, there is an urgency to consider wind power in a more positive way. It is believed that governments and developers play an important role in promoting wind power and that better communication between different stake holders, including the public, has to take place. Improved communication has to be built on trust so that non-rational perspectives can change. As Sections 3 and 8 demonstrate, the siting of a wind farm can have a negative effect on people's sense of place. One rational issue to be argued by proponents of wind energy is the creation of new jobs in the industry. Employment in this new industry could be an escape from poverty in rural and regional Australia, and it could be the catalyst to welcome wind farming. "Farmers are reaping rewards from wind energy" is an article written by Brown (August 21, 2000):

> Farmers and ranchers in the United States are discovering that they own not only land, but also the wind rights that accompany it. A farmer in Iowa who leases a quarter acre of cropland to the local utility has a site for a wind turbine can typically earn $2,000 a year in royalties from the electricity produced. In a good year, that same plot can produce $100 worth of corn.

Just one last thought: Coleby, Miller and Aspinall (2009) find that young people are more accepting of wind turbines than older ones. This then raises the question, do we have to wait another generation to reduce our greenhouse gas emissions? If NIMBYism, as explored in this chapter, persists, Australia may have no option. But that would be a coal-grey future; therefore hope has to prevail that attitudes will change: "We have a larger moral responsibility to reduce our emissions than most other nations and most other people" (Clarke, 2008, p. 9).

> The reasonable man adapts himself to the world; the unreasonable one persists in trying to adapt the world to himself. Therefore all progress depends on the unreasonable man.
> George Bernard Show cited in David Clarke's *The Ramblings of a Bush Philosopher*

11. Acknowledgement

I am grateful for the assistance provided by Ms Alex Zauner-De Ville and Ms Noelene Milliken. I am indebted to Professor Ian Gray for his thoughts on this chapter. And I would also like to thank the referees / editors for their advice and recommendations.

12. References

2008 Australian Coal, *Australian Coal Association*, Retrieved August 5, 2011. Published On-line.

ABC Australian Broadcasting Corporation (25 May, 2011): Acciona scraps wind farm expansion. *ABC Ballarat*. Retrieved August 4, 2011, from http://www.abc.net.au/news/stories/2011/05/25.

ABS Australian Bureau of Statistics (2006). *Regional population growth, Australia and New Zealand*, 2001-02, Cat. No. 3218.0.

Acciona Energy. *Newsletter Waubra Wind Farm*, Edition 15 June 2011.

Acciona (n.d.) Project Snapshot *Waubra Wind Farm*.

Acciona Energy (2006). *Proposed Newfield wind farm: Social impact statement*. Retrieved January 31, 2010, from http://www.nre.vic.gov.au/

Andrews, G. (2001). *Market based instruments: Australia's experience with trading renewable energy certificates*. Paper presented at the Workshop on Good Practices in Policies and Measures. Published On-line.

Berg, C. (2003). *Minimising community opposition to wind farm developments in New Zealand: Opportunities in renewable energy planning*. Wellington: Victoria University.

Birrell, J.R. (n.d.) *Senate wind farm inquiry*. Retrieved August 3, 2011, from http://senate.aph.gov.au/submissions/committees/viewdocument.

Bond, S. (2009). *The tale of two windy cities: Public attitudes towards wind farm development.* Curtin University of Technology: School of Economics and Finance. Published On-line.

Brennan, M.A., Flint, C.G. & Luloff, A.E. (2008). Bringing together local culture and rural development: Findings from Ireland, Pennsylvania and Alaska. *Sociologia Ruralis,* 49(1): 97-112.

Brown, L. (21.8.2000). Farmers are reaping rewards from wind energy. *Grist beta.* Retrieved October 20, 2009, from http://www.grist.org/article/brown-something/.

BTM Consult ApS (2009). *Windfarm management.* Retrieved March 30, 2010, from http://www.btm.dk/special+issues/others+issues/the+wind+power+sector/?s=42+

Clarke, D. (2008). *Greenhouse/Climate Change and Australia.* Retrieved July 7, 2011, from http://www.oocities.org/daveclarkecb/Australia/Greenhouse.html.

Clarke, D. (2008, updated 2010). *Wind farms in New South Wales.* Retrieved May 5, 2010, from http://ramblingsdc.net/Australia/WindNSW.

Clarke, D. (2008, updated 2011). *Australia's wind power potential: Wind in the bush.* Retrieved July 11, 2011, from http://ramblingsdc.net/Australia/WindPotential.html.

Clarke, D. (2009) *Wind power and wind farms in Australia: Wind in the bush.* Retrieved July 3, 2011, from http://www.oocities.org/daveclarkecb/Australia/WindPower.html.

Clarke, D. (2010, updated to 2011). *Wind power and wind farms in Australia: Wind in the Bush.* Retrieved July 2, 2011, from http://ramblingsdc.net.Australia/WindPower.html .

Clarke, D. (2011). *The Ramblings of a bush Philosopher.* Retrieved July 3, 2011, from http://ramblingsdc.net/.

Clean Energy Council (2009) *Clean energy report.* Retrieved March 30, 2010, from http://cleanenergyaustraliareport.com.au/industry-snapshot/

Clean Energy Ideas (2011). *Advantages & disadvantages of wind energy.* Retrieved March 6, 2011, from http://www.clean-energy-ideas.com/articles/advantages_and_ disadvantages _of_wind_energy.html.

Coleby, A.M., Miller, D.R. & Aspinall, P.A. (2009). *Public attitudes and participation in wind turbine development.* Retrieved February 25, 2011, from http://ideas.repect.org/a/wsi/jeapmx/v11y2009i01p69-95.html.

Combet, G., Minister Assisting the Minister for Climate Change and Energy Efficiency (2010). *Renewable Energy (Electricity) Amendment Bill 2010.* Retrieved July 4, 2011, from www.climatechange.gov.au/media/Files/minister.

Commonwealth of Australia, Senate Committee (23 June 2011). *Impact of Rural Wind Farms.* Report Chapter 1. Retrieved July 1, 2011, from http://www.aph.gov.au/senate/committee/clac-ctte/impact_rural_wind_farms/report/cO1.pdf

Courtice, B. (July 30, 2011). Anti-wind power lobby: farming, fear, uncertainty and doubt. *Green Left.* Retrieved August 4, 2011, from http://www.greenleft.org.au.node/4847.

Cross, B. (July 5, 2010). Renewable power from wind farms and the nimby phenomenon. *Aus/NZ/Oceania Affairs*. Retrieved July 2, 2011, from http://www.suite101.com/content/renewable-power-from-wind-farms-and-the -nimby-phenomenon.

Cubby, B. (9.2.2011). *Australia admits CO2 emissions will balloon*. Retrieved February 2, 2011, from http://www.smh.com.au.environment/climite-change/australia-co2-emissions.

CSIRO, Commonwealth Scientific and Industrial Research Organisation (2009). *The science of tackling climate change*. Retrieved October 2010, from www.csiro.au. Published On-line.

CSIRO 2003-2011, Commonwealth Scientific and Industrial Research Organisation, 2003-2011. *Energy from coal*, Retrieved August 5, 2011. Published On-line.

Curry, T., Reiner, D. M., Ansolabehere, S., & Herzog, H. (n.d.). How aware is the public of carbon capture and storage? *Federal Highway Administration*. Retrieved April 4, 2010, from http://knowledge.fhwa.dot.gov/cops/italladdsup.nsf/7ec05a279f17ed

Damborg, S. (1997, updated 2003). *Public attitudes towards wind power*. Danish Wind Industry Association. Retrieved March 7, 2011, from http://www.windpower.org/en/articles/surveys.htm.

Damborg, S. & Krohn, S. (1998). *Public attitudes toward **wind power***: Wind Turbine Manufacturers Association. Published On-line.

Davis, O. (July 2, 2008). *The winds of change*. Retrieved May 4, 2010, from http://www.theage.com.au/environment/the-winds-of-change-20080701-303d.html.

Devine-Wright, P. (2009). Rethinking NIMBYism: The role of place attachment and place identity in explaining place-protective action. *Journal of Community & Applied Social Psychology* (19): 426-441. Published On-line.

Diesendorf, M. (2006). *A sustainable energy future for Australia,* Fact Sheet 5, Issue 2. 5/12/2008. Retrieved February 20, 2011, from www.energyscience.org.au. Published On-line.

Diesendorf, M. (Summer 2003/04). Why Australia needs wind power. *Dissertation*, 13, pp. 43-48. Published On-line.

Equipment and Construction News (2010). *Positives and negatives of wind energy – Facts about wind energy*. Retrieved March 3, 2011, from http://buildernews.galanter.net/2010/09/06/benefits-and-drawbacks-of-wind-energy.

Esaiasson, P. (2010) *Why citizens (sometimes) dispute public facility sitings in their neighborhood – An experimental account of the NIMBY-syndrome*. Presentation at the 2010 Annual Meeting of the American Political Science Association, Washington, September 2-5. Retrieved March 8, 2011, from http://ssm.com/abstract+1644225.

European Commission (1997). *Windenergy - the facts: Volume 5*: Directorate-General for Energy.

Finlay-Jones, R. & Kouzmin, A. (2004) Putting the spin on wind energy projects. Risk management issues in the development of wind energy projects in Australia. *Wind Development Australia Pty. Ltd.* Published On-line.

Firestone, J., Kempton, W., Krueger, A. & Loper, C. E. (2005). *Regulating offshore wind power and aquaculture: messages from land and sea.* University of Delaware: College of Earth, Ocean, and Environment. Published On-line.

Firestone, J., Kempton, W. , & Krueger, A. (2009). Public acceptance of Offshore wind power projects in the USA. *Wind Energy*, 12: 183-202.

Garnaut, R. (2011a) (31.5.2011). *Garnaut Climate Change Review – Update 2011. Australia in the global response to glimate change. Summary.* Retrieved July 7, 2011, from www.garnautreview.org.au. Published On-line.

Garnaut, R. (2011b) (24.2.2011 b). *A re-look at climate change – The Garnaut Review 2011.* Retrieved March 14, 2011, from http://www.carbonneutral.com.au/about-us/news

Gibson, T.A. (2005). NIMBY and the Civic Good. *City & Community*, 4(4):381-401. Published On-line.

Gillard, J. (2011). Gillard reveals carbon price scheme. *ABC News, ABC Broadcasting Corporation.* Retrieved July 12, 2011, from http://www.abc.net.au/news/stories/2011/07/10.

Gipe, P. (2002). Aesthetic guidelines for a wind power future. In M. J. Pasqualetti, P. Gipe & R. W. Righter (Eds.), *Wind power in view.* San Diego: Academic Press.

Glickel, J. (2011). Siting wind turbines: collaborative processes and joint fact finding to resolve NIMBY disputes. *Journal of Social Issues*, 48(4): 39-61, Retrieved June 29, 2011, from http://web.met.edu/dusp/epp/music/pdf/glickel.pdf.

Government Victoria (2007). *Sustainability Victoria. Wind energy. Myths and Facts.* Retrieved March 8, 2011, from http://www.Sustainability_vic.gov.au.

Haggett, C. & Toke, D. (2006). Crossing the great divide – Using multi-method analysis to understand opposition to wind farms. *Public Administration*, 84(1): 103-120.

Hepburn Renewable Energy Association, *Media Release October 2007. Wind farm co-op winding up.* Retrieved December 1, 2010, from http://www.hepburnwind.com.au/windfarm_project.

Hindmarsh, R. (2010). Wind farms and community engagement in Australia: a critical analysis for policy learning. *East Asian Science, Technology and Society: An International Journal.* Published online. Retrieved February 28, 2011, DOI 10.1007/s12280-010-9155-9.

Howden, D. (June 12, 2007). Conservationists fight to keep wind farms off Skyros. *The Independent.* Retrieved July 2, 2011, from http://www.independent.co.uk/news/world/europe/conservationists-fight-to-keep-wind.

Hurst, D. (March 17, 2011). NIMBYs 'a threat to proper planning'. Retrieved July 2, 2011, from http://www.brisbanetimes.com.au/queensland/nimbys-a-threat-to-proper-planning.

Impey, T. (22 June 2011) Man vs wind farm. Australian Broadcasting Corporation, *ABC South East SA.* Retrieved August 4, 2011, from http://www.abc.net/local/stories/2011/06/22/3250488.

International Energy Agency (2009). *IEA Wind Energy Annual Report 2008*, ISBN 0-9786383-3-6. Retrieved August 4, 2011. Published On-line.

Johansson, M. & Laike, T. (2007). Intention to respond to local wind turbines: The role of attitudes and visual perception. *Wind Energy*, 10: 435-451.

Johnson, A. & Jacobsson, S. (2000). *The emergence of a growth industry - a comparartive analysis of the German, Dutch and Swedish wind turbine industries*. Paper presented at The 8th International Joseph A. Schumpeter Society Conference. Reterieved February 10, 2011, from http://www.cric.ac.uk/cric/events/schumpeter/papers/100.pdf.

Jopson, D. (April 2, 2010). Tilting at windmills: why families are at war, *Sydney Morning Herald*. Retrieved April 5, 2010, from http://www.smh.com.au/environment/energy-smart/tilting-at -windmills-why families.a.

Jopson, D. (April 5, 2010). Wind farm approval blows town apart, *Sydney Morning Herald*. Retrieved April 5, 2010, from http://www.smh.com.au/environment/energy-smart/wind-farm-approval-blows-town-a.

Kagkelidou, M. (April 13, 2007). Serifos island opposes gigantic windpark plans. *Athens News*. Retrieved 2 July 2011, from http://www.athensnews.gr/old_issue/13230/15997.

Kampermann, G.W. & James, R.R. (2008a). *The 'how to' guide to siting wind turbines to prevent health risks from sound*. Windaction.org. Retrieved July 7, 2011, from http://www.savethetorontobluffs.com/pdf/ kamperman-james-8-26-08-report-pdf .

Kampermann, G.W. & James, R.R. (2008b). *VIII. Noise-Con 2008 Paper*. Dearborn, Michigan. Windaction.org. Retrieved July 7, 2011, from http://www.savethetorontobluffs.com/pdf/ kamperman-james-8-26-08-report-pdf .

Kann, S. (2009). Overcoming barriers to wind project finance in Australia. *Energy Policy*. Retrieved April 4, 2010, from http://solar.anu.edu.au/pubs/2009/Overcoming%20 barriers%20

Kasperson, R.E., Golding, D. & Tuler, S. (1992). Social distrust as a factor in siting hazardous facilities and communicating risks. *Journal of Social Issues*, 48(4), 161-187. Published On-line.

Knott, M. (7 February 2011). Garnaut to Australia: at least try to keep up. *Crikey*. Retrieved February 13, 2011, from http://www.crikey.com.au/2011/02/07/garnaut-to-oz-stop-blaming-china-for-climate-.

Krohn, S., Morthorst, P-E. & Awerbuch, S. (2007). *The economics of wind energy*. European Wind Energy Association. www.ewea.org. Published On-line.

Leaske, J. (2010). To hatch a Crow. Feeble wind farms fail to hit full power. Retrieved June 29, 2011, from http://tohatchacrow.blogspot.com/2010/03/feeble/wind/farms/fail/to-hit-full.html.

Liebmann, C. (2003). Southern Cross Windpower submission to MRET review. *Submission to the Mandatory Renewable Energy Target Review Panel* Retrieved February 1, 2010, from http://www.mretreview.gov.au/pubs/mret-submission20.pdf.

Lothian, A. (2008). Scenic perceptions of the visual effects of wind farms on South Australian landscapes. *Geographical Research,* 46(2): 196-207. Doi: 10.1111/j.1745-5871.2008.00510.x.

Lynley, M. (May 24, 2011). Main wind could be latest 'not in my backyard' casualty. *Venture Beat and Nokia Conversations on the Global App Economy.* Retrieved July 2, 2011, from http://venturebeat.com/2011/05/24/maine-wind-farm-nimby-casualty.

Macintosh, A. & Downie, C. (2006). *Wind farms: the facts and the fallacies.* The Australia Institute. Retrieved June 29, 2011, from https://www.tai.org.au/documents/downloads/DP91.pdf .

Marquart-Pyatt, S.T. & Petrzelka, P. (2008).Trust, the democratic process, and involvement in a rural community. *Rural Sociology,* 73(2), 250-274. Published On-line.

Mercer, D. (2003). The great Australian wind-rush and the devaluation of lanscape amenity. *Submission to the Mandatory Renewable Energy Target Review Panel.* Retrieved February 1, 2010, from http://www.mretreview.gov.au/pubs/mret-submission22.pdf.

National Health and Medical Research Council (July 2010). *Wind turbines and health. A rapid review of the evidence.* Australian Government. Retrieved July 1, 2011, from http://www.gov.au/_files_nhmrc/attachments/new0048

Nelson, S. (2005). Wind generation and the South Australian economy. *Economic Issues,* 15. Published On-line.

Nolan, T. (February 9, 2011). Australia's emissions projected to rise sharply. *The World Today* with Eleanor Hall, *ABC Local Radio.* Retrieved February 13, 2011, from http://www.abc.net.au/worldtoday/content/2011/s3134150.htm.

Offor, T. (2002). The public and "not in my backyard" - Strategies for engaging angry communities. Paper presented to the *8th Renewable and Sustainable Power Conference,* Alice Springs Convention Centre, 13 August 2002. Published On-line.

Pasqualetti, M. J., Gipe, P. & Righter, R. W. (2002). *Wind power in view: energy landscapes in a crowded world.* San Diego: Academic Press.

Radical Green Watch, (February 13, 2010). Retrieved August 4, 2011, from http://www.radicalgreenwatch.com/main.

Reeves, A. & Beck, F. (2003). *Wind energy for electrical power: a REPP issue brief.* Washington DC: Renewable Energy Policy Project. Published On-line.

Reiner, D., Curry, T., de Figueiredo, M., Herzog, H., Ansolabehere, S., Itaoka, K., et al. (n.d.). *An international comparison of public attitudes towards carbon capture and storage technologies.* Retrieved February 1, 2010, from http://www.geos.ed.ac.uk/ccs/Publications/Reiner1.pdf.

Richards, C. & Bjorkhaug, H. (2009). Comment: The importance of agency in facilitating the transition to a multifunctional countryside. *Journal of Rural Studies,* 25: 249.

Righter, R. W. (2002). Exoskeletal outer-space creations. In M. J. Pasqualetti, P. Gipe & R. W. Righter (Eds.), *Wind power in view.* San Diego: Academic Press.

Rodriguez, J. M., Fernandez, J. L., Beato, D., Iturbe, R., Usaola, J., Ledesma, P., et al. (2002). Incidence on power system dynamics of high penetration of fixed speed and doubly fed wind energy systems: study of the Spanish case. *IEEE Transactions on Power Systems,* 17(4), 1089-1095.

Russell, B. (January 1, 2008). 'Nimbyism' blocking the spread of wind farms. *The Independent.* Retrieved July 2, 2011, from http://www.independent.co.uk/environment/climate/change/nimbyism-blocking-the-spr.

Ryan, A. (2005-2009). *Advantages and disadvantages of wind power.* Retrieved March 6, 2011, from http://www.technologystudent.com/energy1/wind8.htm.

Ryder, T. (April 30, 2011). NIMBYs' whines grist for windmills. *The Australian.* Retrieved July 2, 2011, from http://www.theaustralian.com.au/news/executive-lifestyle/nimbys-whines-grist-for-win.

Seelye, K.Q. (April 28, 2010). Big windfarm off Cape Cod gets approval. *The New York Times.* Retrieved July 8, 2011, from http://www.nytimes.com/2010/4/29/

Siegel, A. (November 27, 2007). NIMBYism - Global Obstacle to a Renewable energy Future. *New York Times.* Retrieved July 2, 2011, from http://www.celsias.com/article/nimbyism-a-global-obstacle-to-a-renewable-energy.

Sims, S. & Dent, P. (2007). Property stigma: wind farms are just the latest fashion. *Journal of Property Investment and Finance,* 25(6), 626-651.

Sobey, E. (April 29, 2010). Daylesford wind farm goes alone. *The Courier.* Retrieved December 1, 2010, from http://www.thecourier/com.au/news/local/news/general/daylesford-wind-farm-goes-alone.

Strong, G. (May 22, 2010). Towns split on which way the wind blows. *The Age.* Ref.70416766

Taylor, R. & Grubel, J. (July 9, 2011) Australia unveils biggest carbon-reduction scheme outside Europe. *Business & Financial News,* Breaking US & International News, Reuters.com. Retrieved July 25, 2011, from http://www.reuters.com/assets/print.

The Border Watch (June 23, 2011). Court blocks $175m wind farm project. Retrieved August 4, 2011, from http://www.borderwatch.com.au/archives/9443.

Walker, C. (January 19, 2011). Nimby-ism to get boost in anti wind campaign? *Business Green,* UK. Retrieved July 2, 2011, from http://yes2renewables.org/2011/01/19/nimby-ism-to-get-boost-in-anti-wind-campaign

United Nations Statistics Division (2008a) *Millennium Development Goals Indicators: Carbon dioxide emissions (CO2), thousand metric tonnes of CO2* (CDIAC). Retrieved August 6, 2011, from http://unstats.un.org/unsd/mdg/SeriesDetail.aspx?srid.

United Nations Statistics Division (2008b) *Millennium Indicators: Carbon dioxide emissions (CO2), metric tons of CO2 per capita* (CDIAC). Retrieved August 6, 2011, from http://unstats.un.org/unsd/mdg/SeriesDetail.

Wikipedia. Windpower in Australia. Retrieved July 13, 2011, from http://en.wikipedia.org/wiki/Wind_power_in_Australia.

Wind Power World Wide. Retrieved March 7, 2011, from www.wwindea.org/home/index2.php?option+com_jce&task=popup&img=imag

Windsor, T. (28.2.2011) Carbon tax 'could create thousands of jobs'. *ABC News,* Climate Change. Retrieved February 28, 2011, from http://www.abc.net.aus/news/stories/2011/02/28/3150493.htm.

Wolsink, M. (2000). Wind power and the NIMBY-myth: institutional capacity and the limited significance of public support. *Renewable Energy*, 21, 49-64.

Woods, M. (2003). Conflicting environmental visions of rural: Windfarm development in Mid Wales. *Sociologia Ruralis*, 43(3): 271-288.

World Future Council, Press Release 24.03.2010. 100% Renewable energy for cities is possible. Retrieved February 1, 2011, from www.worldfuturecouncil.org.

World Wind Energy Association, WWEA (2010). Retrieved February 10, 2011, from http://www.wwindea.org/home/index2.php?option=com_jce&task=popup&img=imag.